MW01640113

GEOMETRY
A UNIFIED COURSE
Expanded Edition

Mackinac Bridge Authority

A precise knowledge of the principles of geometry is required of engineers and men who build structures such as the Mackinac Bridge, which stretches across the Mackinac Strait in northern Michigan. This bridge is suspended from two steel cables, each $24\frac{1}{2}$ inches in diameter. The cables are supported by towers extending 752 feet above bedrock. The central span is 3800 feet in length.

GEOMETRY
A UNIFIED COURSE
Expanded Edition

A. Wilson Goodwin
Supervisor of Mathematics
Des Moines Independent Community Schools
Des Moines, Iowa

Glen D. Vannatta
Chairman, Mathematics Department
Broad Ripple High School
Indianapolis, Indiana

Harold P. Fawcett, Consultant
Ohio State University
Columbus, Ohio

CHARLES E. MERRILL PUBLISHING CO.
A Bell & Howell Company
Columbus, Ohio

MERRILL MATHEMATICS SERIES

for secondary schools

Harold P. Fawcett, Consultant

ALGEBRA ONE: A Modern Course

ALGEBRA TWO: A Modern Course

GEOMETRY: A Unified Course

ADVANCED HIGH SCHOOL MATHEMATICS

ISBN 0-675-05808-2

CHARLES E. MERRILL PUBLISHING CO.
A Bell & Howell Company
Columbus, Ohio 43216

Printed in the United States of America

Preface

This expanded edition of *Geometry: A Unified Course*, is a thorough revision of the highly successful first edition. It has been completely rewritten to reflect the most recent ideas in the teaching of geometry.

The language and concepts of sets are introduced early and are used throughout the book to increase accuracy and clarity of definitions and discussions. Space is considered as the universal set of points, and lines, planes, and other geometric configurations are considered as subsets of space.

The postulational structure of the text has been strengthened and definitions and assumptions have been refined. A distinction has been made between equality (a property of numbers and measures) and congruence (a property of geometric figures). Appropriate symbols have been used to distinguish between a segment and its measure and an angle and its measure. Symbols have also been used to identify lines, rays, and segments.

Deductive logic and conditional statements are carefully discussed in Chapter 4. A section of the appendix using truth tables to prove the Law of Detachment and the Law of Syllogism is keyed to this chapter for enrichment. The logical basis of proof is emphasized.

Concepts of Euclidean geometry are presented so that three-dimensional proofs and exercises follow logically from similar concepts in plane geometry. However, the teacher who wishes to treat solid geometry informally may omit three-dimensional proofs without affecting the continuity of the text. Coordinate geometry is presented as a single unit in the final chapter, and is cross-referenced to corresponding synthetic proofs throughout the book so that it may be fully integrated if desired.

The material on three-dimensional geometry is unusually comprehensive. Solid figures and their measurements are treated in depth, and a number of three-dimensional proofs and exercises are included. The student living in a three-dimensional world will find the study of *Geometry: A Unified Course* much more appealing and meaningful than a course restricted to geometry of the plane.

Student discovery of geometric relationships is encouraged. Informal, intuitive discussions precede many of the proofs. To emphasize that a proposition is not a theorem until proved, the statement of a theorem is placed at the end of the demonstration establishing it. Exercises extend ideas discussed in preceding sections, and lead students naturally and easily into concepts of the following units.

Exercise lists have been greatly expanded, with problems of progressive difficulty providing for varying levels of ability. Comprehensive Review Exercises and a Chapter Test are included at the end of each chapter; questions are original and searching for maximum motivation and reinforcement of learning. A vocabulary list for each chapter brings new terms to attention.

The use of color and type faces is both attractive and functional. In the more complex diagrams, color is used to distinguish the most significant elements; in the text, color emphasizes words and phrases being used for the first time. Other important words in the text are set off by italic type, while theorems and assumptions appear in distinctive boldface type. Photographs illustrate both practical and artistic applications of geometric principles.

For those who desire to teach a modern, unified course in geometry, while retaining the beauty and order of Euclidean geometry, *Geometry: A Unified Course, Expanded Edition* provides material challenging in scope and unusually motivating in approach.

Contents

ALCOA

This is a model of the 25-story Alcoa building, part of the Golden Gateway Center in San Francisco, featuring an exterior structural system, and glass and aluminum sheathing. What geometric forms and patterns do you see?

1

Geometry and the Inductive Process

Geometry is believed to have originated in Egypt about 1350 B.C. In its earliest form it was used largely as a means to measure plane figures such as triangles and rectangles. When land division and measure were instituted under Ramses II in the following century, the usefulness of geometry was permanently established. The annual flooding and channel changes of the Nile River required frequent application of geometry to re-establish proper land boundaries.

Actually the word "geometry" is derived from two Greek words *ge*, meaning earth, and *metrein*, meaning to measure.

Until about 600 B.C. all mathematical knowledge was based upon observation and measurement. The uncertainty of the principles established through this "trial and error" procedure bothered the great scholars of this time. The call for "proof" was probably made by Pythagoras. It is said he asked for evidence that was more reliable than that provided by observation and measurement. Others such as Thales (a teacher of Pythagoras), Plato, Aristotle, and Euclid helped developed a geometry based upon logical deduction.

Euclid (300 B.C.), a teacher of mathematics in the University of Alexandria, collected the geometric knowledge of his day into a volume of thirteen chapters known as *The Elements*. This is believed to be the first systematically arranged and logically sound body of mathematical principles ever produced. It is this "logical pattern of thinking" that we will study as well as the practical applications of geometry.

INDUCTIVE REASONING

The mathematical principles which you will explore in this text were developed from countless experiments performed by mathematicians of recent, as well as ancient, times. If, after a series of experiments using the same set of conditions, a general pattern becomes apparent, a proposition (generalization) may be stated describing the conditions and any tentative conclusions drawn. These tentative conclusions will be called conjectures.

This process is similar to the experimentation done in our science laboratories. In the next few pages, and in many places throughout the text, we want you to experiment with possible geometric relationships, then state the generalization that you feel is warranted from the results of these experiments.

TOOLS OF GEOMETRY

In any laboratory, the scientist—and you, the geometer—needs tools or instruments with which to proceed with his experiments. The tools of geometry historically have been the compasses, protractor, and ruler. These coupled with the language tools developed in previous mathematics courses and some degree of ingenuity will allow us to proceed to our first geometry assignment.

Ruler

The ruler is a straightedge marked with a uniform scale of units. It is useful for finding approximate data about the measure of certain geometric figures. However, many geometric constructions can be made using only an unmarked straightedge and compasses.

Exercises with the Ruler

Experiment 1: Let us measure the length of the line-segment below. Call it segment AB ($\overline{AB}$), since it is labeled A at one end and B at the other.

Figure 1–1

Is your measure the same as that obtained by other members of your geometry class?

Experiment 2: Measure the lengths of the sides of the following figure, ABC, and find their sum.

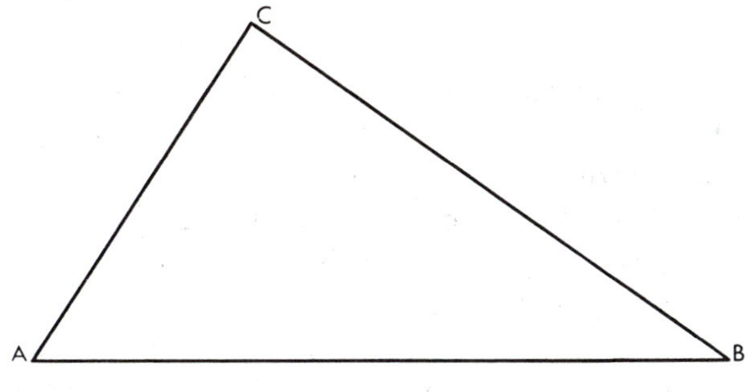

Figure 1–2

How does your result compare with the results obtained by other members of your class?

Experiment 3: The following figures are drawn with two angles whose measures are equal. Measure the lengths of the sides opposite these angles.

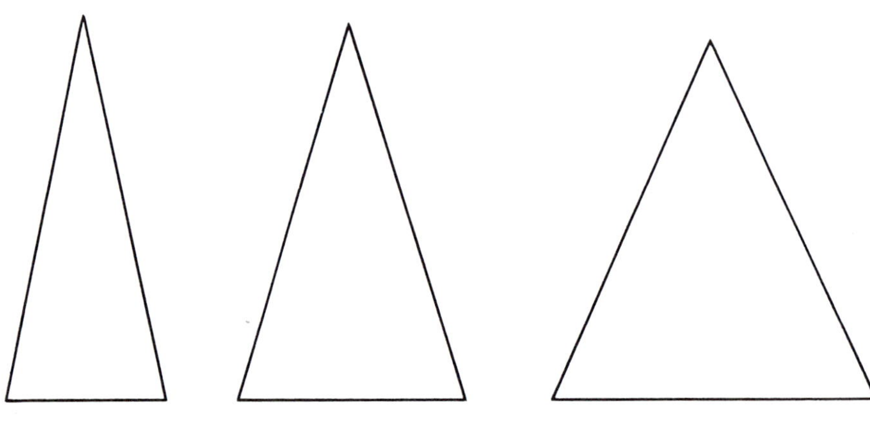

Figure 1–3

Are the measures equal? Exactly equal? Would you be willing to conclude that all such three-sided figures with two angles whose measures are equal must have two sides whose measures are equal? How many such figures would you have to draw to be positive that they all must have two sides with equal measures? If you found only one such figure that did not have two sides of equal measure, would you then have to reject your original conjecture?

Protractor

The protractor is used to measure angles. Though most protractors are in the shape of a semicircle, navigators use a clear, plastic, circular-shaped protractor. The outer edge of the circular type is divided into 360 uniform spaces, and the semicircular type is divided into 180 uniform spaces. Each one of these unit spaces represents a measure of one degree.

Most protractors have one scale marked on the outer edge and a similar scale running in the reverse order on the inner edge of the protractor. The double scale enables you to measure angles from either direction.

To read the measure of an angle, place the center of the circular or semicircular protractor at the intersection of the lines forming the angle. Place the straight edge (diameter) of the semicircle or circle along one side of the angle. Read the number of degrees on the scale whose zero point is on the side of the angle that lies along the straight edge of the protractor. Is the measure of angle *CBA*, (Fig. 1-4), 50 degrees or 130 degrees?

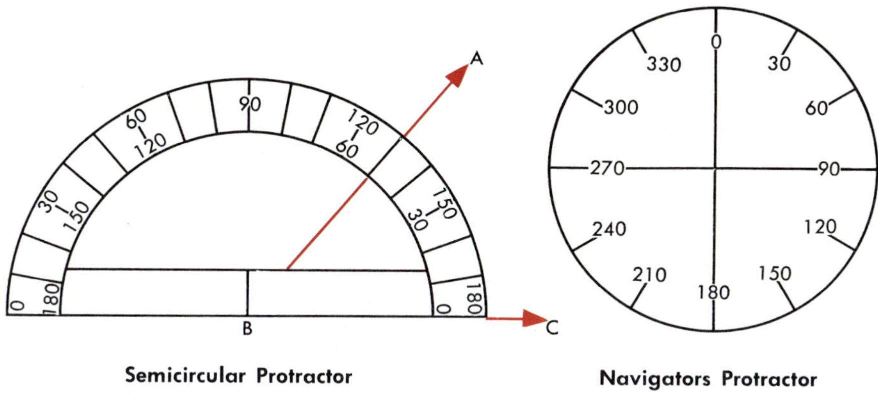

Figure 1–4

Exercises with the Protractor

Experiment 4: Measure the numbered angles below. Draw other triangles showing an external angle and repeat the experiment. If you see a relationship present, write a sentence clearly expressing the relationship.

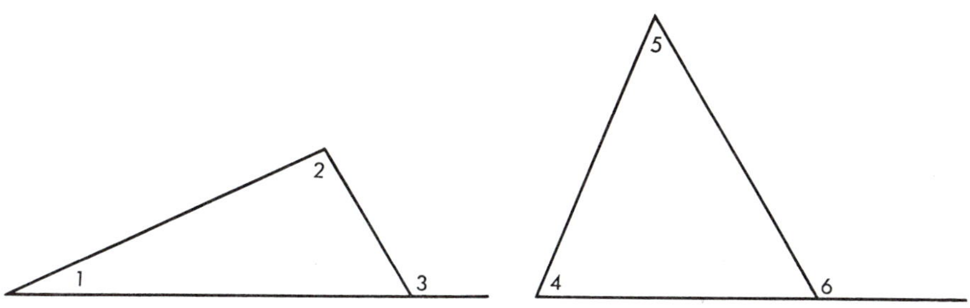

Figure 1–5

Experiment 5: Draw a triangle two of whose angles measure 30 degrees and 60 degrees. Measure the sides. Repeat the experiment. Is it possible to make a conjecture from this experiment? If so, state it clearly.

Experiment 6: Find the sum of the measures of the numbered angles in each of the figures below. Find the sum of the measures of the lettered angles also. Can you make a conjecture from this experiment? If so, write a generalization stating the conditions and the conjecture clearly and concisely.

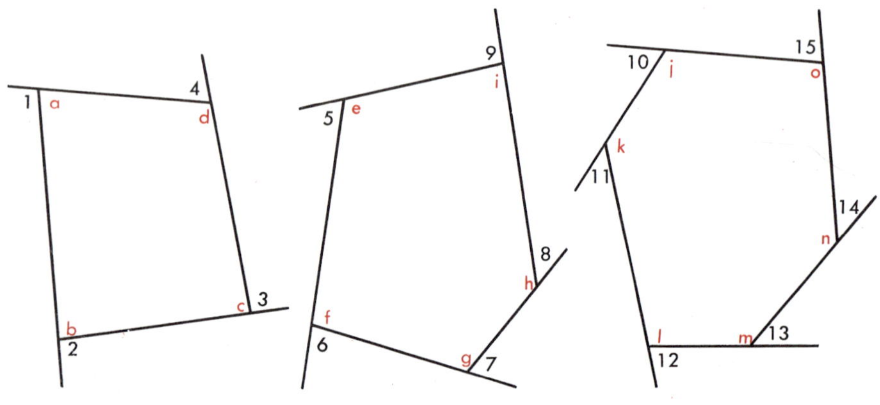

Figure 1-6

Compasses

Compasses are used to draw circles or portions of circles (arcs) and to mark off desired lengths. They may also be used in conjunction with a ruler, or any other linear scale, to find the measure of a part of a line.

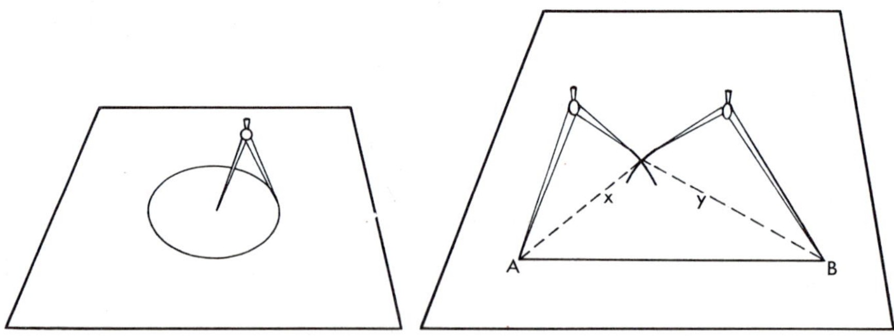

Using compasses to draw a circle (They are also used to draw arcs)

Using compasses to mark off the length *x* and *y*, thus forming a triangle with side *AB*

Figure 1–7

Exercises with Compasses

Experiment 7: By using one setting of your compasses and drawing intersecting arcs from two different points (A and B below), construct a triangle, with two sides ($\overline{AC}$ and $\overline{BC}$) having the same measure. Measure the angles. What is your conjecture? Write a clear and concise sentence stating the conditions and your conjecture. Example: "If a triangle has two sides with the same measure, then—*(Finish the statement with your conjecture)*—."

Does the fact that you have examined one such triangle prove your conjecture? How many triangles with two sides having the same measure would you need to examine before you were certain of your conclusion? How many triangles are there with two sides having the same measure?

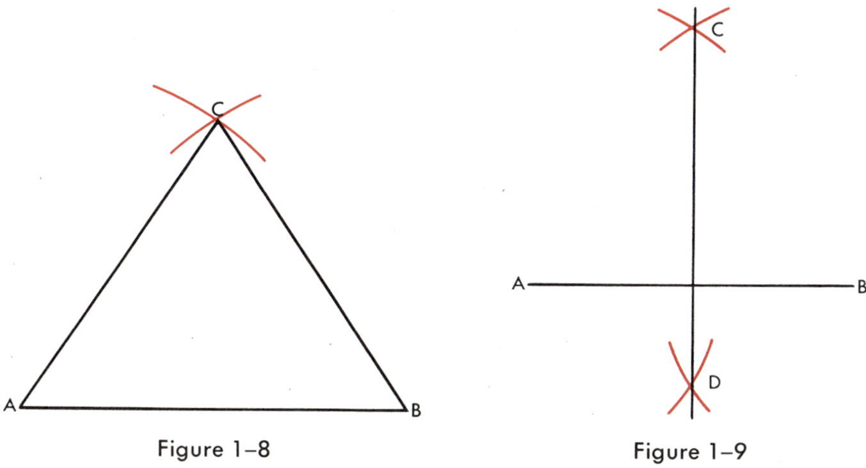

Figure 1–8

Figure 1–9

Experiment 8: (See Figure 1-9.) Draw a line segment such as $\overline{AB}$. Use the compasses to find two points equally distant from the ends of $\overline{AB}$, such as C and D in the accompanying picture. Use your ruler to draw a straight line through C and D. Does the line $\overline{CD}$ divide $\overline{AB}$ into two parts having the same measure? Measure the angles formed by the lines. Do you believe this procedure will always produce the same results? Prepare a statement that expresses clearly and completely the conditions and your conjecture from this experiment. Why is your conjecture only probable and not a *necessary* conclusion?

Experiment 9: Draw a line segment $\overline{AB}$ as shown below. Use one setting of your compasses to draw arcs from the ends of the line segment, one arc above the line and the other below, as shown in Figure 1-10 (left). Use another setting of your compasses and repeat the process (right). Connect the intersections of these arcs with a straight line. Does this second line divide $\overline{AB}$ into two parts having the same measure? Will this procedure always divide a line segment such as $\overline{AB}$ into two parts having the same measure?

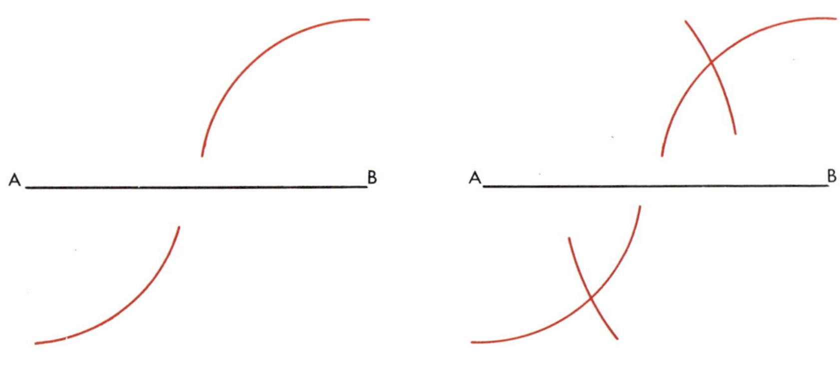

Figure 1–10

Miscellaneous Exercises

Experiment 10: Draw any four-sided figure on your paper. Connect the opposite corners with portions of straight lines. These connecting lines cross each other and are each divided into two parts by the point of intersection. Measure the lengths of the parts of the intersecting lines. Measure the angles formed by these two lines at the point of intersection. Do you see a relationship present that will enable you to make a conjecture? If so, clearly state it and the conditions in a sentence.

Experiment 11: The following list of experiments is intended to check your powers of observation. First use only your eyesight to answer the questions. Then check these answers using the tools of geometry.

(a) Examine a pencil inserted for part of its length in a glass of water. As we observe it from the side of the glass, does it appear to be in one straight piece?

Examine the following diagrams.

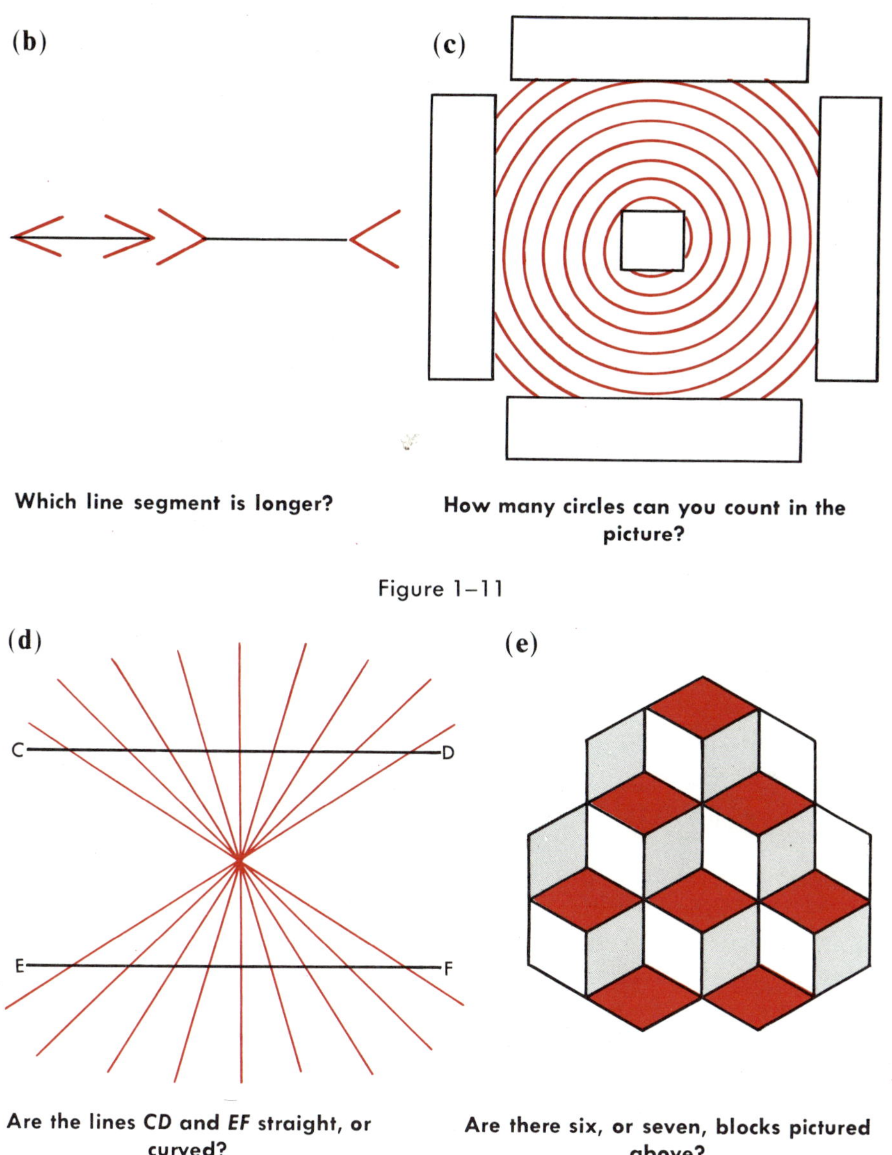

Figure 1–11

Figure 1–12

Experiment 12: Can you conclude from the previous experiment that observation is a satisfactory method of arriving at a conclusion? Is a conclusion based upon observation and measurement only probable, or is it a *necessary* conclusion?

The term "necessary conclusion" may not be clear to you, so we will examine some problems from algebra to illustrate its meaning. (The symbols below represent real numbers.)

EXAMPLE 1. If $4x = 12$, must $x = 3$?

EXAMPLE 2. If $a = b$ and $b = c$, is it necessarily true that $a = c$?

If your answers to the examples are "yes," then you realize that a *necessary* conclusion is a conclusion that is inevitable—one to which no alternative is possible.

Experiment 13: Draw a circle and two radii, forming angle 1. Then draw angles 2 and 3 as shown. Measure angles 1, 2, and 3. State a conjecture, if any, that you can make from this experiment.

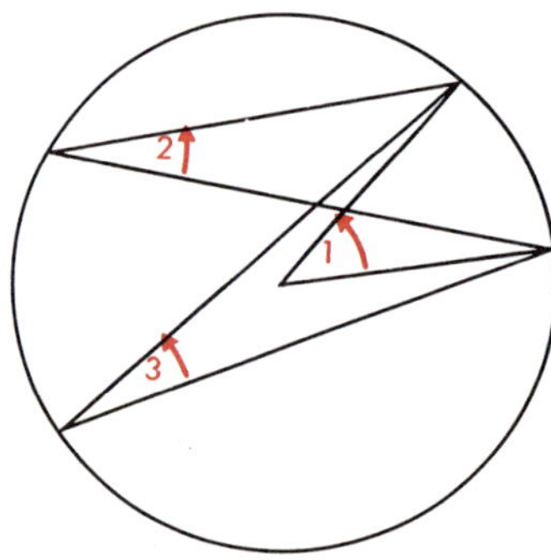

Figure 1–13

Experiment 14: Let us draw a figure with four sides having the same measure.

Step 1. Draw a *base* line and mark off $\overline{AB}$. Use the length of $\overline{AB}$ as the setting of your compasses and draw arcs from each end of the line and on the same side of the line.

Step 2. Using the same setting of the compasses, pick any point on one of the arcs and from this point draw an arc intersecting the remaining arc.

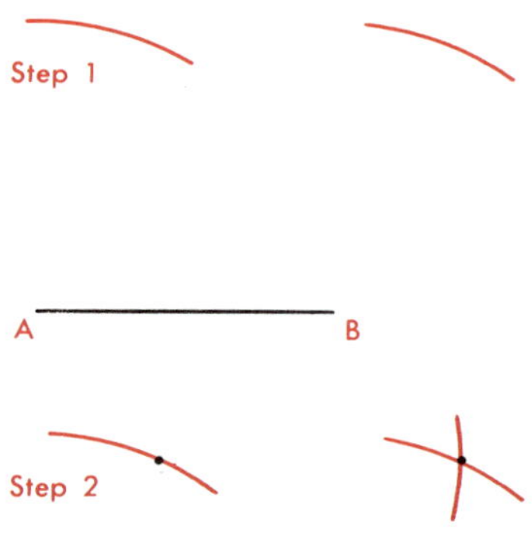

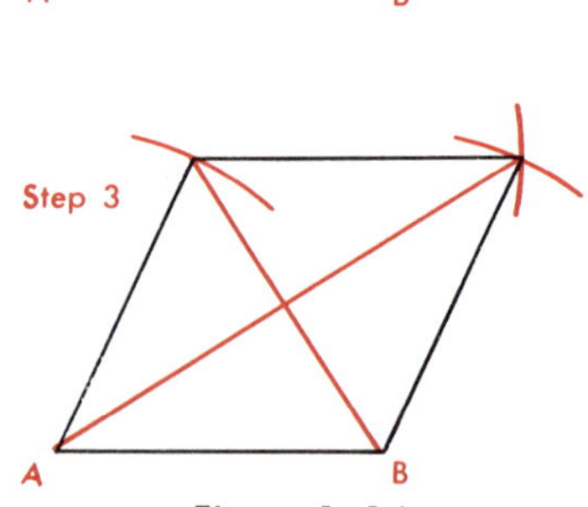

Figure 1–14

Step 3. Connect the points as shown.

Compare the portions of the lines connecting the *corners* of this four-sided figure. Measure the angles formed by these lines at their point of intersection.

Is it possible to make a conjecture from this experiment? If so, write a sentence clearly stating the conditions and your conjecture.

Experiment 15: Draw a triangle. Measure the distance between the midpoints of two sides of the triangle as shown (Figure 1-15). Compare this length with the length of the third side. Shift your ruler to compare the distance between another pair of midpoints with the length of the third side. Repeat the experiment with another triangle of a different shape. Are the results the same? State the conditions completely and clearly and your conjectures, if any.

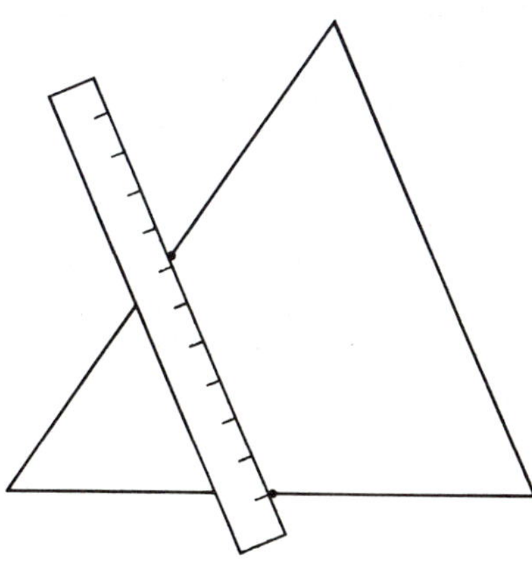

Figure 1–15

If you and your classmates reach the same conjecture using different-shaped triangles does this strengthen, or reduce, the probability of your conjecture's being true.?

Actually it should be clear that our conjecture becomes more probable as we increase the number of supporting experiments with triangles which vary widely in size and shape yet satisfy the basic conditions of the experiment. Can our conjecture ever be more than probable until we have conducted the same experiment upon every triangle it is possible to draw? How many such triangles are there?

Experiment 16: Draw a triangle each of whose sides is a different length.

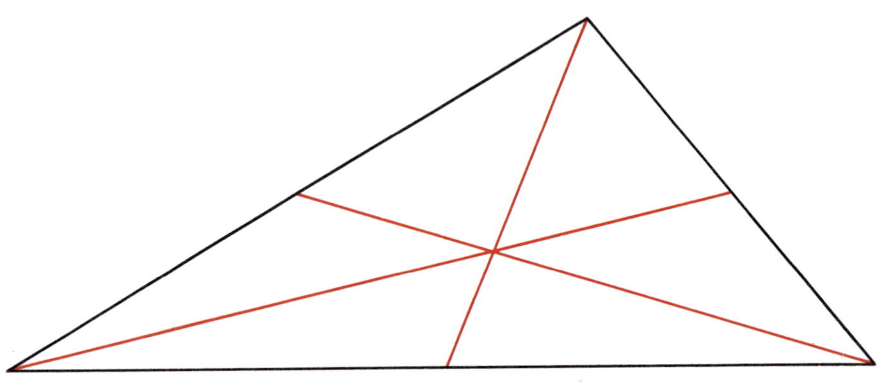

Figure 1–16

Connect each point of intersection of the sides (vertex) of the triangle to the midpoint of the opposite side by a segment of a straight line as shown. Can you be positive that these lines do, or do not, meet at a common point? Compare the length of each of these lines to the length from the vertex to the point they seem to have in common. Is there a conjecture you can make from this experiment? If so, state it and the conditions clearly in a sentence.

Have you and your classmates arrived at the same conjectures from the preceding experiments? If there are differences, how can you reconcile them? If you have all made the same conjecture from a given set of conditions does this prove that it is correct, or only increase the probability that it is correct? Are the conjectures you have made *necessary conclusions* or only *probable conclusions*?

Dr. Albert Einstein once stated, "No amount of experimentation can ever prove me right; a single experiment may at any time prove me wrong." What did he mean by this statement?

Would you want to base all your conclusions on observation and measurement? You will agree that it would be very desirable to be able to *prove* certain statements. This idea of proof is the central idea of geometry.

SUMMARY

We have drawn various figures and measured the lengths of segments and the sizes of angles produced under certain sets of conditions (Experiments 1–7). We arrived at certain conjectures in most of these experiments after observing that repeated trials of the same set of conditions seemed to produce a common result.

This type of reasoning, by which we proceed from particular results, obtained from experimentation, to a generalization called a conjecture, is called *inductive reasoning*. The inductive method is commonly used by scientists to develop new theories (tentative conclusions).

Induction is the process of reasoning by which a conjecture is reached from a study of particular facts. These facts are usually obtained through analysis of several representative samples of the same set of conditions. In this process a common property of several experiments is selected and the tentative conclusion or conjecture is reached that this property fits all such cases, although we did not and usually could not examine *all* cases. A conjecture is proved and becomes a firm conclusion if it is possible to verify every case.

This leads us to see that inductive reasoning has several weaknesses:

1. It depends upon observation and measurement and, as we have determined, neither is absolutely accurate.
2. It usually arrives at a possible conclusion before all possible cases have been studied. We say we "jump to conclusions" by this method. *Any exception to a conjecture proves that it is false.*

We may then state that inductive reasoning merely establishes a degree of probability and never proves the conjecture obtained by it.

This does not mean that reasoning from data secured through observation and measurement should be or could be discarded. The properties of geometric figures were first studied by induction and new ideas of geometry are still being discovered inductively. This process has been and is being used to discover laws and principles in such fields as physics, chemistry, biology, medicine, and economics. However, the tentative or "less than certain" nature of the conclusions drawn by inductive reasoning should leave us with a desire to find a method of proof in which the conclusion is no longer probable but necessary. Mathematics is sometimes defined as "the science of necessary conclusions."

LANGUAGE AND LOGIC

At this point in our study of geometry very few of us could outline a method of reasoning which would always give us a definite and unquestionable conclusion. Most of us would agree that different individuals might reach different conclusions from the same set of conditions if they did not agree upon the meanings of the words used to state the conditions.

To illustrate the importance of the meaning of words and phrases let us examine the following problems in logic.

1. The student council of Leesville High School adopted a regulation that "Each student organization in the school must be sponsored by a teacher." A group of students in this school wished to organize a Reading Club and invited the school librarian to be their sponsor. In view of the regulation adopted by the council, is it possible for her to serve in this capacity?
2. As Peggy Miller left for the junior class party, which she had eagerly anticipated, her mother said, "Now, Peggy, do not stay out late." Peggy replied, "No, Mother, I won't." When she returned to her home following the party, the clock read 12:50 A.M. Had she been faithful to her promise?
3. George Beckwith, a high school student, had been left in charge of the Y.M.C.A. recreation room. When he made an effort to quiet Dick Morrel, who had become noisy and rough, Dick said to him, "You have no right to be in charge here anyway." Then, in support of his statement, Dick quoted a ruling of the local group to the effect that, "Only a person of authority should ever be put in charge of the recreation room." Was Dick correct?
4. The regulations of a certain manufacturing plant state that, "Any employee guilty of chronic absenteeism will be removed from the plant payroll and will not be rehired. A chronic absentee is defined as an employee who missed work without an acceptable excuse for four or more days in two consecutive months or an employee who is consistently absent for short periods of time over a protracted period."

   (a) During the month of October, Bob Nye was absent three days and during the month of November in the same year he was absent four days. He had acceptable excuses for only two absences. According to the rule, should he be dismissed?

   (b) Rose K. returned to her work in the plant one Friday morning after an absence of a day and a half. On examining her record, the Personnel Director found that over a period of ten months she had been absent fifteen times for a total of fourteen days, no single absence being longer than two days. Should she be dismissed?

From a consideration of these problems it is possible to recognize one of the characteristics so essential to precise and careful thinking. One student tried to express this by saying, "I can't think clearly until I know what I am talking about." Another wrote, "People cannot agree on a conclusion when the words they use mean different things to each of them." While these statements reflect some idea of one characteristic of careful and precise thinking neither of them expresses it very clearly. Can you express it more effectively.?

Vocabulary List

geometry	Aristotle	degrees
Euclid	inductive	measure
Pythagoras	induction	conclusion
Thales	compasses	conjecture
Plato	protractor	tentative

Chapter Review

1. Measure the angle determined by segments $\overline{AD}$ and $\overline{AB}$ in the figure below. Measure segments $\overline{AC}$, $\overline{DC}$ and $\overline{BC}$. State any conjectures which follow from this experiment.

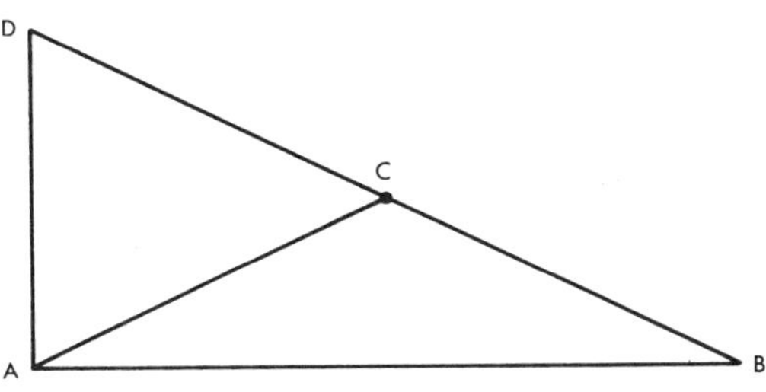

Figure 1–17

2. List the measure of each numbered angle in the figures below. List the sum of the measures of the numbered angles for each figure. How do these sums compare? State any conjectures which follow from this experiment.

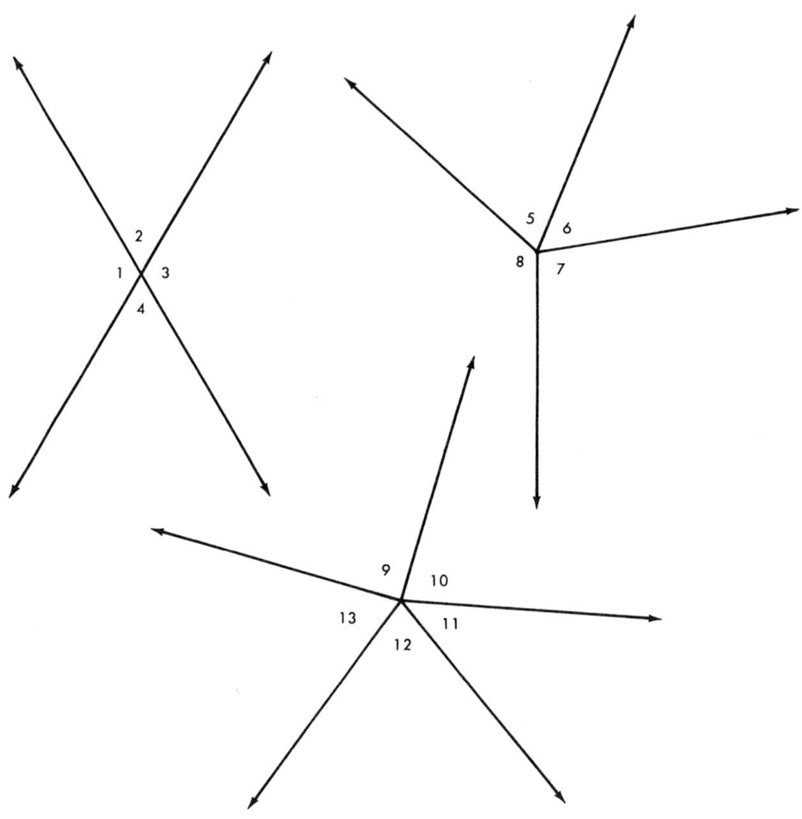

Figure 1–18

3. Draw a circle with your compasses. Draw a diameter of the circle. From any point on the circle draw straight lines through the ends of the diameter. Measure the angle between these two lines. Choose another point on the circle and draw two lines through the ends of the diameter. Measure this angle. Repeat this experiment with other diameters of this circle, and with other circles. State any conjectures which follow from this experiment. Try this with a line through a circle which is not a diameter.

4. List the measures of the numbered angles. What relationship probably exists between segments $\overline{AD}$, $\overline{EF}$ and $\overline{BC}$? Measure segments $\overline{AE}$, $\overline{BE}$, $\overline{DF}$, $\overline{CF}$, $\overline{AD}$, $\overline{BC}$ and $\overline{EF}$. Compare the sum of the measures of $\overline{AD}$ and $\overline{BC}$ to the measure of $\overline{EF}$. State any conjectures which follow from this experiment.

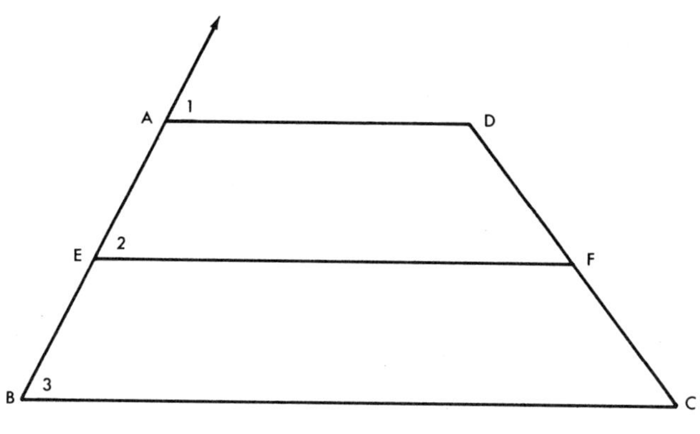

Figure 1–19

5. List the measures of each of the angles of the two figures pictured below. Compare the measures of $\overline{AC}$ to $\overline{DF}$, of $\overline{AB}$ to $\overline{DE}$, of $\overline{BC}$ to $\overline{EF}$. State any conjectures which follow from this experiment.

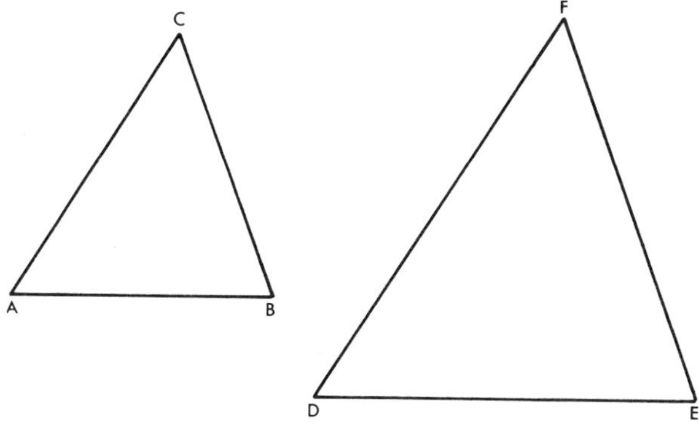

Figure 1–20

6. The picture below represents two lines intersected by a third so that angles 1 and 5 have the same measure. List the measures of the other angles. State any conjectures which follow from this experiment.

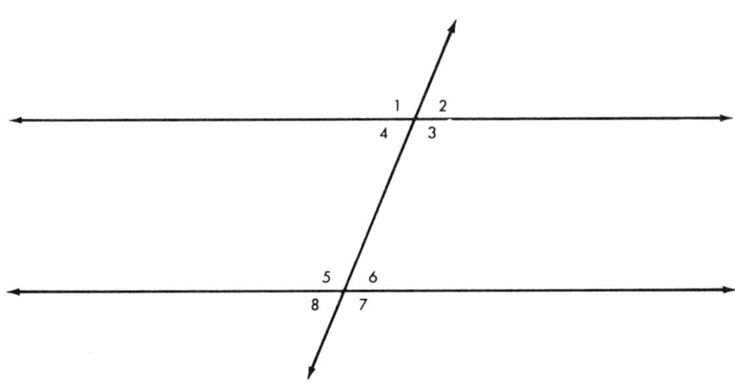

Figure 1–21

Chapter 1 Test

1. In Figure 1-22, measure $\angle A$ and side BC. Measure $\angle B$ and side $\overline{AC}$. Measure $\angle C$ and side $\overline{AB}$. State any conjectures which follow this experiment.

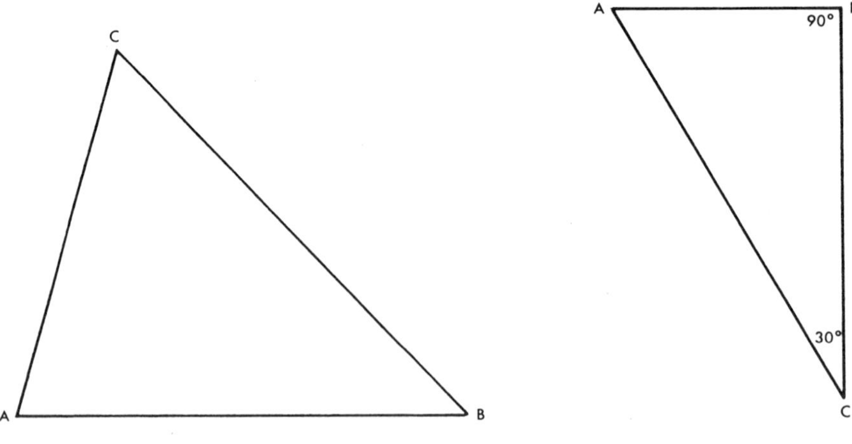

Figure 1–22 Figure 1–23

2. Measure the angles and segments of Figure 1-23. State any reasonable conjectures which follow from this experiment.
3. We have concluded that the meanings of words are very important to our understanding. Attempt to write the definition of the word "segment" used in this test.

True-False

4. Pythagoras is credited with writing the first logically arranged geometry book.
5. Compasses are used to find the measure of angles.
6. The conjectures developed from experimentation are only tentative.
7. The development of conjectures from experimentation is called inductive reasoning.
8. The measures we make of angles are usually exact.

2

Points, Lines, and Planes

The questions posed at the end of Chapter 1 illustrate the need for precise definitions. In geometry we will carefully define the words we use. We must, however, assume your understanding of the meanings of the commonly used words of our language.

To define a term we should use simpler terms which have previously been defined. You should realize that this is not entirely possible. There must be some basic, "simplest terms" which are undefined. For example, let us attempt to define a triangle. Would you say that a triangle is a three-sided figure? What is a *three-sided figure*? (If we say that it is a triangle we have gained nothing.) We need to define the word *side*. Is a *side* a line, or a surface such as the side of a box? If we agree that *side* means *line*, then what does *line* mean? A series of points? What is a point?

UNDEFINED TERMS

In our attempt to define *triangle* we have proceeded through a series of words which might be the basis of a definition. But each new word presented seems to result in more words needing to be defined. We can see that further efforts to define *point* would only lead to other undefined terms or to circular definitions.

For instance, in most dictionaries the word *time* is defined as period or duration, or in terms of itself. Period and duration are defined as time. Examine the definition of *equal* for another example of a circular definition.

Since we cannot define all of the words we use, we must accept some words as primary, or basic, undefined terms.

Let us accept the words *point*, *line*, *surface*, and *plane* as *undefined terms*. Some of the other undefined terms of mathematics are *number*, *order*, *all*, *every*, and *location*.

DEFINITIONS

A definition is merely an agreement as to the meaning of a word or phrase. Examine several definitions from your English text or from the dictionary. For example: (1) a noun is a word which names a person, place, or thing. (2) A fish is an animal, usually with a scaly body and limbs modified into fins, living in water, and breathing through gills and lungs.

What are the characteristics of a good definition?

1. *A good definition names the object being defined and places it in its smallest class.*

   (A noun is a word. A fish is an animal.)

2. *A good definition gives the characteristics which distinguish the object being defined from all other members in the class without being wordy or using unknown terms.* (Explain how this is done in the above definitions.)

3. *A good definition is reversible.*

   That is, the subject phrase and the predicate phrase may be interchanged and this new statement is acceptable. For example: "A word which names a person, place or thing is a noun" is the reverse of the previously stated definition of "noun."

Exercises

Using the criteria for a good definition, analyze the following:

1. A verb is a word used to show action or state of being.
2. A frog is a fish.
3. Snow is frozen water vapor in the form of white, feathery flakes or crystals.
4. Honor is respectful regard.
5. A horse is an animal.
6. A tree is a large, woody plant with a high main trunk, branches and leaves.
7. A pronumeral is used in place of a number.
8. A hood is a covering for the head.
9. Glory is honor given to someone or something by others.
10. Right angles are angles which have the same measure.
11. A magician is an adept of prestidigitation.
12. A triangle is a geometric figure with three equal sides.

Define the following without help from any source:

13. Spoon.
14. Pencil.
15. Sock.
16. Goal.
17. Between.
18. Courage.

SETS

The word "set" has come to be one of the most useful and used words in mathematics. For that reason it is the first word to be defined.

2.00 A set is any collection of objects, ideas or symbols.

2.01 The set with no members is called the empty set or null set.

2.02 A set, all of whose elements are members of a second set, is said to be a subset of the second set. The set is said to be contained in the second set.

EXAMPLES

Sets: The letters a, b, and c are a set. The numerals 6, 8, 10, and 12 are a set. Sets such as those above are commonly shown enclosed in braces: $\{a, b, c\}$, $\{6, 8, 10, 12\}$. Do the members of your basketball team constitute a set? Name other examples of sets.

Null Set: The set of symbols common to the sets in the previous example is the empty or null set. This set is symbolized by $\{\ \}$ or ϕ. We consider there to be *only one empty set* and that *this set is a subset of every set.* Name other examples of the null set.

Subsets: The set $\{6, 8\}$ is a subset of the set $\{6, 8, 10, 12\}$. This is pictured in a Venn diagram below. (These diagrams are named after the mathematician who popularized their use.)

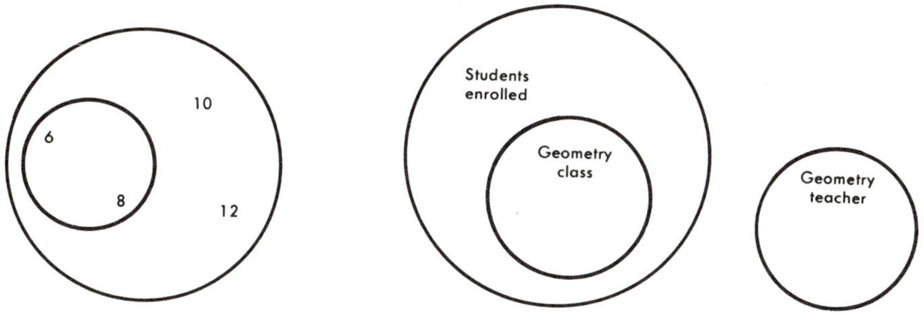

Figure 2–1

Figure 2–2

The student enrollment of your school is a particular set. We will call this the universal set of students. The members of your geometry class constitute a subset of this universal set. Is your geometry teacher also a subset of this universal set?

2.03 Disjoint sets are sets which have no common elements.

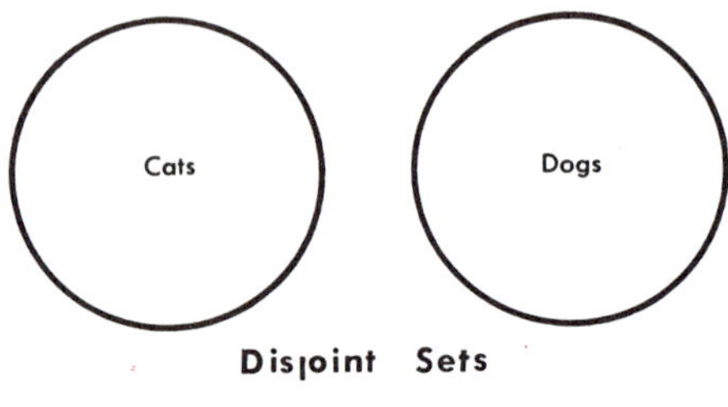

Disjoint Sets

Figure 2–3

2.04 The intersection of two sets is the set of all elements that belong to both of them.

2.05 Two sets intersect if there are one or more elements that belong to both of them.

This means that the *intersection* of two sets may be the empty set; however, if two sets *intersect*, they must have one or more common elements.

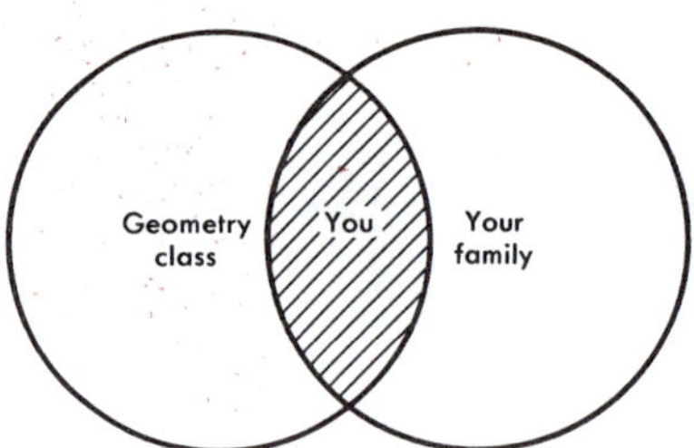

The shaded portion illustrates the intersection of your geometry class and your family

Figure 2–4

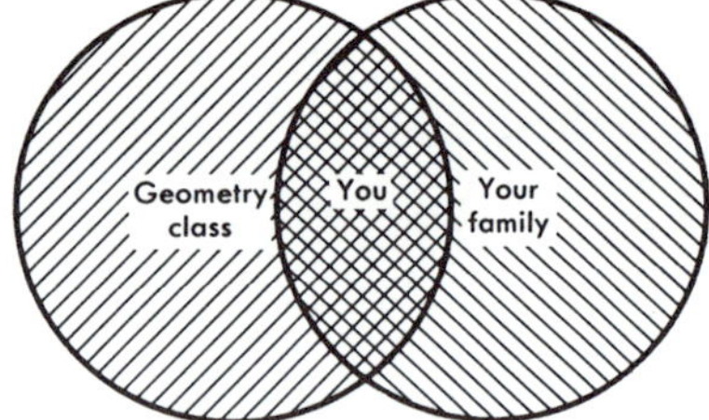

The shaded portion illustrates the union of your geometry class and your family

Figure 2–5

2.06 The union of two sets is the set of all elements that belong to either or both of two sets.

Exercises

Use the following sets to answer questions 1–8.

$A = \{0,1,2,3,4,5\}$ $C = \{2,4,6,8\}$ $E = \{-2,-1,0,1,2\}$
$B = \{1,3,5,7,9\}$ $D = \{0,1\}$

1. List the elements of the union of sets A and C.
2. List the elements of the intersection of sets A and C.
3. List the elements of the intersection of sets B and C.
4. Set D is a subset of which set or sets?
5. Use labeled and shaded circles to represent the intersection of sets A and E.
6. Which of the above sets are disjoint sets?
7. Use labeled and shaded circles to represent the union of sets A and E.
8. Name the pairs of sets that intersect.
9. Use set terminology to describe the intersection of the set of birds and the set of rabbits.
10. List the set of prime numbers between 1 and 10.
11. List the intersection of the sets of real number pairs (x, y) satisfying $x + y = 3$ and $x - y = 2$.
12. Write the reverse of the definition of section 2.03. Of section 2.04.

POINTS

We stated on page 21 that certain terms are going to be undefined, but that does not mean that we should not have some common understanding of the idea the word conveys. For example, point is truly undefined but we can agree that it is a location in space. Location is also undefined. What is a location in space? Does it have any size? Let us agree that it does not have any of the three dimensions (length, width, or height) of our familiar three-dimensional space.

We will think of space as the set of all points and of geometric figures as particular *subsets* of this *universal set.*

Is this dot (.) a point? Is a picture of a horse actually a horse? The dot and the picture are symbols which represent an idea and an object. Man decided that he needed the concept of point so he invented it. Let us understand that a *geometric point* does not exist physically. It is only the idea of a location in space. *Point* in this course will mean geometric point.

We shall use a capital letter to name a point. In Figure 2-6 the dots represent points named by the letters A, B, C and D.

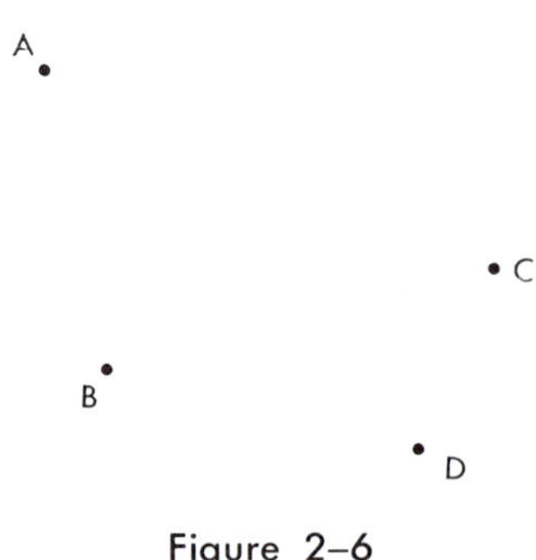

Figure 2–6

2.07 Distinct points are different points. (Not the same point.) Distinct will have the same meaning when applied to all geometric figures (sets of points). That is, one set is not a subset of the other.

LINES

Let us now consider a subset of the universal set of points. Since there are many subsets of the universal set we will start with a subset which might be represented physically by a piece of thread of unlimited length. We will call such a set of points a curved line. If this set is so arranged that it could be represented by unbreakable thread drawn as tautly or as straight as possible, we will name it a straight line. The geometric line has one dimension, length, and, like a point, has no width or height.

Clearly not all continuous one-dimensional figures are straight lines. However, unless otherwise stated, we shall use the word "line" to mean "straight line." We will picture our mental concept of a geometric straight line with figures similar to that below.

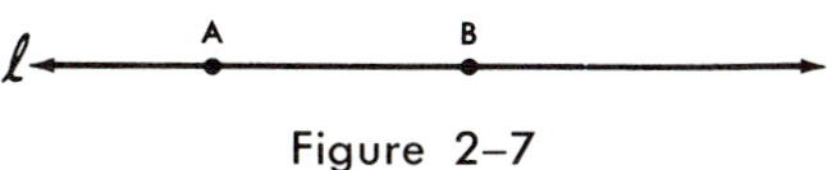

Figure 2–7

The arrowheads indicate that the line represented extends endlessly in both directions. A lower case letter, (*l* in the picture) can be used to name the line. More commonly we name two different points of a line to denote the line. In Figure 2-7, *l*, $\overleftrightarrow{AB}$, or $\overleftrightarrow{BA}$ may be used to identify the line. The double-arrow symbol assures us that we mean a line with unlimited length.

ASSUMPTIONS

Just as sports and games have guiding rules, organizations and institutions usually operate within the framework of certain principles they adopt. These principles are working agreements or assumptions from which other statements may be derived. It is common to regard these assumptions as true, but they are really guiding principles which are accepted as suitable for a particular situation. There are often ground rules which modify regular rules under certain circumstances.

The constitutions of governments, clubs, and business organizations contain statements of certain policies and principles which guide the activities of the group.

Mathematics, too, is based upon certain fundamental rules, (assumptions), concerning the nature of the mathematical objects that we are studying. The assumptions used as the guiding principles in mathematics are called *axioms* or *postulates*. Axioms have universal applicability while postulates refer to specific subject matter. In mathematics an axiom applies to any mathematical situation, but a postulate applies only to geometry. These axioms and postulates represent a convenient set of rules which shall guide our thinking and which we shall accept without proof or the need of proof. In this book we shall refer to all of these fundamental rules as *assumptions* except in the case of a basic list of axioms common to algebra and geometry.

DEVELOPING AN ASSUMPTION INDUCTIVELY

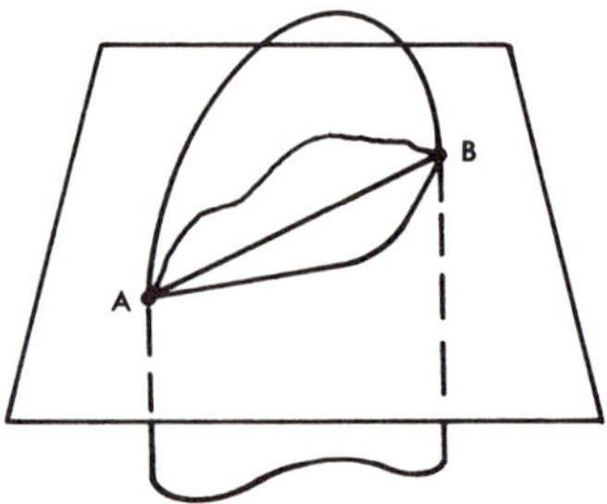

Figure 2–8

How many paths are there between Points A and B? What is the shortest path? How many distinct lines containing point A can also contain point B? How many distinct lines can contain point A? Our answers to these questions are the basis for certain assumptions.

2.08 *Assumption: The number of lines that may contain a given point is unlimited (an infinite number).*

2.09 *Assumption: Every straight line contains at least two distinct points.*

2.10 *Assumption: Given any two distinct points there is exactly one straight line that contains both of them. (This is commonly stated as "two distinct points determine a line.")*

2.11 Points contained in the same straight line are called collinear points.

$\overleftrightarrow{AD}$ and $\overleftrightarrow{EF}$ intersect in point C as represented in Figure 2-9. A, B, C, and D represent collinear points. Do E, C, and F represent collinear points?

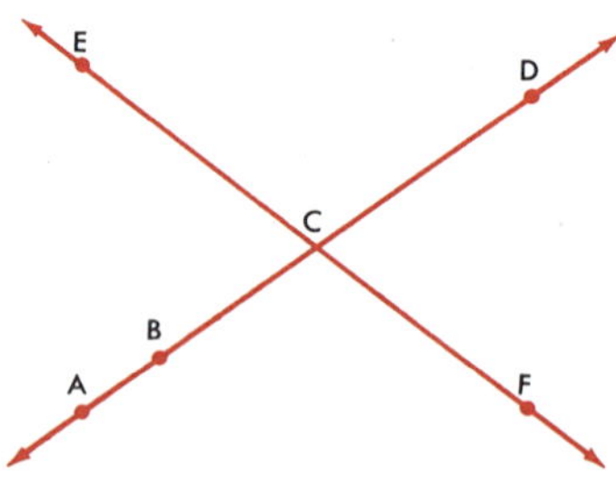

Figure 2–9

2.12 Three or more lines, rays or segments containing the same point are said to be concurrent.

DEVELOPING A THEOREM

The pattern of mathematics is to first define our terms; second, to agree upon a set of assumptions; then third, to develop proofs of conjectures logically related to the basic definitions and assumptions. When proven, these conjectures will be called "theorems," as defined below.

2.13 A statement is a meaningful declarative sentence which is either true or false, but not both.

2.14 A theorem is a statement whose truth has been established by the use of acceptable logic.

Now let us derive a theorem by using some of the definitions and assumptions that we have agreed upon. Let us remember that *if the definitions and axioms are accepted and the method of reasoning is valid, the theorem must also be accepted.*

Reasoning: The definition of intersect (2.05) states that if two lines intersect they have one or more common points (elements). Let us suppose that two lines have two points in common. Assumption 2.10 states that these lines cannot be distinct straight lines. It follows that:

2.15 THEOREM

If two distinct straight lines intersect, their intersection is exactly one point.

Exercises

1. List all possible names for $\overleftrightarrow{EF}$ in Figure 2-9, using only the letters shown.
2. Name the intersection of $\overleftrightarrow{AD}$ and $\overleftrightarrow{EF}$ in Figure 2-9.
3. In Figure 2-9, is every point of $\overleftrightarrow{AD}$ a point of $\overleftrightarrow{BD}$?

4. Draw a picture representing $\overleftrightarrow{FG}$, $\overleftrightarrow{BF}$ and $\overleftrightarrow{EF}$, where B, G, and E are non-collinear points. What name could you apply to this set of lines that would most clearly express their relationship?

5. Must a set of concurrent lines also be a set of intersecting lines?

6. Must a set of intersecting lines also be a set of concurrent lines?

7. A teacher at a university may also be a student of that university. Does this individual represent an intersection or a union of the set of teachers and the set of students?

8. Are $\overleftrightarrow{AB}$ and $\overleftrightarrow{AC}$ disjoint lines? Explain.

9. If we assign to space the three dimensions of length, width, and height, name the number of dimensions of a point. A line.

10. If $\overleftrightarrow{QR}$ and $\overleftrightarrow{ST}$ are distinct, the intersection of $\overleftrightarrow{QR}$ and $\overleftrightarrow{ST}$ could be: (**a**) a point, (**b**) more than one point, (**c**) the null set, (**d**) all points of $\overleftrightarrow{QR}$ and $\overleftrightarrow{ST}$.

Figure 2–10

11. In Figure 2-10, if A and B are points, what authority do we have for stating that both x and y cannot represent straight lines?

12. If l is a line and P is a point of l, what authority allows us to say that there is another point of l?

13. Conditions: Oil is lighter than water. A quantity of oil and water are mixed in a container. State a theorem which necessarily follows from these conditions.

14. Conditions: X and Y are metals. X is heavier than Y; Y has a lower melting temperature than X. State a theorem which logically follows from these conditions.

15. Assumption: There are at least three noncollinear points. Use this assumption together with our present structure of geometry to explain that there are at least two lines containing each point.

PLANES

Plane is one of the undefined terms of geometry. In our minds it is a set of points, a subset of space, of unlimited extent. Physically we associate the idea of a plane with a smooth, flat surface. A desk top or a large plate glass window may suggest planes to us.

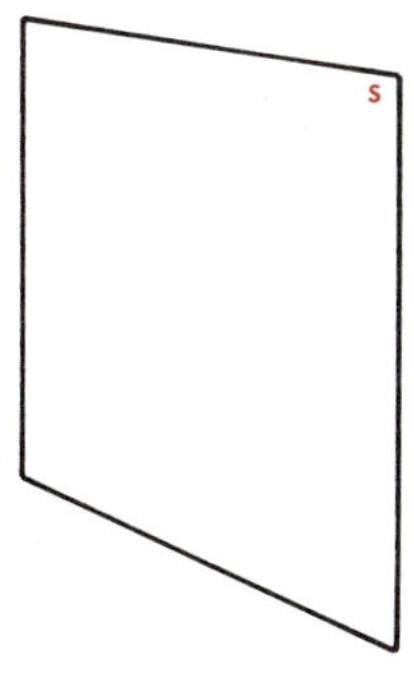

Plane s

Figure 2–11

2.16 *Assumption: A plane separates space into two disjoint sets of points.*

2.17 The two disjoint sets of points separated by a plane in space are called half-spaces. The plane is the face of each half-space but is not a part of either half-space.

2.18 Lines or points contained in the same plane are called coplanar lines or points.

2.19 *Assumption: Every plane contains at least three noncollinear points. (Review 2.09 and exercise 15, page 30.)*

2.20 *Assumption: Space contains at least four distinct noncoplanar points.*

The drawings which represent planes are commonly labeled with a single lower case letter as in Figure 2-11. Sometimes capital letters are placed at opposite "corners" of the drawing. The picture of plane AB in Figure 2-12 is an example.

If M and N are points of plane AB, is $\overleftrightarrow{MN}$ a subset of plane AB? Does $\overleftrightarrow{MN}$ separate plane AB into two disjoint sets of points?

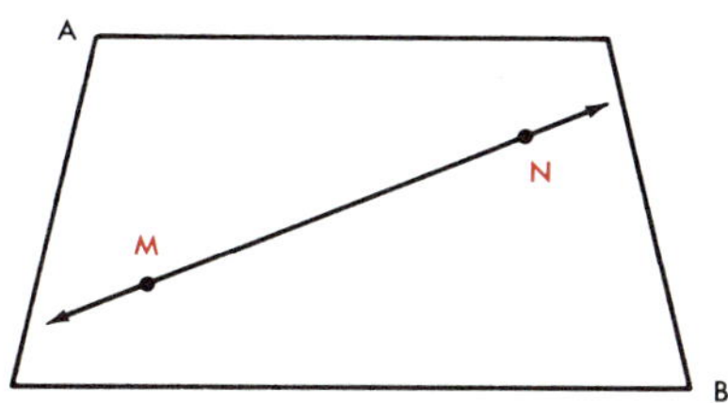

Figure 2–12

2.21 *Assumption: If a plane contains two points of a straight line then all points of the line are points of the plane.*

2.22 *Assumption: A line separates a plane into two disjoint sets of points.*

2.23 The disjoint sets of points of a plane separated by a line are called half-planes. The line is the edge of each half-plane but is not a part of either half-plane.

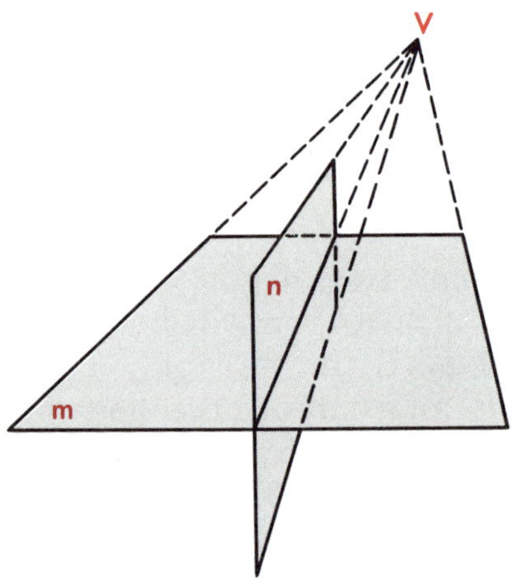

Figure 2–13

To represent horizontal plane m and vertical plane n on your paper, pick a convenient vanishing point (V) as shown in Figure 2-13. Two edges of the drawing recede toward the point while the front and back edges are shown as parallel lines. The nearer lines are made darker than the lines further from the eye to help produce the illusion of three dimensions. Study art books and mechanical drawing books that show perspective drawings having more than one vanishing point.

2.24 *Assumption: If two distinct planes intersect, their intersection is one and only one line.*

EXERCISES

1. In a given plane, what is the union of a line and the two half-planes determined by the line?
2. Name the union of a plane and the two half-spaces separated by the plane.

Figure 2-14 represents a figure formed by portions of planes.

3. In Figure 2-14, are points A, B and E coplanar? Are A, E and C coplanar? Are $\overleftrightarrow{HG}$ and $\overleftrightarrow{AE}$ coplanar?

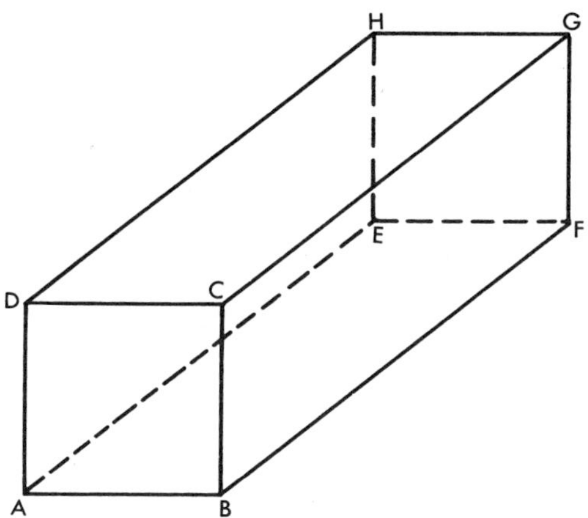

Figure 2–14

4. In Figure 2-14, are $\overleftrightarrow{FG}$, $\overleftrightarrow{BF}$ and $\overleftrightarrow{EF}$ concurrent? Are they coplanar?
5. In Figure 2-14, are $\overleftrightarrow{FG}$ and $\overleftrightarrow{EF}$ coplanar? Are $\overleftrightarrow{AC}$ and $\overleftrightarrow{CG}$ coplanar?
6. Using the conditions for Exercise 9, page 30, name the number of dimensions of a plane. What are they?
7. Make a drawing showing a line intersecting a plane in a point.
8. Make a drawing showing three planes containing the same line.
9. Illustrate two non-coplanar lines.
10. Use the thought process preceding section 2.15 to develop the theorem: "If a line and a plane are distinct and intersect, they have exactly one point in common." (This point is commonly called the *foot* of the line.)
11. Review section 2.19, then discuss the possibility of a plane containing more than three points.

THE NUMBER LINE

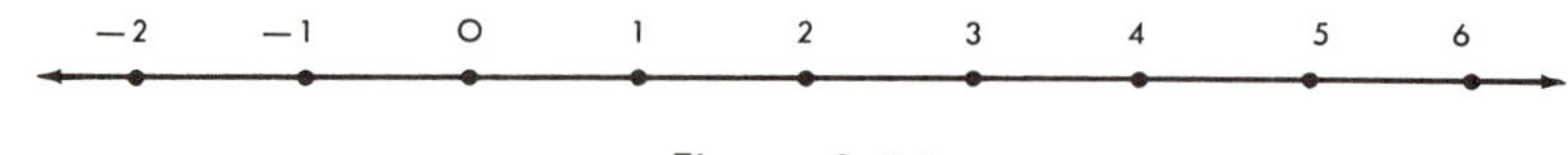

Figure 2–15

Can you think of another way to place a set of numbers (as represented by their numerals) in a one-to-one correspondence with the above set of points (as represented by the dots)? Of course you can. There are many ways.

There are numbers between 0 and 1, between 1 and 2, and so on. Is there a number between $\frac{5}{8}$ and $\frac{6}{8}$? Name it if so. Is there a number between $\frac{3}{153}$ and $\frac{4}{153}$? Name it if so. Is there a number between any two numbers you can name? Explain. If we place the numbers in sequence as in Figure 2-15, is there a different point of the sequence of points that can be paired with each number of this never ending set of numbers? Based upon this thought process and a comprehensive study of the real number system, mathematicians have adopted the following assumption.

2.25 *Assumption: To every point of a straight line there corresponds exactly one real number and to every real number there corresponds exactly one point of the line.*

The assumption of 2.25 establishes what is called a *coordinate system.* That is, each number corresponding to a point of the line is called the coordinate of that point. The point is called the graph of the number.

BETWEENNESS FOR POINTS OF A LINE

Figure 2–16

Following the concept of section 2.25 we can label the points of a line so that 2 corresponds to point A and 4.8 corresponds to point C, as shown in Figure 2-16. Then any point of the line whose number is between 2 and 4.8 is *between A and C.* This is possible only because man created the real numbers with an "order property." That is, 5 is less than 6, 7 is greater than 6, and so forth.

2.26 Any point B of $\overleftrightarrow{AC}$ is said to be between A and C if the numbers corresponding to A, B and C occur in order.

2.27 *Assumption: There is a relation, called between, for collinear points such that:*

(a) For every three distinct collinear points exactly one is between the other two.

(b) If point B is between points A and C, then A, B, and C are three distinct points of a line.

(c) If A and B are two points, then there is at least one point C between A and B, and at least one point D such that B is between A and D. (This is called the line-building assumption.)

2.28 Two points not on a line but in a plane containing the line are said to be on the same side of the line if no point of the line is between the two points. Two points not on a line are said to be on opposite sides of the line if there is a point of the line between the two points.

SEGMENTS, RAYS, AND HALF-LINES

2.29 The set of points of a straight line between, and including, any two designated points of the line is a segment of the line. The two designated points are called the endpoints of the segment.

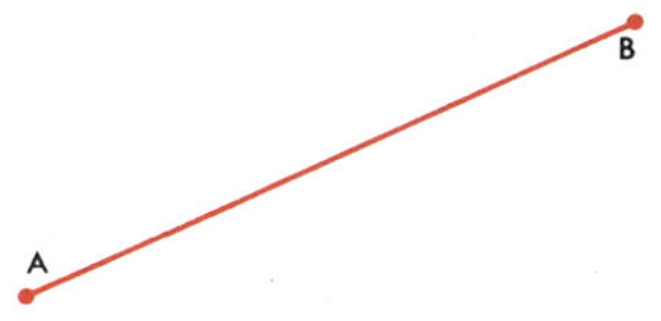

Figure 2–17

In Figure 2-17, the union of points A and B and the points of $\overleftrightarrow{AB}$ between A and B comprise segment $\overline{AB}$. This segment will be named by the symbol $\overline{AB}$ or $\overline{BA}$. The bar above the letters will denote the segment.

2.30 The two disjoint sets of points of a line separated by a point of the line are called half-lines. The point is called the origin of each half-line but it is not a part of either half-line.

2.31 The union of a half-line and its origin is called a ray.

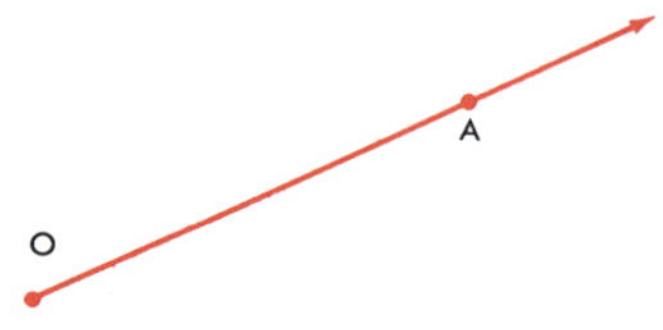

Figure 2–18

Figure 2-18 represents a ray whose origin, sometimes called *endpoint* or *vertex*, is point O. A ray is named by two points of the ray in which the point named first is always the origin. This will be symbolized by the two letters with an arrow placed above the letters; $\overrightarrow{OA}$ in Figure 2-18.

2.32 If B is a point of $\overleftrightarrow{AB}$ between A and P, then $\overrightarrow{BA}$ and $\overrightarrow{BP}$ are called opposite rays.

Exercises

1. If A, B and C are collinear points and B is between A and C, name a pair of opposite rays of $\overleftrightarrow{AC}$.
2. Name the number of endpoints of a segment. A ray. A line.
3. Describe the intersection of $\overrightarrow{BA}$ and $\overrightarrow{AB}$.
4. Describe the union of $\overrightarrow{BA}$ and $\overrightarrow{AB}$.
5. Describe the union of the half-lines determined by opposite rays $\overrightarrow{GR}$ and $\overrightarrow{GT}$. What is the intersection set of these half-lines?
6. Does a ray contain fewer points than a line contains? Explain.
7. Name six geometric figures determined by points M and N.
8. How many points are there between distinct points C and D?
9. State the authority for saying that between any two points of a line there is a third point.

10. Is there a real number between any two given real numbers?
11. Is your answer to Exercise 10 consistent with the assumptions of sections 2.25 and 2.27(c)? Explain.
12. State the authority for saying that a line has no endpoints.
13. Define triangle using the words "union" and "segments."
14. Develop a formula which will give the number of segments that can be drawn between a given number of non-collinear points (3, 4, 5, etc.).

MEASUREMENT OF DISTANCE

Since applications of geometry to our physical world deal with the concept of distance it is necessary to determine that distance. The choice of a unit of measure is arbitrary. We choose that unit most convenient for our purposes whether it be inches, feet, miles, centimeters, meters, or some other unit.

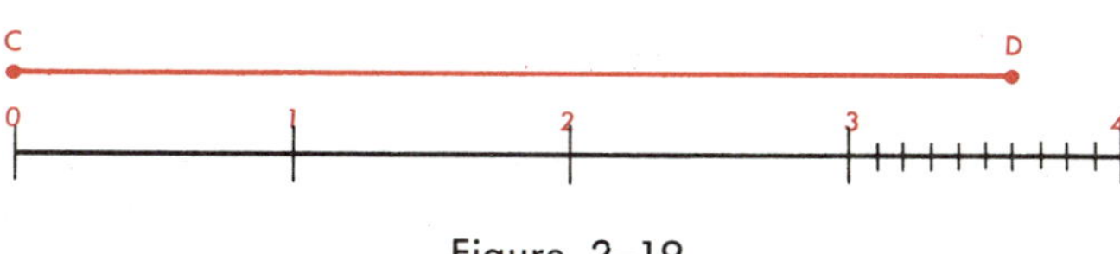

Figure 2–19

To measure a segment ($\overline{CD}$) a number of unit distances are laid off in sequence on the segment, then counted to determine the measure of the segment *in terms of our selected unit.* If the unit we selected is not contained in the segment an integral number of times we then either select a new unit or give an approximate or fractional measure. The fractional measure is found by subdividing the basic unit into smaller units, usually tens (of hundreds, or thousands, etc.), allowing us to write the measure as a decimal fraction. Actually the fractional measure is usually approximate, although more precise than the original unit allowed.

The counting of the unit segments may be accomplished by numbering the starting point (C) zero (as in a coordinate system) and the end of each unit 1, 2, 3, 4 and so on. The number of the last unit (using an integral number of units when this is possible) then names the measure of the segment. This idea coupled with the assumption of section 2.25 prompts us to make the following assumption.

2.33 *Assumption: For every pair of distinct points there is one and only one positive real number called the distance between them as established by a coordinate system of the line determined by the points.*

2.34 The length of a segment is the measure of the distance between the endpoints in terms of a given unit.

The terms "length of $\overline{FD}$" or "measure of $\overline{FD}$ in terms of a given unit" are awkward to use as often as we will need to use them. Since the phrase represents a number, a simple symbol is desired. We will use "m $\overline{FD}$" to mean "the measure of $\overline{FD}$." For example, if M, N and P are collinear points and N is between M and P, then $m\overline{MN} + m\overline{NP} = m\overline{MP}$. (This statement may be proved with the use of coordinate geometry, but we are not in a position to do so at this time.) It does satisfy our intuition, however, so we shall make the following assumption.

2.35 *Assumption: If and only if M, N and P are collinear points and N is between M and P, is* $m\overline{MN} + m\overline{NP} = m\overline{MP}$. *(This is frequently called the Segment Addition Postulate.)*

CONGRUENT SEGMENTS

In general, congruence means correspondence or agreement. In geometry we think of congruent figures (symbol $\cong$) as being duplicates or replicas of each other. To satisfy our intuition that congruent figures do exist we make the following assumption.

2.36 *Assumption: A given geometric figure may be duplicated anywhere.*

2.37 Segments with the same measure are called congruent segments.

Let us remind you that $\overline{AB}$ is the name of a segment while $m\overline{AB}$ denotes a number which is the measure of $\overline{AB}$ in terms of a given unit. The statement $m\overline{AB} = m\overline{CD}$ will mean that the measures of the segments $\overline{AB}$ and $\overline{CD}$ are equal (the same number); while $\overline{AB} \cong \overline{CD}$ refers to the correspondence (congruence) of the two segments. Actually, the definition (2.37) states that $m\overline{AB} = m\overline{CD}$ and $\overline{AB} \cong \overline{CD}$ are equivalent statements; that one may replace the other. We should also comment that if A and B are points, then $A = B$ means that A and B are the same point. The equality sign always implies that the related symbols represent the same thing.

Upon the basis of sections 2.33, 2.36, and 2.37, the following is evident.

2.38 THEOREM

Given $\overline{MN}$ and a ray with endpoint O there is one and only one point P on the ray such that $m\overline{MN} = m\overline{OP}$, or $\overline{MN} \cong \overline{OP}$.

BISECTORS AND MIDPOINTS

2.39 A point M of any segment $\overline{AB}$ is the midpoint if M is between A and B and $m\overline{AM} = m\overline{MB}$.

2.40 The midpoint of a segment is said to bisect the segment. Any geometric figure that contains only the midpoint of the segment is said to bisect the segment.

Exercises

1. If $\overline{AD} \cong \overline{RS}$, is $m\overline{AD} = m\overline{RS}$? Explain.
2. If $\overline{AD} \cong \overline{RS}$, is $\overline{AD} = \overline{RS}$? Explain.
3. If $m\overline{AD} = m\overline{RS}$, is $\overline{AD} \cong \overline{RS}$? Explain.
4. How many lines can bisect $\overline{BC}$? How many planes?
5. If A is not collinear with B and C, how many lines can contain A and bisect $\overline{BC}$? How many planes?
6. Does a ray have a midpoint? Does a line have a midpoint?
7. Does a segment have only one bisector? Explain.
8. Is $\overline{AD} \cong \overline{AD}$? Can you prove your answer? Which axiom of real numbers compares with this congruency statement?
9. If $\overline{AD} \cong \overline{RS}$, is $\overline{RS} \cong \overline{AD}$? Can you prove your answer? With which axiom of real numbers does this compare?

10. If $\overline{AD} \cong \overline{MN}$ and $\overline{MN} \cong \overline{RS}$, is $\overline{AD} \cong \overline{RS}$? Can your prove your answer? With which axiom of real numbers does this compare?

11. Does a segment have only one midpoint? Write your answer to each of the following questions: Assume that $\overline{BC}$ has two midpoints, M and P. Then $m\overline{BM} = \frac{1}{2}(m\overline{BC})$ and $m\overline{BP} = \frac{1}{2}(m\overline{BC})$. Why? Therefore $\overline{BM} \cong \overline{BP}$. Why? It follows that M and P are the same point. Why? State the theorem that you have now proven.

PARALLEL AND SKEW LINES

2.41 Two lines are parallel if they are coplanar and do not intersect. We indicate two lines are parallel by writing "$\overleftrightarrow{AB} \parallel \overleftrightarrow{CD}$."

Segments and rays are said to be parallel when the lines containing them are parallel.

2.42 *Assumption: Two coplanar lines either intersect or are parallel.*

Does this also hold for two lines in space or can two lines in space be other than parallel or intersecting?

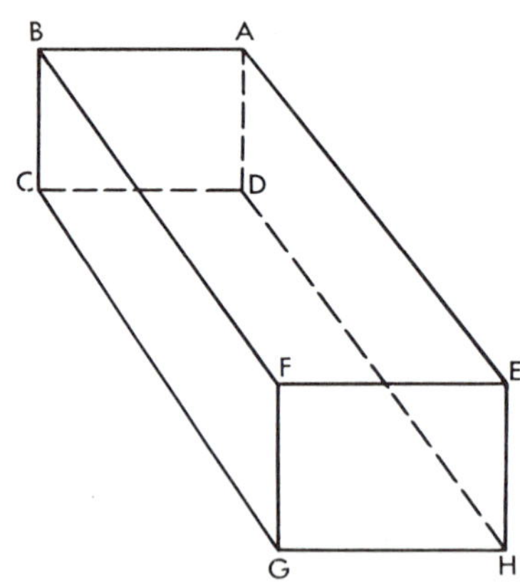

Figure 2–20

Figure 2-20 represents a three-dimensional "box." The edges of the box, $\overline{BC}$ and $\overline{AD}$, represent parallel segments. Also $\overline{BC} \parallel \overline{FG}$,

$\overline{AD} \parallel \overline{EH}$. What other segments seem to be parallel? What is the relationship of $\overline{FE}$ and $\overline{BC}$ of the box? Do they intersect? Are they parallel? Are these segments in the same plane?

2.43 Two non-coplanar lines are called skew lines. This means that *skew lines are not parallel and do not intersect.*

Exercises

1. In Figure 2-20, how many lines can be drawn through point A parallel to $\overleftrightarrow{BC}$? Can you justify your answer?
2. In Figure 2-20, how many lines coplanar with $\overleftrightarrow{BC}$ can contain point A but do not intersect $\overleftrightarrow{BC}$?
3. In Figure 2-20, how many lines can contain point A that are neither parallel to nor intersect $\overleftrightarrow{BC}$? What name describes the relationship of $\overleftrightarrow{BC}$ and one of these lines?
4. Must two planes intersect? Could they be parallel? Could they be skew? State the implied assumption. Compare it to 2.42.
5. Keeping in mind the criteria of a good definition (page 21), define (a) parallel planes; (b) parallel line and plane.
6. Draw a picture representing two parallel planes.
7. Use two letters to name the plane seemingly parallel to plane CF in Figure 2-20.
8. Explain how we might measure the distance between two parallel lines. Does your method fit into the structure of geometry that we have developed to this point? If not, what new terms and concepts must be developed and defined?
9. Suppose that our "space" consisted of the set of points in the interior of a circle. Would this change your answer to Exercise 1? Explain.
10. Suppose that our space consisted of the set of points in the surface of a sphere. Would this change your answer to Exercise 1? Explain.

Are the following statements acceptable in view of the preceding inductive reasoning?

2.44 *Assumption: Two parallel lines are everywhere equidistant.*

2.45 *Assumption: Two parallel planes are everywhere equidistant.*

2.46 *Assumption: Through a given point not on a given line there is one and only one line parallel to the given line.*

Assumption 2.46 is the famous Euclidean Parallel Postulate. This was one of the assumptions accepted by Euclid when he wrote the first geometry book, *The Elements* (See page 2.) We accept this assumption because it appeals to our intuition and because it is the most reasonable for our purposes. Many theorems in this course are proved on the basis of this parallel postulate.

Are you satisfied with the previous assumption? Does it seem consistent with your own experience of actual space? How do you know that the two lines, if extended indefinitely in either direction, will not meet? Euclid called this assumption a "self-evident truth." Would it be acceptable in a finite (limited) space?

NON-EUCLIDEAN GEOMETRIES

Some mathematicians have raised questions concerning the nature of this assumption and do not recognize it as "self-evident." The mathematicians Lobatchevsky of Russia, Bolyai of Hungary, and Gauss of Germany replaced this assumption with one that states that *more than one line can be drawn through the point parallel to the given line.* These men, working independently, produced a geometry just as consistent as that of Euclid, but in which there is the startling conclusion that the sum of the measure of the angles of a triangle is *less* than 180°. This is now called *hyperbolic geometry.*

A little later, Riemann, another German mathematician, replaced the parallel postulate of Euclid by the assumption that there are *no* parallel lines. He proved that in his geometry the sum of the measure of the angles of a triangle exceeds 180°.

Many a student, recognizing the inconsistency of these three conclusions, has asked, "Which is the true geometry?" How would you answer such a question? Would you agree with the student who said, "Even though the conclusions are inconsistent, they are all 'true', (valid), because each of them is consistent with the assumptions from which it was derived"? This episode of mathematical history is an excellent illustration of how a change in only one assumption can have a significant effect on a conclusion derived from it.

CIRCLES AND POLYGONS

As we begin to build more sophisticated geometric figures with the basic building blocks of points, lines, and planes, we immediately think of circles and polygons. We must agree on the definitions of these terms before we can discuss them further. Examine the following definitions and see if they satisfy your notions of circles and polygons.

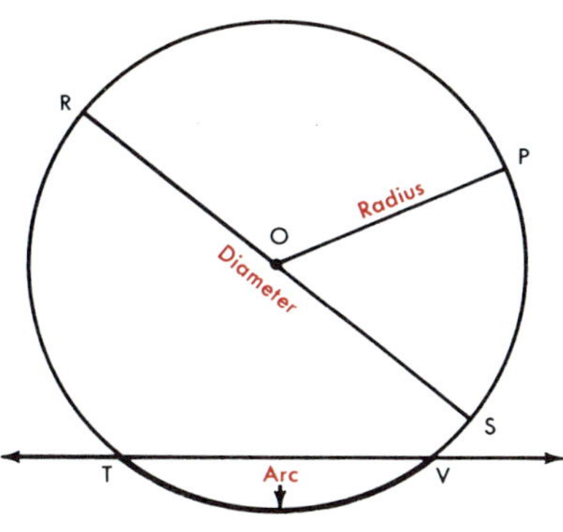

Figure 2–21

2.47 A circle is the set of points in a plane which are at a given distance r from a given point O in the plane. The point O is called the *center* of the circle. The distance r is a positive real number. If P is a point of the circle then $\overline{OP}$ is called a *radius* (plural: radii). If R and S are points of the circle and $\overline{RS}$ contains O, then $\overline{RS}$ is called a *diameter*.

2.48 An arc of a circle is the union of two points of the circle and all points of the circle on one side of the line determined by these two points.

Most of you will see immediately that this definition of arc could be either of two sets of points of the circle. For the moment we need not distinguish between these two sets. It will suffice to know the general meaning of arc.

2.49 A simple polygon is the union of a set of line segments called the *sides* of the polygon so arranged that:

(1) Each segment intersects *exactly two* other segments *only at their endpoints* called the vertices (singular: vertex) of the polygon.

(2) Each vertex of the polygon is the endpoint of *exactly two* segments of the set.

(3) No two intersecting segments lie in the same line.

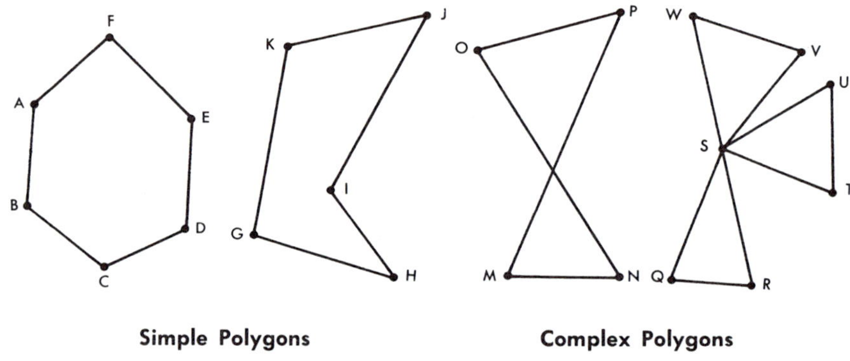

Figure 2—22

We shall restrict our work in this course to simple polygons. The word "polygon" will be understood to mean "simple polygon."

Exercises

1. What name do we apply to the point of intersection of a distinct line and plane?
2. If two straight lines have no common point, what could be their relationship?
3. If two lines have no common point, what must be true for them to be parallel?
4. Draw $\overline{OA}$. Let m$\overline{OA}$ be 1 inch. Rotate $\overline{OA}$ about point O in the plane of the paper. Name the path of point A after one complete rotation. What name could you apply to $\overline{OA}$?

5. In Figure 2-23, are $\overline{AD}$ and $\overline{BE}$ concurrent? Are $\overline{AD}$, $\overline{BE}$, and $\overline{AC}$ concurrent? Is $\overline{AC} \parallel \overline{ED}$? Do $\overrightarrow{ED}$ and $\overrightarrow{AD}$ intersect?

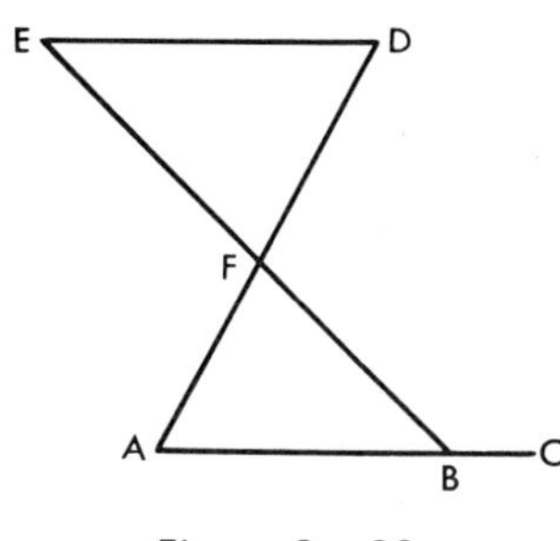

Figure 2—23

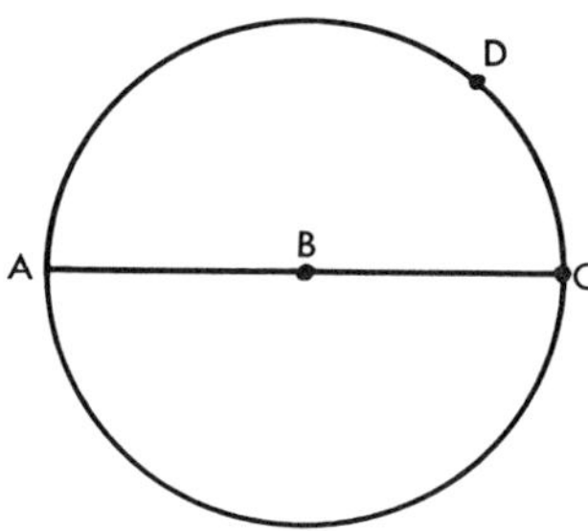

Figure 2—24

6. Figure 2-24 represents a circle with center B and $\overleftrightarrow{AC}$ containing B. Name two radii of the circle. Name two opposite rays of $\overleftrightarrow{AC}$.
7. In Figure 2-24, the intersection of the circle and a line containing D is at most _?_ point(s); is at least _?_ point(s).
8. Describe Figure 2-23 by using set terminology.
9. State the relationship between the number of vertices and the number of sides of a polygon.

Most of your answers to the following questions will be conjectures. You may wish to prove a few of the less obvious and more useful of these conjectures at a later stage in this course.

10. How many planes can contain a given line?
11. Can two different planes contain both of two given parallel lines? Two intersecting lines? Two skew lines?
12. How many planes can contain a given point? Two given points? Three given noncollinear points?
13. If two planes are parallel, is every line in one of the planes parallel to the other plane?
14. Are two lines parallel if they are parallel to the same plane?
15. Are two planes parallel to each other if they are parallel to a third plane?

16. How many lines intersecting at a given point not on a given line can be parallel to the given line?

17. How many lines intersecting at a given point not in a given plane can be parallel to the plane? Are all lines through this point parallel to the plane?

18. Are two lines parallel if they are parallel to the same line?

19. Draw two parallel planes intersected by a third plane.

20. Can three intersecting planes have only one point in common? Explain by means of a drawing.

Vocabulary List

definition	concurrent	length
point	statement	congruent
line	theorem	bisect
plane	assumption	midpoint
set	postulate	parallel
symbol	half-space	skew
null set	coplanar	Lobatchevsky
subset	half-plane	Bolyai
disjoint	space	Gauss
intersect	between	Riemann
intersection	segment	circle
element	endpoint	radius
union	half-line	diameter
distinct	ray	arc
collinear		polygon

Chapter Review

1. Given: $A = \{a, b, c, d, e\}$, $B = \{a, b, c\}$, $C = \{d, e, f\}$, $D = \phi$.

a. Is B a subset of A? Of C?

b. Is D a subset of C? Of A?

c. Name the elements of the intersection of B and C.

d. Name the elements of the union of B and C.

2. Draw a Venn diagram showing the intersection of sets A, B and C, where $A = \{1, 2, 3, 4\}$, $B = \{2, 3, 5, 6\}$, $C = \{3, 4, 6, 7\}$.
3. State several characteristics of a good definition. Illustrate by defining "school."
4. Why must there be undefined terms?
5. Represent three collinear points and label them.
6. Represent a segment and label it.
7. Represent a ray and label it.
8. Represent a line and label it.
9. What is the greatest number of points in which three distinct lines may intersect?
10. Represent a minimum set of concurrent lines.
11. Distinguish between a theorem and an assumption.
12. How many planes can contain two distinct points?
13. How many planes can intersect a line in one point?
14. May two planes have only one point in common? Explain.
15. Draw a picture representing two intersecting planes containing two skew lines.
16. Must the wheels of a bicycle be coplanar? Explain.
17. Draw a picture of a convex polygon having six sides.

Complete the following statements.

18. If point B is between points A and C,
    a. A, B and C are _?_.
    b. $\overrightarrow{BA}$ and $\overrightarrow{BC}$ are called _?_.
    c. $m\overline{AB} + m\overline{BC} = m\overline{AC}$ is justified by the _?_ postulate.
    d. If $\overline{AB} \cong \overline{BC}$, then B is the _?_ of $\overline{AC}$ and B is said to _?_ $\overline{AC}$.
19. If two lines are skew, they are not _?_.
20. If two lines intersect, they are not _?_.
21. A segment has _?_ endpoint(s).
22. A ray has _?_ endpoint(s).

23. A half-line has __?__ endpoint(s).

24. A line has __?__ endpoint(s).

25. The measure of a __?__ of a circle is twice the measure of the __?__. A portion of a circle is called an __?__.

Chapter 2 Test

Select the choice that correctly completes the statement.

1. An example of an undefined geometric term is (**a**) segment (**b**) ray (**c**) half-line (**d**) line.
2. A line has (**a**) 0 (**b**) 1 (**c**) 2 (**d**) 3 dimensions.
3. A polygon may not have (**a**) 2 (**b**) 3 (**c**) 4 (**d**) 5 sides.
4. A plane cannot contain (**a**) concurrent lines (**b**) parallel lines (**c**) intersecting lines (**d**) skew lines.
5. The points of a circle (**a**) are not coplanar (**b**) are not collinear (**c**)are not equidistant from a given point (**d**) are not separated into two disjoint sets by a diameter of the circle.
6. $\overline{AB}$ could be congruent to (**a**) $\overline{CD}$ (**b**) $\overrightarrow{EF}$ (**c**) $\overleftrightarrow{GH}$ (**d**) $\overline{CD}$, $\overrightarrow{EF}$, $\overleftrightarrow{GH}$.
7. Given the sets $W = \{0,1,2,3,4,5\}$, $X = \{3,4,5\}$, $Y = \{3,4,5,6,7\}$, $Z = \phi$, which of the following is not true? (**a**) Z is a subset of X. (**b**) X is a subset of W. (**c**) X is the intersection of sets W and Y. (**d**) W is the union of sets X and Z.
8. Three distinct planes cannot have in common (**a**) exactly one point (**b**) exactly two points (**c**) exactly one line (**d**) more than two points.
9. Parallel lines must be (**a**) coplanar (**b**) intersecting (**c**) skew (**d**) none of these.
10. Of the set of lines containing four non-coplanar points, there are at most (**a**) 2 (**b**) 3 (**c**) 4 (**d**) 5 concurrent in any of these points.

3

Plane Angles

One of the most important of all geometric figures is the angle. Portions of angles are integral parts of many geometric figures such as triangles and parallelograms. We will need a careful definition of "angle" before we proceed further with our geometric structure.

3.00 The union of two distinct rays with a common endpoint is called an angle. The rays are the sides of the angle, and the common endpoint is called the vertex.

If $\overrightarrow{BA}$ and $\overrightarrow{BC}$ are the sides of an angle, the angle is denoted by the symbol $\angle ABC$ or $\angle CBA$. The name of the vertex is always the middle letter of the angle name. Note that in Figure 3-1 $\angle ABC$ and $\angle CBA$ denote the same angle. Either symbol is acceptable in this course. This angle may also be named by the symbol $\angle B$ since B is the vertex of only one angle.

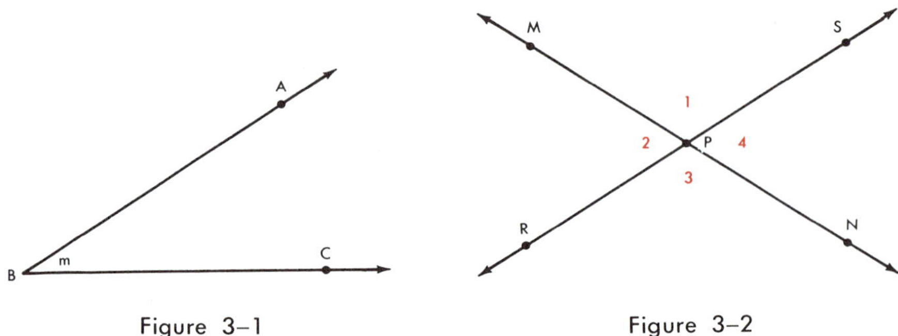

Figure 3–1

Figure 3–2

Sometimes a single lower case letter or numeral is placed between the rays of an angle. See Figures 3-1 and 3-2. In such cases the angle may be named by this letter such as "$\angle m$," "$\angle 2$," and so forth. The use of pictures to represent ideas is so commonly used that we will usually refer to the picture as the idea. For example, the picture in Figure 3-1 will be called an angle rather than "a picture of an angle." However, remember the distinction.

Oral Exercises

1. Can you draw an angle?
2. Using Figure 3-2, denote $\angle 1$ by two other names. Give two other names for $\angle 4$.
3. Use Figure 3-2 to name four different angles with their vertices at point P.
4. Is $\overrightarrow{PS}$ in Figure 3-2 the edge of a half-plane?
5. If N is a point in a half-plane whose edge is $\overleftrightarrow{RS}$, as represented in Figure 3-2, is point P in the same half-plane?
6. What point is the vertex of $\angle RST$?
7. Name the rays which are the sides of $\angle XYZ$.
8. Does an angle separate a plane into two disjoint sets of points?
9. In Figure 3-2 are M, P, and N necessarily collinear?
10. Using Figure 3-2, denote $\angle P$ by another name.

Do you see why $\angle P$ is not a satisfactory name for an angle in Figure 3-2? This is similar to calling a person Mr. Smith at a Smith family reunion. To avoid confusion, where a point is the vertex of more than one angle, use a three-letter symbol which specifies the sides as well as the vertex of the angle. As previously stated, a lower case letter or a numeral placed between the rays and near the vertex in the picture of an angle may also be used to name the angle.

ANGLE MEASUREMENT

The unit of angle measure in common use today is the degree, developed about 2000–4000 B.C. by the Sumerians. A larger unit, the radian, will be used when you study trigonometry and calculus.

The Sumerians lived in Mesopotamia in the valleys of the Tigris and Euphrates rivers and developed a high degree of culture for that time. Their numeration system used a base of 60 with a secondary base of 10. They divided a circle into 360 congruent arcs (a multiple of 60). Each arc represented one day of what they probably considered to be a 360-day year. The measure of one of these arcs was named a degree (1°). Each season thus represented 90 degrees.

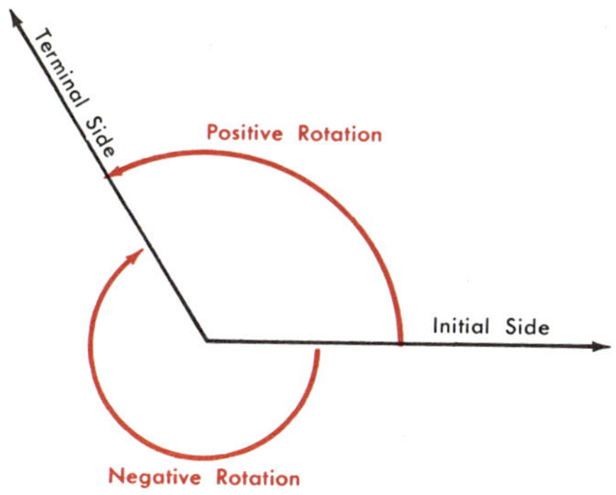

Figure 3–3

Angles and angle measurement are also defined in terms of rotation of a ray in a plane about its endpoint from an initial position (initial side) to a terminal position (terminal side). As with the Sumerians, the degree measure of one rotation is 360, of two rotations is 720, of one-half of a rotation is 180, and so on. Rotation may occur in a clockwise or counterclockwise direction. Counterclockwise rotation is called positive rotation, clockwise rotation is called negative rotation, and the measures of the angles are assigned positive or negative values, respectively.

We shall restrict our degree measure to values equal to or less than 180° and greater than zero. This is necessary so that an angle (two rays with a common endpoint) will have one and only one measure. While angle measures greater than 180° will be used in trigonometry and calculus, it is not necessary to consider such measures in this course.

This is consistent with your experience in using a protractor in Chapter 1 to measure an angle. The concept of angle measure is formalized in the following assumption and definition so that it can be adapted for use in future proofs.

3.01 *Assumption: The set of rays in a half-plane and in its edge whose common endpoint is a point in the edge of the half-plane can be numbered in sequence, starting with a ray in the edge. For each ray there is exactly one real number n such that $0 \leq n \leq 180°$, and for each real number of this set there is exactly one ray of the set.*

3.02 The measure of an angle (written m∠) is the absolute value of the difference between the numbers corresponding to the sides of the angle. The unit of measure is the degree.

> Absolute value is defined in algebra as follows: The absolute value of a number r is denoted by $|r|$. If $r = 0$, then $|r| = 0$. If $r \neq 0$, $|r|$ is the positive number in the pair $r, -r$.

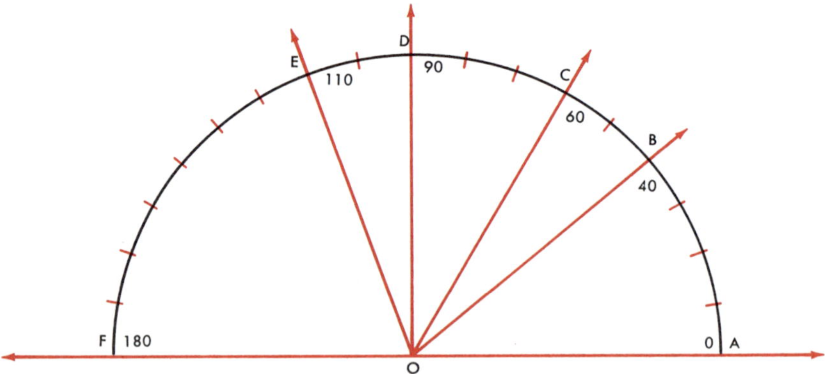

Figure 3–4

EXAMPLE: $m\angle EOB = |110 - 40|$ or $|40 - 110| = 70$
$m\angle COF = |180 - 60|$ or $|60 - 180| = 120$
$m\angle AOD = |90 - 0|$ or $|0 - 90| = 90$

The statement $m\angle ADG = m\angle QRT$ says that the measure of $\angle ADG$ is the same number as the measure of $\angle QRT$. It does not say that these angles are the same angle —although this is possible. If $\angle ADG$ and $\angle QRT$ are the same angle, as shown in Figure 3-5, then we could say $\angle ADG = \angle QRT$ where "equal," as always, means "the same as."

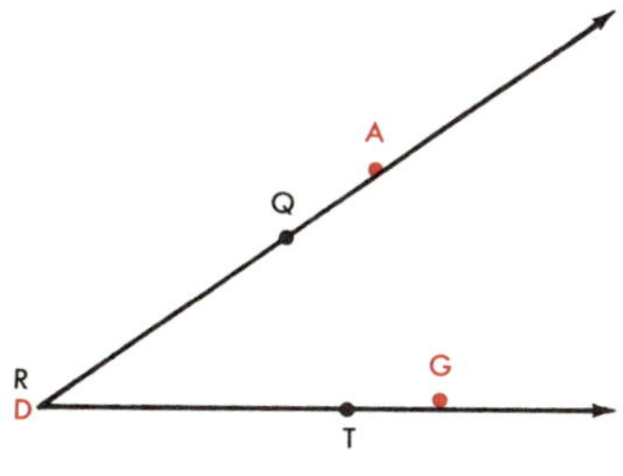

Figure 3–5

It is often necessary to measure angles with greater precision than the whole degree allows. Of course, decimal or fractional parts of a degree can be used. However, it has been agreed to divide a degree into 60 equal parts called minutes ($'$), and each minute into 60 equal parts called seconds ($''$). For example, 54 degrees, 15 minutes, and 47 seconds is written as $54°\,15'\,47''$.

$$1° = 60' \qquad 1' = 60''$$

Below are some examples of degree measures written both in fractional form and in equivalent degree-minute-second form.

$18\frac{1}{3}° = 18°\,20'$, since $\frac{1}{3}$ of a degree is $20'$. $\left(\frac{1}{3} \times 60 = 20\right)$

$21\frac{3}{5}° = 21°\,36'$, since $\frac{3}{5}$ of a degree is $36'$. $\left(\frac{3}{5} \times 60 = 36\right)$

$33.3° = 33°\,18' \left(0.3 \text{ or } \frac{3}{10} \times 60 = 18\right)$

$12\frac{5}{9} = 12°\,33'\,20'' \left(\frac{5}{9} \times 60 = 33\frac{1}{3}; \frac{1}{3} \times 60 = 20\right)$

ORAL EXERCISES

1. Using Figure 3-4, give the measure of $\angle DOC$, $\angle COA$, $\angle AOE$, $\angle AOF$.
2. In Figure 3-4, $m\angle DOC + m\angle COB = m\angle(?)$. Find $m\angle AOB + m\angle DOE$.

3. Name two angles of equal measure in Figure 3-4. Name an angle in Figure 3-4 whose measure is 70°.

4. What is the measure of an angle formed by opposite rays?

5. How many seconds is $1\frac{1}{2}'$? How many minutes is 7°? How many degrees is 10′? How many degrees is 30″?

In the following exercises you will measure "the angles of a polygon." Two consecutive sides of a polygon (2.49) *determine* an angle since the rays containing these sides and having the vertex as the common endpoint *form* an angle.

Exercises

Review page 4 and then measure the lettered angles with your protractor. Do not mark on this book!

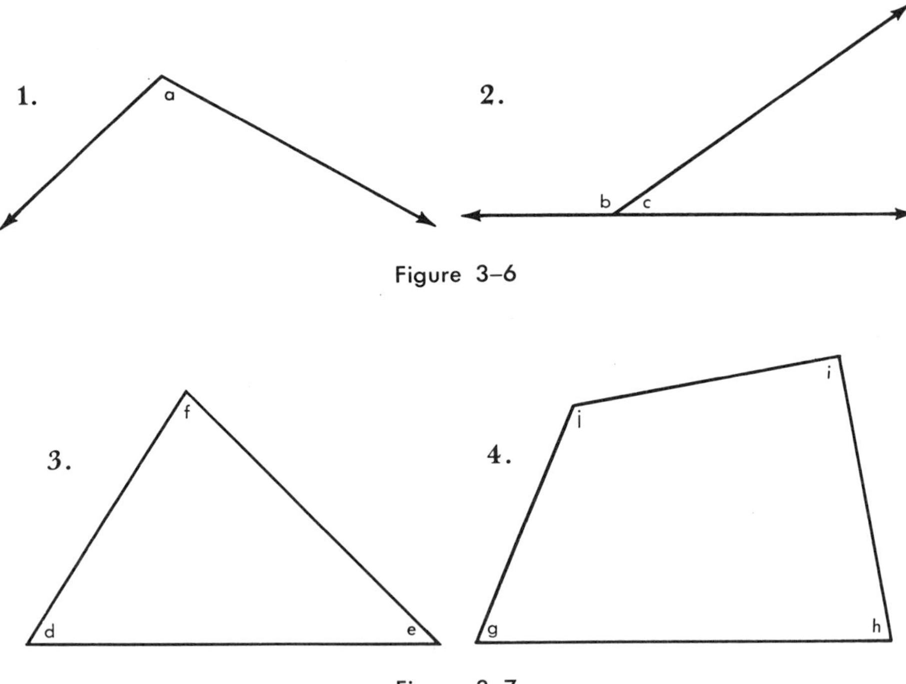

Figure 3–6

Figure 3–7

5. Draw angles whose measures are 15°, 60°, 135°, 210°.

6. Write each of the following as degrees and minutes: $37\frac{1}{2}^\circ$, $18\frac{2}{3}^\circ$, $6\frac{5}{6}^\circ$, $42\frac{3}{4}^\circ$.

7. Write each of the following as degrees, minutes, and seconds, to the nearest second: $18\frac{2}{7}^\circ$, $50\frac{3}{7}^\circ$, $17\frac{3}{11}^\circ$.

Use Figure 3-8 to answer Exercises 8–11.

8. If $m\angle 1 = 30^\circ$ and $m\angle 2 = 35^\circ$, find $m\angle AOC$.
9. If $m\angle 2 = 30^\circ$, $m\angle 3 = 20^\circ$, and $m\angle 4 = 33^\circ$, find $m\angle BOE$.
10. If $m\angle 1 = c^\circ$, $m\angle 2 = b^\circ$, and $m\angle 3 = a^\circ$, find $m\angle AOD$.
11. If $m\angle DOB = 52^\circ$ and $m\angle 3 = 20^\circ$, find $m\angle 2$.

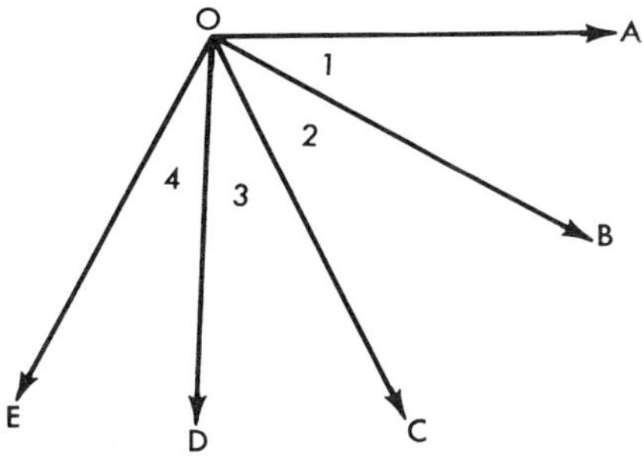

Figure 3–8

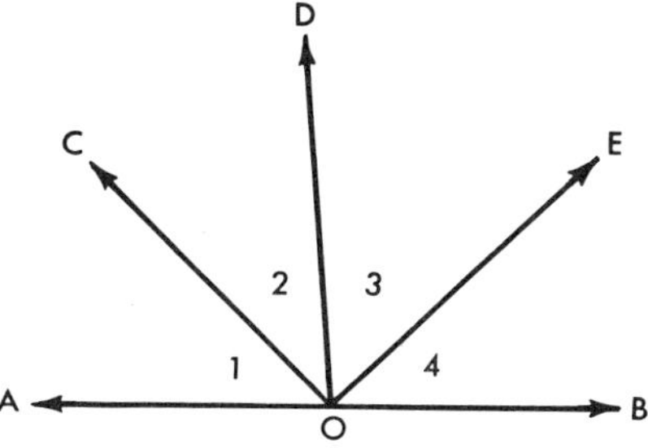

Figure 3–9

Use Figure 3-9 to answer Exercises 12–15; $m\angle AOB = 180^\circ$.

12. If $m\angle 4 = 45^\circ$, find $m\angle AOE$.
13. If $m\angle 1 = a^\circ$, find $m\angle COB$.
14. If $\angle 1 \cong \angle 4$, $m\angle 2 = 30^\circ$, and $m\angle 3 = 50^\circ$, find $m\angle 1$.
15. If $\angle 3 \cong \angle 4$ and $m\angle DOB = a^\circ$, find $m\angle AOE$.
16. Find the degree measure of each of the angles formed by two intersecting straight lines. What relationships do you find?
17. Can two opposite rays form an angle? If so, what is the measure of this angle? State the definitions and assumptions which justify your answers.
18. Devise a definition for "two rays having the same direction."
19. Devise a definition for "two rays having opposite directions."
20. Our structure of geometry does not allow angles whose measure is 0°. What change in our assumptions or definitions would be needed to create such angles?

Using the definitions of 3.00 and 3.02 and assumption 3.01, it follows that:

3.03 THEOREM

If an angle is formed by opposite rays, then its measure is 180°.

REASONING: Two opposite rays form an angle by 3.00. They would be numbered 0 and 180 by assumption 3.01. The measure of this angle is 180° by 3.02.

CONGRUENT ANGLES

3.04 Angles with the same measure are called congruent angles.

We shall use the symbol $\angle ABC \cong \angle DEG$ to represent the statement "$\angle ABC$ is congruent to $\angle DEG$." (Review Sections 2.37 and 2.38.) The congruent statement really refers to the shape of the angles, but as the definition states, $\angle ABC \cong \angle DEG$ and $m\angle ABC = m\angle DEG$ are equivalent statements; that is, one may replace the other.

Congruency is a fundamental concept in geometry. Many times this relationship will be used to develop the structure of geometry. Knowing that angles are congruent will be more important than knowing their measure.

INTERIOR OF AN ANGLE

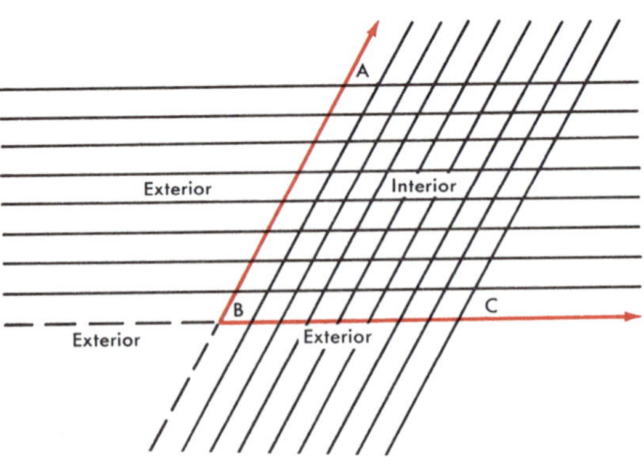

Figure 3–10

3.05 The interior of any angle ABC is the intersection of the half-plane which contains C and whose edge is $\overleftrightarrow{AB}$ and the half-plane which contains A and whose edge is $\overleftrightarrow{BC}$. (See Figure 3-10).

a. The interior of an angle formed by opposite rays is the specified half-plane determined by the line containing the opposite rays.

b. Any point in the plane of the angle which is not a point of the angle or its interior is a point of the exterior of the angle.

BISECTED ANGLES

3.06 An angle bisector is the ray which contains a point in the interior of the angle, has its origin at the vertex, and divides the angle into two congruent angles.

3.07 *Assumption: Every angle whose measure is not zero has one and only one bisector.*

3.08 *Assumption: If E is a point in the interior of any $\angle AOD$, then $m\angle AOE + m\angle EOD = m\angle AOD$. (This is frequently called the Angle-Addition Postulate.)*

ANGLES DEFINED ACCORDING TO SIZE

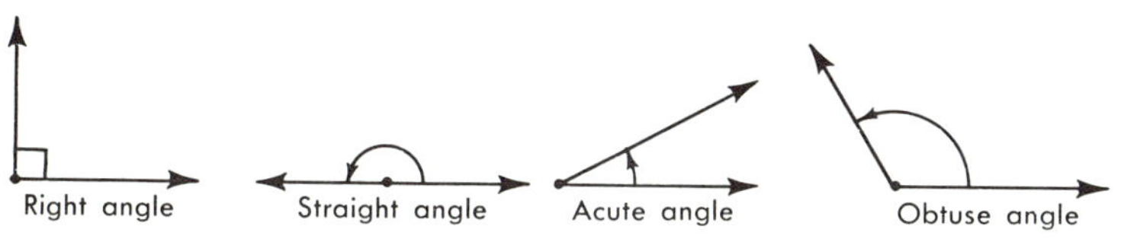

Figure 3–11

3.09 A right angle is an angle whose measure is 90°.

3.10 A straight angle is an angle whose sides are opposite rays. (The measure of a straight angle is 180° by 3.03.)

3.11 An acute angle is an angle whose measure is more than 0° but less than 90°.

3.12 An obtuse angle is an angle whose measure is more than 90° but less than 180°.

Using definitions 3.00, 3.04, 3.09 and 3.10, the following theorem is evident.

3.13 THEOREM

All right angles are congruent to each other; all straight angles are congruent to each other; every angle is congruent to itself.

CLASSIFICATION OF ANGLES BY THEIR POSITION IN RELATION TO ONE ANOTHER

Intersecting lines determine angles, since angles are formed by the rays which are contained in the lines and whose common end-point is the point of intersection.

3.14 Two coplanar angles having a common vertex, a common side, and whose interiors are disjoint sets are called adjacent angles.

Could two nonadjacent angles have a common side? How could this happen? Could two angles share the same vertex and not be adjacent? Could two angles have a common vertex and a common side but not be adjacent?

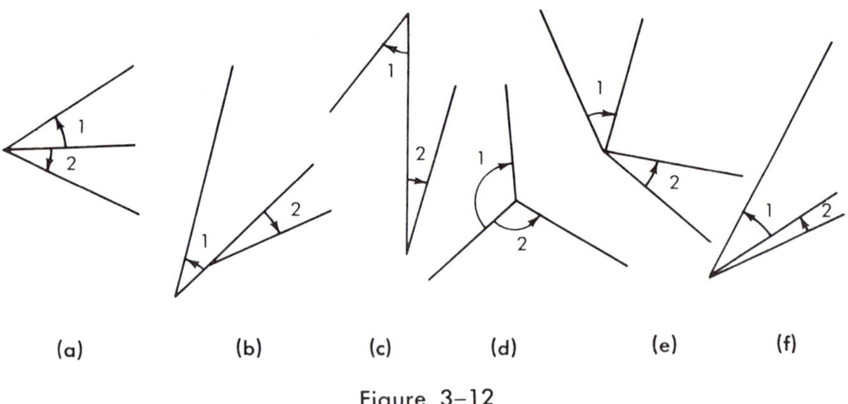

Figure 3–12

In these figures identify the pairs of angles that are adjacent.

3.15 Vertical angles are two angles such that the sides of one are the opposite rays of the sides of the other.

This means that a pair of distinct, intersecting lines forms two pairs of vertical angles. In Figure 3-13, $\angle m$ and $\angle n$, $\angle 1$ and $\angle 2$ are pairs of vertical angles.

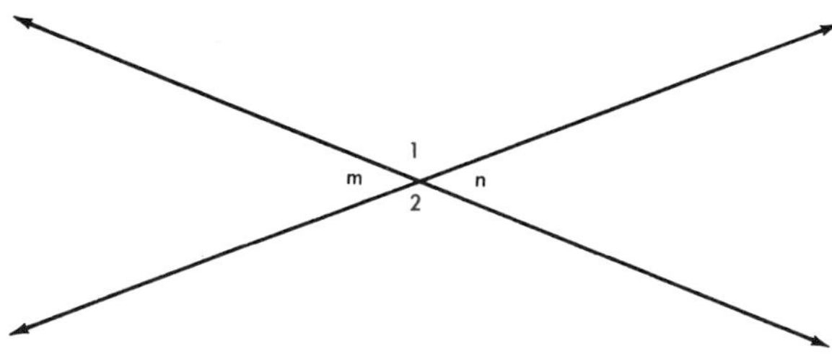

Figure 3–13

Exercises

1. Are all acute angles congruent? All obtuse angles?
2. Can some acute angles be vertical angles? Must all acute angles be vertical angles?
3. Can two right angles be adjacent angles?
4. Can acute angles be right angles?
5. Can a pair of vertical angles also be a pair of adjacent angles?
6. Is the vertex of an angle in the interior of the angle?
7. Does the size of an angle depend upon the length of the sides?

Use Figure 3-14 to answer Exercises 8–11.

8. Name the pairs of vertical angles by naming one letter each; by naming three letters each.
9. Name the pairs of adjacent angles by naming one letter each; by naming three letters each.

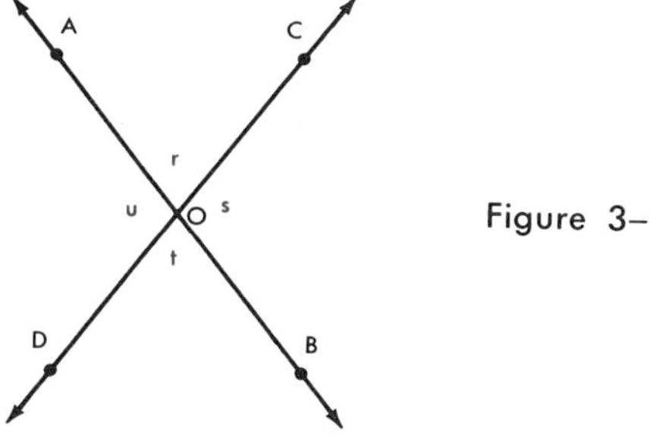

Figure 3–14

10. Name the obtuse angles by a single letter; by three letters.

11. Name the acute angles by a single letter; by three letters.

Use Figure 3-15 to answer Exercises 12–18.

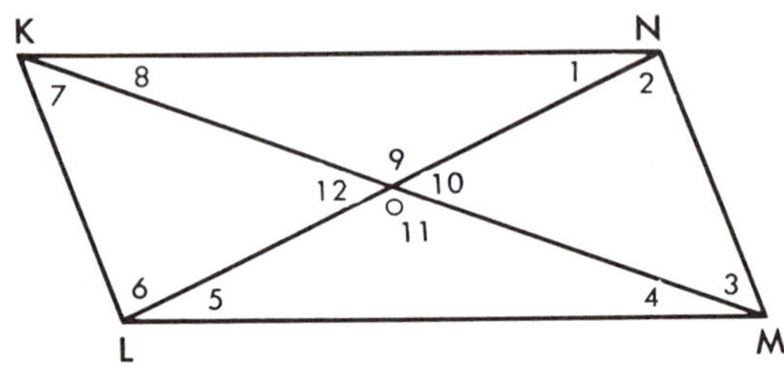

Figure 3–15

12. Use numerals to name the pairs of vertical angles. Use letters.

13. Use numerals to name the pairs of adjacent angles.

14. Use numerals to name the obtuse angles. Use letters.

15. Use numerals to name each of the acute angles.

16. If K, O, and M are collinear, what type of angle is $\angle KOM$?

17. All points of $\overline{LM}$, except L and M, lie in the interior of which angle(s)?

18. If *overlapping* angles have a common vertex and intersecting interiors, name an angle that overlaps $\angle KLM$.

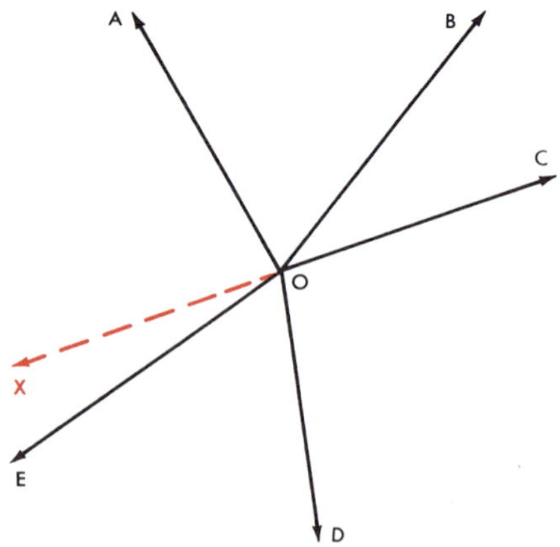

Figure 3–16

19. Given point O with distinct coplanar rays $\overrightarrow{OA}$, $\overrightarrow{OB}$, $\overrightarrow{OC}$, $\overrightarrow{OD}$, and $\overrightarrow{OE}$. See Figure 3-16.

(a) May we draw $\overrightarrow{OX}$ as the opposite of $\overrightarrow{OC}$? Why?

(b) Name $m\angle XOC$. Justify your answer.

(c) Does $m\angle AOB + m\angle BOC = m\angle AOC$? Why?

(d) Does $m\angle XOA + m\angle AOC = 180°$? Why?

(e) $m\angle COD + m\angle DOE + m\angle EOX = ?$ Why?

(f) Does $m\angle EOX + m\angle AOX = m\angle AOE$? Why?

(g) $m\angle AOB + m\angle BOC + m\angle COD + m\angle DOE + m\angle EOA = ?$ Why?

(h) State as a theorem the concept proved in this exercise.

PERPENDICULAR AND OBLIQUE LINES AND PLANES

Most of us have used the terms perpendicular, oblique, vertical, and horizontal, sometimes correctly, but quite often incorrectly. "Correctness" can be determined only if we have an *accepted* meaning or definition for these terms. How would you define them?

Would it be reasonable to define a vertical line as a line which is perpendicular to a horizontal line, and then define a horizontal line as a line which is perpendicular to a vertical line? Explain the weakness of these definitions.

Most of us will realize that neither of the above definitions is possible until we define the word "perpendicular."

3.16 Two intersecting sets of points, each of which is either a line, a ray, or a segment, are perpendicular if the two lines which contain them determine a right angle.

Since a definition is reversible, we can say, "If two lines are perpendicular, they determine a right angle."

Using definitions 3.04, 3.09, 3.14, 3.16, and Theorem 3.13, the following theorems are evident. Remember that definitions are reversible.

3.17 THEOREM

If two intersecting sets of points, each of which is either a line, a ray, or a segment, are perpendicular, they determine congruent adjacent angles.

3.18 THEOREM

If two intersecting sets of points, each of which is either a line, a ray, or a segment, determine congruent adjacent angles, they are perpendicular.

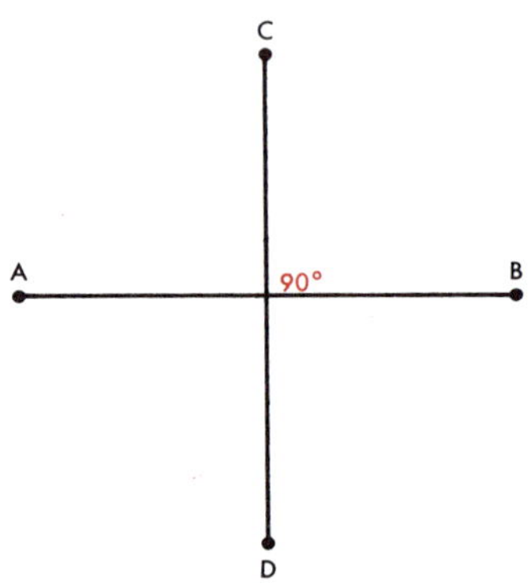

Figure 3–17

Figure 3-17 pictures $\overline{AB} \perp \overline{CD}$, (read "$\overline{AB}$ is perpendicular to $\overline{CD}$"), since the segments determine a right angle.

3.19 THEOREM

The sum of the measures of all the consecutive adjacent angles about a point in a plane is 360°. (This theorem was proved in Exercise 19, page 61.)

In space, how many lines are perpendicular to a given line and contain a given point of the line? What geometric figure is formed by the set of all lines perpendicular to a given line at a given point of the line? Use pencils and cards to help you see the relationships.

If we restrict the lines we are considering to one plane, how many lines are perpendicular to a given line at a point of the line?

We need to make two more assumptions about lines and planes.

3.20 *Assumption: In a plane, there is one and only one line perpendicular to a given line at a given point of the line; there is one and only one line perpendicular to a given line through a given point not in the line.*

3.21 *Assumption: In three-space there is one and only one plane perpendicular to a given line at a given point of the line; there is one and only one plane perpendicular to a given line through a given point not in the line.*

Compare the two assumptions above; what words have been changed?

OBLIQUE LINES

3.22 Two distinct sets of points, each of which is either a line, or ray, or a segment, are said to be oblique if they intersect but are not perpendicular and do not lie in a straight line.

In Figure 3-18 can $\overleftrightarrow{PQ}$ be oblique to $\overleftrightarrow{AB}$? Could $\overleftrightarrow{PQ}$ be perpendicular to $\overleftrightarrow{DE}$? If $\overleftrightarrow{PQ}$ is oblique to $\overleftrightarrow{AB}$ and perpendicular to $\overleftrightarrow{DE}$, can $\overleftrightarrow{PQ}$ be perpendicular to plane RS?

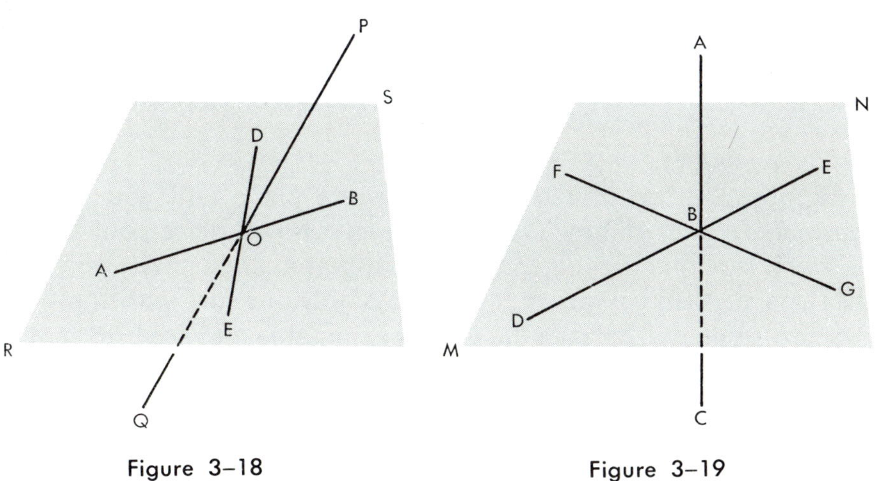

Figure 3–18

Figure 3–19

LINE PERPENDICULAR TO A PLANE

In Figure 3-19 is it possible for $\overleftrightarrow{AC}$ to be perpendicular to both of the two intersecting lines, $\overleftrightarrow{DE}$ and $\overleftrightarrow{FG}$, in the plane MN? Is it possible for $\overleftrightarrow{AC}$ to be perpendicular to all lines in plane MN through point B, the foot of $\overleftrightarrow{AC}$? (The *foot* of a line is the point of intersection of a line and a plane.) This thought process leads to the following definition.

3.23 A line, ray, or segment that is perpendicular to every line in a plane passing through its foot is perpendicular to the plane.

What is the reverse of this definition? Is it true that a line is perpendicular to every line of a plane that passes through its foot if it is perpendicular to the plane?

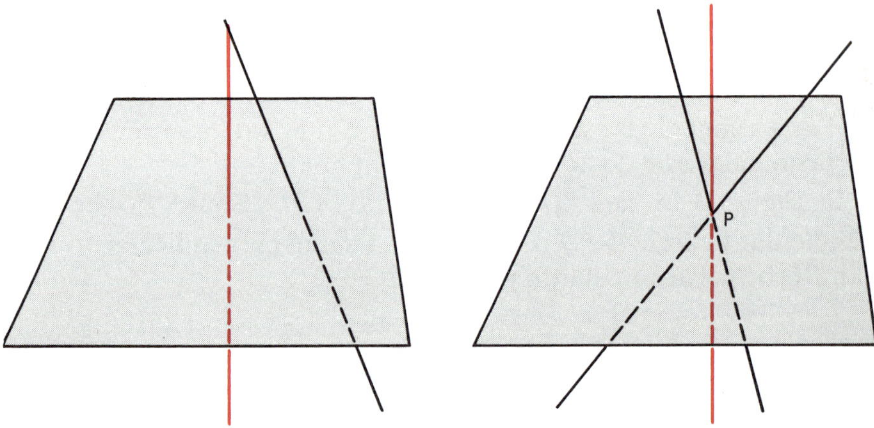

Figure 3–20

Suppose the top of a table represents a plane and your pencil represents a line. In how many positions can you place your pencil so that it is perpendicular to the table top at a single given point of the surface? Can you prove that there is only one such position?

Select some point not in the plane of the table and see how many different lines passing through (containing) the point seem to be perpendicular to the plane. Are you sure that there is one and only one such line? These experiments lead us to make the following assumption.

3.24 *Assumption: There is one and only one line which is perpendicular to a plane at a point in the plane, or from a point not in the plane.*

LINE OBLIQUE TO A PLANE

3.25 A line, ray, or segment that is oblique to a plane intersects the plane in only one point and is not perpendicular to it.

Could a line be oblique to a plane and yet be perpendicular to one line in the plane? Experiment by considering the edge of a table as a line in the plane of the top of the table; can you hold a pencil so that it is perpendicular to the table edge and not perpendicular to the top of the table?

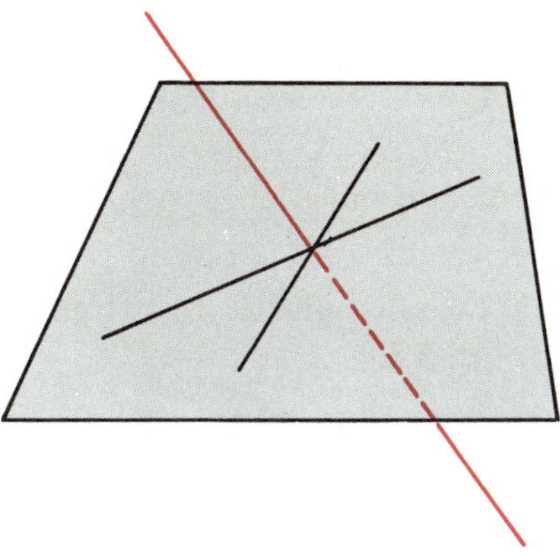

Figure 3–21

Can a line oblique to a plane be perpendicular to two lines in the plane passing through its foot? Make a conjecture about this.

MEASURING DISTANCES

When we wish to find the distance between two points in a plane, we find the measure of the segment connecting the points.

How do we determine the distance from a line to a point not in the line? Could we find the measure of a segment which joins the given point with any point of the line? Explain. Which segment shall we measure to determine the distance? Intuitively most of us accept distance in this case to mean the least distance.

We are not yet able to prove that the measure of a segment from the point perpendicular to and terminated by the line is less than the measure of any other segment from the point to the line. However, we can agree to *use the measure of the distance from a point to a line or to a plane as the length of a segment from the point, perpendicular to and terminated by the line or plane.*

Also, the *distance between two parallel lines or planes* is the length of a segment perpendicular to either and terminated by both of them.

Exercises

Many of your answers to the following questions will be conjectures. You may wish to prove some of these conjectures at a later stage in geometry.

1. If a line does not intersect a plane, is it necessarily parallel to the plane?
2. If a ray and a plane do not intersect, are they necessarily parallel?
3. If two planes are each parallel to a third plane, are they necessarily parallel to each other?
4. If two planes are each parallel to a line, are they necessarily parallel to each other?
5. If two lines are each parallel to a plane, are they necessarily parallel to each other?
6. If a line and plane are perpendicular to the same line, are they necessarily parallel to each other?
7. If a line is parallel to the line of intersection of two planes, is it necessarily parallel to the two planes?
8. How many lines can be perpendicular to a given line from a point not in the line? At a given point of the line?
9. How many lines can be perpendicular to a given plane from a point not in the plane? At a given point of the plane?
10. Are two lines necessarily parallel if they are perpendicular to the same plane?
11. If one of two parallel lines is perpendicular to a plane, is the other also perpendicular to the plane?
12. Can a line be perpendicular to two lines in a plane? Explain.
13. Are two planes parallel to each other if they are perpendicular to the same line?
14. What is the shortest segment that can be drawn from a point to a plane?
15. What relation has the intersection of two walls of your classroom to the plane of the floor?
16. If a line is oblique to a plane, is it oblique to every line in the plane through its foot?

17. Coplanar lines $\overleftrightarrow{AB}$, $\overleftrightarrow{CD}$, and $\overleftrightarrow{EF}$ are concurrent in point O. At O, $\overleftrightarrow{HO}$, not in the plane, is drawn perpendicular to $\overleftrightarrow{AB}$ but oblique to $\overleftrightarrow{CD}$ and $\overleftrightarrow{EF}$. Of the six angles formed with $\overrightarrow{OH}$ as a side, how many are right angles? How many are acute angles? How many are obtuse angles?

18. Assume that two planes perpendicular to the same line are parallel. Can a line be perpendicular to each of two intersecting planes? Justify your response.

VERTICAL AND HORIZONTAL LINES

Review the discussion on page 61 and then compare the following commonly accepted definitions of horizontal and vertical lines with those proposed by you and your classmates.

Some pupils may have confused perpendicular lines with horizontal and vertical lines. If you draw a line through the midpoint of the top of a sheet of paper to the midpoint of the bottom, is the line vertical? Is a line from the midpoint of the left side to the midpoint of the right side a horizontal line? *No, these lines are not necessarily vertical or horizontal.* The words "vertical" and "horizontal" are related to their position with respect to the earth, which we will assume to be a sphere.

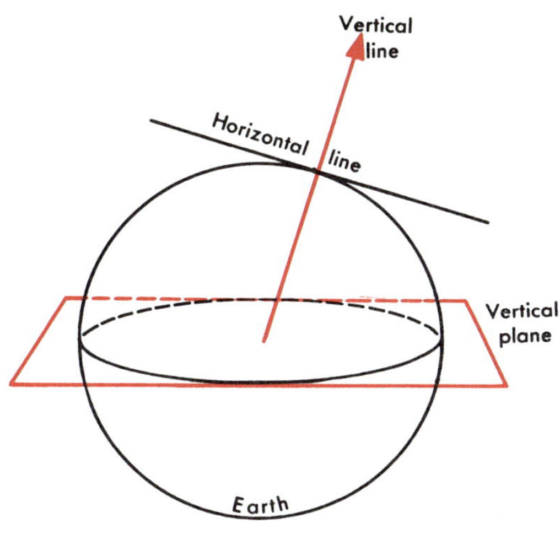

Figure 3–22

3.26 A vertical line is any line which contains the center of the earth. Clearly, a radius or a diameter of the earth is a segment which is part of a vertical line.

3.27 A vertical plane is a plane which contains as one of its points the point at the center of the earth. How many vertical lines are in each vertical plane? Is every line in the plane passing through the center of the earth a vertical line? If the center of gravity of the earth is at the geometric center, then a weight attached to the end of a string will be part of a vertical line. Such a device is called a *plumb line* by surveyors, carpenters, and others. A pendulum at the lowest point of its swing is vertical.

Figure 3–23

Can two vertical lines be parallel? Do two vertical lines have a point in common?

3.28 A horizontal line is any line which does not contain the center of the earth and which is perpendicular to a given vertical line. Is every line, therefore, a horizontal line? Do you see that horizontal is a meaningless term except for a certain point of observation where it is applied? There are an infinite number of horizontal lines through the point of your pencil at this moment. There are also an infinite number of horizontal lines through the point of your neighbor's pencil. Do any of the horizontal lines at the two positions coincide?

Suppose the four corners of a building when built were carefully aligned by a plumb line. Would the vertical lines at the four corners be parallel? Would the area of the second floor of the building be equal to the area of the first floor?

Vocabulary List

angle	minutes	vertical angles
vertex	seconds	perpendicular
interior	right angle	oblique
exterior	straight angle	three-space
numeration	acute angle	vertical
initial	obtuse angle	horizontal
terminal	adjacent angles	plumb line

Chapter Review

1. Draw a picture of two non-congruent angles. Draw an angle whose measure is the sum of the measures of the first two angles.
2. Name the sides of $\angle XYZ$.
3. $\overleftrightarrow{AD}$ and $\overleftrightarrow{GH}$ intersect in F. Name a pair of vertical angles.
4. Opposite rays $\overrightarrow{OD}$ and $\overrightarrow{OC}$ are coplanar with $\overrightarrow{OE}$. If $\angle DOE$ is an acute angle, may $\angle COE$ also be acute? Explain. If $m\angle DOE = 18° \ 12' \ 10''$, find $m\angle COE$.
5. If $\overleftrightarrow{MN} \perp \overleftrightarrow{RS}$ and intersect in P, what is $m\angle RPN$? Name the relationships of $\angle MPS$ and $\angle MPR$.
6. If $\overrightarrow{BD}$ bisects $\angle ABC$ and if $m\angle ABC = 15° \ 22'$, find $m\angle ABD$.
7. If the interiors of $\angle LOM$ and $\angle ROS$ have some points in common, what name do we apply to this pair of angles?
8. What is our authority for saying that the distinct lines $\overleftrightarrow{SR}$ and $\overleftrightarrow{ST}$ are not both perpendicular to plane MN?
9. If two lines are oblique to a plane, may they be parallel to each other? Perpendicular to each other? Draw pictures to illustrate your responses.
10. Point B is in the interior and D is in the exterior of $\angle AOC$. $\overrightarrow{OC} \perp \overrightarrow{OA}$. What type angle is $\angle AOC$? $\angle AOB$? $\angle AOD$? What name could be applied to the pair of angles, $\angle DOC$ and $\angle COB$?
11. Are all lines horizontal which are perpendicular to a vertical line? Explain.
12. Are all lines vertical which are perpendicular to a horizontal line? Explain.

13. Are all planes horizontal which are perpendicular to a vertical line? Explain.

14. Would all planes necessarily be vertical if they are perpendicular to a horizontal line? Explain.

15. Can each of two skew lines be vertical? Each horizontal?

16. Point P is in the interior of $\angle KOM$. Is every point of $\overrightarrow{OP}$ in the interior of $\angle KOM$? Why?

17. If three planes contain a common point, must they intersect in three lines? Explain.

18. What relationships may two planes have?

19. Can four lines be mutually perpendicular? Three lines?

20. Assuming that a line and a point not in the line determine a plane, prove that three non-collinear points determine a plane.

Chapter 3 Test

1. If $\overleftrightarrow{MN}$ intersects $\overleftrightarrow{RT}$ in point P, which of the following is not necessarily true?

(**a**) $\angle MPT$ and $\angle NPT$ are adjacent angles.

(**b**) $\angle MPN$ is a straight angle.

(**c**) $m\angle RPN + m\angle RPM + m\angle MPT + m\angle TPN = 360°$.

(**d**) $\angle RPN \cong \angle RPM$.

2. The sum of the measures of a right angle and an obtuse angle is (**a**) 180° (**b**) more than 180° (**c**) less than 180° (**d**) none of these.

3. Point T is in the exterior of $\angle MON$, $m\angle MON = 72°$.
(**a**) $\overleftrightarrow{MO}$ is oblique to $\overleftrightarrow{NO}$. (**b**) $\angle TOM$ is adjacent to $\angle MON$.
(**c**) $\overrightarrow{OM}$ bisects $\angle TON$. (**d**) $\angle TON$ is obtuse.

4. $\overleftrightarrow{JK}$ intersects $\overleftrightarrow{CD}$ at P so $m\angle JPD = 80°$. Which of the following is not necessarily true?

(**a**) $\angle JPD$ and $\angle CPK$ are vertical angles.

(**b**) If $\overrightarrow{PQ}$ bisects $\angle CPJ$, $m\angle QPJ = 50°$.

(**c**) $\angle KPD$ is an obtuse angle.

(**d**) $\overleftrightarrow{PC} \perp \overleftrightarrow{JK}$.

5. Point R is in the interior of $\angle COD$. If $m\angle COD = 122°\ 48'\ 18''$, which of the following is not true?

   (a) If $\angle COR$ is a right angle, then $\angle ROD$ is an acute angle.

   (b) If $\overrightarrow{OE}$ is opposite to $\overrightarrow{OD}$, $m\angle COE = 57°\ 11'\ 42''$.

   (c) If $m\angle COR = 12°\ 52'$, then $m\angle ROD = 110°\ 4'18''$.

   (d) If $\angle ROD$ is an obtuse angle, then $\angle COR$ is an acute angle.

6. Given: Plane s with $\overleftrightarrow{AB}$ in s, $\overleftrightarrow{CD} \perp \overleftrightarrow{AB}$ at P, F not on $\overleftrightarrow{CD}$, then **(a)** $\overleftrightarrow{CD} \perp s$. **(b)** $\overleftrightarrow{CD}$ is oblique to s. **(c)** $\overleftrightarrow{FP} \perp \overleftrightarrow{AB}$. **(d)** If $\overleftrightarrow{FP} \perp \overleftrightarrow{AB}$, then $\overleftrightarrow{FP}$ and $\overleftrightarrow{CD}$ are not both in s.

7. Given: Plane s with $\overleftrightarrow{AB}$ in s and R not in s.

   (a) If $\overleftrightarrow{RP} \perp \overleftrightarrow{AB}$, then $\overleftrightarrow{RP} \perp s$.

   (b) If $\overleftrightarrow{RT} \perp s$, then $\overleftrightarrow{RM}$ can not be $\perp$ to s, where T and M are different points of plane s.

   (c) If $\overleftrightarrow{RT} \perp s$, then $\overleftrightarrow{RP}$ may be $\perp \overleftrightarrow{AB}$, where T and P are different points of $\overleftrightarrow{AB}$.

   (d) Neither (a), (b), or (c) is correct.

8. If O is the center of the Earth, which of the following is not necessarily true?

   (a) $\overleftrightarrow{TO}$ is a vertical line.

   (b) Plane n containing $\overleftrightarrow{TO}$ is a vertical plane.

   (c) $\overleftrightarrow{CD} \perp n$, is a horizontal line.

   (d) $\overleftrightarrow{ET} \perp \overleftrightarrow{TO}$. $\overleftrightarrow{ET}$ is a horizontal line.

9. Given: $\overleftrightarrow{AB} \parallel \overleftrightarrow{CD}$, $\overleftrightarrow{AC} \perp \overleftrightarrow{CD}$.
   (a) $m\overline{AC}$ is less than $m\overline{BC}$. (b) $\overleftrightarrow{BD} \perp \overleftrightarrow{CD}$.
   (c) $\overline{AB} \cong \overline{CD}$. (d) $\angle CAB \cong \angle ABD$.

10. Distinct planes x and y have points A and B in common. Which one of the following is always false?

    (a) If D is in x and E in y, then $\overleftrightarrow{DE}$ is skew to $\overleftrightarrow{AB}$.

    (b) In (a), D, A, and B are coplanar.

    (c) If A, B, and C are collinear points, C is in x and y.

    (d) Neither (a), (b), or (c) can be true.

4

Deductive Reasoning

To be is to be related.
To know is to see relationships.
—C. J. Keyser

Most of our conclusions in previous chapters were based upon experimental observation and measurement. This process is called *inductive reasoning* and leads only to tentative or probable conclusions called *conjectures.*

To avoid confusion and to enable us to think more precisely, we had to define the words we used. During this process we described some undefined terms and developed the criteria for a good definition.

As we studied the basic building blocks of geometry (points, lines, and planes) and their relationships, we assumed the truth of certain statements. Such statements, called assumptions, plus the definitions and undefined terms led to necessary conclusions concerning geometric figures. These conclusions are called theorems.

The "proofs" of these theorems were acceptable to most of us, but how can we be *certain* that we have actually proven the theorem? This leads us to realize that we need to develop a method of reasoning which will always produce necessary conclusions, not conjectures.

A major aim in the study of geometry is to develop an ability to reason deductively. Man has been developing and refining the rules of logic for many centuries. Those we shall use were essentially established by Aristotle over 2000 years ago. We will make a very elementary study of the basic rules of logic in this chapter. These rules will then be applied to the proofs of the theorems of geometry. We do not mean to minimize the importance of the inductive approach, for both the inductive and the deductive methods must be utilized to create the structure of mathematics.

"Theorem" was defined in Section 2.14 as a statement which has been established by means of acceptable logic. Statement, for the purposes of logic, was defined as a meaningful declarative sentence which is either true or false, but not both. We assume your understanding of declarative sentences. As an example of a declarative sentence which is neither true nor false, we submit "He is a doctor." Until we replace "He" by the name of a specific individual, we can not determine the truth or falsity of the statement.

All statements in the field of logic are either *simple* sentences or *compound* sentences. Simple sentences contain one grammatically independent statement. They do not contain connecting words such as *and*, *or*, and *but*. A compound sentence is formed from two or more clauses that might stand as independent sentences but are joined by connectives such as *and*, *or*, *either . . . or*, and *neither . . . nor*.

Logicians use connectives such as *if . . . then*, and *if, and only if*, called logical connectives, to form compound sentences. These logical connectives are not used as connectives in the field of English grammar.

CONDITIONAL STATEMENTS

Since all theorems, assumptions, and most of our exercises are essentially compound sentences formed by using the logical connective " If . . . , then . . . ," we will confine most of our study to statements of this type. The following is a typical example of a conditional statement.

EXAMPLE: *If* two angles are right angles, *then* they are congruent.

The conditions of the statement are stated in the "if-phrase." The conclusion is found in the "then-phrase."

4.00 A compound sentence formed by using the logical connective "if. . ., then. . . ." is called a conditional sentence or a sentence of implication. Such sentences are usually called *conditionals* or *implications.*

4.01 The "if-phrase" of a conditional states the *conditions* and is called the antecedent. The "then-phrase" states the *conclusion* and is called the consequent.

It is common to represent simple sentences by letters. Thus the general form of a conditional statement is usually represented by "If p, then q." This is further symbolized by "$p \rightarrow q$," which is read "p implies q."

Many times we will find that conditional statements are not written in the classical "If p, then q" form but are written in a shortened form with the same meaning. For example, "If two angles are right angles, then they are congruent" might simply be stated "Right angles are congruent." Even though *if* and *then* are merely implied this latter statement is also a conditional.

EXAMPLES:

1. (a) Congruent segments have equal measures.
   (b) If segments are congruent, then they have equal measures.
2. (a) Women are good drivers.
   (b) If you are a woman, then you are a good driver.
3. (a) A line is perpendicular to a plane if it is perpendicular to every line in the plane through its foot.
   (b) If a line is perpendicular to every line in the plane through its foot, then the line is perpendicular to the plane.

The following exercises will help you to develop skill in identifying the antecedent (conditions) and the consequent (conclusion) of a conditional statement.

Oral Exercises

Restate the following in "if p, then q" form.

1. The sun will shine if there are no clouds.
2. Vertical angles a and b are congruent.
3. Apples are red.
4. Two angles are congruent if their sides are respectively parallel and extend in the same direction.
5. It is warm in the summer.
6. The measure of an obtuse angle is greater than the measure of an acute angle.
7. $x = 3$ since $5x = 15$.
8. $x + 7 = 12$, so $x = 5$.
9. Assuming no misfortune, we will be at the meeting.
10. To gain maturity you must gain self-discipline.

Exercises

Rewrite the following in "if p, then q" form.

1. If it is snowing, it is cold.
2. The humidity is high if it is raining.
3. If two angles of a triangle are congruent, the sides opposite these angles are congruent.
4. Skyscrapers are tall buildings.
5. Multiplying both members of an equation by the same real number produces an equivalent equation.
6. All straight angles are congruent.
7. A good student has good study habits.
8. The opposite sides of a rectangle are parallel.
9. I am ill when my temperature is 102°.
10. The converse of a definition is true.
11. The tangents to a circle at the ends of a diameter are parallel.
12. Any point in the perpendicular bisector of a segment is equidistant from the endpoints of the segment.
13. The diagonals of a parallelogram bisect each other.

14. A group of objects possessing some well-defined common property constitutes a set.

15. A set without members is called the null or empty set.

> We suggest that those classes or individuals, whose interest is aroused by the possibility of establishing truth values for compound sentences in logic, study the section on TRUTH TABLES on pages 568–572 of the appendix at this time.

ARGUMENT

We have advanced a step on the ladder of logic in that we are prepared to recognize conditional statements. We can now proceed to the next step, namely, "how to present an argument." The first thing we must learn is how to determine when certain statements *necessarily* follow others. Consider the following examples.

EXAMPLE 1

(a) If a boy is eligible to play high school football, he must be passing in three subjects.

(b) Bob plays on the high school football team.
It follows necessarily that

(c) Bob is passing in three subjects.

This is an example of a valid deductive argument. In this type of reasoning we must have (a) an accepted generalization or major premise, (b) a specific statement or minor premise satisfying the conditions of the antecedent of the major premise, and (c) the necessary conclusion.

4.02 A premise is a statement as defined in 2.13.

4.03 An argument consists of two or more related premises and a conclusion which follows from these premises.

EXAMPLE 2

(a) If the roads are ice-covered next Friday, the game will be postponed.

(b) It is now Friday and the roads are covered with ice.
It follows that

(c) The game will be postponed.

EXAMPLE 3

Replace (b) in Example 2 with "It is Thursday and the roads are covered with ice."

Does this statement satisfy the conditions of the major premise (a)? Will the game be postponed? If so, will it be due to ice covered roads? If the roads are not ice-covered on Friday, do we know that the game will be played?

There are many valid argument forms. Examples 1 and 2 present one type called the Law of Detachment (or *modus ponens*) which we shall use in this course and which is summarized without proof in 4.04. The Law of Detachment is proved in the truth tables for conditionals on page 569 of the appendix.

4.04 *Law of Detachment.* If we have established (1) that the implication $p \rightarrow q$ is true, and (2) that the antecedent p is true, then we may conclude that the consequent q is true.

The Law of Detachment may be symbolized as "$p \rightarrow q, p, \therefore q$," where $\therefore$ means "therefore."

EXAMPLE 1

$p \rightarrow q$: If a boy is an athlete, then he is healthy.

p : John, a boy, is an athlete.

$\therefore q$: John is healthy.

EXAMPLE 2

$(p \rightarrow q)$ If a boy is an athlete, then he is healthy.

q Jack, a boy, is healthy.

$\therefore p$ Jack is an athlete.

Do you agree with this conclusion? Why? Does this argument satisfy the Law of Detachment? Explain.

The argument form "$p \rightarrow q, q, \therefore p$" in Example 2 above represents an invalid argument since the minor premise does not satisfy the conditions of the antecedent of the implication.

While statements are true or false, arguments are *valid* or *invalid*. Valid arguments follow accepted rules of logic, such as the Law of Detachment and the Law of the Syllogism, to be discussed later.

Oral Exercises

Accept the two statements in each exercise. Decide whether or not they lead to a necessary conclusion. If so, state the conclusion.

1. (a) If a number is greater than 3, it is positive.
   (b) The number 7 is greater than 3.
2. (a) If a number is greater than 3, it is positive.
   (b) The number 2 is less than 3.
3. (a) If it rains on Friday, the football game will be postponed.
   (b) It rained on Thursday.
4. (a) A girl is beautiful if she has blonde hair.
   (b) Ann is a girl with blonde hair.
5. (a) Use 4(a)
   (b) Nancy does not have blonde hair.
6. (a) Use 4(a)
   (b) Toni has blonde hair.
7. (a) Use 4(a)
   (b) Carol is beautiful.
8. Which of the following represents valid reasoning?

(a) $x \rightarrow y$

y

$\therefore \quad x$

(b) $x \rightarrow y$

x

$\therefore \quad y$

Exercises

In the following exercises we are not concerned with the truth of the first two statements; the problem is to determine whether or not these two statements, if accepted, lead necessarily to a third statement. If so, give the third statement (conclusion). Rewrite the statements as conditionals, if necessary.

1. (a) All sophomores are bright.
   (b) I am a sophomore.
2. (a) All basketball players are tall.
   (b) Shorty is a basketball player.

3. (a) September has thirty days.
   (b) This month has thirty days.
4. (a) All moofs are goofs.
   (b) All toofs are moofs.
5. (a) All school houses are red.
   (b) This building is a school house.
6. (a) All marines like to fight.
   (b) Murphy likes to fight.
7. (a) If quantities are equal to the same quantity, they are equal to each other.
   (b) $X = Z$ and $Y = Z$.
8. (a) Halves of equals are equal.
   (b) $A = B$.
9. (a) If equals are subtracted from equals, the results are equal.
   (b) $a = b$ and $c = d$.
10. Give three examples of the Law of Detachment with true conclusions.

SYLLOGISM

If it is raining, then the ground is wet and if the ground is wet, then it is slippery. Does it follow that if it is raining, then the ground is slippery?

4.05 An argument with exactly two premises of the form $p \rightarrow q$ and $q \rightarrow r$ is called a hypothetical syllogism.

4.06 *The Law of the Syllogism.* If $p \rightarrow q$ and if $q \rightarrow r$, then $p \rightarrow r$ *is always true.*
(This is proved in Exercise 2, page 570 of the Appendix.)

When the Law of Detachment is applied to the Law of the Syllogism, the result is a valid argument form.

Law of Detachment	Law of the Syllogism and law of Detachment
$p \rightarrow q$	$[(p \rightarrow q) \text{ and } (q \rightarrow r)] \rightarrow (p \rightarrow r)$
$\underline{p}$	$\underline{(p \rightarrow q) \text{ and } (q \rightarrow r)}$
$\therefore\ q$	$\therefore\ (p \rightarrow r)$

Since the Law of the Syllogism states that $[(p \rightarrow q)$ and $(q \rightarrow r)] \rightarrow (p \rightarrow r)$ is always true, we need only to establish that $[(p \rightarrow q)$ and $(q \rightarrow r)]$ is true to know that $p \rightarrow r$ is true.

Example 1

If a man is successful, he works hard $(p \rightarrow q)$ and if a man works hard, then he is satisfied $(q \rightarrow r)$.

Conclusion: If a man is successful, then he is satisfied. $(p \rightarrow r)$.

Example 2

If two angles are right angles, then the measure of each angle is 90° and if the measure of each of the two angles is 90°, they have the same measure.

Conclusion: If two angles are right angles, they have the same measure.

Example 3

Using the conclusion of Example 2 as a premise, we develop Theorem 3.13.

If two angles are right angles, they have the same measure and if two angles have the same measure, they are congruent.

Conclusion: If two angles are right angles, they are congruent.

The sequence of implications involved in Examples 2 and 3, leading to the proof of a theorem, is typical of the chain of logic we shall use to develop future theorems.

To summarize, *true premises and valid reasoning will always produce true conclusions.* (This does not mean that invalid reasoning will never produce true conclusions, or that valid reasoning will produce true conclusions if the premises are not true.)

Examples

1. Valid reasoning, false conclusion.

$p \rightarrow q$	(a) If you are a student, you are not lazy.
$q \rightarrow r$	(b) If you are not lazy, you will receive good grades.
$p \rightarrow r$	(c) If you are a student, you will receive good grades.

2. Invalid reasoning, true conclusion.

$p \rightarrow q$ (a) People are interesting.

$r \rightarrow q$ (b) All high school seniors are interesting.

$r \rightarrow p$ (c) All high school seniors are people.

Exercises

Using valid reasoning, write the necessary conclusions in the following exercises.

1. If Ruff is a dog, then Puff is a cat.
   If Puff is a cat, then he has fur.
2. If the corn is to grow, we must have rain.
   If we have rain, the ball game will be cancelled.
3. $\triangle ABC$ is isosceles, if $\overline{AB} \cong \overline{BC}$.
   If $\overline{AB} \cong \overline{BC}$, $\angle A \cong \angle C$.
4. If not-B, then A.
   Not-C, if A.
5. If $x = y$, then $r = t$.
   If $a = b$, then $x = y$.
6. If $\angle A \cong \angle D$, then $m\angle 1 = m\angle 2$.
   If $\triangle ABC \cong \triangle DEF$, then $\angle A \cong \angle D$.
7. If two angles are straight angles, they have the same measure.
   If angles have the same measure, they are congruent.
8. If two intersecting lines determine a right angle, they are perpendicular.
   If two lines are perpendicular, they form congruent adjacent angles.
9. If the exterior sides of two adjacent angles are opposite rays, they are supplementary.
   If two angles are supplementary, the sum of their measures is 180°.
10. If the sky is clear, then I will carry an umbrella. If I carry an umbrella, then it will not rain.
11. If the weather is warm, then the birds sing. If the birds sing, then the dog barks.
12. Devise a hypothetical syllogism with a true conclusion.

PROOF

The previous pages have presented an elementary background in logic which justifies our belief that "a theorem is proved" when we follow accepted rules of logic.

4.07 A statement is proved when the statement is the conclusion of a valid argument in which the premises are assumed to be true.

The essence of the *deductive method* of proof is the drawing of conclusions from given premises. Compare this with the *inductive method* of proof summarized on page 13.

EULER'S CIRCLES AND VENN DIAGRAMS

4.08 Leonhard Euler (1701–1783), a brilliant Swiss mathematician, is usually given original credit for using circles to visualize deductive reasoning. John Venn (1835–1923), an English mathematician, more recently popularized the use of diagrams in set algebra and logic so that most recent literature now calls them "Venn diagrams."

EXAMPLE 1

Premises: (a) All undergraduates are geniuses. $(q \rightarrow r)$
(b) All juniors are undergraduates. $(p \rightarrow q)$

Conclusion: All juniors are geniuses. $(p \rightarrow r)$

Let us try to prove this syllogism by Euler's Method.

1. First, we draw a circle representing all juniors.
2. According to statement (b), all juniors are undergraduates. We can, then, draw a larger circle representing all undergraduates. This circle is drawn so that it includes all of the smaller circle, because all juniors are included in the group of undergraduates.

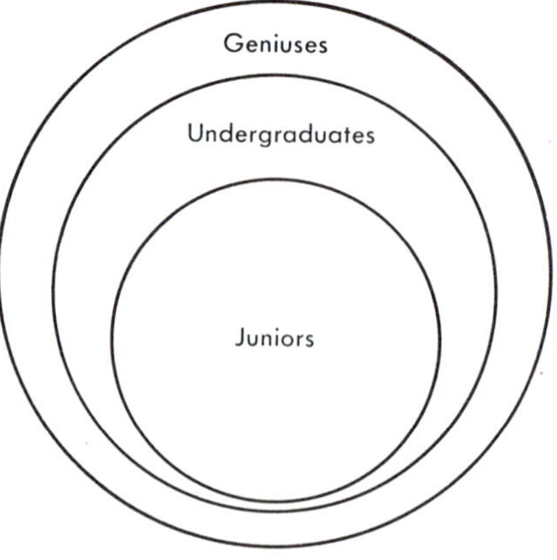

Figure 4–1

3. Finally we draw a still larger circle to represent all geniuses. This circle is drawn so that it includes all of the circle representing undergraduates. This is justified by statement (a), which says, "All undergraduates are geniuses."
4. Thus we show visually the conclusion that juniors are geniuses, for the circle representing juniors falls completely within the circle representing geniuses.
5. Is this correct according to the Law of Detachment?

EXAMPLE 2

Premises: (a) All sophomores are human.
(b) All undergraduates are human.

Conclusion: All sophomores are undergraduates.

The premises do not require that you place the set of sophomores within the set of undergraduates, so the argument is invalid.

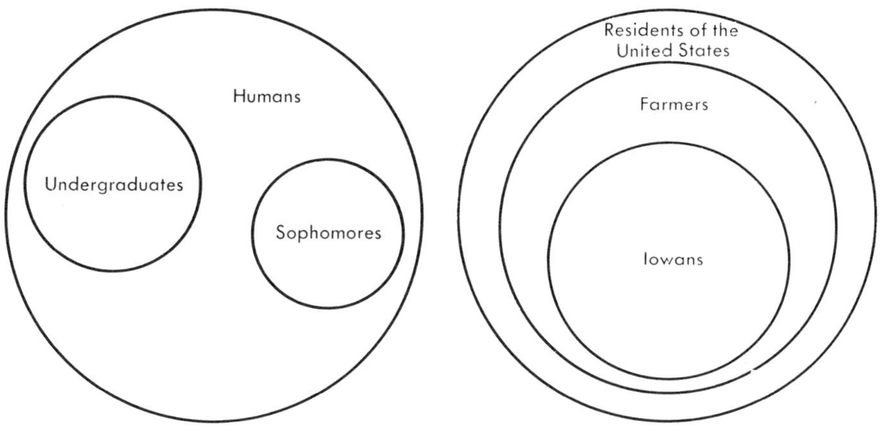

Figure 4–2

Figure 4–3

EXAMPLE 3

Premises: (a) All farmers are residents of the United States.
(b) All Iowans are farmers.

Conclusion: All Iowans are residents of the United States.

The circles show that the conclusion is inescapable if the premises are accepted. This shows how a false premise may yield a true conclusion. *The truth of the conclusion does not imply the truth of the premises.* Many theories about the physical world and why things happen as they do have been discarded because the premises were found to be false, even though the conclusions were true.

EXAMPLE 4

Premises: (a) No interesting people are tiresome.
(b) Some interesting people are seniors.

Conclusion: Some seniors are not tiresome.

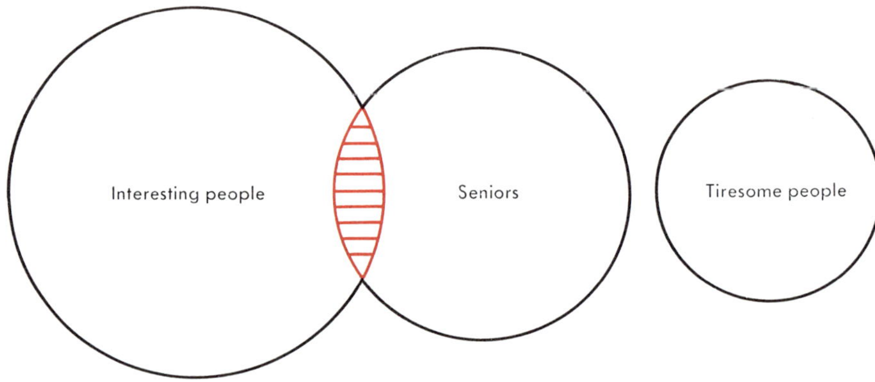

Figure 4–4

This is a valid argument since those seniors who are interesting people are not tiresome people, by premise. The shaded area represents some interesting people who are seniors.

Exercises

Test the validity of the following arguments by using circles. (Note: *Only the premises, or conditions, are to be represented by circles.*)

1. *Premises:* (a) All absent-minded people are odd.
(b) All mathematics teachers are absent-minded people.

   Conclusion: All mathematics teachers are odd.

2. *Premises:* (a) All intelligent students can pass mathematics.
(b) Todd can pass mathematics.

   Conclusion: Todd is an intelligent student.

3. *Premises:* (a) All clever people are students.
(b) All freshmen are clever people.

   Conclusion: All freshmen are students.

4. *Premises:* (a) All rabbits are timid.
(b) Sylvester is timid.

Conclusion: Sylvester is a rabbit.

5. *Premises:* (a) All careless drivers are foolish.
(b) Some careless drivers are speed demons.
(c) Some intelligent people are careless drivers.

Conclusion: (a) Some speed demons are foolish.
(b) Some intelligent people are foolish.
(c) Some intelligent people are foolish speed demons.

6. *Premises:* (a) All a's are b's.
(b) All c's are b's.

Conclusion: Some a's are c's.

7. *Premises:* (a) Some a's are b's.
(b) All b's are c's.

Conclusion: Some a's are c's.

8. Assuming that the conclusion follows from the premises by valid reasoning, complete each of the following sentences with either "must be true," "must be false," or "may be true or false."
(a) If the premises are true, then the conclusion ____.
(b) If the premises are false, then the conclusion ____.
(c) If the conclusion is true, then the premises ____.
(d) If the conclusion is false, then the premises ____.

CONVERSE OF A CONDITIONAL STATEMENT

One of the troublesome errors that occur in deductive proofs by high school students is the tendency to assume that if a conditional statement $p \rightarrow q$ is true, then the conditional statement $q \rightarrow p$ is true.

EXAMPLE 1

Theorem 3.13 stated that "All right angles are congruent." This might be rewritten as "If two angles are right angles, they are congruent."
Now, *interchanging the antecedent and consequent*, the original $p \rightarrow q$ statement becomes $q \rightarrow p$, or "If two angles are congruent, they are right angles."
Is the second statement always true?

4.09 The converse of a conditional statement $p \rightarrow q$ is $q \rightarrow p$.

EXAMPLE 2

Statement: If it is a dog, then it is an animal.
Converse: If it is an animal, then it is a dog.
Is the original statement true? Is the converse true?

EXAMPLE 3

Statement: If it is raining, then the grass is wet.

Converse: If the grass is wet, then it is raining.
Is the original statement true? Is the converse true?

EXAMPLE 4

Statement: If $a = b$, then $b = a$.

Converse: If $b = a$, then $a = b$.

Is the original statement true? Is the converse true? Can we say that all converse statements are false? Can we say that all converse statements are true?

If a conditional statement contains multiple distinct conditions and conclusions, converses are obtained by interchanging any number of distinct conditions with an *equal number* of distinct conclusions of the original implication.

EXAMPLE 5

If a ray bisects the vertex angle of an isosceles triangle, it is perpendicular to the base and bisects the base.

Let p = a ray bisects the vertex angle,
q = an isosceles triangle,
r = is perpendicular to the base, and
s = bisects the base.

The implication is then represented by

$$p \text{ and } q \rightarrow r \text{ and } s$$

Converse A (p and $r \rightarrow s$ and q)

If a ray bisects the vertex angle and is perpendicular to the base of a triangle, it bisects the base and the triangle is isosceles.

Converse B (p and $s \rightarrow r$ and q)

If a ray bisects the vertex angle and the base of a triangle, it is perpendicular to the base and the triangle is isosceles.

Converse C (r and $q \rightarrow p$ and s)

If a ray is perpendicular to the base of an isosceles triangle, it contains the ray which bisects the vertex angle and the base of the triangle.

Converses D and E are left for the student to complete.

Exercises

Write each statement in "if-then" form. Identify the antecedent by the letter p and the consequent by the letter q.

Write the converse of each of the following statements and determine its truth if you can.

1. A farmer is a hard worker.
2. A senator is a politician.
3. Good students have desire and determination.
4. Circles having equal radii are congruent.
5. An equilateral triangle is equiangular.
6. The sum of the measures of two sides of a triangle is greater than the measure of the third side.
7. The diagonals of a rectangle are congruent.
8. The opposite angles of a parallelogram are congruent.
9. When I am asleep my eyes are closed.
10. When I drink coffee I can't sleep.
11. The altitude of an isosceles triangle bisects the base.
12. Two angles are congruent if they have the same measure.
13. If two quantities are equal to the same quantity, they are equal to each other.
14. A point on the perpendicular bisector of a line segment is equidistant from the endpoints of the segment.
15. If two sides of a triangle are congruent, the angles opposite those sides are congruent.
16. A rectangle is a geometric figure with four sides.
17. A resident of Missouri is a resident of the United States.

INVERSE OF A CONDITIONAL STATEMENT (Optional)

Inserting the phrase "It is false that" preceding a statement negates the statement. The negation of a statement p is written $\sim p$ and is read "not p."

EXAMPLE 1

p: It is raining.

$\sim p$: It is false that it is raining.

$\sim p$: It is not raining.

EXAMPLE 2

p: A and B are equal.

$\sim p$: It is false that A and B are equal.

$\sim p$: A and B are not equal.

$\sim p$: A and B are unequal.

It may happen that an antecedent or consequent is already stated in a negative form.

EXAMPLE 3

$\sim p$: Two sides of triangle ABC are not congruent.

$\sim\sim p$: It is false that two sides of triangle ABC are not congruent. (Which means they are congruent)

$\sim\sim p$: Two sides of triangle ABC are not, not congruent. (Which means they are congruent)

We shall agree that *the negative of a negative form is a positive form.* That is, the negative of "Two sides of triangle ABC are not congruent" is "Two sides of triangle ABC are congruent."

4.10 The inverse of an implication is written by negating both the antecedent and the consequent. Given the implication $p \rightarrow q$, $\sim p \rightarrow \sim q$ is the inverse of the implication.

EXAMPLE 1

$p \rightarrow q$: If two angles are right angles, they are congruent.

$\sim p \rightarrow \sim q$: If two angles are not right angles, they are not congruent.

Is the original statement true? Is the inverse statement true?

EXAMPLE 2

$p \rightarrow q$: If it is not snowing, then it is warm.

$\sim p \rightarrow \sim q$: If it is snowing, then it is not warm.

Is the original statement true? Is the inverse statement true?
Can we say that all inverse statements are true? Can we say that all inverse statements are false?

Exercises

Write the inverse of each statement in the Exercises on page 87. If possible, determine the truth of each inverse. How does the truth of the converse and the inverse of each statement compare?

CONTRAPOSITIVE OF A CONDITIONAL STATEMENT
Optional

The results of our examination of the ways to write and to rewrite an implication are summarized below.

Statement: $p \rightarrow q$

Converse: $q \rightarrow p$

Inverse: $\sim p \rightarrow \sim q$

Obviously this does not exhaust the ways to rewrite an implication. Suppose we rewrite $p \rightarrow q$ as $\sim q \rightarrow \sim p$, the *inverse of the converse.*

EXAMPLE 1

($p \rightarrow q$) If two angles are right angles, they are congruent.

($\sim q \rightarrow \sim p$) If two angles are not congruent, then they are not right angles.
Are both statements true?

EXAMPLE 2

($p \rightarrow q$) If it is water, then it can freeze.

($\sim q \rightarrow \sim p$) If it can not freeze, then it is not water.
Are both statements true?

EXAMPLE 3

($p \rightarrow q$) If it is summer, then it is cold.

($\sim q \rightarrow \sim p$) If it is not cold, then it is not summer.
Are both statements true? Both false?

EXAMPLE 4

($p \longrightarrow q$) If it is a flower, then it eats meat.

($\sim q \rightarrow \sim p$) If it does not eat meat, then it is not a flower.

Are both statements true? Both false?

Suppose we rewrite $p \longrightarrow q$ as the *converse of the inverse.*

Inverse: $\sim p \rightarrow \sim q$

Converse of the Inverse: $\sim q \rightarrow \sim p$

Is this different than the inverse of the converse of $p \longrightarrow q$?

4.11 The contrapositive of $p \longrightarrow q$ is $\sim q \rightarrow \sim p$.

The truth or falsity of a conditional statement and its contrapositive was examined in Examples 1-4. Your conclusions can be proved by means of truth tables. (See Appendix, page 571.)
We therefore state the following theorem.

4.12 THEOREM

Any conditional statement and its contrapositive are simultaneously true or false.

4.13 A conditional statement and its contrapositive are said to be logically equivalent statements.

Exercises

1. Write the contrapositive of each statement in the list on page 87.
2. Can you justify each of the following statements? Explain.
   (a) If two segments are not congruent, then they do not have the same measure.
   (b) If the measure of an angle is not 180°, then it is not a straight angle.
   (c) If two distinct straight lines do not contain exactly one point in common, then they do not intersect.
   (d) If $m\angle AOE + m\angle EOD \neq m\angle AOD$, then E is not a point of the interior of $\angle AOD$.
   (e) If two intersecting sets, each of which is either a line, a ray or a segment, are not perpendicular, then they do not determine congruent adjacent angles.
3. Show that the inverse of any statement is true if the converse is true.

COUNTEREXAMPLE

We have developed the machinery we will need to prove the truth of the conjectures (conditionals) we will make and that others have made. Since we attempt to form conjectures that we hope are true, we will usually be successful in proving them true. If a conjecture is not true, must it be false? Review section 2.13. What if we make a conjecture that is false? How can we disprove it (prove it to be false)?

One method of disproof is to assume that the conjecture is true and then to derive theorems from this. If we arrive at a theorem which contradicts a theorem known to be true, we then show the conjecture to be false. This is called the *contradiction* method. This is the basis for Indirect proof in Chapter 10.

For our purpose at the moment, there is a simpler method. Recall that a conditional statement says that a certain property is true for *every member* of a certain set within our structure of geometry. If we can find *one member* of the specified set *for which the property is not true, then the conditional statement is false.* This one example of a member of the set for which the specified properties do not hold is called a counterexample of the statement.

Stating a counterexample of a conditional statement will thus disprove the statement.

EXAMPLES:

1. Two angles are congruent if their sides are parallel.

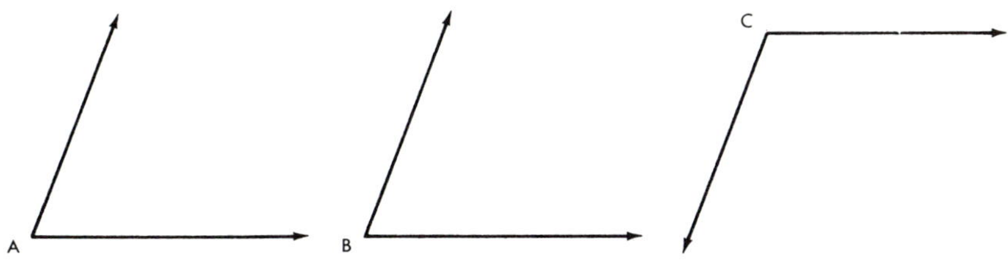

Figure 4–5

While the statement is true for many angles of the set of angles we can find at least one example ($\angle C$ in the above set) where it is not true.

2. If a line bisects a segment, it is perpendicular to the segment. The counterexample is illustrated in Figure 4-6, where $\overline{DE} \simeq \overline{EF}$. Since the lines do not form a right angle, the lines are not perpendicular (3.16).

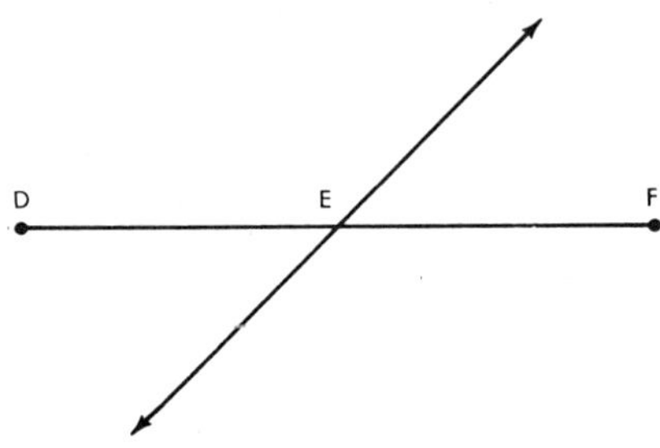

Figure 4–6

3. Every prime number is odd.

Are 2, 3, 5, and 7 prime numbers? Are all of these odd numbers? What is the counterexample?

The use of a counterexample to disprove a statement is simple and easy, if a counterexample can be found, but this is not always possible. Goldbach (1690–1764) developed the conjecture that "every even number except two could be represented as the sum of two primes." No one has yet proved or disproved this statement.

Exercises

Prove or disprove the following:

1. All animals with feathers can fly.
2. All right angles are congruent.
3. All congruent angles are right angles.
4. The intersection of a line and a plane is exactly one point.
5. For every real number a, $-a < a$.
6. For every real number a, $a + 1 > 1$.
7. If two lines do not intersect, then they are parallel.
8. If two angles are not congruent, they do not have the same measure.

9. If a line is perpendicular to $\overline{AB}$, it bisects $\overline{AB}$.

10. If a line is perpendicular to a line in a plane it is perpendicular to the plane.

Vocabulary List

deductive
simple sentence
compound sentence
conditional statement
implication
antecedent
consequent
implies
negation
counterexample
premise
argument
Law of Detachment
modus ponens
valid
invalid
syllogism
Euler
Venn
converse
inverse
contrapositive

Chapter Review

1. Is "$2 \cdot 2 = 5$" a statement? Why? (Use definition 2.13.)

2. Is "$x + 3 = 7$" a statement? Why?

3. Is "This represents an irresistible force and an immovable object" a statement? Why?

4. Demosthenes, a Greek, wrote that "All Greeks are liars." Is this a statement? Why?

5. Which of the following are simple sentences and which are compound sentences?
   (a) The Sun is hot.
   (b) The sky is blue.
   (c) The Sun is hot and the sky is blue.
   (d) The Sun is hot or the sky is blue.
   (e) If the Sun is hot, then the sky is blue.
   (f) If and only if the sky is blue, is the Sun hot.

6. Write the (a) converse, (b) inverse, and (c) contrapositive of the following conditional: Dry weather means poor crops.

7. Prove that the following is false. Name the type of "disproof" you use.
   Since moss grows on the north side of trees, I can tell directions without the aid of a compass, Sun, or stars.

8. Complete each of the following:

   (a) If a statement can be proved, it is _?_.

   (b) Any conditional sentence and its _?_ are logically equivalent sentences. That is, when one is true the other is true, or when one is false the other is false.

   (c) For every argument there is a corresponding conditional sentence, whose _?_ represents the premise or conjunction of premises and whose _?_ represents the conclusion of the argument.

   (d) The converse of a true conditional is _?_ true.

   (e) The inverse of a true conditional is _?_ true.

   (f) The contrapositive of a false conditional is _?_ true.

9. All Kats have tails.
   All dogs are Kats.

   Use the above sentences to answer the following:

   (a) Write the sentences in "If . . . , then . . ." form.

   (b) Denote the antecedents and consequents with p, q, and r.

   (c) Using the sentences as the major and minor premises of a Hypothetical Syllogism, state the conclusions of the argument.

   (d) Show that your conclusion is the result of a valid argument (satisfies the Law of Detachment).

   (e) Name the major premise of the argument.

   (f) Picture the argument using Eulers circles or Venn diagrams.

10. Vic said to Ric, "I will hire you only if our company receives the construction contract." Vic's company received the contract but refused to hire Ric. Can Ric logically win a breach of promise suit? Explain.

11. Lee was captured by the Logics, a cannibal tribe. The tribe sacrifices victims to the Idol of Truth or to the Idol of Falsity, depending upon the captives answer to a question posed by a Chief. The question asked was, "To which Idol will you be sacrificed?" What answer allowed Lee to flee, free?

Chapter 4 Test

1. Rewrite the following in "If . . . , then . . ." form.
The measure of an angle formed by opposite rays is 180°.
2. Name the antecedent of the conditional of question 1.
3. The statement, "Salmon are not mammals" disproves and is a _?_ of, "All animals are mammals."
4. The symbols $p \rightarrow q$, _?_, $\therefore$ _?_ represent the Law of Detachment.
5. The symbols $a \rightarrow b, b, \therefore a$ represent _?_ argument.
6. The symbols $[(p \rightarrow q)$ and $(q \rightarrow r)] \rightarrow (p \rightarrow r)$ represent the Law of the Syllogism.
   (a) $[(p \rightarrow q)$ and $(q \rightarrow r)] \rightarrow (p \rightarrow r)$ is always _?_, and thus satisfies the first rule of the Law of Detachment.
   (b) To satisfy the second rule of the Law of Detachment we need to show _?_ is true.
   (c) The conclusion of the above argument would be represented by the symbols _?_.

Using valid reasoning in problems 7, 8, and 9, write the conclusion if one exists. If none exists, write, "no conclusion."

7. If it rains, we will stay inside. It is not raining.
8. All cows eat grass. Bossie is a cow.
9. If two angles are right angles, then they are congruent. If two angles are vertical angles, then they are congruent.
10. Write the converse of the conditional, "If it is snowing, then it is cold."
11. Write the (a) inverse and (b) contrapositive of the following conditional: "If 15 is a factor of a number, then 3 is a factor of the number."
12. If the contrapositive of a statement is true, the statement is (never, sometimes, always) true.

5

Parallel Lines, Axioms, and Angle Relationships

Proof is the key concept in the study of geometry. If we follow the rules of deductive logic, as developed in Chapter 4, we are assured that our proofs will be valid. To proceed further with the proof of various conjectures concerning the relationships of geometric figures, we need to introduce additional definitions and to review the axiomatic structure of the real number system as a source of "true" premises. This we shall do in this chapter.

COMPLEMENTARY AND SUPPLEMENTARY ANGLES

5.00 Two acute angles are complementary if the sum of their degree measures is 90.

In Figure 5-1, $\angle x$ and $\angle y$ represent complementary angles. Angle x is the complement of $\angle y$, and $\angle y$ is the complement of $\angle x$. Using Sections 3.09, 3.14 and 3.16, the following is evident.

5.01 THEOREM

If the exterior sides of two adjacent angles are perpendicular, then the angles are complementary.

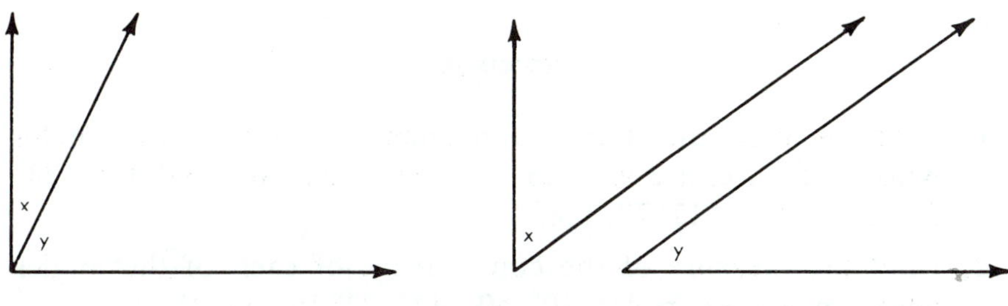

Figure 5–1

EXAMPLE: If the measure of an angle is $22°33'$, find the measure of its complement.

Solution: Since $90° = 89°\ 60'$, then $89°\ 60' - 22°\ 33' = 67°\ 27'$.

5.02 Two angles are supplementary if the sum of their degree measures is 180.

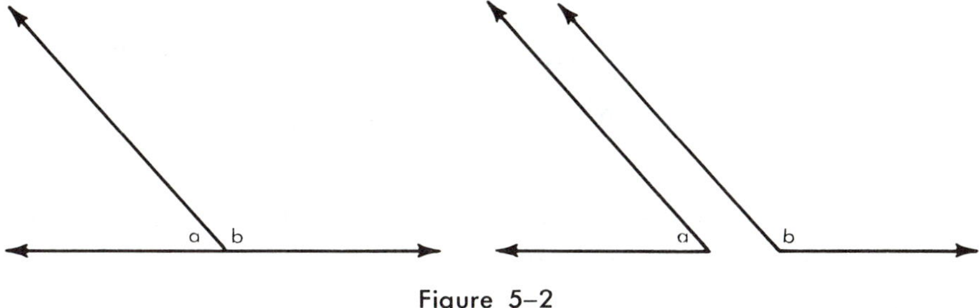

Figure 5–2

In Figure 5-2, $\angle a$ is the supplement of $\angle b$, and $\angle b$ is the supplement of $\angle a$. Using 3.03 and the definition of supplementary angles, the following is evident.

5.03 THEOREM

If the exterior sides of two adjacent angles are opposite rays, the angles are supplementary.

This theorem means that a pair of adjacent angles formed by two distinct intersecting lines are supplementary.

EXAMPLE: If the measure of an angle is $37^\circ\ 12'\ 17''$, find the measure of its supplement.

Solution: Since $180^\circ = 179^\circ\ 59'\ 60''$, then $179^\circ\ 59'\ 60'' - 37^\circ\ 12'\ 17'' = 142^\circ\ 47'\ 43''$.

Exercises

1. Find the measure of the supplement of each of the angles whose measures are 30°, 125°, 73°, 90°, 130°, 67°, $37^\circ\ 15'$, $123^\circ\ 53'$, $51^\circ\ 29'$, $77^\circ\ 17'\ 7''$, $148^\circ\ 11'\ 12''$.
2. Find the measure of the complement of each of the angles whose measures are 45°, 30°, 50°, 17°, $22^\circ\ 15'$, $38^\circ\ 18'\ 20''$.
3. Write the measures of the supplements of the angles whose measures are the following. Leave the answers in simplest form.

(a) x° (b) $\frac{y^\circ}{2}$ (c) $x^\circ - y^\circ$

(d) $x^\circ + y^\circ$ (e) $\frac{x^\circ + y^\circ}{2}$ (f) $\frac{x^\circ - y^\circ}{2}$

(g) $60^\circ + x^\circ$ (h) $90^\circ - y^\circ$ (i) $180^\circ - y^\circ$

(j) $45^\circ + \frac{a^\circ}{2}$ (k) $90^\circ - \frac{a^\circ}{2}$

4. Find the measures of the complements of the angles whose measures are given in Exercise 3.
5. Name the angle that is congruent to its supplement.
6. Find the measure of an angle that is twice the measure of its complement.
7. Find the measure of an angle that is three times the measure of its supplement.
8. In some of the figures on page 99 there are one or more pairs of numbered angles that are supplementary because they are formed when one straight line meets another. List the figures by the appropriate Roman numerals and identify the supplementary angles in each. If a figure has no supplementary angles, write the word "none."

9. Complete the following statements concerning Figures 5-3 and 5-4. If the answer cannot be found, write the words "not known." The Roman numerals refer to Figures 5-3 and 5-4.

I. If $m\angle 3 = 50°$, $m\angle 2 = \underline{?}$

II. (a) If $m\angle 1 = 40°$, $m\angle 2 = \underline{?}$

(b) If $m\angle 3 = 72°$, $m\angle 5 = \underline{?}$

III. If $m\angle 1 = 28°$, $m\angle 2 = \underline{?}$

IV. (a) If $m\angle 10 = 115°$, $m\angle 9 = \underline{?}$

(b) If $m\angle 8 = 85°$, $m\angle 7 = \underline{?}$

V. If $m\angle 3 = 90°$, $m\angle 4 = \underline{?}$

VI. If $m\angle 4 = 120°$, $m\angle 5 = \underline{?}$

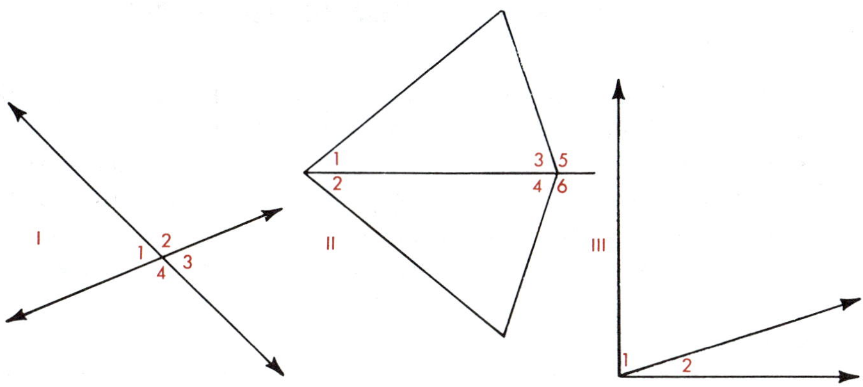

Figure 5–3

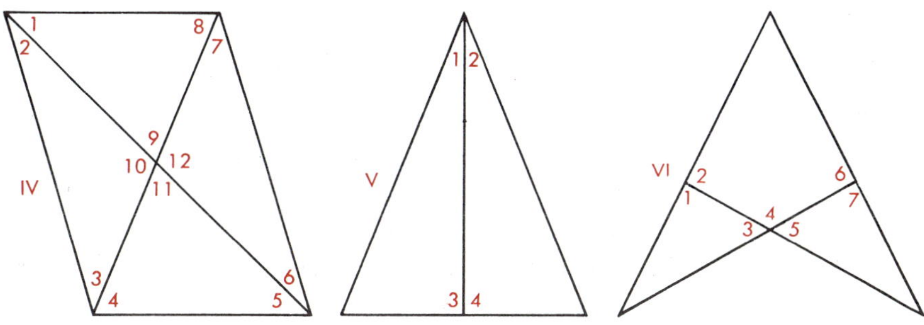

Figure 5–4

10. Angles CDE and ADB are right angles. Name the pairs of complementary angles. Can you show without measuring that $\angle 1 \cong \angle 3$? If $m\angle 2 = x°$, find $m\angle 1$ and $m\angle 3$.

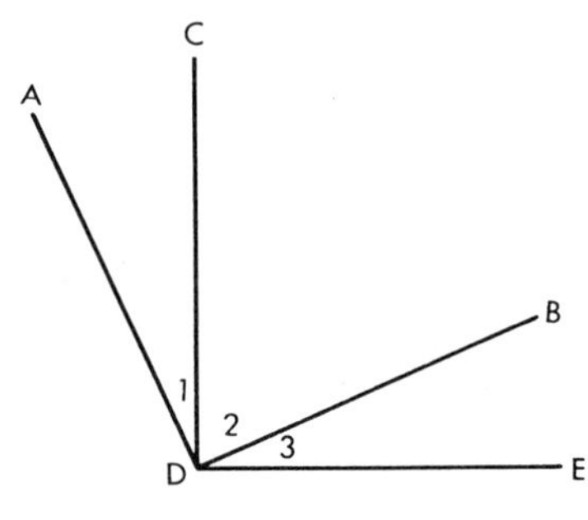

Figure 5–5

THE MEANING OF GEOMETRIC PROOF

Most of the foundation we shall use to develop deductive proofs of conjectures about geometric figures has now been laid. We have done a considerable amount of experimental work which caused you to think about the relationships between geometric figures such as points, lines, planes, and angles. By experimentation, observation, measurement, and sometimes intuition, you arrived at certain conjectures which were admittedly tentative.

Next we defined many terms used in geometry. These definitions made use of undefined terms, some commonly used words of our language, and some previously defined geometric terms.

Then we made certain assumptions about the geometric figures which we had defined. These *assumptions* (often called *postulates*) are statements about relationships which seem to be true from observation of physical, but imperfect, representations of geometric figures. Additional assumptions will be required from time to time as we progress in the study of geometry.

There remains a set of other assumptions (often called *axioms*) which are used throughout mathematics and, indeed, have many applications to everyday life. These are the axioms of equality for real numbers and you should be familiar with them from your study of algebra. They will be used to provide justification for those theorems dealing with measure.

THE AXIOMS OF EQUALITY FOR REAL NUMBERS

Let a, b, and c represent real numbers.

5.04 Reflexive Axiom. $a = a$.

5.05 Symmetric Axiom. If $a = b$, then $b = a$.

5.06 Transitive Axiom. If $a = b$ and $b = c$, then $a = c$.

5.07 Substitution Axiom. If $a = b$, then b can replace a in any numerical (mathematical) expression.

5.08 Addition Axiom. If $a = b$, then $a + c = b + c$.

5.09 Subtraction Axiom. If $a = b$, then $a - c = b - c$.

5.10 Multiplication Axiom. If $a = b$, then $ac = bc$.

5.11 Division Axiom. If $a = b$, $c \neq 0$, then $\frac{a}{c} = \frac{b}{c}$.

A relation $\otimes$, defined on a set is called an equivalence relation if the following conditions hold.

1. Reflexivity. $a \otimes a$, for every a.
2. Symmetry. If $a \otimes b$, then $b \otimes a$.
3. Transitivity. If $a \otimes b$ and $b \otimes c$, then $a \otimes c$.

Congruence is an equivalence relation as is equality; therefore, it has the properties listed below.

THE AXIOMS OF CONGRUENCY FOR GEOMETRIC FIGURES

Let a, b, and c represent geometric figures.

5.12 Reflexive Axiom. $a \cong a$.

5.13 Symmetric Axiom. If $a \cong b$, then $b \cong a$.

5.14 Transitive Axiom. If $a \cong b$ and $b \cong c$, then $a \cong c$.

EXAMPLES OF THE USE OF THE AXIOMS

Examples of the use of the reflexive, symmetric and transitive axioms of equality might seem trivial, but the axioms are powerful and necessary concepts in the structure of mathematics.

Substitution Axiom Examples

The letters a, b, c, d, x, y and z represent real numbers.

(1) If $a + b = 180$ and $b = c$, then $a + c = 180$.

(2) If $x + y = z$ and $x = a + b$, then $a + b + y = z$.

(3) If $a + b = 90$ and $c + d = 90$, then $a + b = c + d$. (Note: We could use the Axiom of Symmetry, since $90 = c + d$, and the Axiom of Transitivity to show $a + b = c + d$.)

(4) The Segment-Addition Postulate (2.35) allows us to state that if N is between M and P, then $m\overline{MN} + m\overline{NP} = m\overline{MP}$. Thus $m\overline{MN} + m\overline{NP}$ is another name for $m\overline{MP}$, or $m\overline{MN} + m\overline{NP}$ may replace or be replaced by $m\overline{MP}$ using the Substitution Axiom.

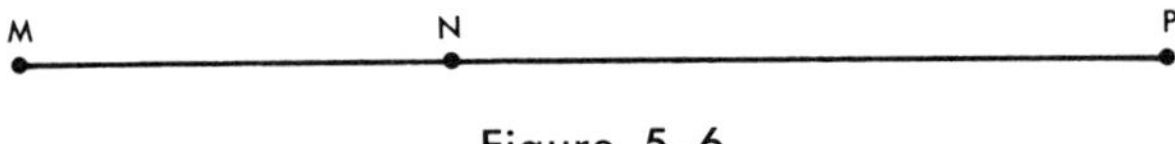

Figure 5–6

(5) The Angle-Addition Postulate (3.08) allows us to state that if $\angle ADB$ is adjacent to $\angle BDC$, then $m\angle ADB + m\angle BDC = m\angle ADC$. Under these circumstances $m\angle ADB + m\angle BDC$ may replace or be replaced by $m\angle ADC$, using the Substitution Axiom.

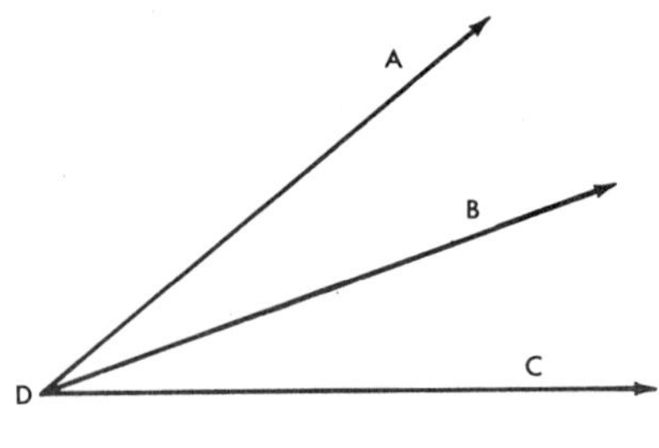

Figure 5–7

Figure 5-8 illustrates certain statements in the examples of the Addition and Subtraction Axioms.

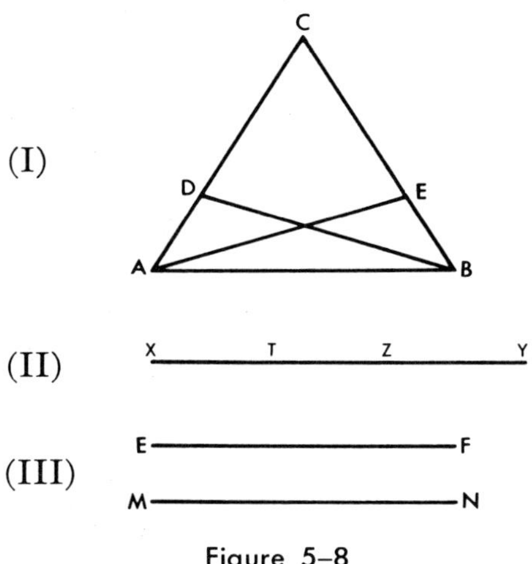

Figure 5–8

Addition Axiom Examples

(1) If $x = y$, then $x + 3 = y + 3$.

(2) If $m\overline{XT} = m\overline{ZY}$, then $m\overline{XZ} = m\overline{TY}$. What did we add? (Figure 5-8.)

(3) If $x - 2 = 15$, then $x = 15 + 2$.

(4) In Figure 5-8, if $m\overline{AD} = m\overline{BE}$ and $m\overline{DC} = m\overline{EC}$, then $m\overline{AD} + m\overline{DC} = m\overline{BE} + m\overline{EC}$. (Note: It follows that $m\overline{AC} = m\overline{BC}$. Why?)

Subtraction Axiom Examples

(1) If $a = b$, then $a - 5 = b - 5$.

(2) If $a + 4 = b$, then $a = b - 4$.

(3) If $m\overline{XZ} = m\overline{TY}$, then $m\overline{XT} = m\overline{ZY}$. (Figure 5-8.) What did we subtract?

(4) If $m\overline{AD} + m\overline{DC} = m\overline{BE} + m\overline{EC}$ and $m\overline{AD} = m\overline{BE}$, then $m\overline{DC} = m\overline{EC}$. (Figure 5-8)

(5) If $m\overline{AC} = m\overline{BC}$ and $m\overline{AD} = m\overline{BE}$, then $m\overline{DC} = m\overline{EC}$. (Figure 5-8)

Multiplication Axiom Examples

(1) If $c = d$, then $4c = 4d$. (2) If $\frac{x}{5} = y$, then $x = 5y$.

(3) If $\text{m}\overline{AD} = \frac{1}{2}\text{m}\overline{AC}$, then $2\text{m}\overline{AD} = \text{m}\overline{AC}$.

Division Axiom Examples

(1) If $\text{m}\overline{EF} = \text{m}\overline{MN}$, then $\frac{\text{m}\overline{EF}}{2} = \frac{\text{m}\overline{MN}}{2}$.

(2) If $a = b$, then $\frac{a}{3} = \frac{b}{3}$.

Reflexive Axiom of Congruency

(1) $\angle C \cong \angle C$ (2) $\angle ACB \cong \angle BCA$ (3) $\overline{AB} \cong \overline{AB}$ (4) $\overline{AB} \cong \overline{BA}$

Symmetric Axiom of Congruency

(1) If $\angle CAB \cong \angle CBA$, then $\angle CBA \cong \angle CAB$.

(2) If $\overline{EF} \cong \overline{MN}$, then $\overline{MN} \cong \overline{EF}$.

Transitive Axiom of Congruency

(1) If $\overline{XT} \cong \overline{TZ}$ and $\overline{TZ} \cong \overline{ZY}$, then $\overline{XT} \cong \overline{ZY}$.

(2) If $\angle CAB \cong \angle O$ and $\angle DEF \cong \angle O$, then $\angle CAB \cong \angle DEF$.

(Note: We used the Axiom of Symmetry to realize that $\angle O \cong \angle DEF$ before we could apply the Transitive Axiom. It might be simpler to develop a Substitution Axiom for Congruency but it is not necessary to do so.)

USING AXIOMS TO JUSTIFY EQUATION-SOLVING OPERATIONS

5 Using axioms to justify the steps in solving equations will to remember them.

Solve the equation $3x - 5 = x + 3$.

Cite an axiom to justify each operation.

Statements	*Reasons*
1. $3x - 5 = x + 3$	1. Given
2. $3x - 5 - x = x + 3 - x$ or $2x - 5 = 3$	2. Subtraction Axiom
3. $2x - 5 + 5 = 3 + 5$ or $2x = 8$	3. Addition Axiom
4. $\frac{2x}{2} = \frac{8}{2}$ or $x = 4$	4. Division Axiom
5. *Check*: $3(4) - 5 = 4 + 3$ or $7 = 7$	5. Substitution Axiom

Exercises

Solve each equation by following the form of the preceding example.

1. $2x - 5 = 13$
2. $7x + 4 = -3$
3. $7x + 14 = 2x - 6$
4. $\frac{x}{3} + \frac{x}{7} = 4$
5. $ax + b = c$
6. $\frac{x}{a} - b = c$
7. $ax + b = cx$
8. $a(x - b) = c$
9. $7x - 55 = 18 - 1 - 2x$
10. $2(x + 3) = 4x + 6 - 2x$

ELEMENTS OF FORMAL PROOF

In Chapter 4 it was established that a statement is *proved* when the statement is the conclusion of a valid argument in which the premises are assumed to be true. In geometry, most valid arguments will follow the rules of the *Law of Detachment* and the *Law of the Syllogism.* The premises will be true because they will consist of theorems and accepted definitions, assumptions and the conditions proposed in the conjecture.

We will usually follow a prescribed pattern or form in presenting a proof (argument). The form consists of a series of related statements accompanied by an acceptable reason for each, resulting in a necessary conclusion. The completed proof, sometimes called a demonstration, is called a formal proof because it follows a prescribed pattern or form. This pattern is an aid to clear and logical thinking because it helps you to organize your thinking.

Actually, a formal proof as described by logicians (experts in the study of logic), would proceed from step to step using the Law of Detachment and the Law of the Syllogism much more rigorously than we shall do. (See Example, page 111) We shall use an abbreviated form of a formal proof so that the length will be reduced and the clarity improved. We still seek full understanding, so clarity should not be sacrificed for brevity.

Our approach to the formal proof of a statement will usually begin inductively or intuitively. Let us develop an argument and then summarize our thinking into standard form.

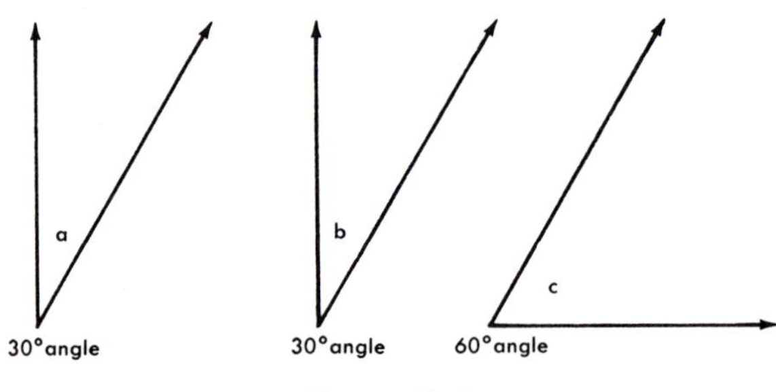

Figure 5–9

PROBLEM: If two angles are each complementary to the same angle, is there any relationship between the two angles? Let us examine two angles, $\angle a$ and $\angle b$, each of which is complementary to a third angle, $\angle c$.

If we make an arbitrary (unrestricted) choice of $\angle c$, stipulating only that it be acute, and if we draw $\angle a$ and $\angle b$ each complementary to $\angle c$, have we represented the situation in completely general form?

Usually we would need to measure $\angle a$ and $\angle b$ in several different but similar experiments to determine their relationship, if any. Our intuition tells us that $\angle a$ and $\angle b$ must have the same measure if each is to be complementary to $\angle c$. Since neither experimentation nor intuition is proof, we shall make a conditional statement, and then attempt to prove it by the rules of deduction.

United Redevelopment Corporation

Parallel and perpendicular lines have been used in the design of this apartment building. Geometric forms are often evident in modern architecture.

"If two angles are each complementary to the same angle, then the angles have equal measures." Let us investigate this relationship to determine whether it is consistent with the assumptions and definitions which have been accepted.

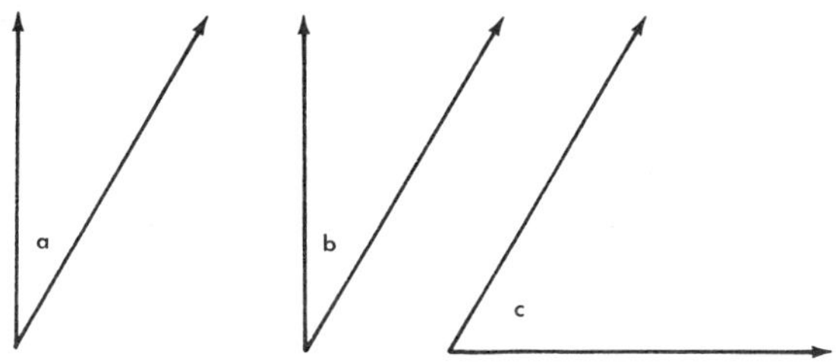

Figure 5–10

A figure is drawn to represent the conditions in general form. The antecedent is translated into specific form in terms of the letters in the figure and labeled "Given."

Given: $\angle a$ and $\angle c$ are complementary; $\angle b$ and $\angle c$ are complementary.

Then the tentative conclusion is translated into specific form using the letters of the figure and is labeled "Conjecture."

Conjecture: $m\angle a = m\angle b$.

We must now plan the *chain of reasoning* which we will follow from the given statements to our conclusion (the conjecture), supporting each statement in the chain by a reason. The acceptable reasons include given data, a definition, an assumption, an axiom, or a previously proved theorem.

Plan: The sum of the measures of two complementary angles is 90°. Therefore, $m\angle a + m\angle c = 90°$ and $m\angle b + m\angle c = 90°$. By the Transitive Axiom, $m\angle a + m\angle c = m\angle b + m\angle c$. Then by the Subtraction Axiom, $m\angle a = m\angle b$. (NOTE: Later, when you have gained more experience, you may abbreviate the plan by simply citing the principal idea or reason to be used.)

Proof:

Statements	*Reasons*
1. $\angle a$ and $\angle c$ are complementary; $\angle b$ and $\angle c$ are complementary.	1. Given
2. $m\angle a + m\angle c = 90°$ $m\angle b + m\angle c = 90°$	2. Definition of complementary angles (5.00)
3. $m\angle a + m\angle c = m\angle b + m\angle c$	3. Transitive Axiom
4. $m\angle c = m\angle c$	4. Reflexive Axiom
5. $m\angle a = m\angle b$	5. Subtraction Axiom (subtract equation 4 from equation 3.)
6. $\angle a \cong \angle b$	6. Definition of congruent angles (3.04)

We have now proved deductively the following theorem:

5.16 THEOREM

Angles complementary to the same angle are congruent.

5.17 The following outline gives the elements of a formal proof. Whenever you are making such a proof, you should observe the form very carefully. Of course, the most important part is the proof itself, in which the *statements form an unbroken chain of logic* and *each statement is supported by an acceptable reason.* Correct form is worthless if the argument is faulty or invalid.

PATTERN OF FORMAL PROOF

1. **Given:** The conditions of the antecedent expressed in terms of the Figure and its labels.
2. **Figure:** A drawing which is general so that it represents all conditions of the antecedent and nothing more.
3. **Conjecture:** The conclusion of the consequent expressed in terms of the Figure and its labels.
4. **Plan:** An analysis of the problem with key reasons cited.
5. **Proof:**

Statements	*Reasons*
Statements arranged as a deductive chain.	Given, definition, assumption, axiom, or previously proved theorem.

6. **Theorem:** The proved statement expressed in best form.

The theorem is not proved until completion of the demonstration. However, a few theorems which require knowledge beyond the scope of this book will be accepted without proof. Also, many theorems will be stated without proof with the understanding that the student will furnish the proof. Theorems which follow easily from main theorems are sometimes called *corollaries*. A corollary usually requires only one or two additional statements to be proved and is frequently a special case of the main theorem.

We shall now present a formal proof of a theorem closely related to the one just completed.

Given: $\angle a$ and $\angle c$ are complementary; $\angle b$ and $\angle d$ are complementary; $\angle c \cong \angle d$.

Figure: See Figure 5-11, page 110.

Conjecture: $\angle a \cong \angle b$

Plan: Show $m\angle a + m\angle c = m\angle b + m\angle d$ because each is equal to 90°. By subtraction, $m\angle a = m\angle b$.

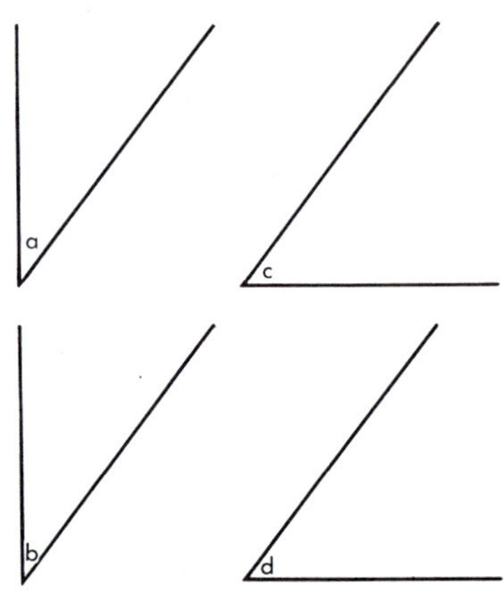

Figure 5–11

Proof:

Statements	*Reasons*
1. $\angle a$ comp. to $\angle c$; $\angle b$ comp. to $\angle d$	1. Given
2. $m\angle a + m\angle c = 90°$; $m\angle b + m\angle d = 90°$	2. Why?
3. $\therefore m\angle a + m\angle c = m\angle b + m\angle d$	3. Transitive Axiom
4. $m\angle c = m\angle d$	4. Given ($\angle c \cong \angle d$), and 3.04
5. $\therefore m\angle a = m\angle b$	5. Subtraction Axiom
6. So $\angle a \cong \angle b$	6. Why?

5.18 THEOREM

Angles complementary to congruent angles are congruent.

Alternate Proof

The previous proof was, as stated on page 106, an abbreviated formal proof. A more nearly complete formal proof of the theorem of Section 5.18 would proceed somewhat as follows:

(Using the conditions of the previous proof.)

Statements	*Reasons*
1. If two angles are complementary then the sum of their measures is 90°.	1. Definition 5.00.
2. $\angle a$ is complementary to $\angle c$; $\angle b$ is complementary to $\angle d$.	2. Given.
3. $\therefore m\angle a + m\angle c = 90°$, $m\angle b + m\angle d = 90°$.	3. Law of Detachment.
4. If $a = b$, then b may replace a in any mathematical expression.	4. Substitution Axiom 5.04.
5. $m\angle a + m\angle c = 90°$; $m\angle b + m\angle d = 90°$.	5. Step 3.
6. $\therefore m\angle a + m\angle c = m\angle b + m\angle d$.	6. Law of Detachment.
7. If two angles are congruent, they have the same measure.	7. Definition 3.04.
8. $\angle c \cong \angle d$.	8. Given.
9. $\therefore m\angle c = m\angle d$.	9. Law of Detachment.
10. If $a = b$ and $c = d$, then $a - c = b - d$.	10. Subtraction Axiom 5.04.
11. $m\angle a + m\angle c = m\angle b + m\angle d$.	11. Step 6.
12. $m\angle c = m\angle d$	12. Given.
13. $\therefore m\angle a = m\angle b$.	13. Law of Detachment.
14. If two angles have the same measure, they are congruent.	14. Definition 3.04.
15. $m\angle a = m\angle b$	15. Step 13.
16. $\therefore \angle a \cong \angle b$	16. Law of Detachment.

Sections 5.19 and 5.20 involve propositions that are stated here as theorems even though we have not proved them in this course. Consider the following theorems and develop proofs for each.

5.19 THEOREM

Angles supplementary to the same angle are congruent.

5.20 THEOREM

Angles supplementary to congruent angles are congruent.

The proof of 5.19 leads to a theorem concerning vertical angles.

Given: $\overleftrightarrow{AB}$ and $\overleftrightarrow{CD}$, forming vertical angles, $\angle 1$ and $\angle 3$

Figure: 5-12

Conjecture: $\angle 1 \cong \angle 3$

Plan: Use Section 5.19

Proof: (The proof is left to the student.)

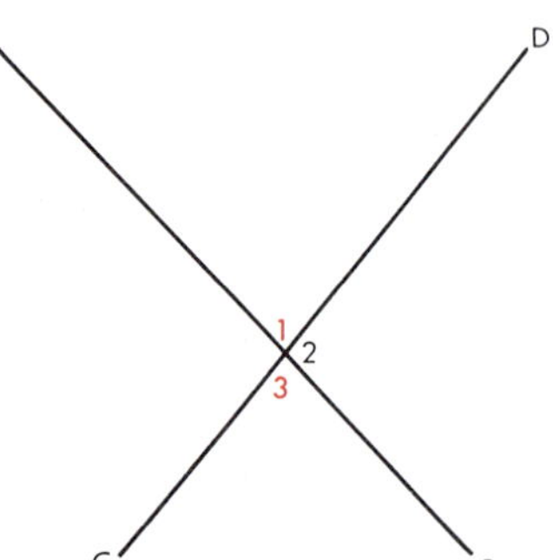

Figure 5–12

5.21 THEOREM

The two angles in a pair of vertical angles are congruent.

Exercises

In each of the following exercises, state the authority (theorem, axiom, etc.) for your answer.

1. If $\angle a$ is complementary to an angle whose measure is 50°, and $\angle b$ is complementary to another angle whose measure is 50°, what do you know about $\angle a$ and $\angle b$?

Use Figure 5-13 for Exercises 2-10.

2. If $\angle 5$ and $\angle 1$ are each supplementary to $\angle 3$, what do you know about $\angle 5$ and $\angle 1$? Explain.
3. If $\angle 8$ and $\angle 4$ are each supplementary to $\angle 6$, what do you know about $\angle 8$ and $\angle 4$? Could you answer this without looking at the figure? Explain.
4. If $\angle 2$ is supplementary to $\angle 8$, could you show that $\angle 2$ is congruent to $\angle 6$? Explain.
5. If $\angle 1 \cong \angle 5$, is $\angle 1 \cong \angle 8$? Explain.

6. If $\angle 3 \cong \angle 6$, is $\angle 3 \cong \angle 7$? Explain.
7. If $\angle 7$ is supplementary to $\angle 1$, is $\angle 7 \cong \angle 2$? Explain.
8. If $\angle 3$ is supplementary to $\angle 5$, is $\angle 2 \cong \angle 6$? Explain.

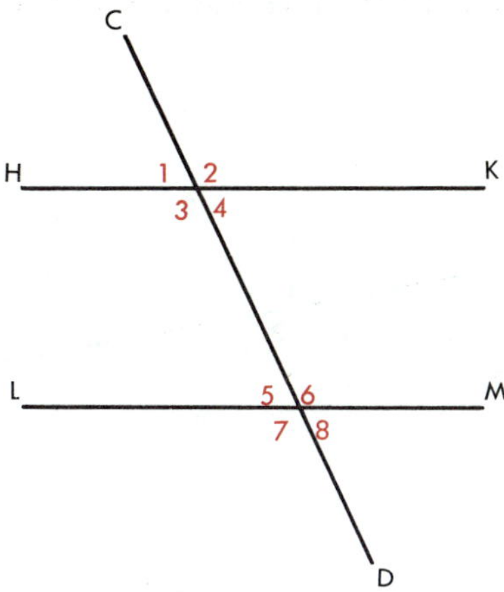

Figure 5–13

9. If $\angle 3$ is supplementary to $\angle 8$, is $\angle 3 \cong \angle 6$? Explain.
10. If $\angle 1$ is supplementary to $\angle 7$, is $\angle 4 \cong \angle 8$? Explain.

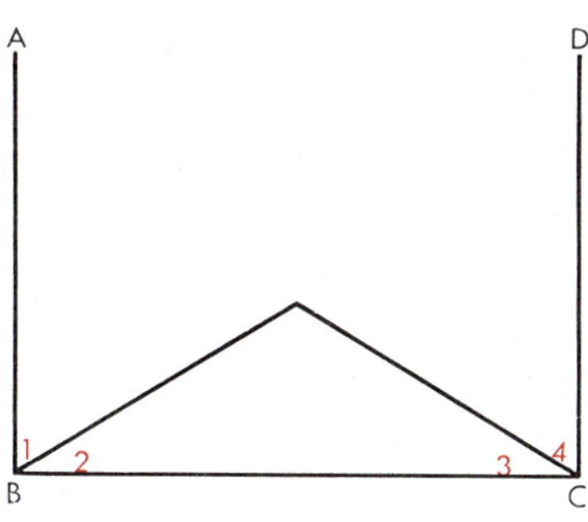

Figure 5–14

11. In Figure 5-14, $\overline{AB} \perp \overline{BC}$ and $\overline{DC} \perp \overline{BC}$. $\angle 1 \cong \angle 4$. Is $\angle 2 \cong \angle 3$? Explain.

PARALLEL LINES

5.22 A transversal is a line which intersects two or more lines in distinct points.

In Figure 5-15, $\overleftrightarrow{AB}$ is a transversal of $\overleftrightarrow{CD}$ and $\overleftrightarrow{EF}$.

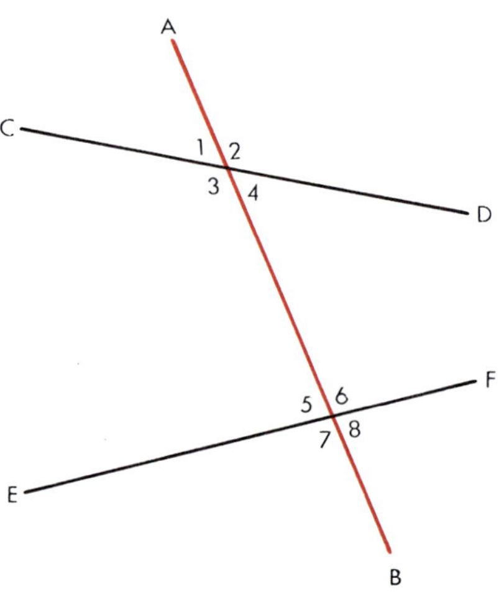

Figure 5–15

If $\overleftrightarrow{AB}$, $\overleftrightarrow{CD}$, and $\overleftrightarrow{EF}$ are coplanar, then the angles 1, 2, 7, and 8, "outside" $\overleftrightarrow{CD}$ and $\overleftrightarrow{EF}$, are called exterior angles. Angles 3, 4, 5, and 6 are interior angles. Angles 3 and 6 are alternate interior angles, as are angles 4 and 5.

Angles 1 and 8 are a pair of alternate exterior angles, as are angles 2 and 7. Pairs of corresponding angles are angles 1 and 5, 3 and 7, 2 and 6, 4 and 8. Angles 4 and 6 are called consecutive interior angles, as are angles 3 and 5.

In Figure 5-15 what relationship would $\overleftrightarrow{CD}$ and $\overleftrightarrow{EF}$ have if a pair of corresponding angles are congruent?

5.23 *Assumption: If two lines in the same plane are cut by a transversal so that the corresponding angles are congruent, the lines are parallel.*

5.24 *Assumption: If two parallel lines are cut by a transversal, the corresponding angles are congruent.*

It is possible that in another development of parallel lines, a different set of propositions would be accepted as assumptions and the assumptions made here would be proved as theorems. If, for instance, the assumptions had been made concerning the alternate interior angles, then the propositions concerning the corresponding angles would be open to proof or disproof.

In the drawing of parallel lines cut by a transversal (Figure 5-16), do any pair of angles, other than corresponding angles, appear to be congruent? Could you prove that they are congruent?

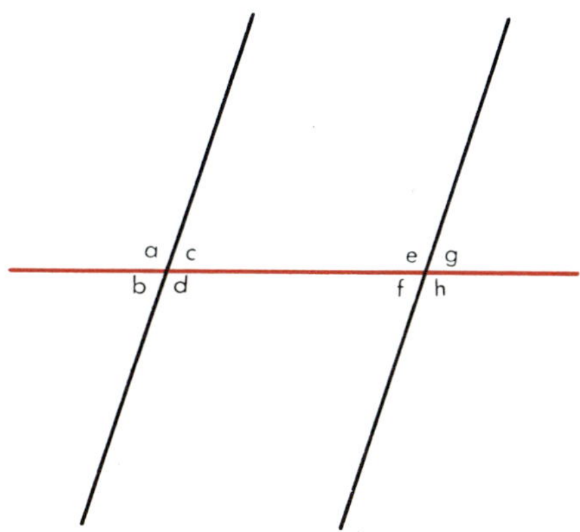

Figure 5–16

Name the pairs of corresponding angles in Figure 5-16. Name the pairs of alternate interior angles. Name the pairs of alternate exterior angles. Name the pairs of consecutive interior angles. Are there any nonadjacent, supplementary angles? Can you prove your answer?

Two students have attempted to prove the proposition, "If two parallel lines are cut by a transversal, the alternate interior angles are congruent." Are both demonstrations correct?

STUDENT A

Given: $\overleftrightarrow{MN} \parallel \overleftrightarrow{RS}$, $\overleftrightarrow{XY}$ is a transversal.

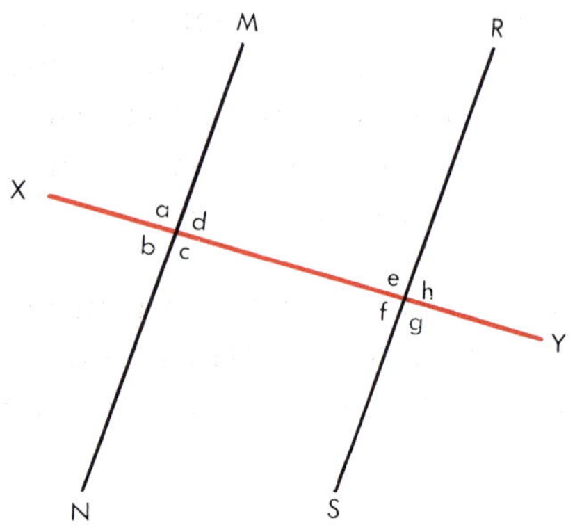

Figure 5–17

Conjecture: $\angle d \cong \angle f$ and $\angle c \cong \angle e$

Plan: To show $\angle d$ and $\angle f$ are supplementary to the same angle and $\angle c$ and $\angle e$ are supplementary to the same angle.

Proof:

Statements	*Reasons*
1. $\overleftrightarrow{MN} \parallel \overleftrightarrow{RS}$ and cut by transversal $\overleftrightarrow{XY}$	1. Given.
2. $\angle d$ supplementary to $\angle a$	2. If two lines meet to form adjacent angles, then the angles are supplementary.
3. $\angle f \cong \angle b$	3. If parallel lines are cut by a transversal, then the corresponding angles are congruent.
4. $\angle b$ supplementary to $\angle a$	4. Same as reason 2.
5. $\therefore \angle f$ supplementary to $\angle a$	5. Substitution Axiom.
6. so $\angle f \cong \angle d$	6. Supplements of the same angle are congruent (both are supp. to $\angle a$).

The student used a similar procedure to prove $\angle c \cong \angle e$.

STUDENT B

Given: $\overleftrightarrow{AB} \parallel \overleftrightarrow{CD}$. $\overleftrightarrow{EF}$ is a transversal.

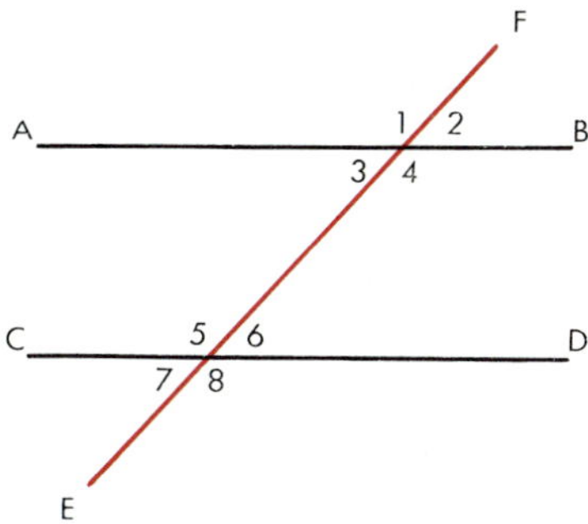

Figure 5–18

Conjecture: $\angle 3 \cong \angle 6$, and $\angle 4 \cong \angle 5$

Plan: Use 5.06, 5.23, and 5.21

Proof: *Statements*	*Reasons*
1. $\overleftrightarrow{AB} \parallel \overleftrightarrow{CD}$ and cut by transversal $\overleftrightarrow{EF}$	1. Given.
2. $\angle 1 \cong \angle 5$	2. If parallel lines are cut by a transversal the corresponding angles are congruent. (5.23)
3. $\angle 1 \cong \angle 4$	3. Vertical angles are congruent. (5.21)
4. $\angle 4 \cong \angle 5$	4. Transitive Axiom (5.06)

The student used a similar procedure to prove $\angle 3 \cong \angle 6$.

Have either or both of the demonstrations proved the following theorem?

5.25 THEOREM

If two parallel lines are cut by a transversal, the alternate interior angles are congruent.

$\overline{MN} \parallel \overline{RS}$. Measure angles 1 and 2. Measure angles 3 and 4. What is your conclusion? Can you prove your conclusion?

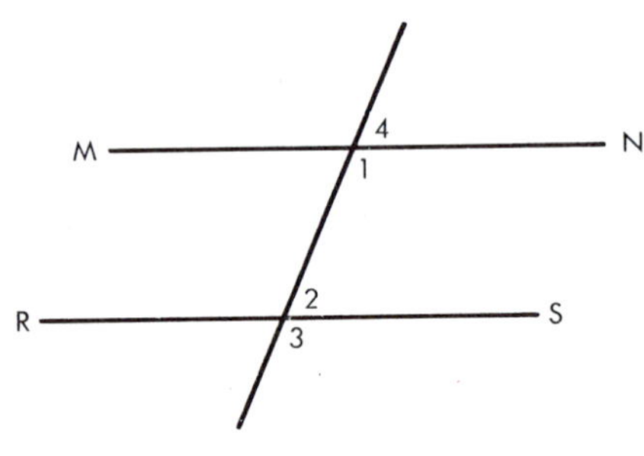

Figure 5–19

Given: $\overleftrightarrow{AB} \parallel \overleftrightarrow{CD}$. Transversal $\overleftrightarrow{EF}$

Conjecture: $\angle 3$ supplementary to $\angle 5$, $\angle 4$ supplementary to $\angle 6$

Plan: Use 5.07 (Substitution Axiom)

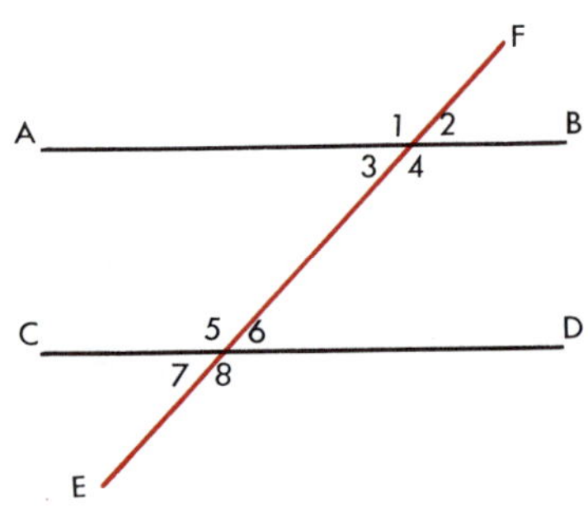

Figure 5–20

Proof: *Statements*	*Reasons*
1. $\overleftrightarrow{AB} \parallel \overleftrightarrow{CD}$ with transversal $\overleftrightarrow{EF}$.	1. Given
2. $\angle 1 \cong \angle 5$	2. If parallel lines are cut by a transversal, the corresponding angles are congruent. (5.24)

3. $\angle 1$ supplementary to $\angle 3$.	3. If two straight lines meet to form adjacent angles, the angles are supplementary. (5.03)
4. $\angle 5$ supplementary to $\angle 3$	4. Substitution Axiom

Use a similar procedure to prove $\angle 4$ supplementary to $\angle 6$.

5.26 THEOREM

If two parallel lines are cut by a transversal, the consecutive interior angles are supplementary.

The following two theorems are converses of the previous two theorems (5.25 and 5.26). Remember that converses are not always true. See if you can prove or disprove them. (*Hint:* Attempt to prove that the corresponding angles are congruent.)

5.27 THEOREM

If two lines in the same plane are cut by a transversal so that the alternate interior angles are congruent, the lines are parallel.

5.28 THEOREM

If two lines in the same plane are cut by a transversal so that the consecutive interior angles are supplementary, the lines are parallel.

Exercises

1. Does the relation "parallel" have the Reflexive Property? Symmetric Property? Transitive Property? Explain.

2. State the inverse of the statement of Section 5.23. Is it true? Explain.

3. State the contrapositive of the statement of Section 5.23. Is is true? Explain.

4. Figure 5-21 represents two parallel lines cut by a transversal. If m$\angle a$ = 130°, find the measure of each of the remaining angles.

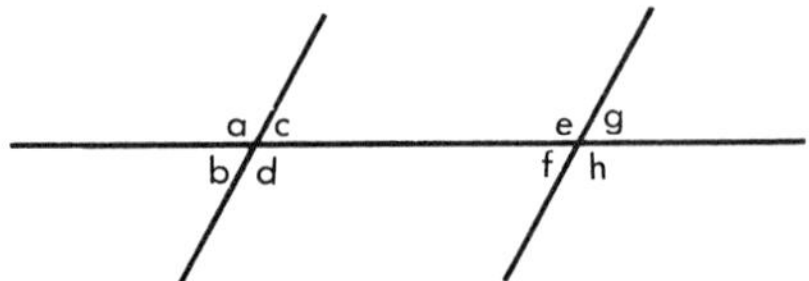

Figure 5–21

5. Assuming m$\angle g$ = 48°, find the measure of the remaining angles. (Figure 5-21)

6. If $\angle a \cong \angle d$, which lines are parallel? (Figure 5-22)

7. If $\angle c \cong \angle b$, which lines are parallel? (Figure 5-22)

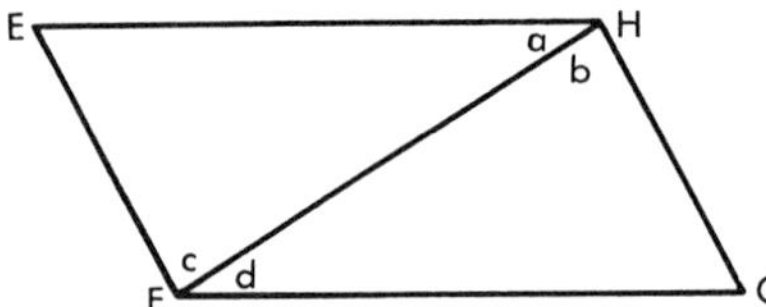

Figure 5–22

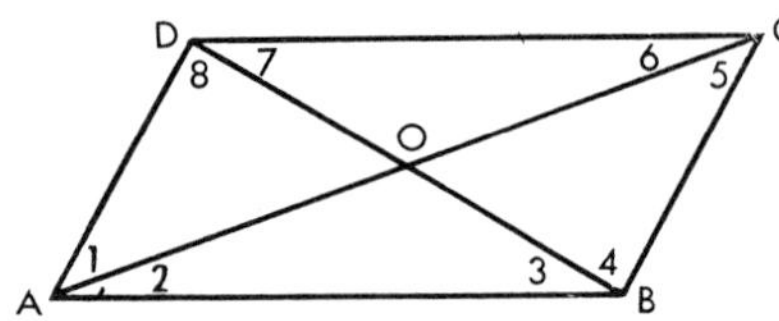

Figure 5–23

8. If $\overleftrightarrow{DC} \parallel \overleftrightarrow{AB}$, which angles are congruent? Which angles are supplementary? (Figure 5-23)

9. If $\overleftrightarrow{AD} \parallel \overleftrightarrow{CB}$, which angles are congruent? Which angles are supplementary? (Figure 5-23)

10. If $\overleftrightarrow{AB} \parallel \overleftrightarrow{CD}$, which angles are congruent? Which angles are supplementary? (Figure 5-24)

11. If $\overleftrightarrow{AB} \parallel \overleftrightarrow{CD}$, prove m$\angle 1$ + m$\angle 2$ + m$\angle 3$ = m$\angle 5$ + m$\angle 2$ + m$\angle 6$. (Figure 5-24)

12. If $\overleftrightarrow{AC} \parallel \overleftrightarrow{BD}$, name the pairs of congruent angles. (Figure 5-25)

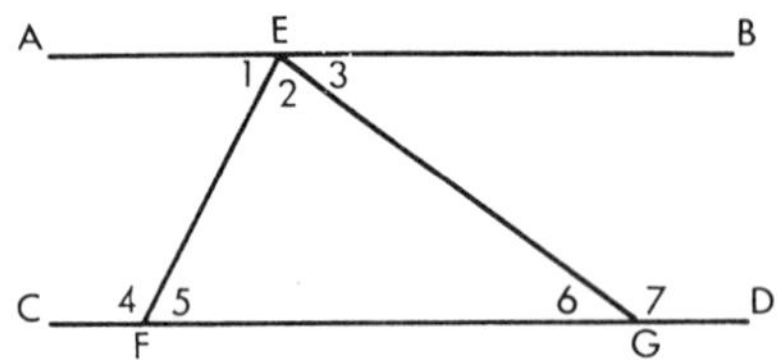

Figure 5–24

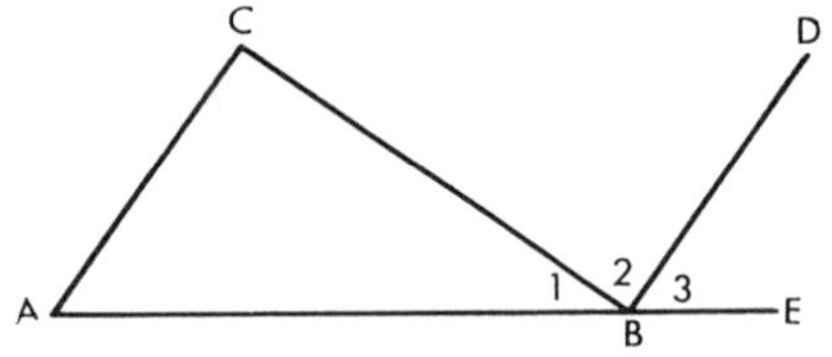

Figure 5–25

13. If $\angle A \cong \angle 1$, what could you prove about Figure 5-26?

14. If $\overleftrightarrow{AB} \parallel \overleftrightarrow{DE}$, what could you prove about Figure 5-26?

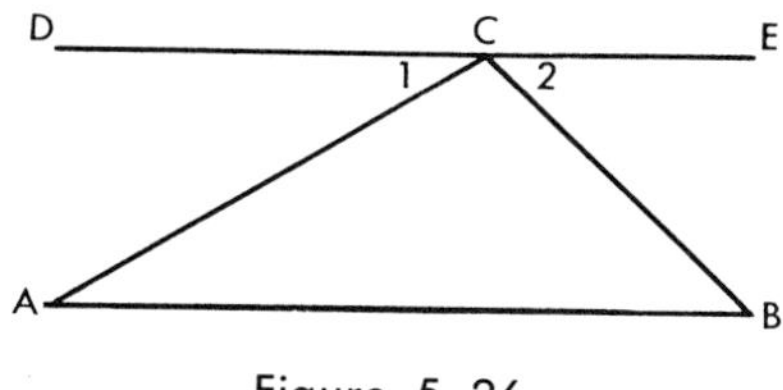

Figure 5–26

15. Prove: If a line is in the plane of two parallel lines and intersects one of the lines, then it intersects the other line. (*Hint*: Assume that it does not intersect the second line. Show that this assumption is false, then use 2.42. This is called Indirect Proof. See page 91 and Chapter 10.)

USING OUR INTUITION

In geometry, and in our environment, "lines" and objects are arranged in patterns that are both pleasing and useful. One of these "patterns" we now wish to study.

Suppose $\overleftrightarrow{AB} \parallel \overleftrightarrow{DE}$ and $\overleftrightarrow{AC} \parallel \overleftrightarrow{DF}$. What is the relationship of $\angle A$ to $\angle D$? What is their relationship to $\angle 3$? Would this help prove the relationship of $\angle A$ to $\angle D$?

If $\overleftrightarrow{DF} \parallel \overleftrightarrow{AC}$ and $\overleftrightarrow{EF} \parallel \overleftrightarrow{BC}$, is $\angle F$ related to $\angle 1$? How? Can you prove this?

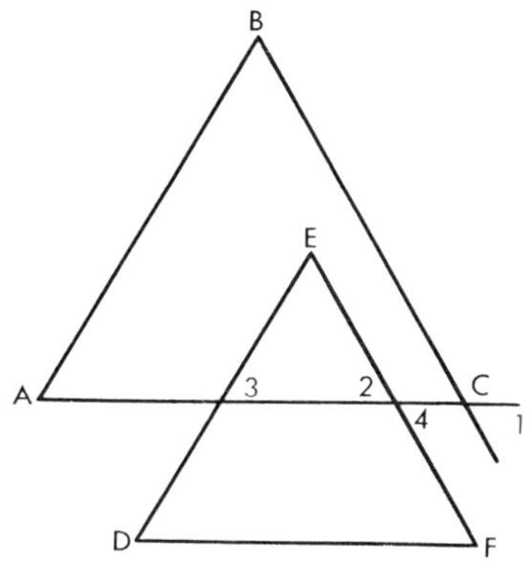

Figure 5–27

Given: $\angle ABC$ and $\angle DEF$ with $\overrightarrow{BC} \parallel \overrightarrow{EF}$ and $\overrightarrow{BA} \parallel \overrightarrow{ED}$

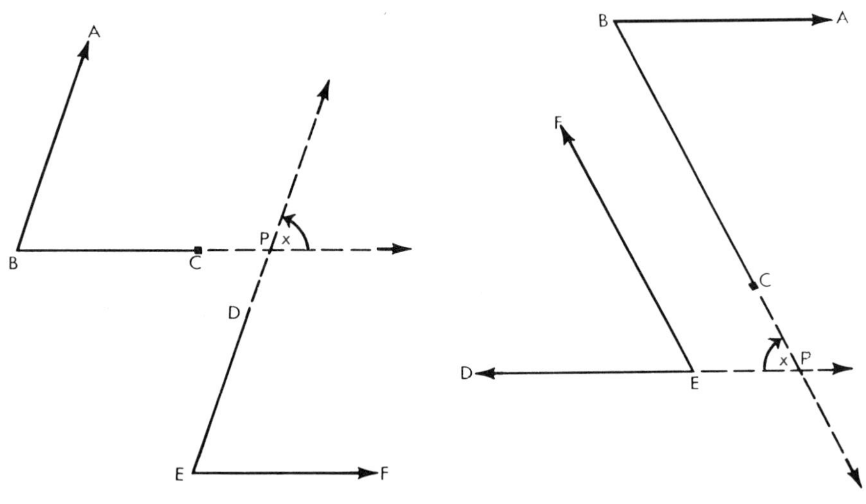

Figure 5–28

Conjecture: $\angle B \cong \angle E$

Plan: Prove $\angle ABC$ and $\angle DEF$ are both congruent to $\angle x$.

Proof:

Statements	*Reasons*
1. $\angle ABC$ and $\angle DEF$ with $\overrightarrow{BC} \parallel \overrightarrow{EF}$ in the same or opp. directions and $\overrightarrow{BA} \parallel \overrightarrow{ED}$ in the same or opp. directions	1. Given
2. $\overleftrightarrow{BC} \nparallel \overleftrightarrow{ED}$	2. Exercise 15, p. 121
3. $\overleftrightarrow{BC}$ will intersect $\overleftrightarrow{ED}$ at some point P.	3. 2.42
4. $\overrightarrow{BA} \parallel \overrightarrow{EP}$	4. Why?
5. Therefore, $\angle ABC \cong \angle x$	5. Give reason for each figure.
6. $\overrightarrow{BC} \parallel \overrightarrow{EF}$	6. Why?
7. $\angle E \cong \angle x$	7. Why?
8. $\therefore \angle ABC \cong \angle DEF$	8. Why?

5.29 THEOREM

If two angles in the same plane have sides which are parallel and which extend respectively in the same direction*, or in opposite directions, the angles are congruent.

Application of Geometry

The following drawing illustrates an instrument used by carpenters to draw parallel lines.

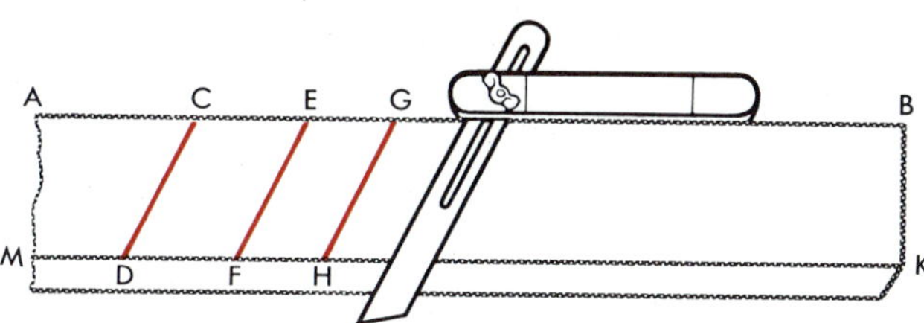

Figure 5–29

The arm of the instrument is set at a fixed position, the bevel is moved along the plane $ABMK$, and the parallel segments $\overline{CD}$, $\overline{EF}$, and $\overline{GH}$ are drawn along the edge of the arm. Is $\overleftrightarrow{AB}$ a transversal? Why? Is $\angle ACD \cong \angle AEF \cong \angle AGH$? What proposition can you apply to prove that these angles are congruent?

Certain relationships seem too obvious to warrant stating them formally; at least, they do not warrant proof. These same "obvious" relationships may also be needed frequently to justify statements in the proof of less obvious conjectures. The following three theorems fall in this "obvious, but important" class. The proofs are left to the student.

5.30 THEOREM

Two lines in the same plane perpendicular to the same line are parallel. (Use 5.23)

---

*See Exercises 18 and 19, p. 55.

5.31 THEOREM

In a plane, a line perpendicular to one of two parallel lines is perpendicular to the other line also.

5.32 THEOREM

In a plane, two lines parallel to a third line are parallel to each other.

Exercises

1. If $\overline{AC} \perp \overline{AB}$ and $\overline{DE} \perp \overline{AB}$, what can you prove about the angles in Figure 5-30?

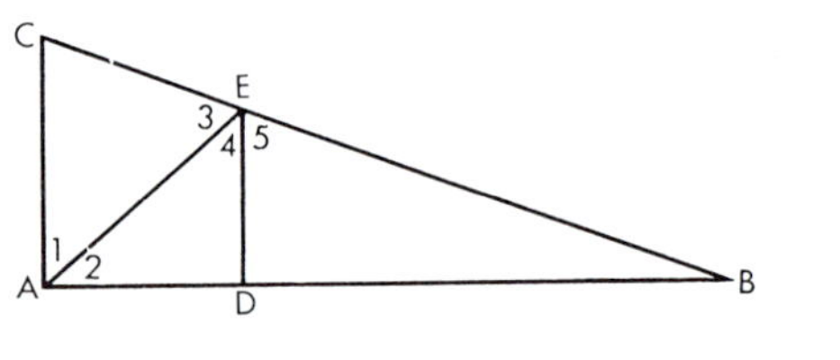

Figure 5–30

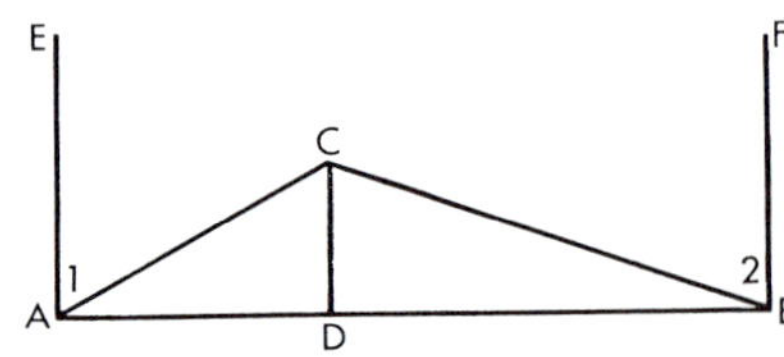

Figure 5–31

2. $\overline{AE} \perp \overline{AB}$, $\overline{CD} \parallel \overline{AE}$, and $\overline{BF} \parallel \overline{AE}$, $m\angle 1 = 60°$, $m\angle 2 = 70°$. Evaluate all other angles of Figure 5-31.
3. In Figure 5-32, if $m\angle 1 = 150°$, find $m\angle 2$.

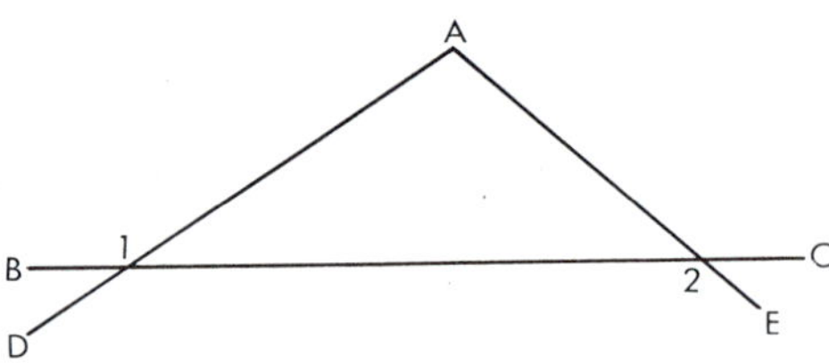

Figure 5–32

4. If the sides of $\angle A$ and $\angle B$ are parallel and one angle is acute while the other is obtuse, what is the relationship of the angles? Prove your response.
5. Which of the following properties does Theorem 5.32 represent for parallel lines: reflexive, symmetric, or transitive?
6. Does perpendicularity of lines have reflexive, symmetric, and transitive properties? Explain.

7. In Figure 5-33, $\overrightarrow{OD} \parallel \overrightarrow{BC}$, $\overrightarrow{OE} \parallel \overrightarrow{CA}$, and $\overrightarrow{OF} \parallel \overrightarrow{AB}$. Find $m\angle 1 + m\angle 2 + m\angle 3$. Prove your response.

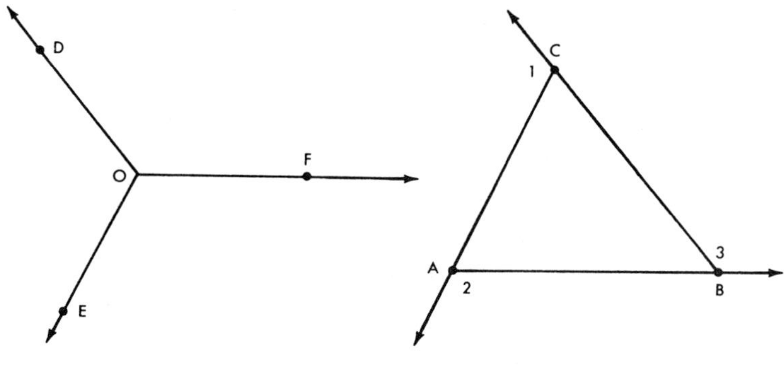

Figure 5–33

8. In Figure 5-34, if $\overline{BE} \parallel \overline{AD}$, prove that $m\angle A + m\angle D = m\angle CBD$.

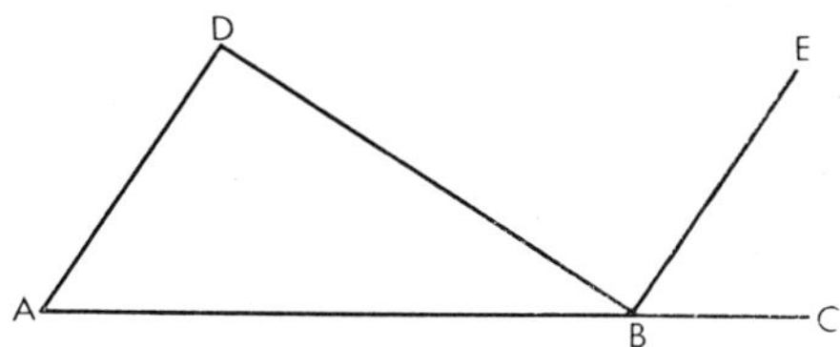

Figure 5–34

9. Why are the lines, which were drawn with a T-square as shown in Figure 5-35, parallel? How could you draw parallel "vertical" lines with the T-square? Find the name of the drawing instrument used to draw oblique parallel lines.

Figure 5–35

Vocabulary List

complementary	corollary
supplementary	transversal
axiom	alternate interior
reflexive	alternate exterior
symmetric	corresponding
transitive	consecutive

Chapter Review

State the axioms which justify the statements of Exercises 1–10.

1. $\angle ABC \cong \angle CBA$.
2. If $m\angle A = m\angle B$ and $m\angle B = m\angle C$, then $m\angle A = m\angle C$.
3. If $\overline{AD} \cong \overline{CD}$ and $\overline{CD} \cong \overline{EF}$, then $\overline{AD} \cong \overline{EF}$.
4. If $m\overline{RS} = m\overline{QT}$, then $m\overline{QT} = m\overline{RS}$.
5. If $a + b = c$ and $b = d$, then $a + d = c$.
6. If $m\overline{RN} = m\overline{ST}$, then $m\overline{RN} + m\overline{AB} = m\overline{ST} + m\overline{AB}$.
7. If $m\angle C = m\angle A$, then $\frac{1}{2} m\angle C = \frac{1}{2} m\angle A$.
8. If $x + y = a + b$ and $x = a$, then $y = b$.
9. If $3x = 21$, then $x = 7$.

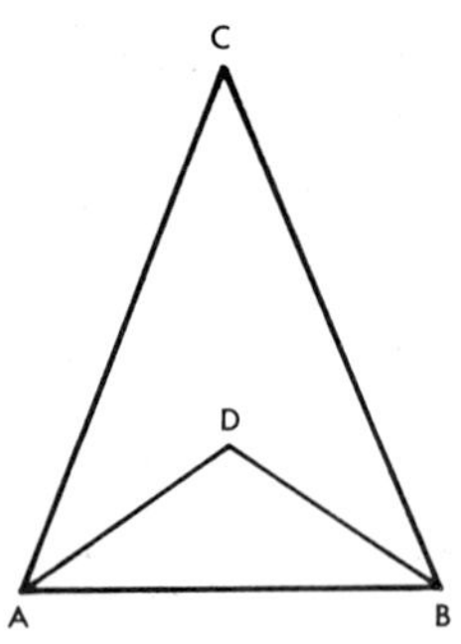

Figure 5–36

10. In Figure 5-36, if $m\angle CAB = m\angle CBA$ and $m\angle DAB = m\angle DBA$, then $m\angle CAD = m\angle CBD$.

11. If $m\angle A = 17°18'22''$, name the measure of the complement of $\angle A$. The supplement of $\angle A$.

12. Using Figure 5-37, is $\overleftrightarrow{AB} \parallel \overleftrightarrow{CD}$ if

(a) $\angle 4 = 60°$ and $\angle 6 = 60°$?

(b) $\angle 1 = 100°$ and $\angle 5 = 110°$?

(c) $\angle 2 = 80°$ and $\angle 8 = 80°$?

(d) $\angle 3 = 110°$ and $\angle 8 = 60°$?

(e) $\angle 3 = 110°$ and $\angle 6 = 70°$?

(f) $\angle 1 = 100°$ and $\angle 8 = 100°$?

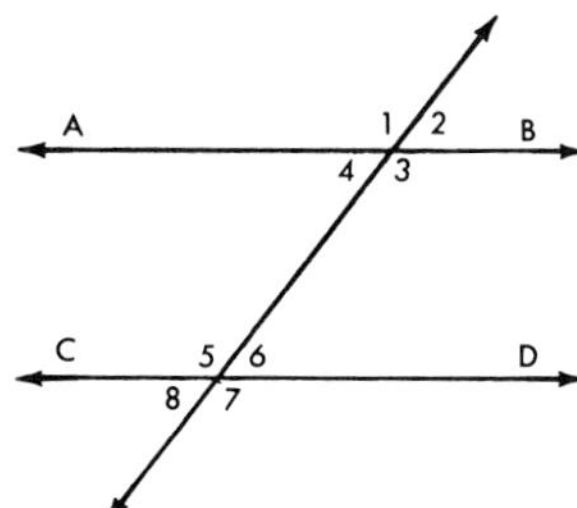

Figure 5–37

13. Using the conditions of Figure 5-38, is $\overleftrightarrow{MN} \parallel \overleftrightarrow{RT}$?

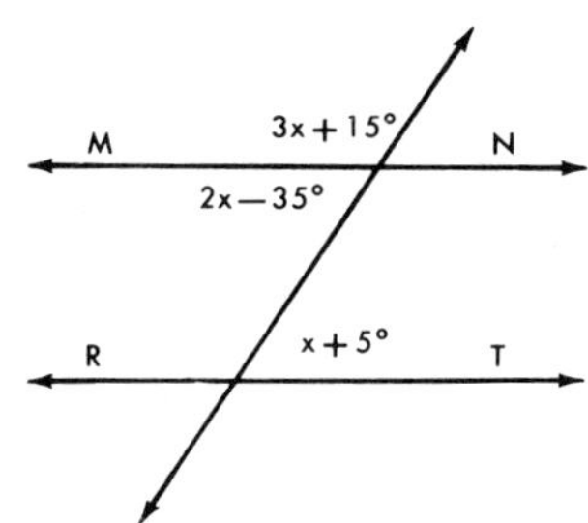

Figure 5–38

14. Given: $\overleftrightarrow{AB}$ with $\angle 2 \cong \angle 3$, in Figure 5-39.

(a) What is the relationship of $\angle 1$ and $\angle 2$? Why?

(b) What is the relationship of $\angle 1$ and $\angle 4$? Why?

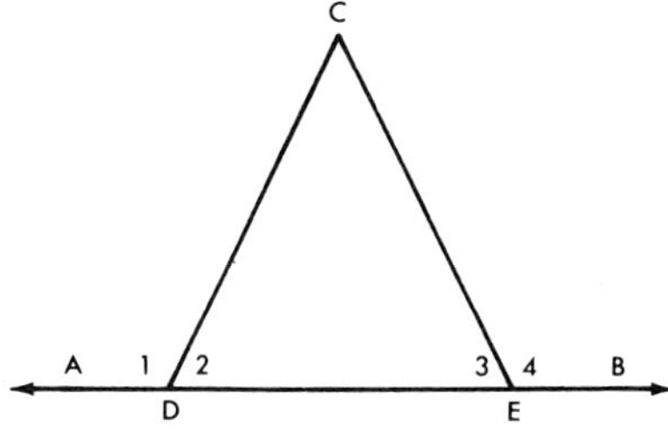

Figure 5–39

15. **Given:** In Figure 5-40, $\overleftrightarrow{AB}$, $\overrightarrow{GH} \parallel \overrightarrow{FE}$, $\overrightarrow{GH} \perp \overrightarrow{GA}$, $\overrightarrow{GH} \parallel \overrightarrow{CD}$, $m\angle C = 90°$. Justify your response to each of the following by quoting theorems, assumptions, etc.

 (a) $\overrightarrow{GA} \perp \overrightarrow{FE}$. (c) $\angle 6 \cong \angle 7$. (e) $\overrightarrow{GA} \parallel \overrightarrow{CB}$.

 (b) $\overrightarrow{FE} \parallel \overrightarrow{CD}$. (d) $\angle 8 \cong \angle 10$.

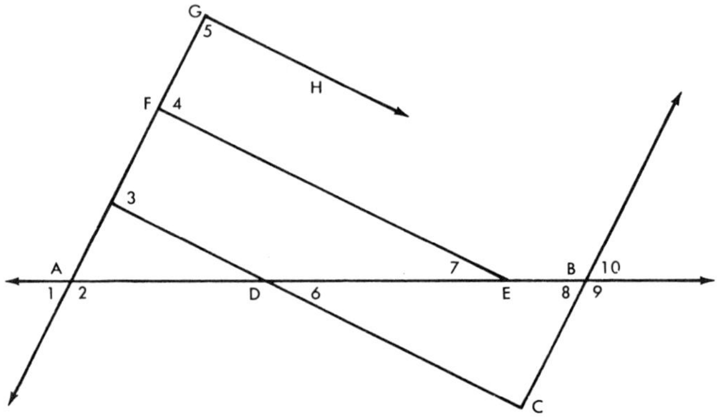

Figure 5–40

16. If $\overrightarrow{OC} \perp \overrightarrow{OA}$ and $\overrightarrow{OB} \perp \overrightarrow{OD}$ where B is in the interior of $\angle COA$, why is $\angle COB \cong \angle DOA$? ($D$ and B are on opposite sides of $\overrightarrow{OA}$.)

17. In Figure 5-41, $\overrightarrow{AM} \perp \overrightarrow{MN}$, $\overrightarrow{CN} \perp \overrightarrow{MN}$, $\angle 1 \cong \angle 3$. Is $\overrightarrow{AM} \parallel \overrightarrow{CN}$? Why? Is $\angle 2 \cong \angle 4$? Why? Is $\overrightarrow{MB} \parallel \overrightarrow{ND}$? Why?

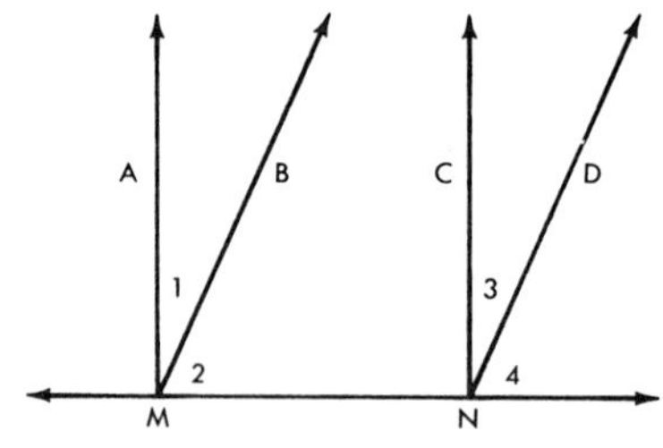

Figure 5–41

18. **Given:** Distinct lines $\overleftrightarrow{AB}$, $\overleftrightarrow{CD}$ and $\overleftrightarrow{EF}$ in plane r concurrent in O, so that $\angle DOF$ and $\angle DOB$ are congruent adjacent angles.

 Prove: $\angle COA \cong \angle COE$.

19. In Figure 5-42, $\overleftrightarrow{AB} \parallel \overline{DE}$, $\angle D \cong \angle E$. Prove $\angle CAB \cong \angle CBA$.

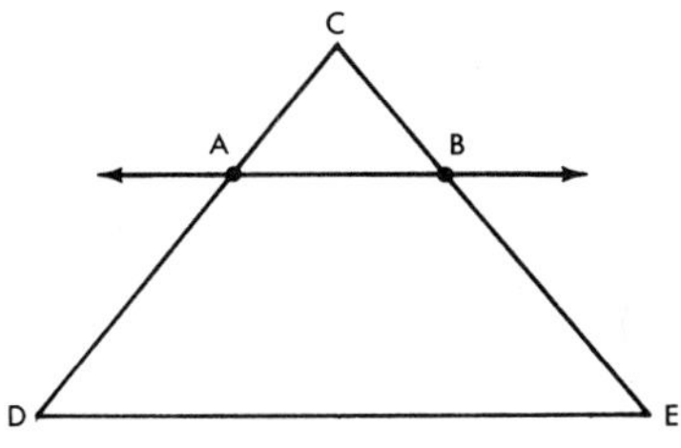

Figure 5–42

20. Given: In Figure 5-43, $\overleftrightarrow{AB}$ with O between A and B. $\overrightarrow{OC}$ bisects $\angle AOD$. $\overrightarrow{OE}$ bisects $\angle BOD$.
Prove: $\overrightarrow{OC} \perp \overrightarrow{OE}$.

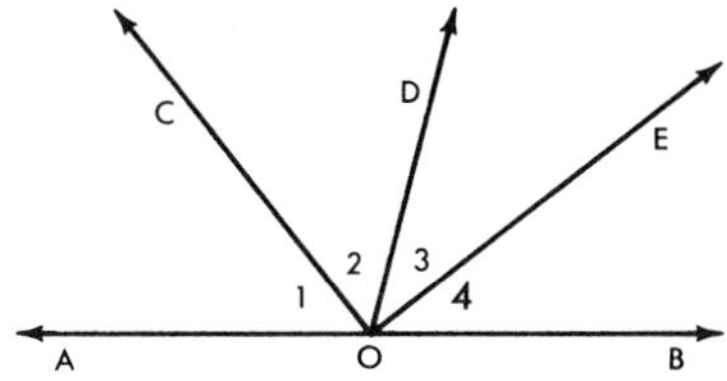

Figure 5–43

Chapter 5 Test

Complete the statements of problems 1–10.

1. The measure of an angle which is congruent to its complement is __?__.
2. The measure of an angle which is congruent to its supplement is __?__.
3. The measure of an angle which is half the measure of its complement is __?__.
4. The sum of the measures of two adjacent angles formed by two intersecting lines is __?__.
5. The measure of an angle supplementary to the angle whose measure is 73° is __?__.

6. The bisectors of two complementary adjacent angles form an angle whose measure is _?_ degrees.
7. The bisectors of a pair of vertical angles form a _?_ angle.
8. If $\angle A$ is complementary to $\angle C$ and if $\angle B$ is complementary to $\angle C$, then _?_.
9. If $\angle A$ is supplementary to $\angle C$, $\angle B$ is supplementary to $\angle D$, and $m\angle C = m\angle D$, then _?_.
10. "If $m\angle A = m\angle B$ and $m\angle B = m\angle C$, then $m\angle A = m\angle C$" is an example of the _?_ axiom of _?_.

Use Figure 5-44 to answer questions 11–25.

Given: $\overline{DG} \parallel \overline{EF}$ and $\overline{AB} \parallel \overline{EF}$. $\overline{BD}$, $\overline{AG}$, and $\overline{CH}$ concurrent at O.

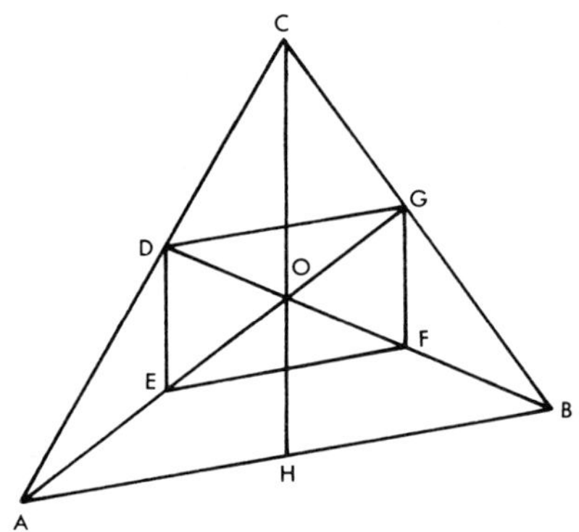

Figure 5–44

If the statements of 11–20 are true, state the theorem or assumption which justifies it. If it is false, write "False."

11. $\overline{DG} \parallel \overline{AB}$.
12. $\angle CDG \cong \angle CAB$.
13. $\angle DOC \cong \angle BOH$.
14. $\angle OEF \cong \angle DGO$.
15. $\angle GDE$ is supplementary to $\angle DEF$.
16. $\angle CDG \cong \angle ABD$.
17. If $\overline{DE} \parallel \overline{CH}$, then $\angle DEF \cong \angle OHB$.
18. if $\overline{DE} \perp \overline{EF}$ and $\overline{CH} \perp \overline{EF}$, then $\overline{DE} \parallel \overline{CH}$.
19. Using the conditions of question 18, then $\overline{DG} \perp \overline{CH}$.
20. $\overline{CH} \cong \overline{HC}$.

21–25. If $\angle EOH \cong \angle FOH$, prove $\angle DOH \cong \angle GOH$. (Write a complete formal proof.) Use Figure 5-44.

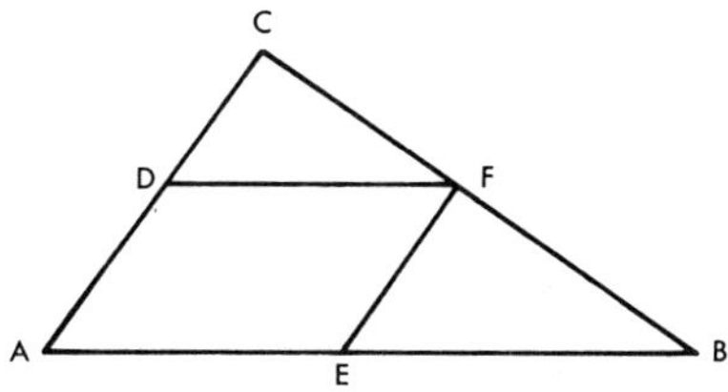

Figure 5–45

26-30. Given: Using Figure 5-45, $\overline{DF} \parallel \overline{AB}$, $\overline{EF} \parallel \overline{CA}$.

Prove: $\angle CDF \cong \angle FEB$. (Write a complete formal proof.)

6

Triangles
A Subset of the Set of Polygons

You will remember that we defined polygons in Section 2.49, page 44. In this and succeeding chapters we shall examine several of the subsets of the set of polygons. It is rather obvious that the most basic subset is the set having the least number of sides. This, as you know, is called a *triangle*. Each polygon with more than three sides can be subdivided, by drawing the proper segments, into a set of distinct triangles. It follows that if we know the geometric relationships in the set of triangles, we will know much about the more complex polygons.

The triangle is widely used in art, architecture, and engineering, and for this reason, it is probably the most important figure studied in geometry. Since the shape of a triangle cannot be changed without changing the length of a side, the triangle is used to give rigidity in construction work.

6.00 Let us classify several important subsets of the set of polygons by the number of segments (sides) forming them.

Triangle: three sides
Quadrilateral: four sides
Pentagon: five sides
Hexagon: six sides
Heptagon: seven sides
Octagon: eight sides
Nonagon: nine sides
Decagon: ten sides
Dodecagon: twelve sides
Icosagon: twenty sides

The prefixes used to describe various polygons will also be used in a future chapter for naming three-dimensional geometric figures.

6.01 Two sides of a polygon with a common vertex are called consecutive or adjacent sides. Two vertices which are the endpoints of the same side are consecutive vertices. Two angles of a polygon with a common side are consecutive angles.

6.02 The sum of the measures of the sides of a polygon is called the perimeter of the polygon. A "perimeter" will be a positive number.

6.03 Every simple polygon separates the points of its plane, excluding itself, into two sets called the interior and exterior of the polygon.

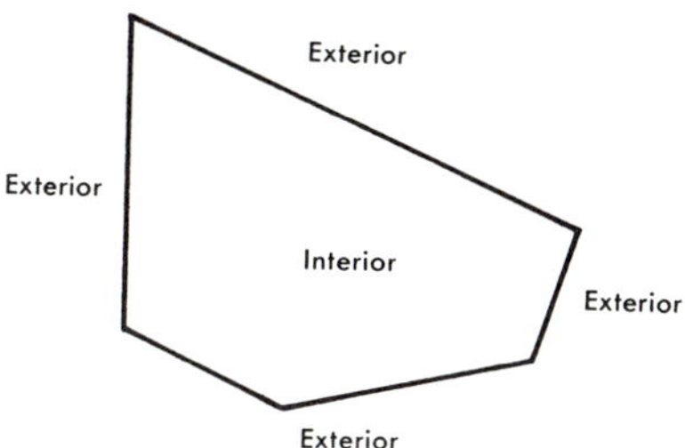

Figure 6–1

6.04 The union of a polygon and its interior is called a polygonal region. The names of specific regions will follow the classifications of Section 6.00, that is, *triangular region*, *quadrangular region*, and so on.

6.05 A polygon with all angles congruent is equiangular.

6.06 A polygon with all sides congruent is equilateral.

6.07 Polygons that are both equiangular and equilateral are called regular polygons.

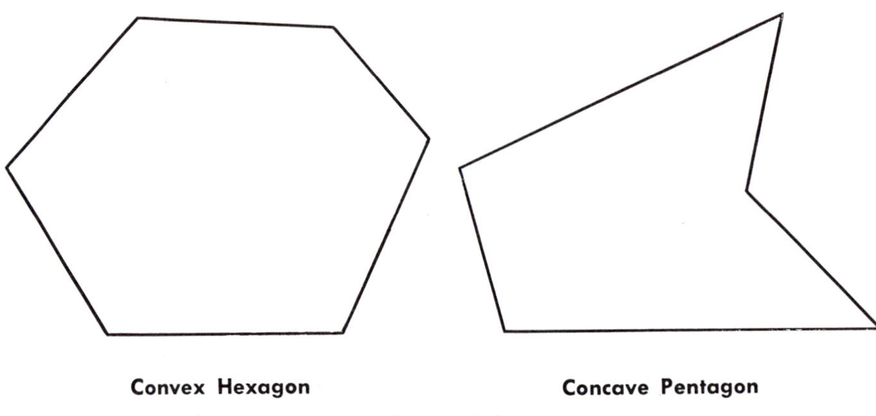

Figure 6–2

6.08 A simple polygon is called a convex polygon if a segment connecting any two points of the interior lies entirely in the interior of the polygon; otherwise, it is called a concave polygon. Since we will primarily study convex polygons in this course, the word "polygon" hereafter will mean "convex polygon."

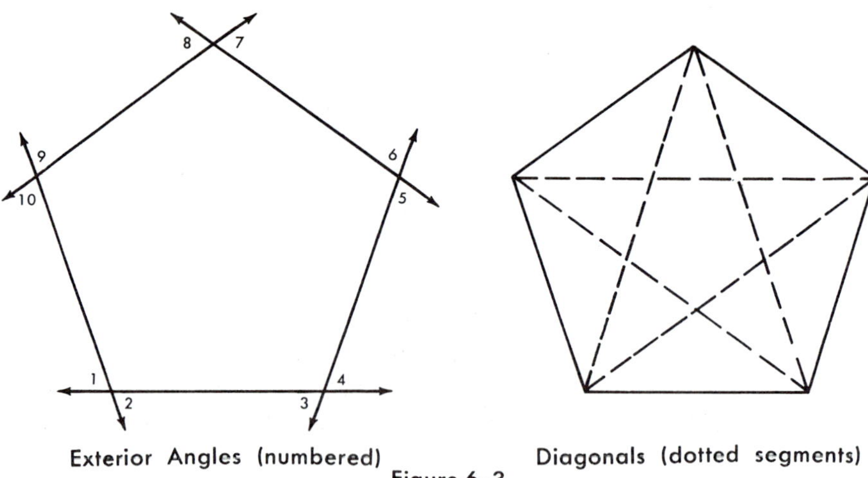

Figure 6–3

6.09 A diagonal of a polygon is a segment whose endpoints are non-consecutive vertices of the polygon.

6.10 An exterior angle of a polygon is the angle determined by one side and the opposite ray of an adjacent side of an angle of the polygon.

Exercises

1. The measures of the sides of a triangle are $6\frac{1}{2}$, $7\frac{3}{16}$, and $8\frac{5}{16}$ inches. What is the perimeter of the triangle?
2. What is the class (subset) name of polygon $ABCDE$?
3. Does a triangle contain three angles or does it determine three angles? Explain.
4. Is a regular polygon equiangular? Equilateral?
5. Is an equilateral polygon necessarily equiangular? Explain.
6. Is an equiangular polygon necessarily equilateral? Explain.
7. An octagon has how many sides?
8. In polygon $ABCD$, name two consecutive sides. Two consecutive angles.
9. What is the least number of sides a concave polygon may have?
10. What is the name of the union of a polygonal region and the exterior of the polygon?
11. A quadrilateral has how many diagonals? How many exterior angles?
12. What is the relationship of an exterior angle of a polygon to its adjacent angle of the polygon?
13. Define "interior of a convex polygon" in terms of union or intersection of the interiors of the angles of the polygon.
14. Devise a formula which will determine the number of distinct diagonals of a polygon of n sides.

TRIANGLES CLASSIFIED

Just as there are distinguishable subsets of the set of polygons, there are also various recognizable subsets of the set of triangles. The distinguishing features represent variations in the relationships of the sides and angles. Before reading further, attempt to develop your own classifications. The commonly accepted names for these subsets are given below.

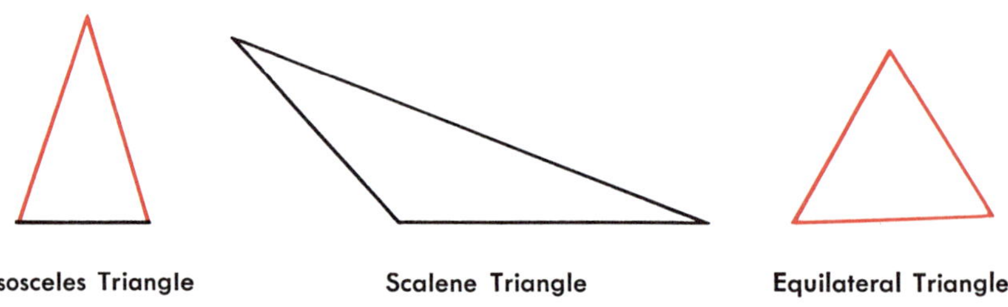

Figure 6–4

6.11 A scalene (general) triangle is a triangle with no two sides congruent.

6.12 An isosceles triangle is a triangle with two congruent sides.

Sometimes the congruent sides are called the legs of a triangle. The angle determined by the congruent sides is commonly called the vertex angle and the third side is called the base of the triangle. The angles of the triangle containing the base as a common side are called base angles. Note that all triangles have three "bases" and an isosceles triangle is no exception.

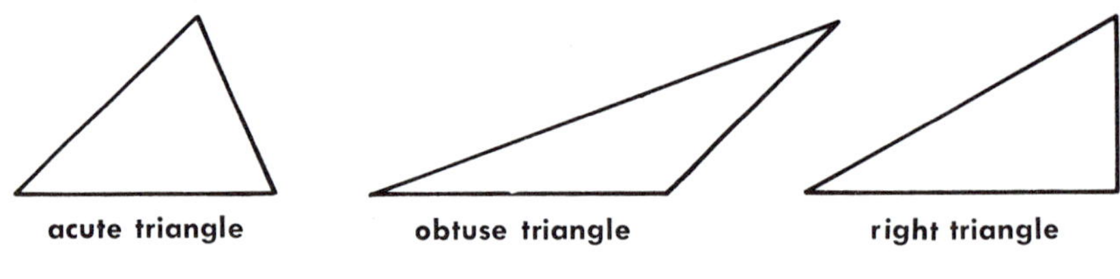

Figure 6–5

6.13 An acute triangle is a triangle each of whose angles is acute.

6.14 An obtuse triangle is a triangle with an obtuse angle.

6.15 A right triangle is a triangle with a right angle. The side opposite the right angle is called the hypotenuse.

Warning! Just as *John E. Smith, Jr.*, distinguishes one individual in the John E. Smith family from all other members of the family, or from the Smith clan in general, so must the names we use distinguish the designated members of the set of geometric figures from all other members of the set.

EXAMPLE: "Name the three-sided polygon with two and only two congruent sides." *Isosceles triangle* is correct. Triangle or equilateral triangle would not be acceptable.

ALTITUDES, MEDIANS, AND BISECTORS

In the study of triangles we will find it necessary to examine the relationships of certain segments and of the lines and rays containing these segments. The definitions of these terms follow, and will be used throughout your study of geometry.

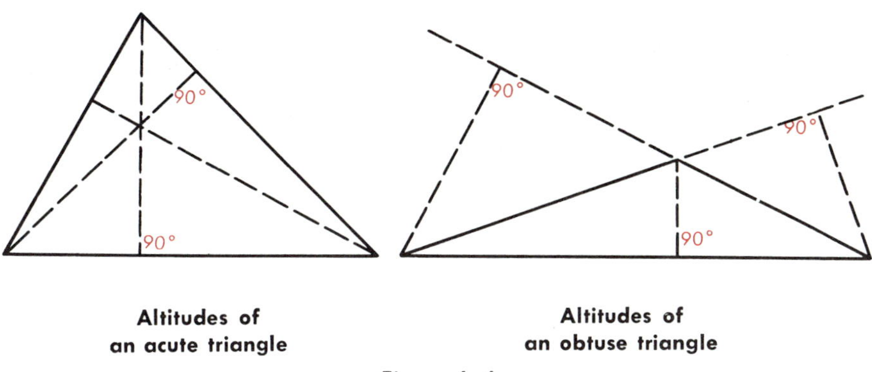

Figure 6–6

6.16 An altitude of a triangle is a segment which joins a vertex to a point of the line containing the opposite side, and is perpendicular to that line. The measure of an altitude is called the height of the triangle with respect to the base to which the altitude is drawn.

6.17 A median of a triangle is a segment whose endpoints are a vertex and the midpoint of the opposite side.

6.18 An angle bisector of a triangle is a segment contained in the ray bisecting an angle and whose endpoints are the vertex of the angle and a point of the opposite side.

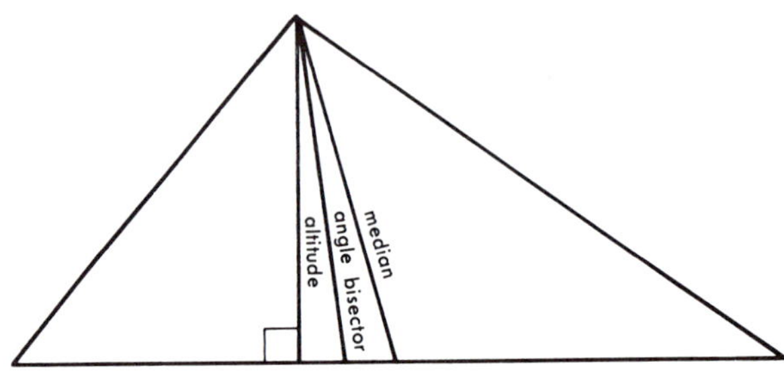

Figure 6–7

Exercises

1. A triangle has how many altitudes? How many medians? How many angle bisectors?
2. A triangle has how many diagonals?
3. A triangle has how many exterior angles?
4. Name the segment in Figure 6-8 which could represent the altitude, the segment which could represent the median, and the segment which could represent the angle bisector of $\triangle DEF$. (The symbol $\triangle$ represents the word "triangle.") Use a ruler and protractor to help you confirm your statement.

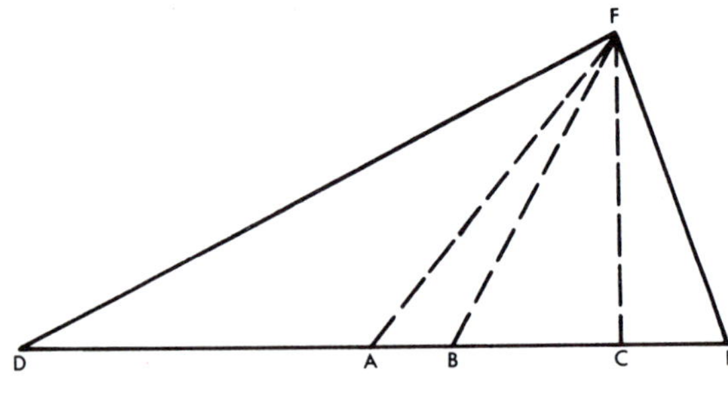

Figure 6–8

5. Are all equilateral triangles isosceles triangles? Explain.
6. Are all isosceles triangles equilateral triangles? Explain.

7. Define "equiangular triangle."
8. Define "equilateral triangle."
9. Define " regular triangle."
10. Draw right triangle ABC, where $\angle C$ is the right angle. Draw and label the altitudes of the triangle.
11. Can an isosceles triangle be a scalene triangle? Explain.
12. Could a right triangle be an isosceles triangle? Could it be a scalene triangle?
13. Could an obtuse triangle be an acute triangle? Explain. Is a right triangle an acute triangle? Explain.
14. Could a right triangle be an obtuse triangle? Explain.
15. Have we established that an obtuse triangle can exist? Explain.

INDUCTIVE REASONING WITH TRIANGLES

The questions of Exercises 13, 14, and 15 in the preceding list make us realize that we need to establish the sum of the measures of the angles of a triangle. Let us proceed with some experimentation in this area.

1. Use a straight edge and a protractor to draw a scalene acute triangle, a scalene obtuse triangle, and a scalene right triangle. Label the vertices of the triangles. Use your protractor to determine the sum of the measures of the angles of each triangle. Are your results the same as other members of your class?
2. Have you now proved that the sum of the measures of the angles of a (any) triangle is 179°, or 180°, or 181°, or $179\frac{1}{2}°$, or any other constant value? Explain.
3. Which value do you believe to be correct? Why?
4. Does it seem possible that the sum would change with different shaped triangles?
5. Draw several non-planar "triangles" on a sphere. Attempt to measure the "angles" of these figures. Can you use your protractor? What is the sum of the measures you found? How does this value or values compare with the values determined by other members of your class?

MEASURE OF THE ANGLES OF A TRIANGLE

Your work in the previous experiments may have enabled you to state a conjecture concerning the sum of the measures of the angles of a plane triangle. Since an experimental or inductive procedure merely establishes probability, we must develop a deductive proof before accepting the conjecture as a theorem.

The only angle-measure relationships we now have available are those concerning parallel lines intersected by a transversal and those concerning complementary, supplementary, and vertical angles. It is obvious that any proof of our conjecture must rest upon these concepts. As a basis for the deductive proof, review Exercise 11, page 120. In anticipation of your ability to prove the usual conjecture, we state it below as a theorem.

6.19 THEOREM

The sum of the measures of the angles of a plane triangle is 180°.

(The proof is left to the student.)

6.20 THEOREM

If two angles of one triangle are congruent to two angles of another triangle, the remaining angles are congruent.

MEASURE OF THE ANGLES OF A POLYGON

Draw any polygon of more than three sides. Draw all possible diagonals from one vertex. Does the sum of the measures of the angles of all these triangles equal the sum of the measures of the angles of the polygon? How does the number of triangles compare with the number of sides (or angles) of the polygon?

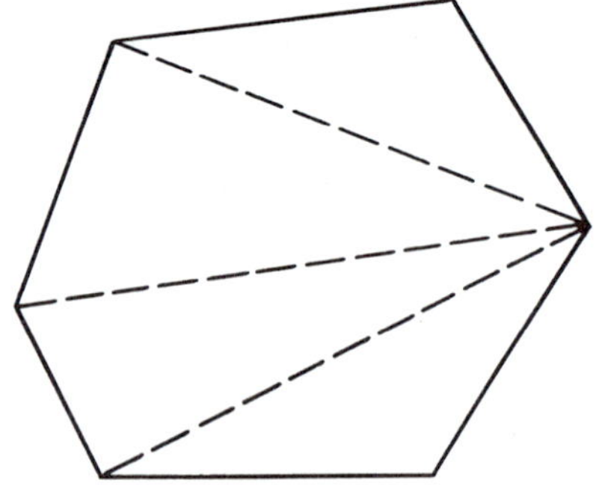

Figure 6–9

The answers to the previous questions lead us to the following proposition which is stated as a theorem in anticipation of your proof. The formal proof of Theorem 6.21 and Theorem 6.22 is left to the student.

6.21 THEOREM

In a polygon of n sides, the sum of the measures of the angles is $(n-2)180°$.

6.22 THEOREM

In a regular polygon of n sides, the measure of one angle is $\frac{(n-2)180°}{n}$.

We have established a formula for finding the sum of the measures of the angles of a polygon. This leads us naturally to ask whether the sum of the measures of the *exterior* angles is equally significant.

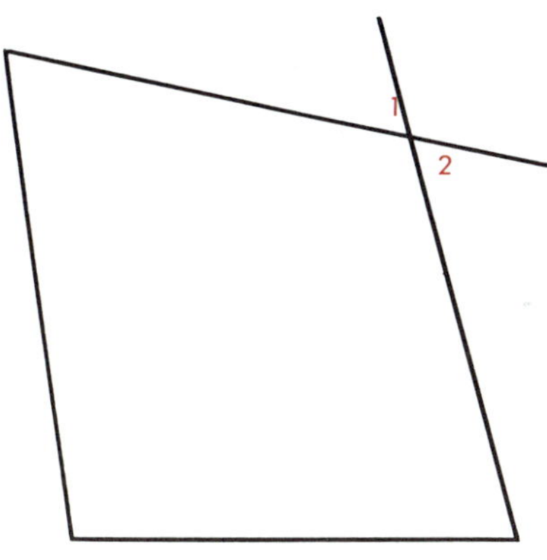

Figure 6–10

There are two exterior angles possible at each vertex of a polygon, as shown in Figure 6-10. Since these are vertical angles and therefore are congruent, let us consider only one exterior angle at each vertex in our experimentation.

Draw several different types of polygons and find the sum of the measures of the exterior angles (one at each vertex). Write any conjecture that you are able to reach by inductive reasoning. If we now have a tentative conclusion, we should test it by attempting a deductive proof.

Given: Polygon $ABCDE$ with exterior angles k, l, m, n, and o

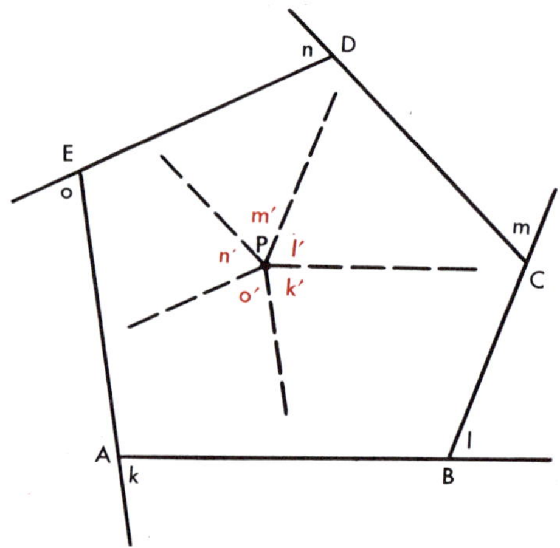

Figure 6–11

Conjecture: $m\angle k + m\angle l + m\angle m + m\angle n + m\angle o = 360°$

Plan: Through any point P draw lines parallel to the sides of the polygon, then show that the sum of the measures of the exterior angles is 360°.

Proof:

Statements	Reasons
1. Through some point P, draw lines parallel to the sides of the polygon.	1. Why possible?
2. $m\angle k' + m\angle l' + m\angle m' + m\angle n' + m\angle o' = 360°$	2. 3.19
3. $m\angle k = m\angle k', m\angle l = m\angle l', m\angle m = m\angle m', m\angle n = m\angle n', m\angle o = m\angle o'$	3. 5.29
4. $\therefore m\angle k + m\angle l + m\angle m + m\angle n + m\angle o = 360°$	4. Why?

Is the following proposition now definitely established?

6.23 THEOREM

The sum of the measures of the exterior angles of a polygon, one at each vertex, is 360°.

Exercises

1. Determine the sum of the measures of the angles of a quadrilateral; of a pentagon.
2. What is the measure of each angle of a regular hexagon? Of a regular octagon?
3. Name the polygon having the sum of the measures of its angles totaling 720°; 900°; 1440°.
4. Name the regular polygons each of whose angles have the following measure: 144°; 162°; 135°. (*Hint:* Divide 360° by the measure of the supplementary exterior angle.)
5. As the number of sides of a regular polygon increases, the measure of an angle of the polygon _?_ and the measure of an exterior angle _?_.
6. How many sides has a regular polygon if the measure of each exterior angle is 40°? If it is 45°?
7. The measure of one angle of a triangle is 40°. What is the measure of the angle formed by the bisectors of the other two angles? (*Hint:* Draw a diagram.)
8. What is the measure of the largest exterior angle formed by any regular polygon? What is the name of the polygon?
9. The measure of an exterior angle of a regular polygon is twice the measure of an angle of the polygon. Name the polygon.
10. Given triangles ABC and DEF, such that $\angle A \cong \angle D$ and $\angle B \cong \angle E$. Is $\angle C \cong \angle F$? Justify your response.

Vocabulary List

triangle	perimeter	isosceles
quadrilateral	region	legs
pentagon	equiangular	base
hexagon	equilateral	acute triangle
heptagon	regular	obtuse triangle
octagon	convex	right triangle
nonagon	concave	hypotenuse
decagon	diagonal	altitude
dodecagon	scalene	median
icosagon		angle bisector

Chapter Review

1. Name the perimeter of an equilateral triangle if the measure of a side is $3\frac{3}{4}$ yards.
2. What is the name of the "base" of an isosceles right triangle?
3. Can a triangle have two obtuse angles? Explain.
4. What is the relationship of the acute angles of a right triangle? Explain.
5. If the measure of an acute angle of a right triangle is 40°, what is the measure of the other acute angle?
6. Name the sets of triangles that cannot be obtuse triangles.
7. Name the sets of triangles that cannot be scalene triangles.
8. Name the sets of triangles that cannot be right triangles.
9. Name the sets of triangles that cannot be regular triangles.
10. Draw a diagram to represent the relationships of the subsets of the set of triangles in Sections 6.11–6.15 and in Exercise 9, page 139. (*Hint*: Use simple closed curves to represent each subset; then the intersections of the curved regions will indicate the relationships.)
11. If a triangle is equiangular, name the measure of each angle.
12. Name the measure of each angle of a regular triangle.
13. If $m\overline{CD} = 10''$ of regular $\triangle CDE$, name $m\overline{DE}$ and $m\overline{CE}$.
14. In $\triangle ABC$, $m\overline{AB} = 2x + 5$, $m\overline{BC} = 3x - 8$, and $m\overline{AC} = 4x - 12$. If the perimeter of the triangle is 57, what type triangle is it?
15. In $\triangle RST$, the altitude $\overline{RQ}$ is drawn with Q between S and T. $m\angle SRT = 90°$. (a) Name the relationship of $\angle S$ to $\angle T$. (b) Name the relationship of $\angle S$ to $\angle SRQ$. (c) Name the relationship of $\angle T$ to $\angle SRQ$. (d) State a theorem or assumption that justifies your response to (c).
16. If the median $\overline{AD}$ of $\triangle ABC$ is drawn, we know that $\underline{?} \cong \underline{?}$.
17. If the angle bisector $\overline{BE}$ of $\triangle ABC$ is drawn, we know $\underline{?} \cong \underline{?}$.
18. If the altitude $\overline{AF}$ of $\triangle ABC$ is drawn, we know that $\overline{BF} \cong \overline{CF}$ only if the triangle is $\underline{?}$.

19. $\overline{AE}$ and $\overline{BD}$ intersect at C. $\overline{AB}$ and $\overline{DE}$ are drawn so that $\angle B \cong \angle D$. Formally prove that (1) $\overline{AB} \parallel \overline{DE}$, and (2) $\angle A \cong \angle E$. (You may prove both in the same proof.)

20. How many diagonals may be drawn from one vertex of a pentagon? From one vertex of an octagon? Would the union of the triangular regions determined by these diagonals and the sides of the polygons completely "fill" the polygonal region?

21. Which regular polygonal regions can be placed so that only their sides and/or vertices intersect, thus "filling" a plane region? (Consider first the region about a point in a plane.)

22. Can we "fill" a plane region using a combination of different and non-overlapping regular polygons? Explain by a diagram.

23. Carefully draw a regular pentagon, a regular hexagon, and a regular octagon. Using a different colored pencil, draw the lines containing the sides of the polygons until they meet. What name would you apply to the resulting figures?

24. Develop a formula that will give the sum of the measures of the "star" angles of the concave polygons of Exercise 23.

Chapter 6 Test

Completion: On your paper opposite the number of the question place the word or phrase that best completes the statement.

1. The measure of a side of an equilateral triangle is $2\frac{1}{3}$ inches. Its perimeter is __?__ inches.
2. If the sum of the measures of two sides of a triangle is 10 inches, the measure of the third side is __?__ than 10 inches.
3. A convex quadrilateral has __?__ diagonals.
4. The name of the side opposite the right angle of a right triangle is the __?__.
5. The segment whose endpoints are a vertex and the midpoint of the opposite side of a triangle is named the __?__ of the triangle.
6. The __?__ of a triangle is the segment whose endpoints are a vertex and a point of the line containing the opposite side and which is perpendicular to this line.
7. If the measure of one exterior angle of a regular polygon is 12°, the polygon has __?__ sides.

8. The measure of an acute angle of a right triangle is 27° 29′. The measure of the other acute angle is _?_ .

9. The measures of two angles of a triangle are 65° and 50°. The measure of the angle formed by the bisectors of the 65° angle and the third angle is _?_ .

10. A polygon has _?_ sides if the measure of each of its angles is 150°.

Correctly complete the statements of questions 11–15 by inserting the word "sometimes," "always," or "never" for the missing word in each.

11. An isosceles triangle is _?_ an acute triangle.

12. A right triangle _?_ has an obtuse angle.

13. An equilateral triangle is _?_ an isosceles triangle.

14. A convex polygon is _?_ a concave polygon.

15. The measure of an exterior angle of a regular dodecagon is _?_ 30°.

7

Congruency as a Method of Proof

Huge metal presses in Detroit form thousands of duplicate car bodies, stamping machines cut thousands of duplicate hub caps, large cutting machines cut many layers of fabric into identical shapes. Duplication is important in our economy and it is important in our study of geometry. Mathematicians have found the concept of duplicate sets of points, or congruency (Sections 2.36, 2.37, 3.04), to be a very useful tool. We now wish to apply the concept of congruency to the study of triangles.

7.00 Congruent triangles are triangles whose corresponding sides are congruent and whose corresponding angles are congruent.

For convenience, this definition will be symbolized by *c.p.c.t.c.* (corresponding parts of congruent triangles are congruent). Remember that the words *congruent* and *equal measure* are interchangeable when they refer to plane angles or line segments. It will be helpful to review the Axioms of Congruency, Sections 5.12–5.14, at this time.

INDUCTIVE APPROACH TO CONGRUENCY

Since we must learn to recognize congruent figures, let us work with triangles, for they are probably the most commonly used geometric figure, and any polygon can be divided into a number of triangles by drawing certain diagonals.

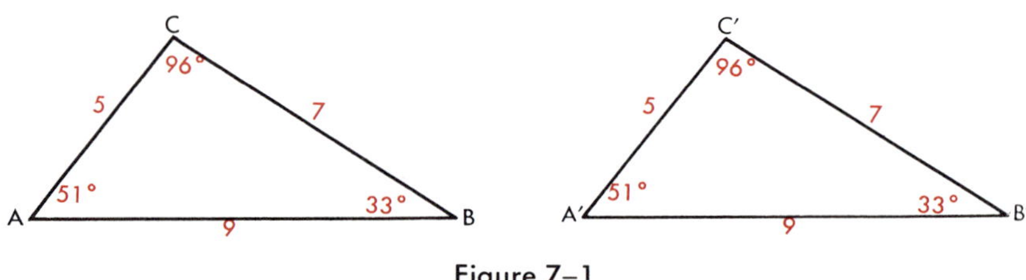

Figure 7–1

In Figure 7-1, is $\triangle ABC \cong \triangle A'B'C'$?
($A'B'C'$ is read A-prime, B-prime, C-prime. $A''B''C''$ is read A-double prime, B-double prime, C-double prime. This designation is used to show the close correlation between figures.)

Must we know that all three sides and all three angles of one triangle are congruent to the corresponding three sides and three angles of the second triangle to know that the triangles are congruent?

Let us examine the following sets of conditions:

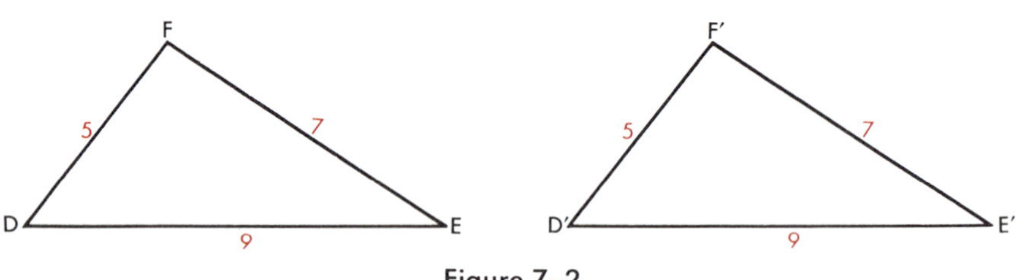

Figure 7–2

Is $\triangle DEF \cong \triangle D'E'F'$? (Figure 7-2)
Are two triangles congruent if three sides of one are congruent to the three sides of the other?

Is $\triangle GHI \cong \triangle G'H'I'$? (Figure 7-3)
Are two triangles congruent if two sides and the included angle (angle determined by these sides) of one are congruent to two sides and the included angle of the other?

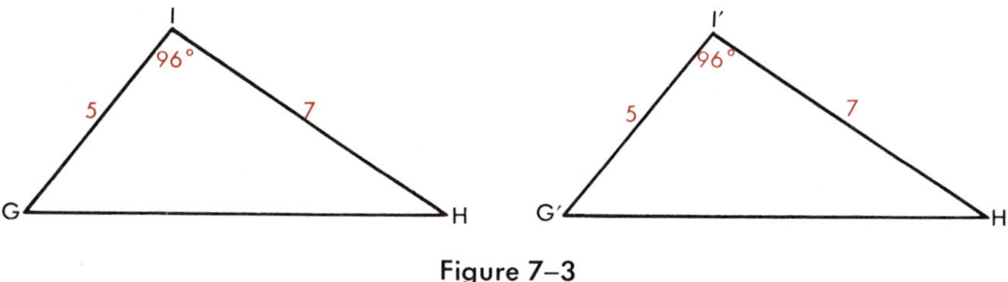

Figure 7–3

Is $\triangle JKL \cong \triangle J'K'L'$? (Figure 7-4)
Are two triangles congruent if two angles and the included side (side joining the vertices of the two angles) of one are congruent to two angles and the included side of the other?

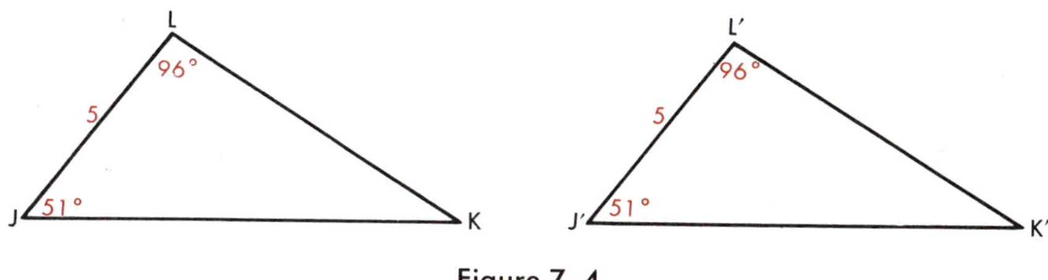

Figure 7–4

Is $\triangle MNO \cong \triangle M'N'O'$? (Figure 7-5)
Are two triangles congruent if two angles and a side opposite one of the angles in one triangle are congruent to two angles and a side opposite the corresponding angle in the other triangle?

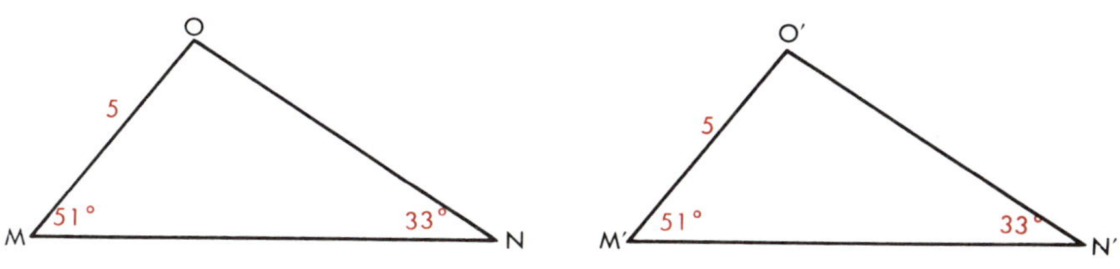

Figure 7–5

Are two triangles congruent if two sides and an angle opposite one of the sides in one triangle are congruent to two sides and an angle opposite the corresponding side in the other triangle?

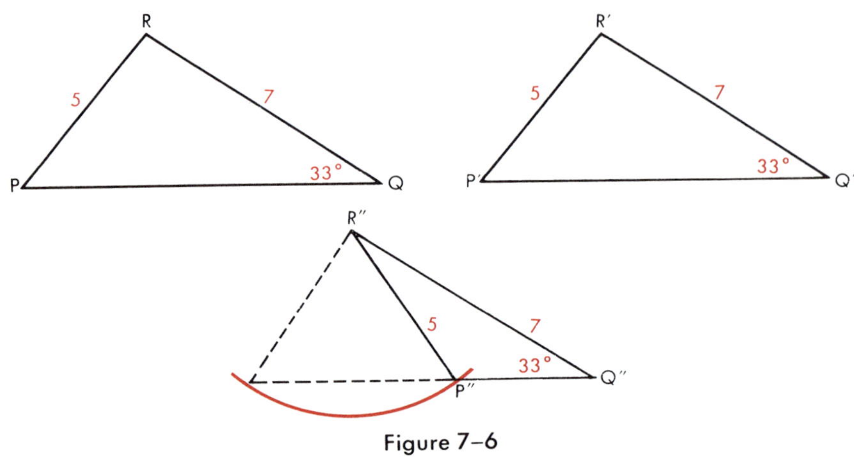

Figure 7–6

Is $\triangle PQR \cong \triangle P'Q'R'$?
Is $\triangle PQR \cong \triangle P''Q''R''$?
Do both $\triangle P'Q'R'$ and $\triangle P''Q''R''$ satisfy the conditions given with $\triangle PQR$?

Are two *right* triangles congruent if the hypotenuse and a side of one are congruent to the hypotenuse and a side of the other?
Is this the same set of conditions used for triangles PQR, $P'Q'R'$, and $P''Q''R''$?

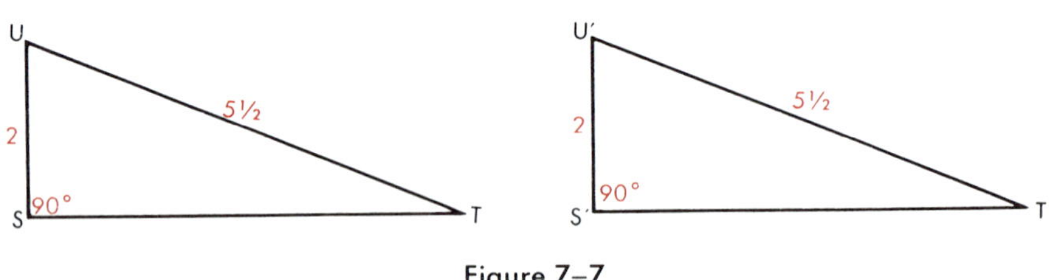

Figure 7–7

Is $\triangle STU \cong \triangle S'T'U'$?

Is $\triangle VWX \cong \triangle V'W'X'$? (Figure 7-8).
Are two triangles congruent if the three angles of one are congruent to the three angles of the other?

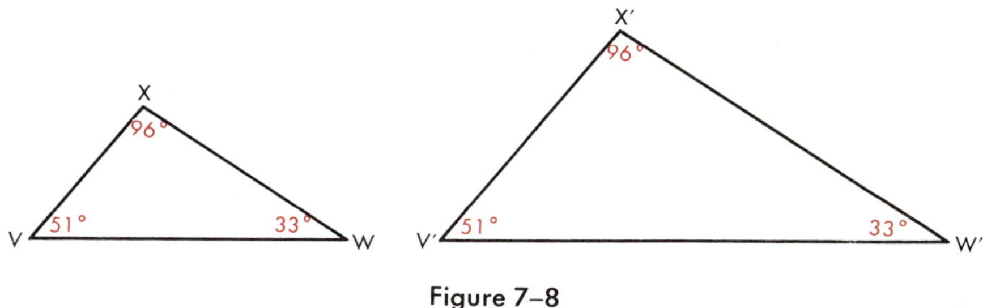

Figure 7–8

We used three elements of each triangle in the previous examples. Could we use only two elements of each to prove congruency? Only one element of each?

CONSTRUCTION OF CONGRUENT TRIANGLES

To fortify the conjectures developed intuitively on pages 148–151, draw a triangle on your paper. Now, using a ruler or compasses and protractor, attempt to construct duplicates of the triangle using the conditions implied on pages 148–151, and any others which may occur to you. Check their congruency by measuring their corresponding parts, or by placing the constructed triangle upon the original triangle (superimposing).

Summary of Inductive Reasoning

The inductive reasoning on pages 148–151 leads us to the following assumptions for congruency. After postulating any one of the first three assumptions, it is possible to prove the others. Proofs are listed in the appendix.

7.01 *Assumption: Two triangles are congruent if three sides of one are congruent to the three sides of the other. (s.s.s.)*

7.02 *Assumption: Two triangles are congruent if two sides and the included angle of one are congruent to two sides and the included angle of the other. (s.a.s.)*

7.03 *Assumption: Two triangles are congruent if two angles and the included side of one are congruent to two angles and the included side of the other. (a.s.a.)*

7.04 THEOREM

Two triangles are congruent if two angles and a side opposite one of these angles in one triangle are congruent to two angles and a side opposite the corresponding angle of the other triangle. (*s.a.a.*) (This can be proved by using 6.20 and 7.03)

7.05 *Assumption: Two right triangles are congruent if the hypotenuse and a side of one are congruent to the hypotenuse and a side of the other. (h.s.)*

RECOGNITION OF CONGRUENT TRIANGLES

We must be able to recognize congruent figures on the basis of 7.01–7.05. Since some congruent triangles are overlapping parts of geometric figures and are confusing to the eye, it is often helpful to separate the figures as shown in the following examples, or to draw the congruent parts of the congruent figures with colored pencils.

EXAMPLE 1: Given $\triangle ABC$, with $\overline{AD} \cong \overline{BE}$ and $\angle DAB \cong \angle EBA$. Is $\triangle ABE \cong \triangle BAD$?

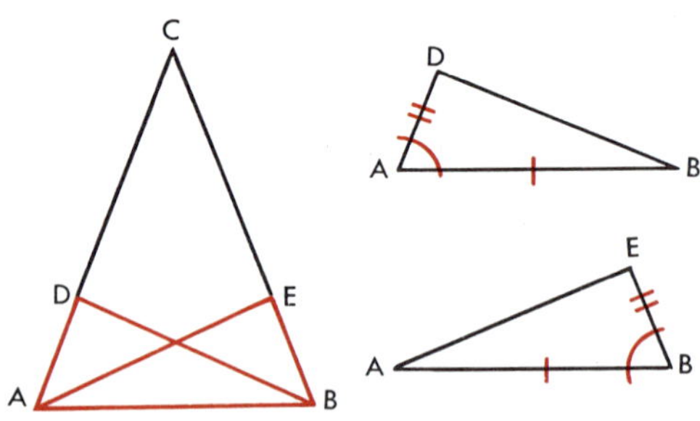

Figure 7–9

Answer: Yes, by *s.a.s.* (7.02).

EXAMPLE 2: Given quadrilateral $ABCD$, with $\angle BAC \cong \angle DCA$ and $\angle BCA \cong \angle CAD$. Is $\triangle ABC \cong \triangle CDA$?

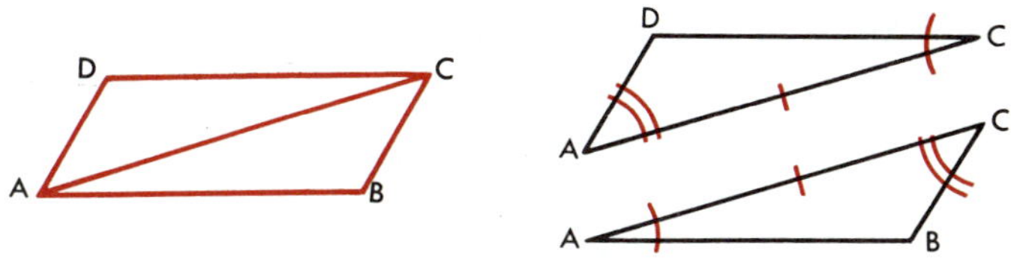

Figure 7–10

Answer: Yes, by *a.s.a.* (7.03).

Exercises

Exercises 1–5 may be done orally.

1. In $\triangle ABC$, which sides include $\angle A$? $\angle B$? $\angle C$? Which angles include $\overline{AB}$? $\overline{BC}$? $\overline{AC}$?

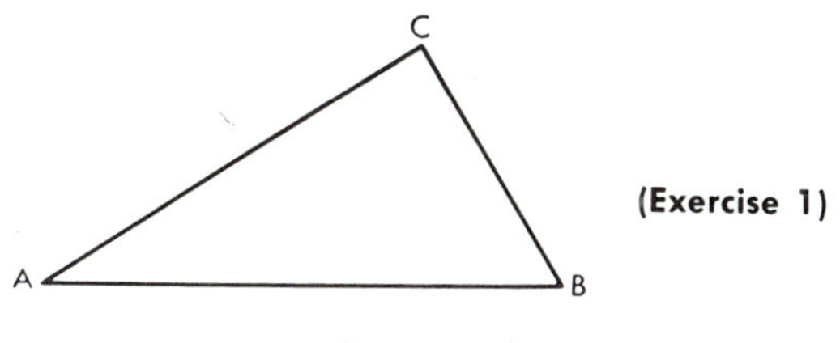

Figure 7–11

2. In $\triangle CDE$, $\angle CDE$, $\angle DFE$, $\angle DFC$ are right angles. Name the hypotenuse of $\triangle CDE$; of $\triangle DFC$.
3. Name the sides which include $\angle C$; $\angle E$; $\angle CDF$; $\angle CFD$. In $\triangle CDF$, name the angles which include $\overline{CF}$; $\overline{CD}$; $\overline{FD}$.

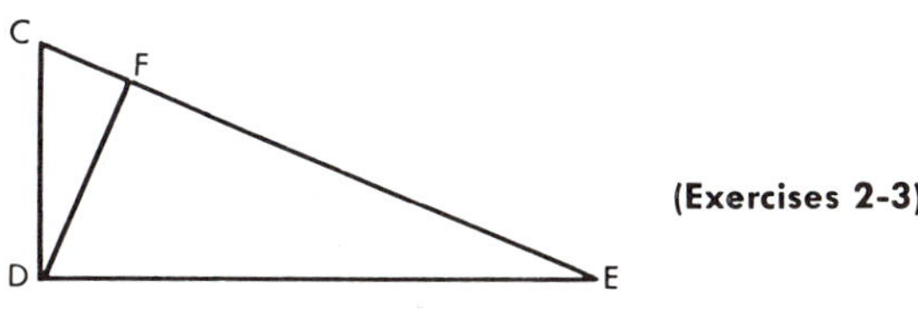

Figure 7–12

4. In $\triangle DOA$, name the sides which include $\angle 1$; $\angle 9$; $\angle 8$. In $\triangle DOC$ name the angles which include $\overline{DC}$; $\overline{DO}$; $\overline{CO}$.

5. In $\triangle ABC$, name the sides which include $\angle 2$; $\angle 5$; $\angle ABC$. In $\triangle ADC$, name the angles which include $\overline{AD}$; $\overline{DC}$; $\overline{AC}$.

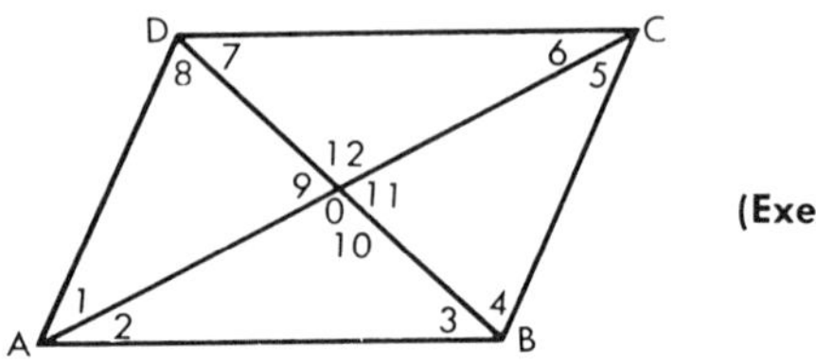

(Exercises 4-5)

Figure 7–13

Decide whether the given conditions in the following exercises are sufficient to prove the triangles congruent. Study the conditions and then write s.s.s., s.a.s., a.s.a., s.a.a., h.s., or No, as the case may be. Do not draw conclusions from the appearance of the figure.

Use Figure 7-14 for Exercises 6-11.

	Given:	**Prove:**
6.	$\overline{AC} \cong \overline{BC}$, $\overline{AD} \cong \overline{BD}$	$\triangle ADC \cong \triangle BDC$
7.	$\angle A \cong \angle B$, $\overline{AD} \cong \overline{BD}$	$\triangle ADC \cong \triangle BDC$
8.	$\overline{CD} \perp \overline{AB}$, $\overline{AD} \cong \overline{BD}$	$\triangle ADC \cong \triangle BDC$
9.	$\overrightarrow{CD}$ bisects $\angle C$, $\overline{AC} \cong \overline{BC}$	$\triangle ADC \cong \triangle BDC$
10.	$\overrightarrow{CD}$ bisects $\angle C$, $\overline{CD} \perp \overline{AB}$	$\triangle ADC \cong \triangle BDC$
11.	$\overline{AC} \cong \overline{BC}$, $\overline{CD} \perp \overline{AB}$	$\triangle ADC \cong \triangle BDC$

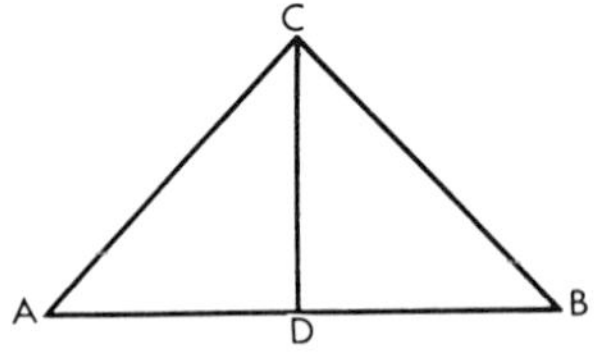

(Exercises 6-11)

Figure 7–14

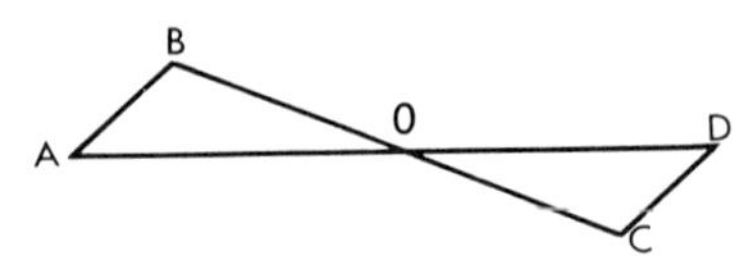

(Exercises 12-13)

Figure 7–15

12. $\overline{AD}$ intersects $\overline{BC}$ at O, $\angle A \cong \angle D$, $\overline{AB} \cong \overline{DC}$ — $\triangle ABO \cong \triangle DCO$

13. $\overline{AB} \cong \overline{DC}$, $\overline{BO} \cong \overline{CO}$ — $\triangle ABO \cong \triangle DCO$

14. $\overline{EA}$ and $\overline{DC} \perp \overline{AC}$, B the midpoint of $\overline{AC}$, $\overline{EB} \cong \overline{DB}$ — $\triangle EAB \cong \triangle DCB$

15. $\overline{EA} \cong \overline{DC}$, B the midpoint of $\overline{AC}$ — $\triangle EAB \cong \triangle DCB$

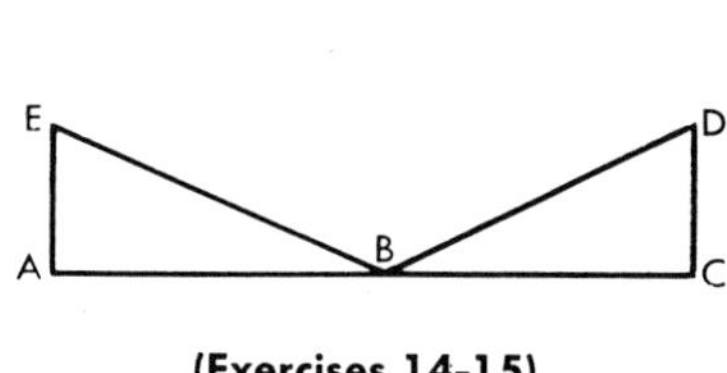

(Exercises 14-15)

Figure 7–16

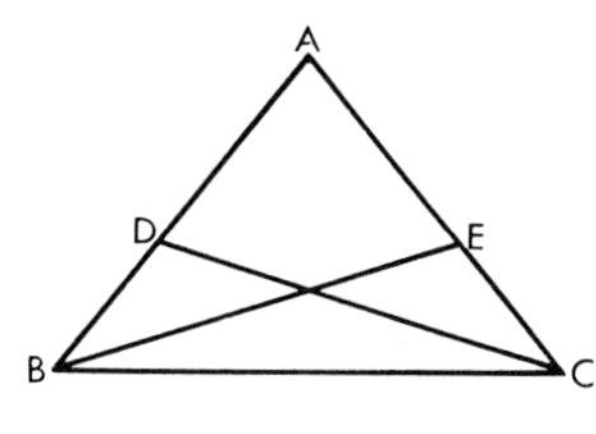

(Exercises 16-17)

Figure 7–17

16. $\overline{AB} \cong \overline{AC}$, $\overline{DA} \cong \overline{EA}$ — $\triangle BEA \cong \triangle CDA$

17. $\overline{BD} \cong \overline{CE}$, $\overline{BE} \cong \overline{CD}$ — $\triangle BEC \cong \triangle CDB$

18. $\overline{AD} \cong \overline{BC}$, $\angle 8 \cong \angle 4$ — $\triangle ADB \cong \triangle CBD$

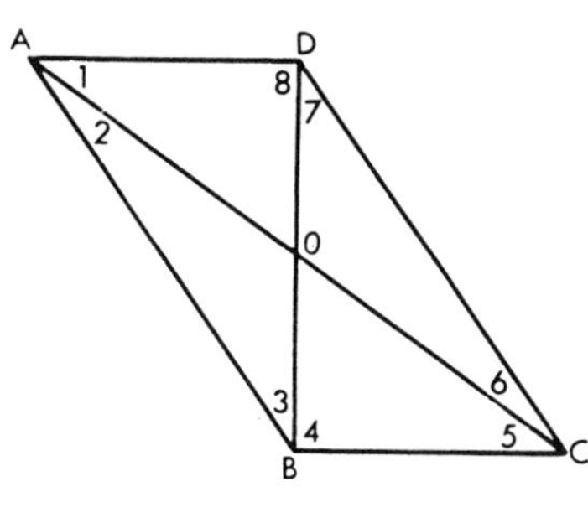

(Exercise 18)

Figure 7–18

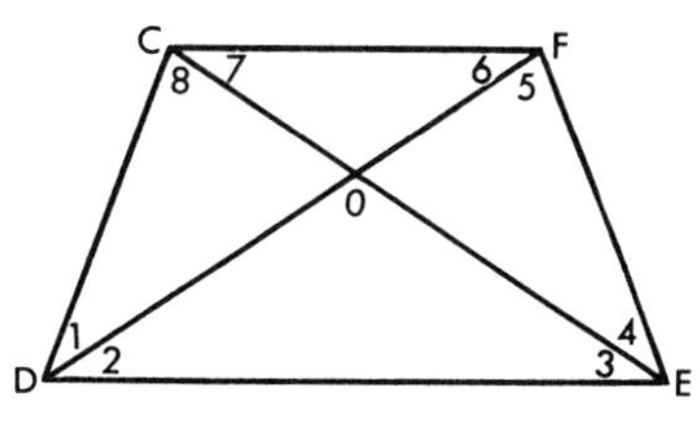

(Exercises 19-20)

Figure 7–19

19. $\overline{CD} \cong \overline{FE}$, $\overline{DF} \cong \overline{EC}$ — $\triangle CDE \cong \triangle FED$

20. $\angle 2 \cong \angle 6$, $\angle 7 \cong \angle 3$ — $\triangle FOC \cong \triangle DOE$

ANALYSIS OF PROPOSITIONS

Review 4.01 and then read the statements carefully and, in terms of the letters of the accompanying figure, state the antecedent (given conditions) and consequent (that which is to be proved).

EXAMPLE: The segments which bisect the angles at the base of an isosceles triangle and are terminated by the congruent sides are congruent.

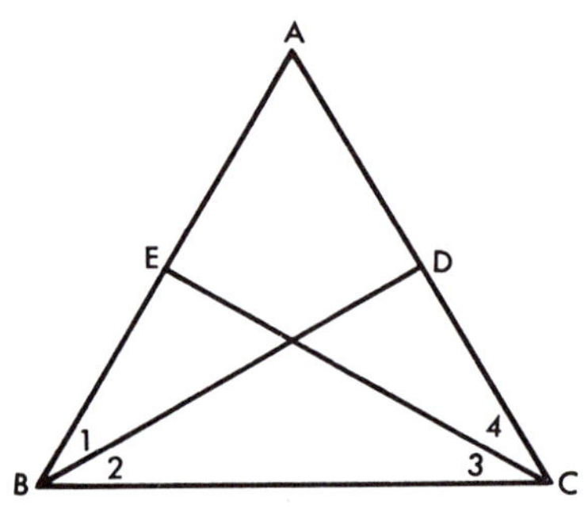

Figure 7–20

Given: $\triangle ABC$, $\overline{AB} \cong \overline{AC}$, $\overrightarrow{BD}$ bisects $\angle B$, and $\overrightarrow{CE}$ bisects $\angle C$

Conjecture: $\overline{BD} \cong \overline{CE}$

Exercises

Follow the instructions preceding the example given above. In addition, construct the appropriate figure for Exercises 4-10.

1. If the bisector of an angle of a triangle is perpendicular to the opposite side, the triangle is isosceles.

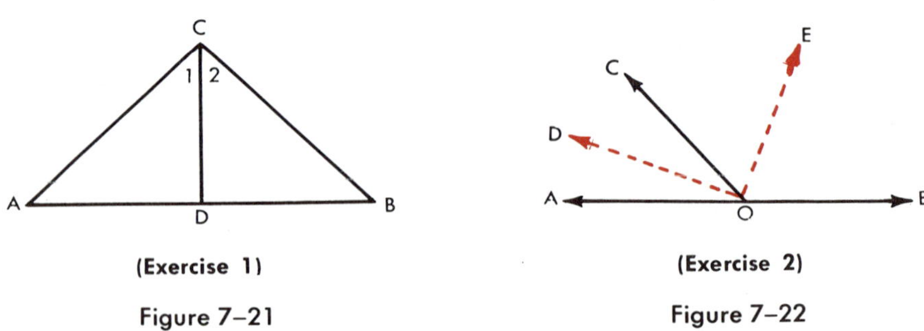

(Exercise 1)

Figure 7–21

(Exercise 2)

Figure 7–22

2. The bisectors of two supplementary adjacent angles are perpendicular to each other.

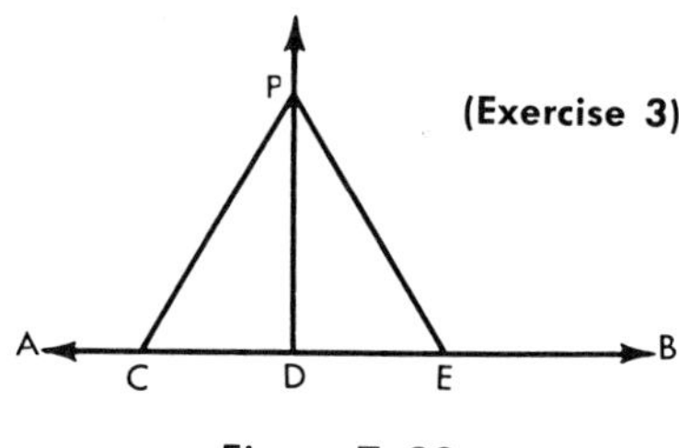

Figure 7–23

3. Two oblique segments drawn from a point on the perpendicular to a line so that they cut off congruent segments from the foot of the perpendicular are congruent.
4. The bisector of the vertex angle of an isosceles triangle is perpendicular to the base.
5. The segments joining the midpoints of the sides of an equilateral triangle form another equilateral triangle.
6. If a segment connects the midpoints of two sides of a triangle, its measure is equal to one-half the measure of the third side.
7. If segments are drawn from any point on the perpendicular bisector of a segment to the endpoints of the segment, they are congruent.
8. Segments drawn from any point on the bisector of an angle perpendicular to the sides of the angle and terminated by the sides, are congruent.
9. If a line is perpendicular to the bisector of an angle, it forms an isosceles triangle with the sides of the angle.
10. The altitudes to the congruent sides of an isosceles triangle are congruent.

AUXILIARY LINES

In this and in future chapters, you will be asked to *draw* certain lines, angles, and arcs. Remember that your diagrams are merely representations of the geometric figures and that these figures exist only in our minds.

7.06 For purposes of proof it is often necessary to draw auxiliary lines, segments, or rays. Auxiliary lines are extra (helping) lines which are not given in the antecedent. They are usually represented by dotted lines.

A common error is to assume too many conditions for an auxiliary line. For example, a student might say, "Draw $\overleftrightarrow{CD}$ as the perpendicular bisector of side $\overline{AB}$ in $\triangle ABC$." If he draws $\overleftrightarrow{C'D}$ so that it is the perpendicular bisector of $\overline{AB}$, it may or may not pass through point C. He may choose to draw the median $\overline{CD}$ and the perpendicular to $\overline{AB}$ from point C realizing that two distinct segments may result.

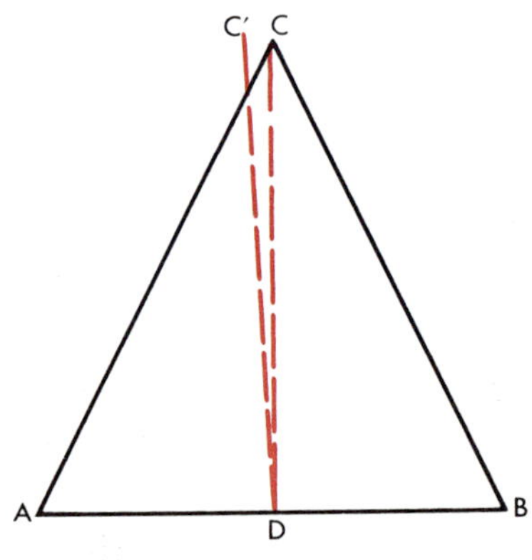

Figure 7–24

A line, segment, or ray is *determined* by two points, *or* a point and a direction. We can not use both conditions simultaneously with an auxiliary line.

PURPOSES OF CONGRUENCY

Our purpose in this chapter is to develop congruency as a method of proof so that, in addition to our definitions, assumptions, and theorems, congruency may be used to discover relationships in geometric figures. Remember that once a theorem has been proved, it may be used to prove other theorems and exercises.

PRACTICE WITH CONGRUENCY

We urge you to follow the formal method of proof outlined in 5.17. We can think of each proof as a brief prepared by a lawyer. This brief must be written in the phraseology of geometry. The judge (teacher) will throw out arguments (reasons) that are not stated in the proper form and that do not use proper terms, as well as those that are incorrect.

EXAMPLE 1: If $\overline{AD} \cong \overline{AB}$ and $\overrightarrow{AC}$ bisects $\angle BAD$ in the accompanying figure, prove $\triangle ADC \cong \triangle ABC$.

Given: $\overline{AD} \cong \overline{AB}$, $\overrightarrow{AC}$ bisects $\angle BAD$.

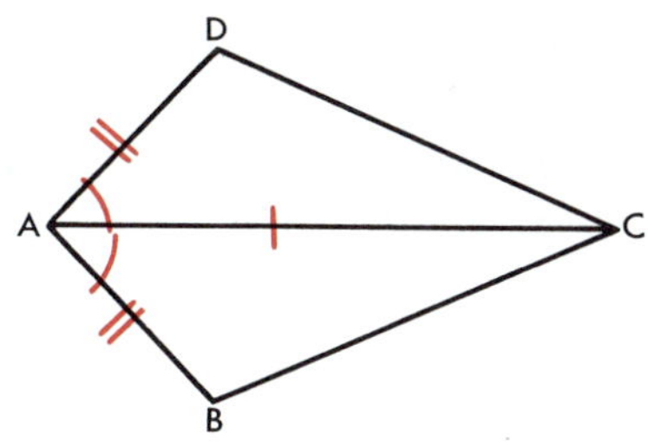

Figure 7–25

Conjecture: $\triangle ADC \cong \triangle ABC$

Plan: Use *s.a.s.* as the method for proof of congruency.

Proof:

Statements	*Reasons*
1. $\overline{AD} \cong \overline{AB}$	1. Given
2. $\overrightarrow{AC}$ bisects $\angle BAD$	2. Given
3. $\therefore \angle DAC \cong \angle BAC$	3. If an angle is bisected it is divided into two congruent parts. (3.06)
4. $\overline{AC} \cong \overline{AC}$	4. Reflexive Axiom
5. $\therefore \triangle ADC \cong \triangle ABC$	5. If two sides and the included angle of one triangle are congruent to two sides and the included angle of another triangle, then the triangles are congruent (*s.a.s.*).

We remind you that a more nearly complete formal proof might be written as follows, using the conditions of Example 1:

Plan: (To establish $[(p \rightarrow q)$ and $(q \rightarrow r)]$ of the Law of the Syllogism.)

If two sides and the included angle of one triangle are congruent respectively to two sides and the included angle of a second triangle, then the triangles are congruent; and if two triangles are congruent, then the corresponding parts are congruent.

Conjecture: $\overline{DC} \cong \overline{BC}$ (Note the change from Example 1.)

Proof:

Statements	*Reasons*
1. If an angle is bisected, it is divided into two congruent parts.	1. Definition 3.06.
2. $\overrightarrow{AC}$ bisects $\angle BAD$	2. Given.
3. $\therefore \angle DAC \cong \angle BAC$	3. Law of Detachment.
4. If two segments have the same measure, they are congruent.	4. Definition 2.37.
5. $m\overline{AC} = m\overline{AC}$	5. Reflexive Axiom 5.04.
6. $\therefore \overline{AC} \cong \overline{AC}$	6. Law of Detachment.
7. $\overline{AD} \cong \overline{AB}$	7. Given.
8. If two sides and the included angle of one triangle are congruent respectively to two sides and the included angle of a second triangle, the triangles are congruent.	8. Assumption 7.02.
9. $\overline{AC} \cong \overline{AC}$, $\overline{AD} \cong \overline{AB}$, $\angle DAC \cong \angle BAC$.	9. Steps 3, 6, 7.
10. $\therefore \triangle ADC \cong \triangle ABC$.	10. Law of Detachment.
11. $\overline{DC}$ and $\overline{BC}$ are corresponding parts of $\triangle ADC$ and $\triangle ABC$.	11. Agreement concerning undefined term.

12. If two triangles are congruent, then their corresponding parts are congruent.	12. Definition 7.00.
13. $\triangle ADC \cong \triangle ABC$, of which $\overline{DC}$ and $\overline{BC}$ are corresponding parts.	13. Steps 10 and 11.
14. $\therefore \overline{DC} \cong \overline{BC}$.	14. Law of Detachment.

EXAMPLE 2: A student's proof corrected by the teacher.

Given: $\triangle ACD$ with $\overline{DB} \perp \overline{AC}$, $\angle A \cong \angle C$, and $\overline{AB} \cong \overline{BC}$

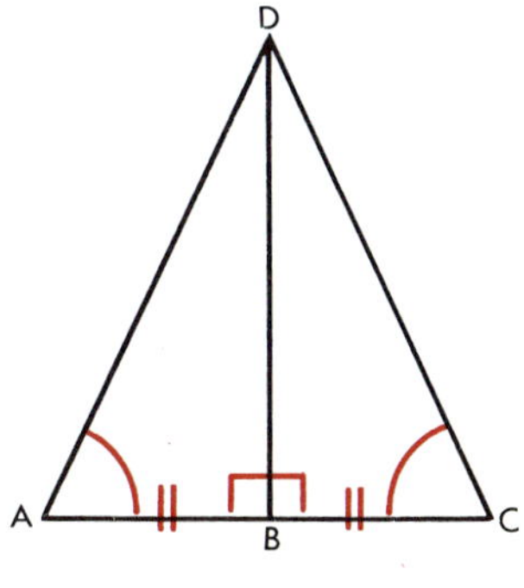

Figure 7–26

Conjecture: $\triangle ABD \cong \triangle CBD$.

Analysis: Use *a.s.a.* as the method of congruency.

Proof: *Statements*	*Reasons*
(A) 1. $\angle A \cong \angle C$	1. Given
(S) 2. $\overline{AB} \cong \overline{BC}$	2. Given
Insert → If $\overline{DB} \perp \overline{AC}$	→ Given
(A) 3. then $\angle ABD \cong \angle CBD$	3. (Given) See 3.17
4. ($\overline{DB} \cong \overline{DB}$) A useless statement. Do you see why?	4. Reflexive Axiom
5. $\therefore \triangle ABD \cong \triangle CBD$	5. *a.s.a.*

EXAMPLE 3: Another student's proof of Example 2. Is it correct? Would you change it in any way?

Given: $\triangle ACD$ with $\overline{DB} \perp \overline{AC}$, $\angle A \cong \angle C$, and $\overline{AB} \cong \overline{BC}$

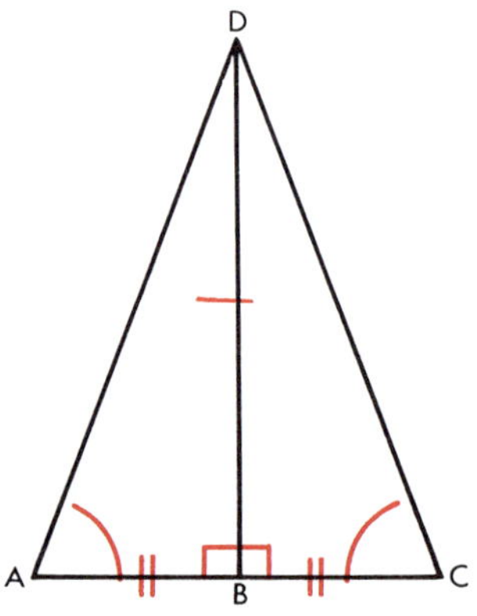

Figure 7–27

Conjecture: $\triangle ABD \cong \triangle CBD$

Analysis: The conditions of the problem satisfy *s.a.s.*

Proof:

Statements	*Reasons*
1. $\overline{DB} \perp \overline{AC}$	1. Given
2. $\therefore \angle ABD \cong \angle CBD$	2. Perpendicular lines form equal adjacent angles.
3. $\overline{DB} \cong \overline{DB}$	3. Reflexive Axiom
4. $\angle A \cong \angle C$ and $\overline{AB} \cong \overline{BC}$	4. Given
5. $\therefore \triangle ABD \cong \triangle CBD$	5. *s.a.s.*

Is reason number two correct as it is written, or should the student have used the converse of this statement?

NOTE: Nearly every reason in a proof is an example of an if-then relationship. This follows since most "reasons" are theorems, or assumptions, which are necessarily if-then propositions. Review 4.01.

EXAMPLE 4: A model proof.

Given: $\overline{AE} \cong \overline{DE}$, $\angle x \cong \angle y$, $\angle A \cong \angle D$

Conjecture: $\angle 1 \cong \angle 2$

Plan: Prove $\triangle ACE \cong \triangle DBE$

(An alternate plan would be to prove $\angle ABE \cong \angle DCE$ by proving $\triangle ABE \cong \triangle DCE$ and then use 5.20.)

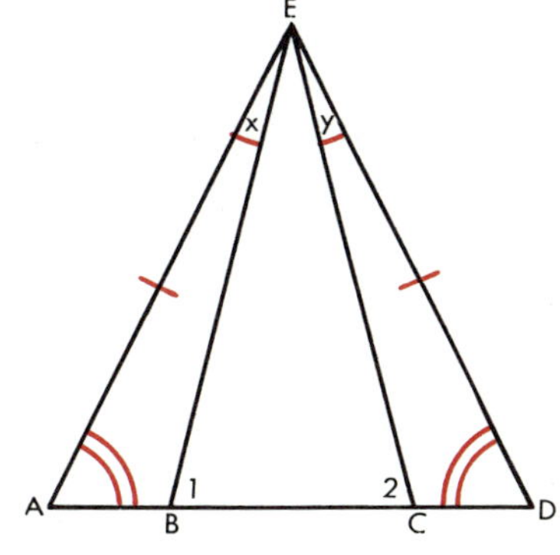

Figure 7–28

Proof: *Statements*	*Reasons*
1. $\overline{AE} \cong \overline{DE}$, $\angle A \cong \angle D$, $\angle x \cong \angle y$	1. Given
2. $\angle BEC \cong \angle BEC$	2. Reflexive Axiom
3. $m\angle x + m\angle BEC = m\angle y + m\angle BEC$	3. Addition Axiom
4. $m\angle AEC = m\angle x + m\angle BEC$, $m\angle DEB = m\angle y + m\angle BEC$	4. Angle Addition Axiom (3.08)
5. $\angle AEC \cong \angle DEB$	5. Transitive Axiom of Congruency.
6. $\therefore \triangle AEC \cong \triangle DEB$	6. If two angles and the included side of one triangle are congruent to two angles and the included side of another triangle, the triangles are congruent (*a.s.a.*).
7. Then $\angle 1 \cong \angle 2$	7. Corresponding parts of congruent triangles are congruent. *c.p.c.t.c.* (7.00)

If the alternate plan above were used, (proving $\triangle ABE \cong \triangle DCE$ and then applying Theorem 5.20), which assumption for congruency would be used to prove that $\triangle ABE \cong \triangle DCE$?

Exercises

Prove the following exercises in proper form.

1. If $\overline{AC}$ and $\overline{BD}$ bisect each other, prove $\triangle ABE \cong \triangle CDE$.

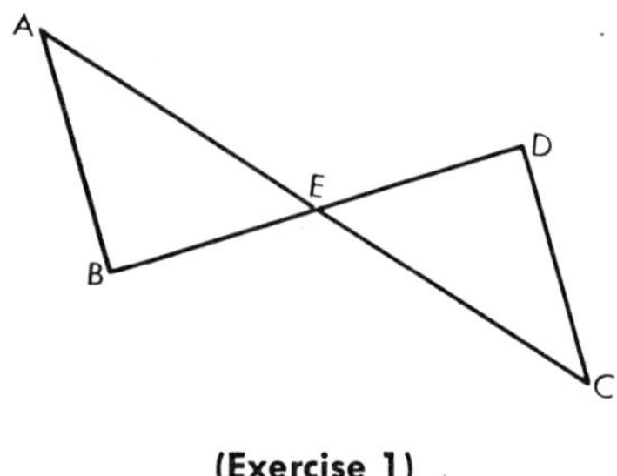

(Exercise 1)

Figure 7–29

2. If $\angle A \cong \angle B$ and $\overline{AD} \cong \overline{BD}$, prove $\triangle ACD \cong \triangle BED$. Does it follow that $\overline{AC} \cong \overline{BE}$? Explain.

3. If $\overline{CD} \perp \overline{AB}$ and $\overrightarrow{CD}$ bisects $\angle C$, prove $\triangle ACD \cong \triangle BCD$. Does it follow that $\angle A \cong \angle B$? Explain.

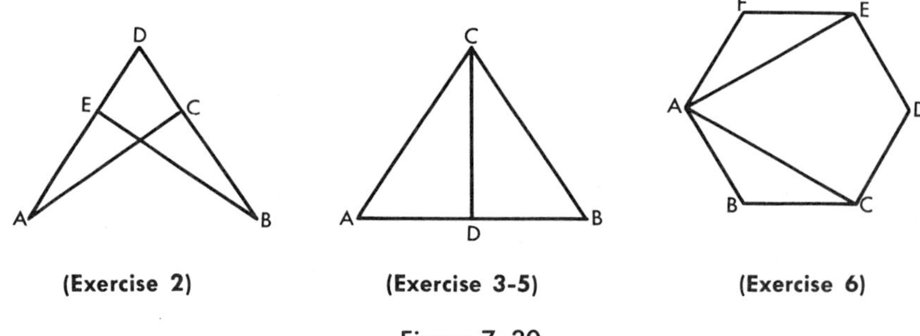

(Exercise 2) **(Exercise 3-5)** **(Exercise 6)**

Figure 7–30

4. If $\overline{CD} \perp \overline{AB}$ and $\overline{AC} \cong \overline{BC}$, prove $\triangle ACD \cong \triangle BCD$. Is $\overline{AD} \cong \overline{BD}$? Explain.

5. If $\angle ACD \cong \angle BCD$, and $\angle A \cong \angle B$, prove that $\triangle ACD \cong \triangle BCD$. Is $\overline{CD} \perp \overline{AB}$? Explain.

6. If $ABCDEF$ is a regular polygon, prove $\triangle ABC \cong \triangle AFE$. Is $\overline{AE} \cong \overline{AC}$? Explain.

7. Given: $\overline{AD} \cong \overline{CD}$, $\overline{AB} \cong \overline{CB}$. Prove $\angle A \cong \angle C$.
8. Given: $\overline{OL} \cong \overline{ON}$, $\overline{LM} \cong \overline{MN}$. Prove $\angle OLM \cong \angle ONM$.

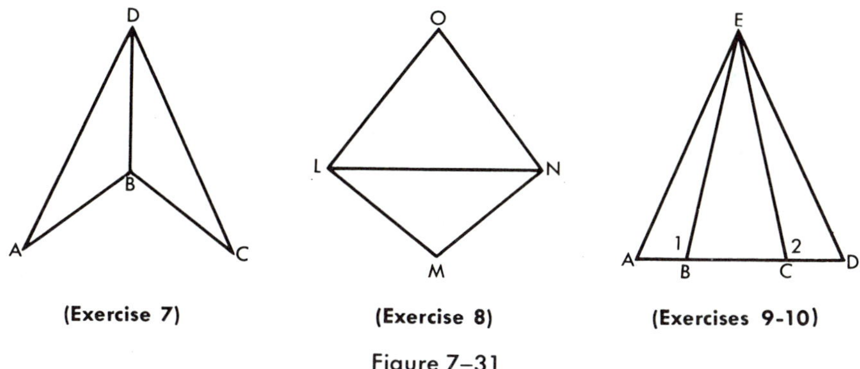

(Exercise 7) (Exercise 8) (Exercises 9-10)

Figure 7–31

9. If $\overline{AE} \cong \overline{DE}$, $\angle 1 \cong \angle 2$, and $\angle A \cong \angle D$, prove $\overline{BE} \cong \overline{CE}$.
10. In Exercise 9, prove $\triangle ACE \cong \triangle DBE$.
11. Given: $\angle NKL \cong \angle KLM$, $\overrightarrow{KM}$ bisects $\angle NKL$, and $\overrightarrow{LN}$ bisects $\angle KLM$. Prove $\overline{NK} \cong \overline{ML}$.

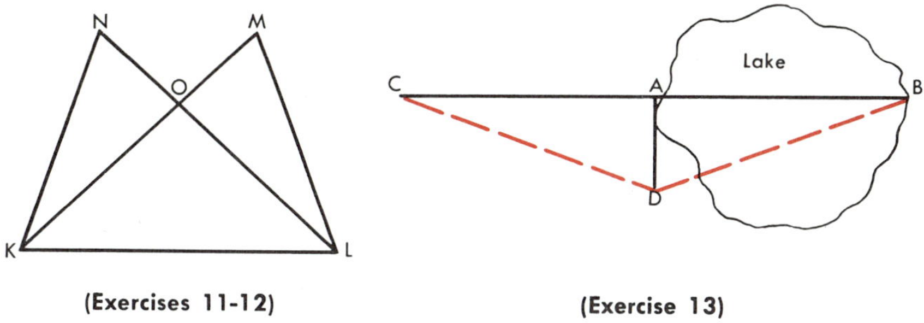

(Exercises 11-12) (Exercise 13)

Figure 7–32

12. If $\overline{NL} \cong \overline{MK}$ and $\angle MKL \cong \angle NLK$, prove that $\overline{ON} \cong \overline{OM}$. (*Hint*: Use corresponding parts from one set of congruent triangles to prove a second set of triangles congruent.)
13. Explain how you could find the width of the lake from A to B, without crossing the lake, by measuring $\overline{AC}$.

14. Explain how you could measure the distance from A to D, without crossing the river, by using the accompanying figure.

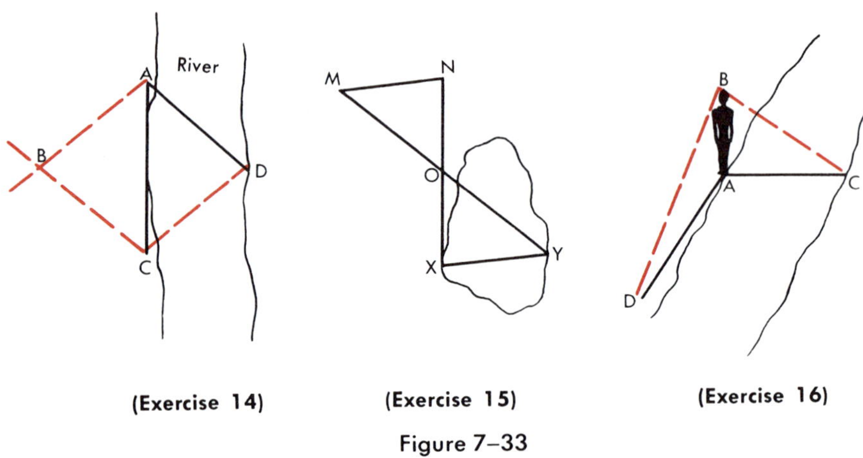

(Exercise 14) (Exercise 15) (Exercise 16)

Figure 7–33

15. Explain how you could measure the distance from X to Y, without crossing the lake, using the accompanying figure.

16. A person standing at A and sighting across the river to C, turns to sight along the same angle to D. Is $\overline{AD} \cong \overline{AC}$? Explain.

17. Explain how you could bisect $\angle ABC$ by using a carpenter's square as shown.

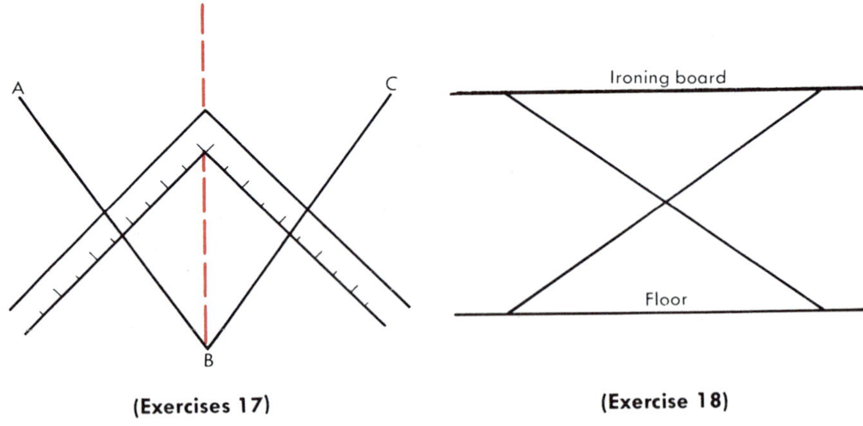

(Exercises 17) (Exercise 18)

Figure 7–34

18. Explain why the plane of the ironing board is always parallel to the floor plane if the legs bisect each other.

19. Draw a line perpendicular to a plane. From any point of this line draw two segments to the plane whose other endpoints are equally distant from the foot of the perpendicular line. Are these segments oblique to the plane? Do these two segments represent any and all segments fulfilling the above conditions? What is the relationship of these segments? Prove your conjecture and state it as a theorem.

20. State the converse of your theorem in Exercise 19. Prove or disprove it.

REASONING BY ANALOGY

7.07 Reasoning by analogy is closely related to the inductive method (page 13). Induction investigates many cases; *analogy compares two.* When we examine two cases and, upon finding them similar in many ways, infer that what is true of one must therefore be true of the other, we are using an analogy. Analogies are more often false than true, so we should use them as arguments only with caution. It is very difficult to find two cases that are alike in all respects, so attempts to reason by analogy frequently result in unfair or invalid comparisons.

EXAMPLE A. Tom looks good with his "crew cut," so I'm sure I would look good with the same type of haircut.

EXAMPLE B. Mr. Smith plays the stock market and is rich, so I'm going to buy stocks and become rich.

EXAMPLE C. *Advertisement:* "Daisy cough drops stopped my cough. Use Daisy cough drops to stop your cough."

Exercises

Criticize the following analogies.

1. "If Mary is allowed to go to the party, I should be permitted to go."
2. "Bob should be a good athlete since he is bigger than his brother Tom, who was a good athlete."

3. Fertilization of plants and grass seems to stimulate their growth, so we should "feed" our hair to stimulate its growth.

4. $\dfrac{12 \times 15}{3 \times 4} = \dfrac{\overset{3}{\cancel{12}} \times \overset{5}{\cancel{15}}}{\cancel{3} \times \cancel{4}} = 15$, so $\dfrac{12 + 15}{3 + 4} = \dfrac{\overset{3}{\cancel{12}} + \overset{5}{\cancel{15}}}{\cancel{3} + \cancel{4}} = 8$.

5. Since $\dfrac{5}{5} = 1$ and $\dfrac{\frac{1}{2}}{\frac{1}{2}} = 1$, then $\dfrac{0}{0} = 1$.

6. We can compare an electric current with the flow of water in a pipe.

7. "A flatterer is like a distorted mirror."

8. Studying geometry is much like constructing a building.

9. Write an analogy to 5.25 in three dimensions, that is, use planes instead of lines. Is this analogy valid? Is its converse valid?

10. If two triangles are congruent by *s.s.s.*, it must be true that two quadrilaterals are congruent by *s.s.s.s.*

11. All equilateral triangles are regular; therefore, all equilateral polygons will be regular.

12. Since a triangle has three angles and a hexagon has six angles, the sum of the measures of the angles of the hexagon should be twice the sum of the measures of the angles of the triangle.

Vocabulary List

auxiliary
congruent
elements of a triangle
corresponding parts
analogy
inductive
deductive
proposition

Chapter Review

Use Figure 7-35 for Exercises 1-6. $\overline{DC} \parallel \overline{AB}$, $\overline{DB}$ *and* $\overline{AC}$ *intersect at E so that* $\overline{DE} \cong \overline{CE}$ *and* $\overline{AE} \cong \overline{BE}$.

1. Name the angle included by $\overline{AC}$ and $\overline{BA}$.

2. Name the side included between $\angle DCA$ and $\angle DAC$.

3. Is $\angle DCA \cong \angle BAC$? Explain.

4. From the conditions given, is $\triangle AED \cong \triangle BEC$? Explain.

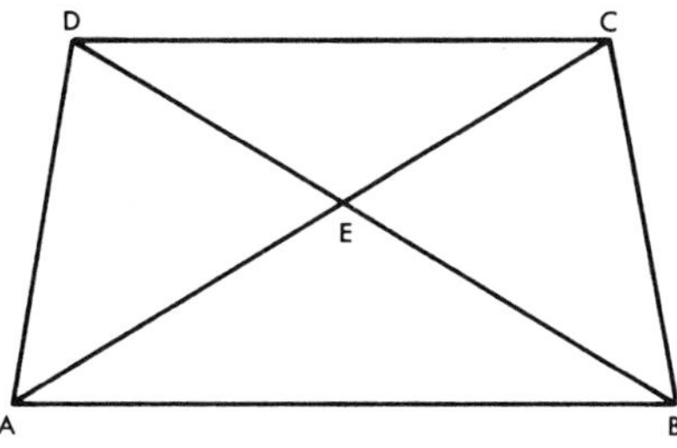

Figure 7–35

5. From the conditions given, is $\triangle DEC \cong \triangle AEB$? Explain.
6. From the conditions given, is $\triangle ADC \cong \triangle BCD$? Explain.
7. In plane s, A and B are on the same side of $\overline{CD}$, E is the midpoint of $\overline{CD}$, $\overline{AE} \cong \overline{BE}$, $\overline{AC} \perp \overline{CD}$, and $\overline{BD} \perp \overline{CD}$. Prove $\triangle ACE \cong \triangle BDE$.
8. In plane quadrilateral $ABCD$, $\overline{AC}$ bisects $\angle A$ and $\angle C$. Prove $\triangle ABC \cong \triangle ADC$.

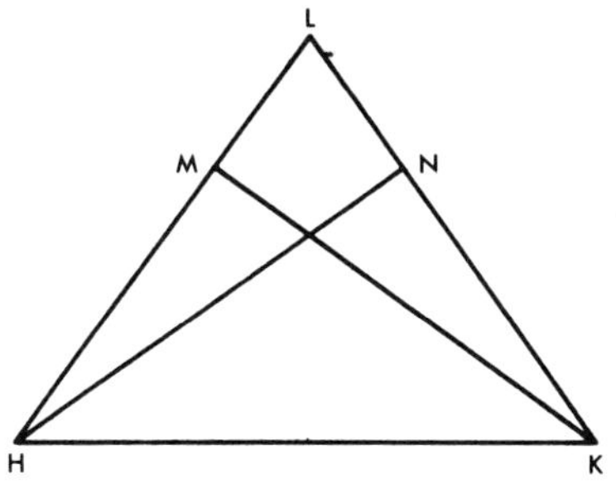

Figure 7–36

9. In Figure 7-36, $\angle LHK \cong \angle LKH$, $\angle HMK \cong \angle KNH$. Prove $\triangle MHK \cong \triangle NKH$.
10. **Given:** $\triangle ABC$ with D the midpoint of $\overline{AC}$ and E the midpoint of $\overline{BC}$. E is between D and F so that $\overline{DE} \cong \overline{EF}$. $\overline{BF}$ is drawn.
Prove: $\overline{BF} \parallel \overline{AD}$ and $\overline{BF} \cong \overline{AD}$.

Chapter 7 Test

1. In $\triangle RST$, sides $\overline{RS}$ and $\overline{RT}$ include angle ___?___.

2. In $\triangle KHN$, side $\overline{HK}$ is included between angles ___?___ and ___?___.

3. In Figure 7-37, $\triangle BAF$ appears to be congruent to $\triangle$ ___?___. (Name the vertices in the corresponding order.)

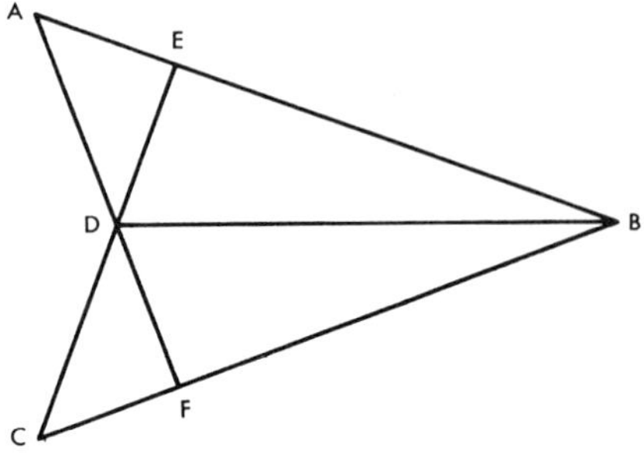

$\overline{AF}$ and $\overline{CE}$ intersect in D

(Exercises 3-8)

Figure 7–37

4. Write the assumption or theorem you would use to prove $\triangle ADE \cong \triangle CDF$ if $\angle A \cong \angle C$ and $\overline{AD} \cong \overline{CD}$.

5. Write the assumption or theorem you would use to prove $\triangle AFB \cong \triangle CEB$ if $\overline{AB} \cong \overline{CB}$ and $\overline{EB} \cong \overline{FB}$.

6. Write the assumption or theorem you would use to prove $\triangle BED \cong \triangle BFD$ if $\overline{BE} \cong \overline{BF}$ and $\overline{DE} \cong \overline{DF}$.

7. Write the assumption or theorem you would use to prove $\triangle BDE \cong \triangle BDF$ if $\overline{DE} \perp \overline{AB}$, $\overline{DF} \perp \overline{CB}$, and $\overline{DE} \cong \overline{DF}$.

8. Write the assumption or theorem you would use to prove $\triangle ADB \cong \triangle CDB$ if $\overline{BD}$ bisects $\angle ABC$ and $\angle A \cong \angle C$.

9. Present a formal proof of the statement "If $\overline{AB}$ and $\overline{CD}$ bisect each other at point M, then $\overline{AC} \cong \overline{BD}$."

10. **Given:** $\overline{AB} \perp$ plane x at B.
    Points C and D are in x.
    $\overline{CB} \cong \overline{DB}$.
    Prove: $\angle ACD \cong \angle ADC$

Figure 7–38

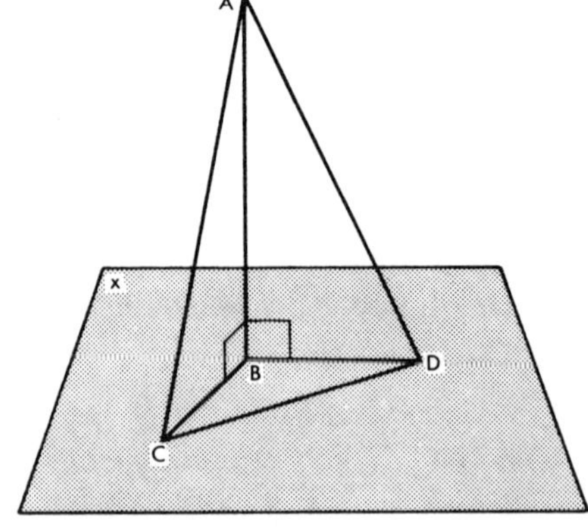

8

Basic Constructions and Isosceles Triangles

In geometry the words *draw* and *construct* have different meanings. When you are told to construct a figure (meaning a picture of a geometric figure) you are to use only compasses and straight edge. When drawing a figure you may reproduce it by the best means available, which may include tracing an object or use of a marked rule or protractor.

8.00 Use of Compasses to Construct Congruent Segments

Suppose that you are asked to construct a segment 3 inches long. First, draw the ray $\overrightarrow{AB}$. Second, place one point of your compasses at the 1 inch mark on a ruler and the other point of your compasses at the 4 inch mark on the ruler. Tighten your compasses to keep the correct distance between the points. Now place the sharp point of the compasses at A and draw an arc intersecting $\overrightarrow{AB}$. Call the point of intersection C. The m$\overline{AC}$ is 3 inches.

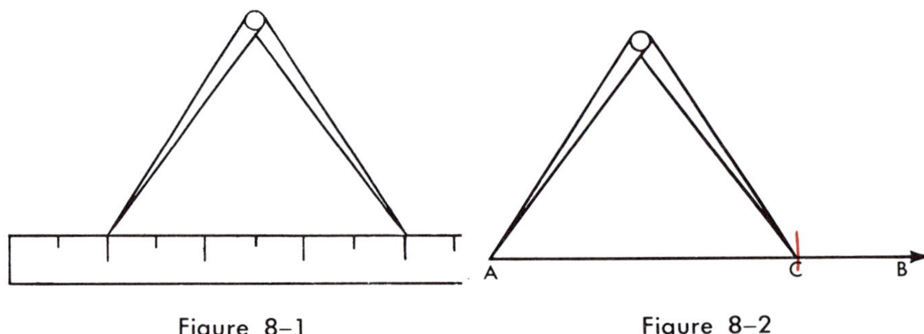

Figure 8–1 Figure 8–2

Exercises

Do not erase your arcs of construction.

1. Draw a segment of length x on your paper. Construct a segment whose length is $x + 1.5$ inches.
2. Given segments whose measures are a and b. Construct a segment whose measure is equal to $2a + 3b$.
3. Given segments whose measures are a and b. Construct a segment whose measure is equal to $a - b$. (a is larger than b)

PROOF OF CONSTRUCTION METHODS

You may remember the procedures used in seventh and eighth grade arithmetic classes to bisect an angle or to construct a perpendicular bisector of a segment. We now have the background to prove that our methods do produce the desired results. The completion of the proofs of Sections 8.01–8.05 provide this justification. After reviewing each construction procedure, verify it by proof.

8.01 Bisecting an angle

Given: $\angle ABC$

Required: To construct the bisector of $\angle ABC$

Construction Procedure:

(1) Place the point of the compasses at the vertex (B) of the angle.

(2) Draw an arc intersecting the sides of the angle.

(3) Using the points of intersection (F and D) as centers, draw arcs of congruent radius to intersect at E.

(4) $\overrightarrow{BE}$ is the angle bisector.

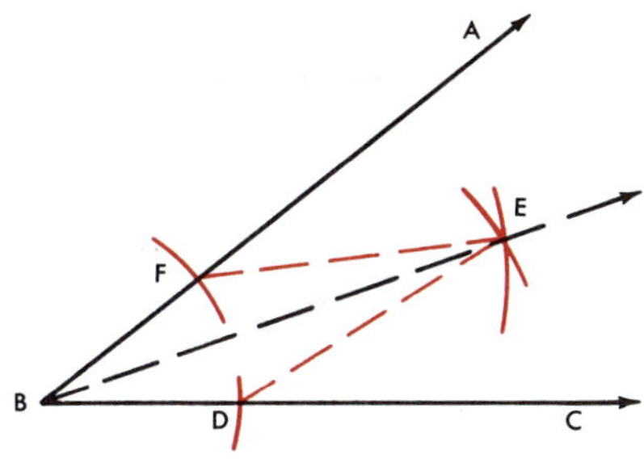

Figure 8–3

Verification:

Given: $\angle ABC$

Construction: $\overline{BF} \cong \overline{BD}, \overline{FE} \cong \overline{DE}$

Conjecture: $\angle ABE \cong \angle CBE$

Exercises

1. Draw and bisect an acute angle.
2. Draw and bisect an obtuse angle.
3. Draw and bisect a right angle. What is the measure of each of the resulting angles?
4. Draw and bisect a straight angle. What is the measure of each of the resulting angles?
5. Bisect each of the angles formed at the intersection of two straight lines. What can you prove about the figure you have just constructed?

8.02 Perpendicular bisector of a segment

Given: $\overline{AB}$

Required: To construct the perpendicular bisector of $\overline{AB}$

Construction Procedure:

(1) Use your compasses to construct two sets of intersecting arcs, each intersection point being equally distant from A and B, and on opposite sides of the segment for the best results. It is obvious that the measures of the radii of the arcs must be greater than $\frac{1}{2}\text{m}\overline{AB}$. Why?

(2) Label the points of intersection C and D. $\overleftrightarrow{CD}$ is the perpendicular bisector of $\overline{AB}$ at E.

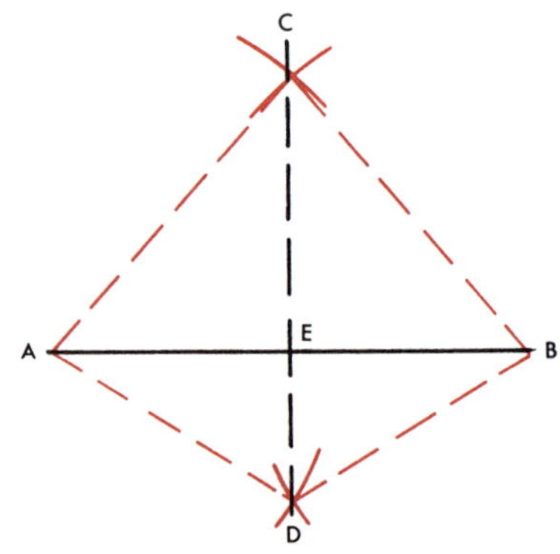

Figure 8–4

Verification:

Given: $\overline{AB}$

Construction: $\overline{AC} \cong \overline{BC}, \overline{AD} \cong \overline{BD}$

Conjecture: $\overline{AE} \cong \overline{BE}, \overleftrightarrow{CD} \perp \overline{AB}$

Exercises

1. Draw $\overline{CD}$ so that $\text{m}\overline{CD} = 4$ inches. Find the midpoint by constructing the perpendicular bisector.
2. Use your protractor to measure the angles determined by the segments constructed for Exercise 1.

3. Divide a segment into four congruent parts by using the construction method outlined above.
4. How many bisectors can a segment have? How many perpendicular bisectors? How many midpoints?
5. Develop a method of constructing the bisector of a segment so that it will not be perpendicular to the segment.

8.03 Construction of a line perpendicular to a given line at a given point of the line.

Given: $\overleftrightarrow{AB}$ and point P on $\overleftrightarrow{AB}$

Required: To construct a line perpendicular to $\overleftrightarrow{AB}$ at point P.

Construction Procedure:

(1) Place the point of your compasses at P, the desired point of $\overleftrightarrow{AB}$.

(2) Mark equal distances on each side of P on $\overleftrightarrow{AB}$. Label these points C and D.

(3) Consider $\overline{CD}$. Is P equidistant from C and D? Can you find another point equidistant from C and D? (Use 8.02).

(4) Draw a line connecting P and the point found in step 3.

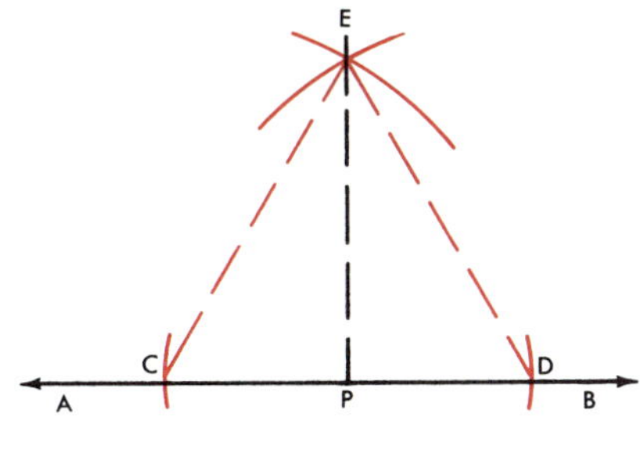

Figure 8–5

Verification:

Given: $\overleftrightarrow{AB}$ with point P on $\overleftrightarrow{AB}$.

Construction: $\overline{PC} \cong \overline{PD}$, $\overline{CE} \cong \overline{DE}$

Conjecture: $\overline{EP} \perp \overleftrightarrow{AB}$

8.04 Construction of a line perpendicular to a given line from a point not on the line.

Given: $\overleftrightarrow{AB}$ and point P not on $\overleftrightarrow{AB}$

Required: To construct a line perpendicular to $\overleftrightarrow{AB}$ from P.

Construction Procedure:

(1) Place the point of your compasses at the given point P.

(2) Using some convenient radius draw arcs intersecting $\overleftrightarrow{AB}$ at points C and D.

(3) Consider $\overline{CD}$. Is P equidistant from C and D? Can you find another point (E) equidistant from C and D? (For best results place this point on the opposite side of the line from point P.)

(4) $\overleftrightarrow{PE}$ is $\perp$ to $\overleftrightarrow{AB}$.

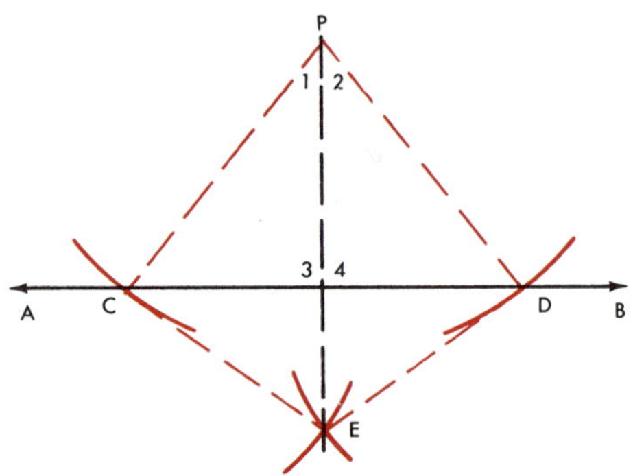

Figure 8–6

Verification:

Given: $\overleftrightarrow{AB}$ with P not on $\overleftrightarrow{AB}$.

Construction: $\overline{PC} \cong \overline{PD}$, $\overline{CE} \cong \overline{DE}$

Conjecture: $\overleftrightarrow{PE} \perp \overleftrightarrow{AB}$

Exercises

1. Construct a line perpendicular to line $\overleftrightarrow{MN}$ at point Q on $\overleftrightarrow{MN}$.
2. Construct a line perpendicular to line $\overleftrightarrow{RS}$ from a point P which is not on $\overleftrightarrow{RS}$.
3. Draw a segment 2 inches long and construct perpendiculars to the segment at both ends.

4. Draw an acute angle and, from any point on one side of the angle, construct a perpendicular to the other side of the angle.

5. Draw $\overline{AB}$ and bisect it, calling the midpoint C. At A and B construct lines perpendicular to $\overline{AB}$. Complete the figure by drawing any line through C, meeting one perpendicular at D and the other at E.

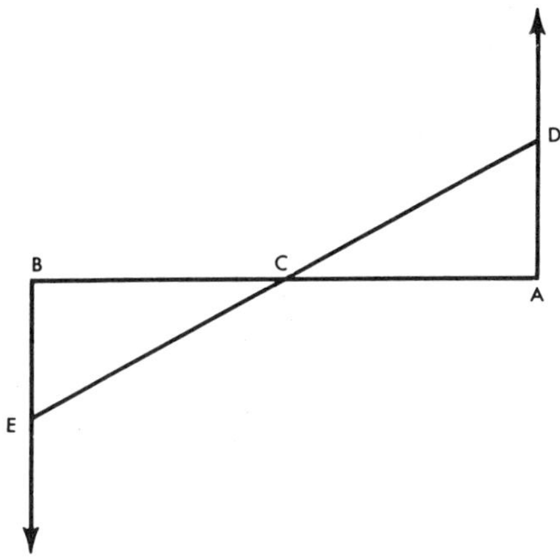

Figure 8–7

6. Measure $\overline{AD}$ and $\overline{BE}$. Measure $\overline{DC}$ and $\overline{EC}$. Did we attempt to construct these as congruent segments? Which of the following statements describes Exercise 5?

   (a) If $\overline{AD} \cong \overline{BE}$, $\overline{AD} \perp \overline{AB}$, and $\overline{BE} \perp \overline{AB}$, then $\overline{AC} \cong \overline{BC}$.

   (b) If $\overline{DE}$ bisects $\overline{AB}$, $\overline{DA} \perp \overline{AB}$, and $\overline{EB} \perp \overline{AB}$, then $\overline{AD} \cong \overline{BE}$ and $\overline{DC} \cong \overline{CE}$.

7. Draw an angle whose measure is 45°. From a point two inches from the vertex of the angle on one side of the angle, construct a perpendicular to the other side of the angle. Measure the angles of the triangle thus formed. How do they compare?

8. How would you measure the least distance from a point to a line? From a point to a plane?

9. How many perpendiculars to a line can be drawn at a given point on the line? How many can be drawn to the line from a point not on the line? How many can be drawn to a plane at a given point on the plane? How many can be drawn to a plane from a point not on the plane?

8.05 Construction of an angle congruent to a given angle.

Given: $\angle ABC$

Required: Construct $\angle FDE \cong \angle ABC$

Construction Procedure:

(1) Draw $\overrightarrow{DE}$.

(2) Place the point of your compasses at B. Draw an arc any convenient distance from B to intersect $\overrightarrow{BA}$ at R and $\overrightarrow{BC}$ at M.

(3) Place the point of your compasses at D and with the same compass setting, draw arc z intersecting $\overrightarrow{DE}$ at N.

(4) Set the points of your compasses on M and R. Using this compass setting place the point of your compasses at N and draw an arc intersecting arc z at P.

(5) Draw $\overrightarrow{DP}$. Check the measures of the angles with a protractor.

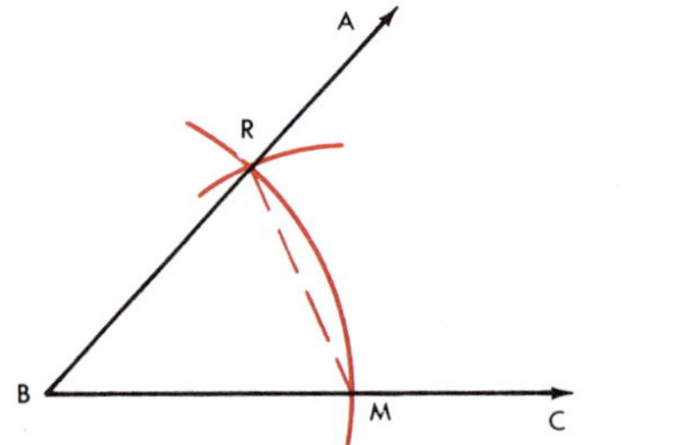

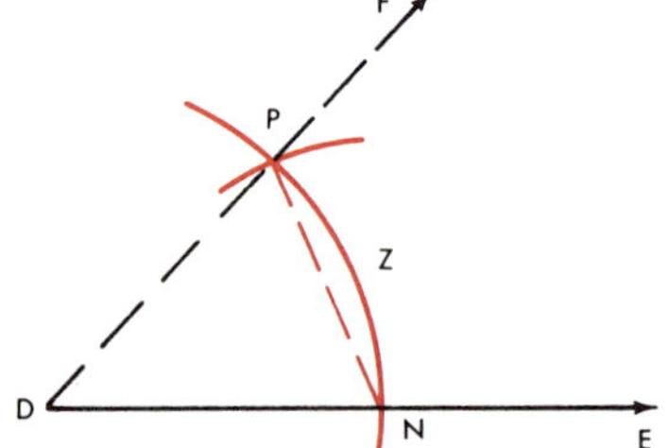

Figure 8–8

Verification:

Given: $\angle ABC$

Construction: On $\overleftrightarrow{DE}$ make $\overline{DN} \cong \overline{BM}$.
Make $\overline{DP} \cong \overline{BR}$ and $\overline{NP} \cong \overline{MR}$.

Conjecture: $\angle FDE \cong \angle ABC$

Exercises

1. Draw an obtuse angle ABC. Construct $\angle DEF \cong \angle ABC$.
2. Given an angle whose measure is $x°$, construct an angle whose measure is $1\frac{1}{2}x°$.

3. Construct a line perpendicular to a given line to obtain a right angle. By constructing three other angles congruent to this one, form a regular quadrilateral (square) $1\frac{1}{2}$ inches on a side.

4. Through a point P not on $\overleftrightarrow{AB}$, construct $\overleftrightarrow{CD} \parallel \overleftrightarrow{AB}$. (*Hint:* Draw $\overleftrightarrow{PO}$, O on $\overleftrightarrow{AB}$, and form congruent corresponding angles.)

5. Draw any acute angle CDE. Construct $\angle FCD \cong \angle CDE$ in the position indicated by the dotted line. Measure $\overline{DF}$ and $\overline{CF}$. Which of the two statements below, *a* or *b*, is the given condition and which is the conclusion?
   (a) $\overline{DF} \cong \overline{CF}$ (b) $\angle D \cong \angle C$

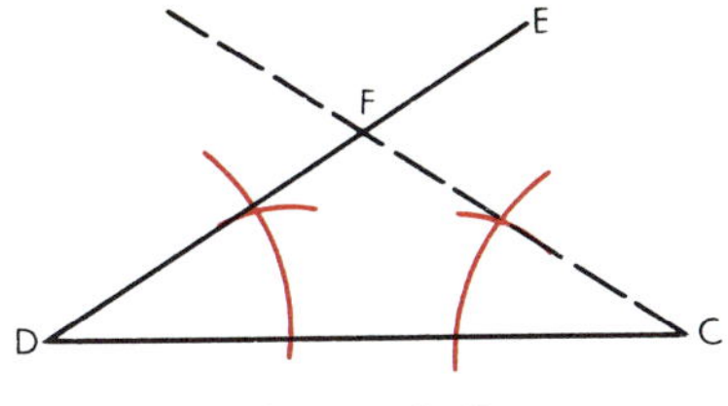

Figure 8–9

6. Repeat the problem above with the given angle slightly different in size. Measure $\overline{DF}$ and $\overline{CF}$ again. Which of the two statements below is best illustrated by this construction?
   (a) If $\angle C \cong \angle D$, then $\overline{DF} \cong \overline{CF}$. (b) If $\overline{DF} \cong \overline{CF}$, then $\angle C \cong \angle D$.

The skills we have developed in construction may be used to construct congruent polygons. The restrictions we have placed upon construction (use only compasses and straight edge—not the marks on a ruler) may be artificial but they do force you to use, and thus know, more geometry. The next few pages will serve as a review of geometry to this point and an extension of the use of your construction skills.

UNITS OF MEASUREMENT

Most of us are familiar with the use of inches as a unit of measurement. Another unit in international use is the centimeter. One inch equals about 2.54 centimeters (cm). A portion of a rule containing both units of measurement is shown in Figure 8-10. Each centimeter is divided into ten equal parts called millimeters (mm). Because centimeters are smaller than inches, they will sometimes be more convenient to use in constructions.

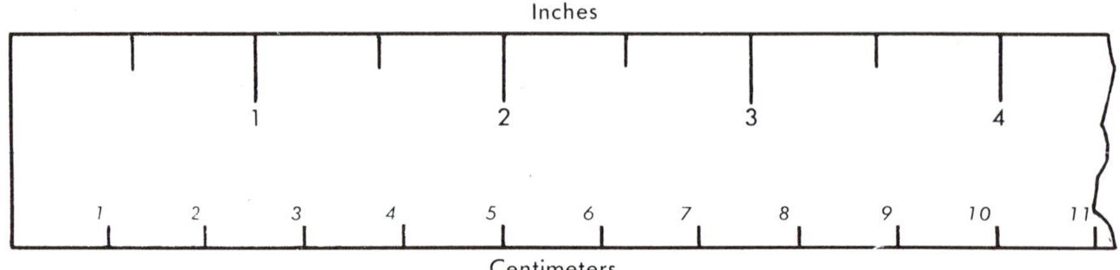

Figure 8–10

8.06 Constructing a Triangle

Construct a triangle:

I. Having three sides given.

PROBLEM: Construct the triangle whose sides are 5 cm, 4 cm, and 3 cm.

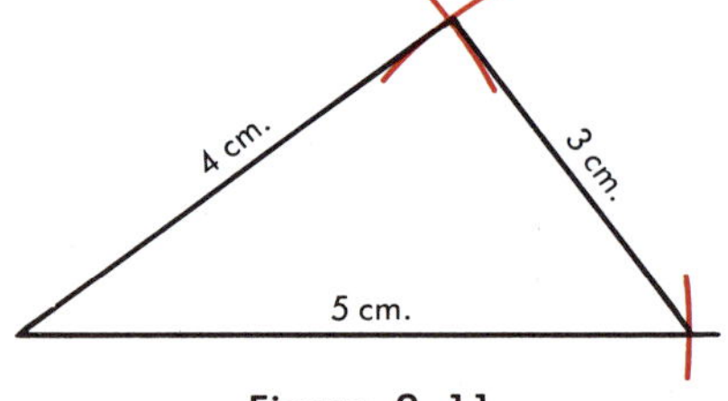

Figure 8–11

Solution: Mark off 5 cm with your compasses on a "base" line. Then, from the ends of the 5 cm segment, draw intersecting arcs of 3 cm and 4 cm radius. How many constructions are possible with the same base?

PROBLEM: Construct the triangle which has sides congruent to those shown below.

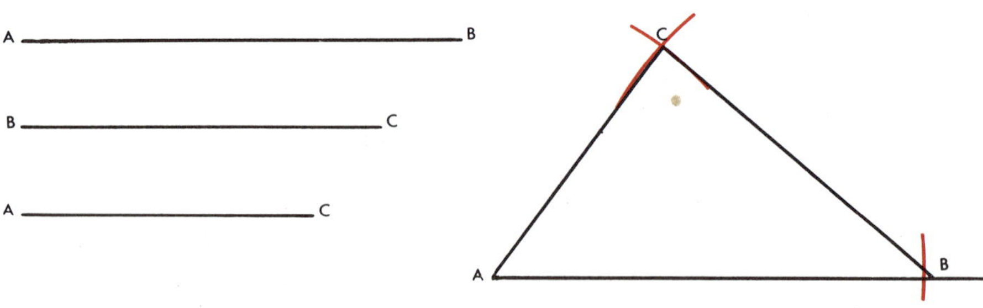

Figure 8–12

Solution: Construct $\overline{AB}$ on a base line with your compasses. From the ends of $\overline{AB}$, draw intersecting arcs whose radii are congruent to $\overline{AC}$ and $\overline{BC}$.

II. Having given two sides and the angle included by those sides.

PROBLEM: Construct the triangle having $\overline{AB}$, $\overline{AC}$, and $\angle A$ given as shown below.

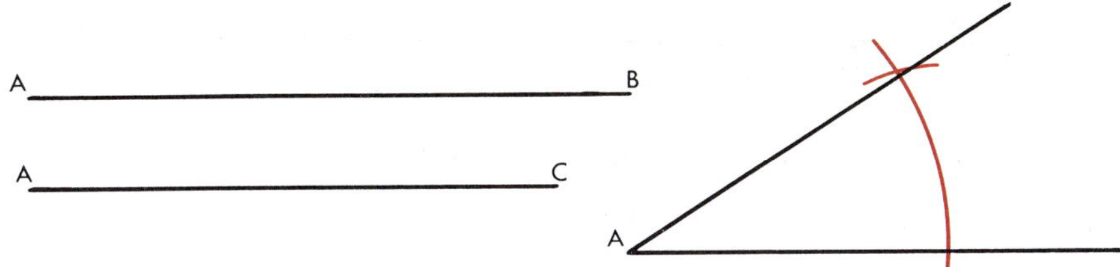

Figure 8–13

Solution: Construct $\overline{AB}$ (or $\overline{AC}$) on a base line. Construct an angle at A on this line congruent to the given $\angle A$. From A mark off the length $\overline{AC}$ and complete the triangle as shown.

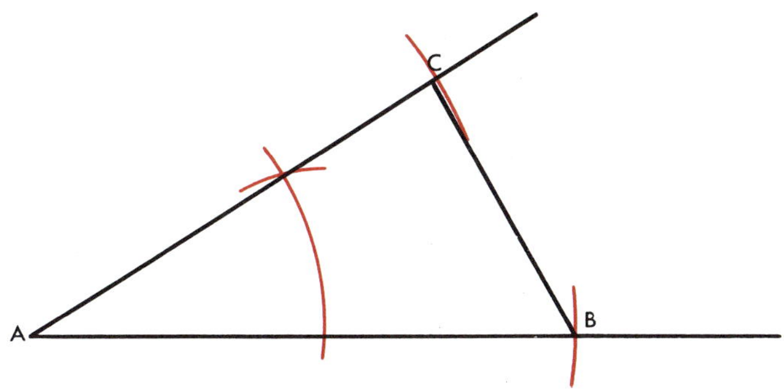

Figure 8–14

Could you construct a triangle if the angle given above were not included by the two given sides? Could you construct more than one triangle? Can you prove triangles congruent by *s.s.a.*?

III. Having two angles and the included side given.

PROBLEM: Construct $\triangle ABC$ given $\angle A$, $\angle B$, and $\overline{AB}$ as shown below.

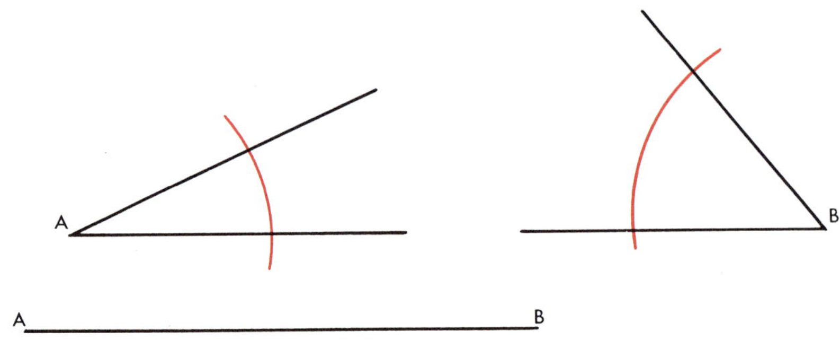

Figure 8–15

Solution: On a base line mark off $\overline{AB}$ with your compasses. Then, at A and B, construct angles A and B, extending the sides of the angles until the triangle is completed.

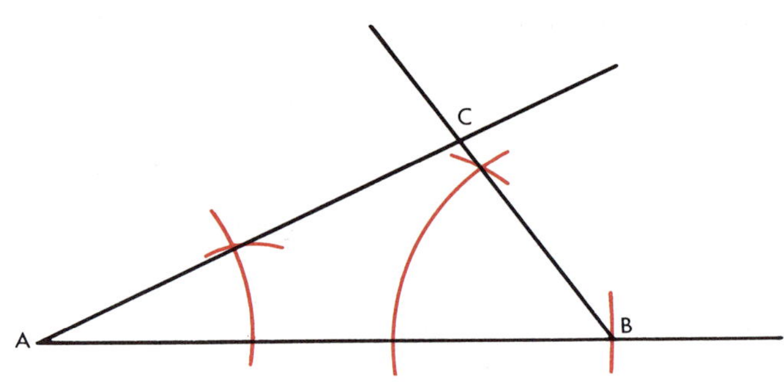

Figure 8–16

Exercises

1. What is the perimeter of a triangle the lengths of whose sides are 4 in., 8 in., and 10 in.?
2. If the measure of the base of an isosceles triangle is 5 inches and the measure of each leg is 8 inches, what is the perimeter?

3. Find the perimeter of the triangles the lengths of whose sides are as follows:

(a) 55 ft, 38 ft, 47 ft

(b) $61\frac{1}{2}$ in., $47\frac{3}{4}$ in., $87\frac{5}{8}$ in.

(c) 2.67 ft, 9.83 ft, 7.75 ft

(d) $4\frac{3}{8}$ yd, $5\frac{1}{3}$ yd, $3\frac{1}{2}$ yd

4. What is the perimeter of an equilateral triangle the measure of whose side is:

(a) 5 ft (b) 23 in. (c) $3\frac{3}{4}$ yd (d) 2.23 ft

5. Construct triangles the lengths of whose sides are as follows:

(a) 2 in., 3 in., $1\frac{1}{2}$ in. (b) 3 cm, 4 cm, 6 cm (c) 1 in., 2 in., 3 in.

(d) 1 in., 2 in., $3\frac{1}{2}$ in. What difficulty do you encounter with (c) and (d)?

6. Complete: The sum of the measures of _?_ sides of a triangle must be _?_ than the measure of the third side.

7. Construct the quadrilateral the lengths of whose sides are 1 in., $1\frac{1}{2}$ in., 2 in., $2\frac{1}{2}$ in. Is your quadrilateral congruent to those constructed by your neighbors? Do the given elements determine a definite quadrilateral? What is the perimeter of this quadrilateral?

8. Draw two acute angles and a line segment about $1\frac{1}{2}$ inches long on your paper. Construct a triangle using these elements as two angles and their included (common) side.

9. Draw two line segments and an obtuse angle on your paper. Construct a triangle using these elements as two sides and their included angle of the triangle.

10. Construct the quadrilateral whose diagonals are the perpendicular bisectors of each other and have lengths of 3 cm and 4 cm. Measure the angles and the sides of the quadrilateral. State a conjecture concerning your construction.

U. S. Forest Service

Since the shape of a triangle cannot be changed without changing the length of one of its sides, triangular structures like this are often used. Such a pattern makes a strong framework with the least possible weight.

ISOSCELES TRIANGLES—INDUCTIVE REASONING

1. Using ruler and compasses, draw models of isosceles triangles of various shapes.
   (a) Construct a segment from the vertex perpendicular to the base. Measure the segments of the base.
   (b) Construct a segment from the vertex to the midpoint of the base. Measure the angles this segment forms with the base. Does it bisect the vertex angle?
   (c) Construct the bisector of the vertex angle. Measure the angles it forms with the base. Measure the segments of the base.
   (d) Measure the angles opposite the congruent sides of an isosceles triangle.
2. Construct several triangles having two congruent angles. Measure the sides opposite those congruent angles.

3. State clearly the conjectures derived from the preceding inductive reasoning.

Our objective now is to test by deductive methods the conjectures suggested by the inductive procedures. They appear to be true in the few cases we have studied. Are they true in general? Deductive proof can determine this for you.

Draw a model of an isosceles triangle to represent the set of all isosceles triangles. Do not make it equilateral. Make $\overline{AC} \cong \overline{BC}$. How can we show deductively that $\angle A \cong \angle B$? Can we use the method of congruency?

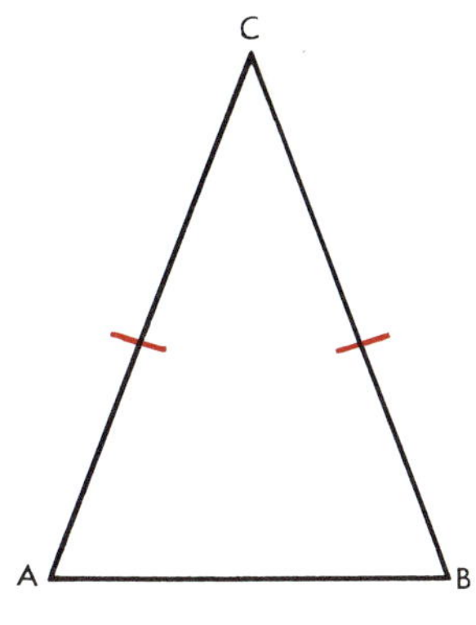

Figure 8–17

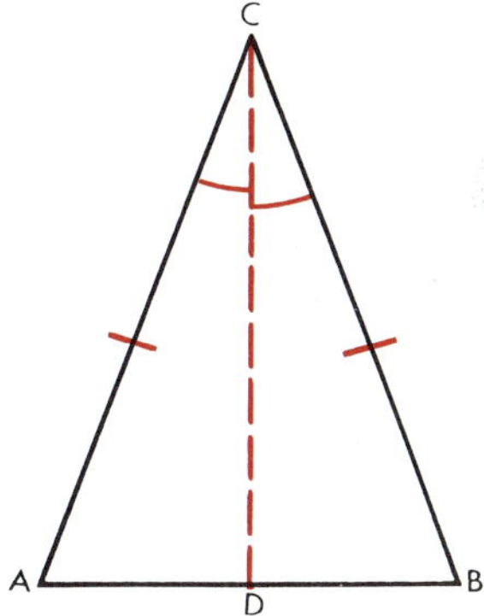

Figure 8–18

Do we have two congruent triangles—or even two triangles? Can we use an auxiliary line (7.06) to form two triangles that would be congruent? (Figure 8-18).

Copy Figure 8-17 carefully, and draw auxiliary ray $\overrightarrow{CD}$ bisecting $\angle C$. Are the triangles now congruent? Why? Does it then follow that $\angle A \cong \angle B$? Have we now proved the following proposition? (Figure 8-18.)

If two sides of a triangle are congruent, then the angles opposite these sides are congruent.

Could we have accomplished the same result if $\overrightarrow{CD}$ had been constructed to bisect $\overline{AB}$? What if the auxiliary segment had been constructed perpendicular to $\overline{AB}$ from point C?

The following propositions and their converses (8.07–8.12) have been selected for proof from the inductive conjectures on the preceding pages. The formal proofs, with the exception of 8.11 are left to the student to complete. The proof of 8.11 is included as a guide and reminder of the formal method.

8.07 THEOREM

If two sides of a triangle are congruent, the angles opposite those sides are congruent.

8.08 THEOREM

An equilateral triangle is also equiangular.

8.09 THEOREM

If two angles of a triangle are congruent, the sides opposite those congruent angles are congruent.

8.10 THEOREM

An equiangular triangle is also equilateral.

Groups of exercises are inserted at intervals to further your understanding and ability to use the previously developed theorems.

Exercises

1. The measure of each angle of an equilateral triangle is _?_ degrees.
2. Construct an isosceles right triangle. Write the measure of each angle within the angle.
3. An altitude of an equilateral triangle divides it into two _?_ _?_ triangles. The measures of the acute angles are _?_ and _?_ degrees.
4. Is the theorem of 8.09 the converse of the theorem of 8.07?
5. If the measure of the vertex angle of an isosceles triangle is 50°, find the measure of each of the remaining angles.
6. If the measure of an angle opposite one of the congruent sides of an isosceles triangle is 40°, state the measures of the other two angles.
7. If the measure of the vertex angle of an isosceles triangle is 90°, find the measure of an exterior angle at the base of the triangle.

Use Figure 8-19 for Exercises 8-12.

8. If $\overline{AB} \cong \overline{BC}$ and $\overline{AD} \cong \overline{EC}$, prove $\overline{BD} \cong \overline{BE}$.

9. If $\overline{AB} \cong \overline{BC}$ and $\angle 1 \cong \angle 2$, prove $\angle 3 \cong \angle 6$.

10. If $\overline{AB} \cong \overline{BC}$ and $\overline{AD} \cong \overline{CE}$, prove $\angle 4 \cong \angle 5$.

11. If $\overline{AB} \cong \overline{BC}$ and $\angle 1 \cong \angle 2$, prove $\angle 4 \cong \angle 5$.

12. If $\angle 4 \cong \angle 5$ and $\angle 1 \cong \angle 2$, prove $\overline{AB} \cong \overline{CB}$.

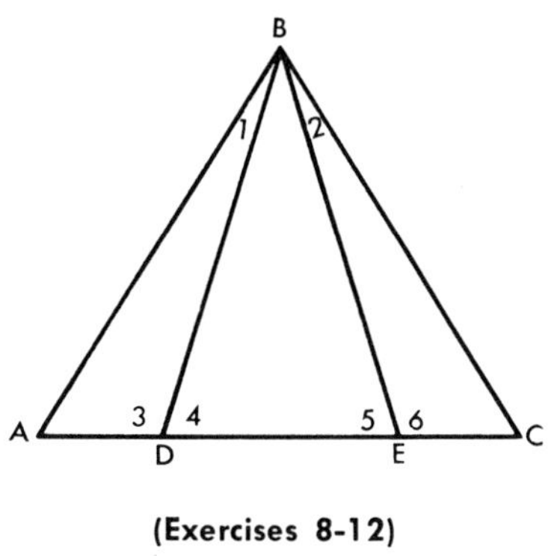

(Exercises 8-12)

Figure 8–19

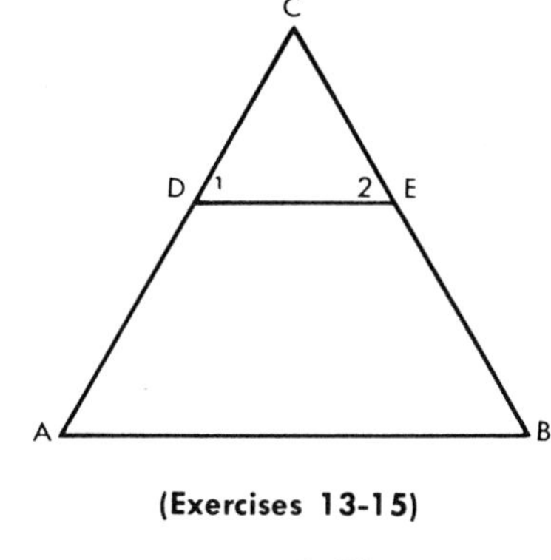

(Exercises 13-15)

Figure 8–20

Use Figure 8-20 for Exercises 13–15.

13. If $\overline{AC} \cong \overline{BC}$ and $\overline{AD} \cong \overline{BE}$, prove $\angle 1 \cong \angle 2$.

14. If $\overline{AD} \cong \overline{BE}$ and $\overline{DC} \cong \overline{EC}$, prove $\angle A \cong \angle B$.

15. If $\overline{AC} \cong \overline{BC}$, $\angle 1 \cong \angle A$ and $\angle 2 \cong \angle B$, prove $\overline{CD} \cong \overline{CE}$.

16. What type of triangle is formed by the bisectors of the congruent angles of an isosceles triangle and the base? Prove your conjecture.

17. Construct a line perpendicular to the bisector of an angle in the plane of the angle. Prove any relationships evident in this figure.

ISOSCELES TRIANGLES—PROOF

What conclusions did you reach after constructing the bisector of the vertex angle of an isosceles triangle on page 184? What was the relation of the angle bisector to the base of the triangle? Can you prove these conjectures?

STUDENT A developed the following "proof."

Given: Isosceles triangle ABC with $\overline{AC} \cong \overline{BC}$, and $\overline{CD}$ bisecting angle C

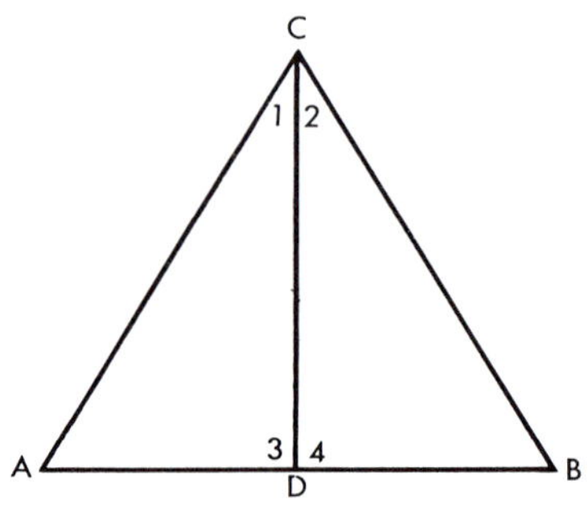

Figure 8–21

Conjecture: $\overline{AD} \cong \overline{BD}$ and $\overline{CD} \perp \overline{AB}$

Plan: Prove $\triangle ADC \cong \triangle BDC$.

Proof:

Statements	*Reasons*
1. $\overline{AC} \cong \overline{BC}$	1. Given
2. $\overline{CD}$ bisects $\angle C$	2. Given
3. $\angle 1 \cong \angle 2$	3. If an angle is bisected, it is divided into two congruent angles.
4. $\overline{CD} \cong \overline{CD}$	4. Reflexive Axiom
5. Then $\triangle ADC \cong \triangle BDC$	5. *s.a.s.*
6. $\overline{AD} \cong \overline{BD}$	6. *c.p.c.t.c.* (Corresponding parts of congruent triangles are congruent.)
7. $\angle 3 \cong \angle 4$	7. *c.p.c.t.c.*
8. $\overline{CD} \perp \overline{AB}$	8. If two segments determine congruent adjacent angles, they are perpendicular.

STUDENT B "proved" the same proposition as follows (using the same figure as Student A):

Given: Isosceles triangle ABC with $\overline{AC} \cong \overline{BC}$, and $\overline{CD}$ bisecting angle C

Conjecture: $\overline{AD} \cong \overline{BD}$ and $\overline{CD} \perp \overline{AB}$

Plan: Prove $\triangle ADC \cong \triangle BDC$.

Proof:

Statements	*Reasons*
1. $\overline{AC} \cong \overline{BC}$	1. Given
2. $\angle A \cong \angle B$	2. If two sides of a triangle are congruent, the $\angle s$ opposite those sides are congruent.
3. $\overrightarrow{CD}$ bisects $\angle C$	3. Given
4. $\angle 1 \cong \angle 2$	4. If an angle is bisected, it is divided into two congruent angles
5. $\therefore \triangle ADC \cong \triangle BDC$	5. *a.s.a.*
6. so $\angle 3 \cong \angle 4$	6. *c.p.c.t.c.*
7. $\therefore \overline{CD} \perp \overline{AB}$	7. If two segments determine congruent equal adjacent angles, the lines are $\perp$.
8. also $\overline{AD} \cong \overline{BD}$	8. *c.p.c.t.c.*
9. $\therefore \overline{CD}$ bisects the base	9. If a segment is divided into two congruent parts, it is bisected.

Are both proofs correct?

Has the following theorem been proved?

8.11 THEOREM

The bisector of the vertex angle of an isosceles triangle bisects the base and is perpendicular to it.

The preceding theorem made conclusions concerning the angle bisector of the vertex angle of an isosceles triangle. Can any conclusions be made about the median from the vertex angle?

A student's attempt to prove his conjectures was corrected by his teacher as follows. Be sure that you understand why the corrections were made.

Given: Isosceles $\triangle EFG$, with $\overline{EG} \simeq \overline{FG}$ and $\overline{EH} \simeq \overline{FH}$

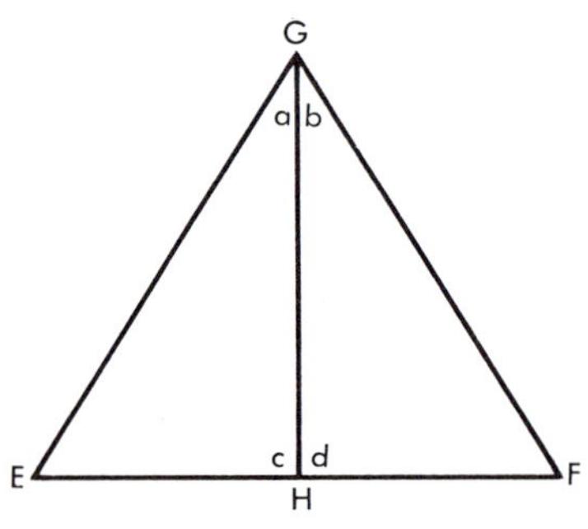

Figure 8–22

Conjecture: $\angle a \simeq \angle b$ and $\overline{GH} \perp \overline{EF}$

Plan: Prove that $\triangle EHG \simeq \triangle FHG$; then show $\angle c \simeq \angle d$ in order to prove $\overline{GH} \perp \overline{EF}$.

Proof:

	Statements		*Reasons*
(S)	1. $\overline{EG} \simeq \overline{FG}$	Do the pairs of congruent sides "include" the angles?	1. Given Shouldn't you have used the converse of th
(A)	2. $\angle E \simeq \angle F$		2. If two angles of a triangle are congruent, the sides opposite are congruent.
(S)	3. $\overline{GH} \simeq \overline{GH}$		3. Reflexive Axiom
	use $\overline{EH} \cong \overline{FH}$, then		
	4. $\triangle EGH \simeq \triangle FGH$	4. s.s.s.	4. *s.a.s.*
	5. $\overline{GH} \perp \overline{EF}$ and $\angle a \simeq \angle b$		5. *c.p.c.t.c.*
	This is not established by c.p.c.t.c. Show $\angle c \cong \angle d$. Then use #3.18.		

Can you rewrite this argument so that it forms an acceptable formal proof?

Another student, using the same figure and conditions, but omitting his plan of approach, attempted to prove the conjecture above and was corrected as follows.

Proof: *Statements* Redo! *Reasons*

Statements	Reasons
1. $\overline{EG} \cong \overline{FG}$	1. Given
2. $\angle C \cong \angle D$	2. If two sides of (a) triangle are congruent, the $\angle$s opposite are congruent.
3. $\overline{GH} \perp \overline{EF}$	3. 3.18

The reason does not apply $\angle c + \angle d$ are $\angle$s of two different $\triangle$.

Do you mean $\angle c \cong \angle d$? Use proper labels.

Why is $\angle a \cong \angle b$?

Write out your reason. We don't attempt to memorize section numbers.

Is this proof now acceptable?

If the student proofs were corrected, would the following theorem be established?

8.12 THEOREM

The segment that connects the vertex angle of an isosceles triangle with the midpoint of the base bisects the vertex angle and is perpendicular to the base.

Exercises

1. Is theorem 8.12 the converse of theorem 8.11?
2. Justify the statement: "The median to the base of an isosceles triangle bisects the vertex angle."
3. Specify the conditions where the bisector of an angle of a triangle and a median are the same segment.
4. Specify the conditions where the median and the altitude of a triangle are the same segment.
5. Prove or disprove the statement: "A diagonal of a quadrilateral forms an isosceles triangle with two sides of the quadrilateral if the diagonal bisects the angles whose vertices it joins."
6. Prove or disprove the statement: "Two isosceles triangles are formed by segments of transversals intersecting between parallel lines."
7. Prove or disprove the statement: "Two diagonals from the same vertex of a regular pentagon form an isosceles triangle with the opposite side."
8. If $\overline{AB}$ and $\overline{CD}$ intersect at E so that $\overline{AC} \cong \overline{BC}$ and $\overline{AD} \cong \overline{BD}$, is $\overline{AE} \cong \overline{BE}$? Prove your answer.

9. In $\triangle ABC$, $\overline{AC} \simeq \overline{BC}$, X is a point of $\overline{AB}$ and O is a point of $\overline{CX}$ such that $\overline{AO} \simeq \overline{BO}$. Is $\overline{CX} \perp \overline{AB}$? Prove your answer.
10. In $\triangle DEF$, $\overline{DF} \simeq \overline{EF}$, $\overline{FR}$ bisects $\overline{DE}$, R is a point of $\overline{DE}$ and P is a point of $\overline{FR}$. Is $\triangle DFP \simeq \triangle EFP$? Prove your answer.
11. Prove or disprove the statement: "The diagonals of an equilateral quadrilateral are the perpendicular bisectors of each other."
12. Prove or disprove the statement: "The bisector of an angle of a regular pentagon is the perpendicular bisector of the opposite side."
13. Construct $\overleftrightarrow{LM}$ as the perpendicular bisector of $\overline{RT}$. If P is a point of $\overleftrightarrow{LM}$, what is the relationship of $\overline{PR}$ and $\overline{PT}$? Prove it. State the theorem you proved.
14. Can a plane be the perpendicular bisector of a segment? Justify your answer by quoting applicable definitions, assumptions, and theorems.
15. Replace $\overleftrightarrow{LM}$ of Exercise 13 by plane LM. Is the proof you developed for Exercise 13 still correct? Should the theorem be rewritten? If so, write it.
16. In a given plane select any two points A and B such that each point is equally distant from the endpoints of $\overline{GH}$ in the plane. What is the relationship of $\overleftrightarrow{AB}$ to $\overline{GH}$? Prove it. State the relationship as a theorem.
17. Would the theorem of Exercise 16 be true if A, B, G and H are not coplanar? Explain.

Because the theorems you developed in Exercises 13 through 17 are so commonly used we will formalize then below.

8.13 THEOREM

Any point on the perpendicular bisector of a segment is equidistant from the endpoints of the segment.

8.14 THEOREM

In a given plane, two distinct points equidistant from the endpoints of a segment determine the perpendicular bisector of the segment.

Proof by Algebra

Some of you will be interested to know that several of the proofs in this chapter and many of the proofs throughout the remainder of the text can be accomplished by an algebraic approach to geometry. In some cases this simplifies the proof considerably, but in other cases the proofs become more difficult because of the complexity of the algebra involved. Chapter 21 outlines this approach to geometry.

If you are interested in pursuing the coordinate geometry in Chapter 21 at this point, it will be necessary to assume the Pythagorean Theorem (page 440) or to prove the theorem as shown in the Appendix (page 576) upon the basis of the understanding of area you have obtained from previous mathematics courses.

As we proceed through the proofs of the theorems in Chapters 9–20, references will be listed for the alternate algebraic proof in Chapter 21.

THE PROBLEM OF TRISECTING AN ANGLE

It has been proved that the problem of trisecting an angle cannot be solved using only the compasses and the straightedge. However, we often read of an alleged solution of the problem. The following is one of the most common. See if you can find the fallacy.

Given $\angle AOB$. Place two marks C and D on a straightedge. With O as center and a radius congruent to $\overline{CD}$, construct a circle. Place the straightedge so that it contains point B; C is on the circle; and D is on $\overrightarrow{AO}$. Now $\overline{OB} \cong \overline{OC} \cong \overline{CD}$. Therefore $m\angle ADB \cong \frac{1}{3} m\angle AOB$. Why?

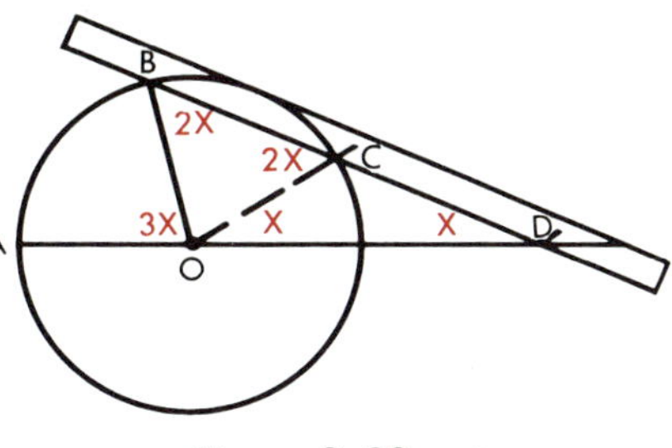

Figure 8–23

Exercises

1. Are two isosceles triangles congruent if they have congruent bases? Congruent vertex angles? Explain.
2. Does the bisector of any angle of an equilateral triangle bisect the opposite side? Explain.

Use Figure 8-24 for Exercises 3–6.

3. If a vertical telephone pole $\overline{AO}$ is braced with two congruent guy wires $\overline{AB}$ and $\overline{AC}$, which are fastened to the horizontal plane MN, are $\overline{BO}$ and $\overline{CO}$ congruent? Explain. (Figure 8-24.)
4. $\overline{AD} \cong \overline{BE}$, $\overline{AE} \cong \overline{BD}$. Prove that $\angle 1 \cong \angle 2$.

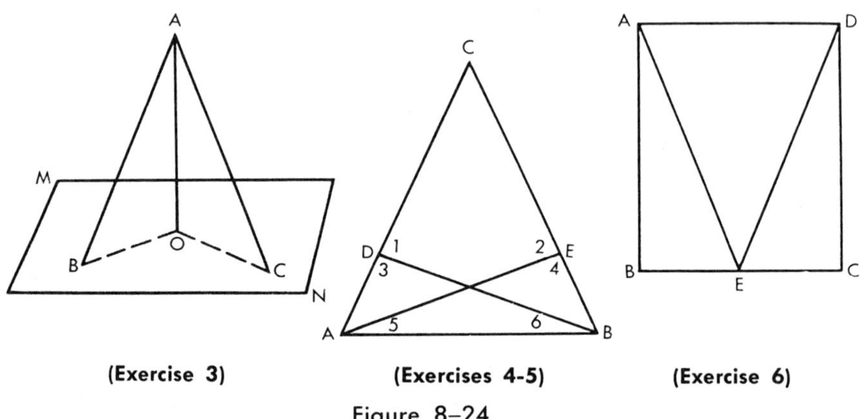

(Exercise 3) (Exercises 4-5) (Exercise 6)

Figure 8–24

5. $\overline{AE}$ and $\overline{BD}$ determine congruent angles with $\overline{AB}$, and $\angle CAB \cong \angle CBA$. Prove that $\angle 1 \cong \angle 2$ and $\angle 3 \cong \angle 4$.
6. Prove: If in quadrilateral $ABCD$, $\overline{AB} \cong \overline{BC} \cong \overline{CD} \cong \overline{AD}$, and $\angle A$, $\angle B$, $\angle C$ and $\angle D$ are right angles, and E is a point on $\overline{BC}$ such that $\overline{AE} \cong \overline{ED}$, then m$\overline{EC}$ is one-half of m$\overline{BC}$.

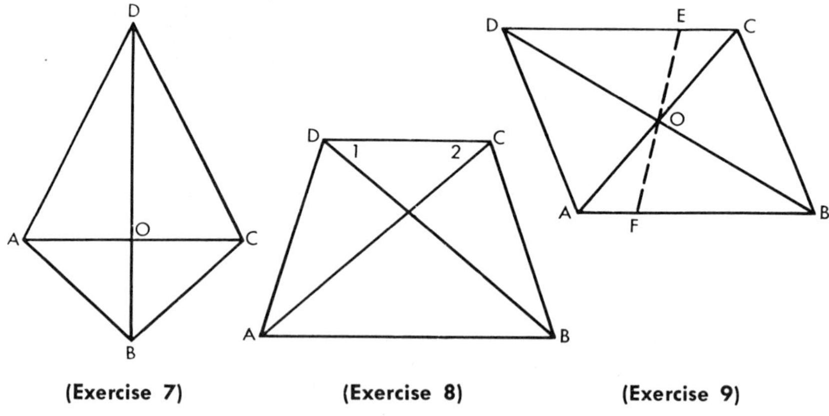

(Exercise 7) (Exercise 8) (Exercise 9)

Figure 8–25

Use Figure 8-25 for Exercises 7–9.

7. Prove: If $\overline{AD} \cong \overline{CD}$ and $\overline{AB} \cong \overline{CB}$, then $\overline{AO} \cong \overline{CO}$.
8. Prove: If $\overline{AD} \cong \overline{BC}$ and $\overline{AC} \cong \overline{BD}$, then $\angle 1 \cong \angle 2$.
9. Prove: In quadrilateral $ABCD$, if $\overline{AC}$ and $\overline{BD}$ bisect each other at O, then any segment $\overline{EF}$ through O with endpoints on $\overline{AB}$ and $\overline{CD}$ is bisected by O.

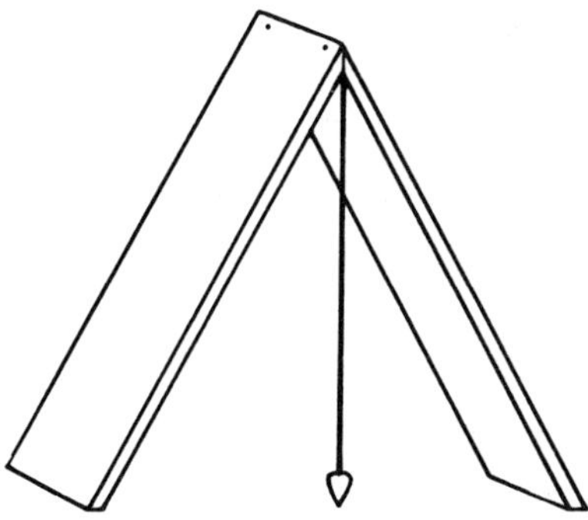

Figure 8–26

10. Figure 8-26 shows a partially completed level using a *plumb line* as the major part of the figure. Devise a method to complete the figure so that it can be used as a level. Explain.

Vocabulary List

equidistant	construct
midpoint	compasses
bisect	centimeter
perpendicular	perimeter
isosceles	

Chapter Review

1. Draw a segment, and call its measure x. From this, construct a segment whose measure is $2x$, and a segment whose measure is $\frac{1}{2}x$.
2. Draw any angle, and call its measure $x°$. From this, construct an angle whose measure is $3x°$.

3. Name the segments constructed in Figures I, II, and III.

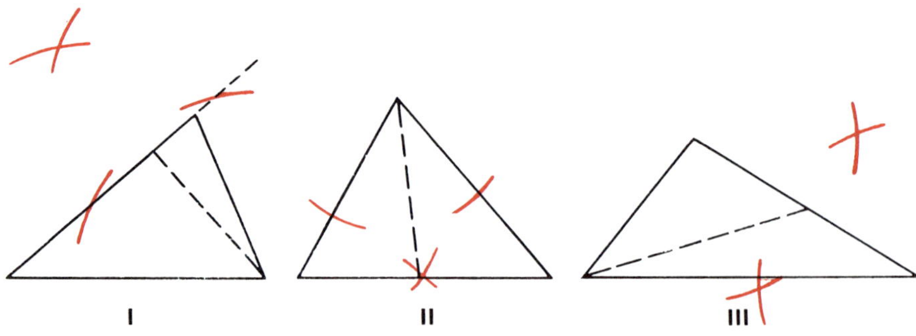

Figure 8–27

4. Draw a scalene triangle. Bisect its three angles and continue the bisectors until they meet. Do they seem to meet at some point within the triangle? Does this point seem equidistant from the sides? From the vertices?
5. Draw an acute triangle and construct the perpendicular bisectors of the three sides. Do they seem to meet at some point P within the triangle? Are these concurrent lines? Is this point equidistant from the sides? From the vertices?
6. Construct a right triangle. Construct the perpendicular bisectors of the three sides. Where do the bisectors seem to meet?
7. Repeat Exercise 6 using an obtuse triangle.
8. Draw an acute scalene triangle and construct its altitudes. Continue these lines until they meet.
9. Repeat Exercise 8 using an obtuse scalene triangle.
10. Draw a scalene triangle and construct its medians.
11. What sets of concurrent lines, rays, or segments have you discovered inductively?
12. Construct an isosceles triangle and through its vertex construct a line parallel to the base.
13. Construct a line through the midpoint of the base of a triangle parallel to one of the sides.
14. Construct angles whose measures are 60°; 30°; 75°.
15. Construct $\triangle ABC$ so that $m\angle A = 60°$, $m\angle C = 45°$, and $m\overline{AB} = 6.3$ cm.

16. Construct an isosceles right triangle whose hypotenuse is 3 inches in length.

17. Construct an isosceles triangle, having a leg and the vertex angle given. (Draw a line segment and an angle on your paper and label them to represent the given elements. Then use these elements to construct the isosceles triangle.)

18. Construct an equilateral triangle, having an altitude given. (Draw a line segment on your paper to represent the altitude, label it, and then, elsewhere on your paper, construct an equilateral triangle whose altitude is congruent to the segment.)

Determine whether the following statements are true or false. If true, prove them formally. Remember that the "truth" of these statements can only be determined within the framework of the mathematical system we are now developing.

19. A median of a triangle bisects the angle whose vertex it contains.

20. If the base of an isosceles triangle is trisected and from each point of division a segment is drawn perpendicular to the nearer side, the segments are congruent.

21. If a median of a triangle is also an altitude, the triangle is isosceles.

22. A pair of corresponding altitudes of two congruent triangles are congruent.

23. A pair of corresponding medians of two congruent triangles are congruent.

24. A pair of corresponding angle bisectors of two congruent triangles are congruent.

25. Any two points equidistant from the end points of a line segment determine the perpendicular bisector of the segment.

26. The altitudes to the legs of an isosceles triangle are congruent.

27. Write the converse of the above and determine its truth.

28. The exterior angle at the base of an isosceles triangle is congruent to the angle formed by the bisectors of the base angles.

29. If from the vertex of one of the congruent angles of an isosceles triangle a segment is drawn perpendicular to the opposite side, it makes an angle with the base whose measure is one-half the measure of the vertex angle of the triangle.

Chapter 8 Test

Use compasses and straight edge to construct a triangle, the measures of whose sides are 2, 3, and 4 inches, for exercises 1–4. Do not erase your "construction arcs."

1. Construct the bisector of the angle with the least measure.
2. Construct the altitude to the side whose measure is 4 inches.
3. Construct a perpendicular bisector of the side whose measure is 3 inches.
4. Construct a line parallel to the 3 inch side through the opposite vertex.

5. Draw a segment whose measure is $1\frac{1}{2}$ inches. Construct a perpendicular to this segment at one of its endpoints.

Correctly complete the statements of questions 6–10 by inserting the word "sometimes," "always," or "never" for the missing word in each.

6. Isosceles triangles are __?__ congruent.
7. The bisector of the vertex angle of an isosceles triangle __?__ contains the midpoint of the base.
8. The perpendicular bisector of a side of a scalene triangle __?__ bisects the angle opposite that side.
9. Any point in the perpendicular bisector of a segment is __?__ equally distant from the endpoints of that segment.
10. Two isosceles triangles are __?__ congruent if a side and an adjacent angle of one are congruent to the corresponding parts of the other.

11. **Given:** $\overline{AB}$ is the common base of coplanar isosceles triangles ABC and ADB.

    One of the following statements is correct. Prove it.

    (a) $\overleftrightarrow{AB}$ is the perpendicular bisector of $\overline{CD}$.

    (b) $\overleftrightarrow{CD}$ is the perpendicular bisector of $\overline{AB}$.

12. **Given:** ΔRST with $\overline{RT} \cong \overline{ST}$, M on $\overline{RT}$ and N on $\overline{ST}$ such that $\overline{RM} \cong \overline{SN}$.

    Prove: $\overline{RN} \cong \overline{SM}$.

9

Quadrilaterals

In this chapter we will study another subset of the set of simple polygons, the *quadrilaterals*. Following our established procedure, we will define the more common members of this set and then develop geometric relationships concerning them.

9.00 A parallelogram is a quadrilateral whose opposite sides are parallel. It is one of the most important quadrilaterals.

9.01 An altitude of a parallelogram is a segment whose endpoints lie in opposite sides and which is perpendicular to one of these sides. The measure of an altitude is called the *height*.

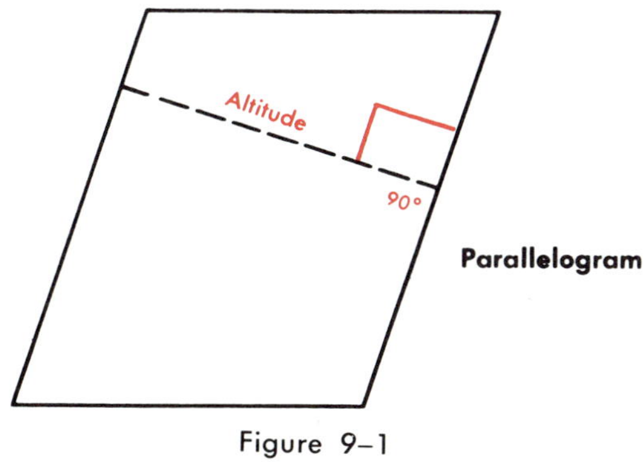

Figure 9–1

LEARNING ABOUT PARALLELOGRAMS BY INDUCTIVE REASONING

1. Construct several parallelograms of varying shapes. (a) Measure the angles and the sides of each figure. (b) Compare the lengths of the diagonals and the segments of the diagonals.
2. State the various relationships suggested to you as a result of your experimentation.
3. Consider the converses of these relationships. Do you believe these converses would be true?

Our objective now is to test by deductive methods the conjectures suggested by the preceding inductive procedure. They appear to be true in the special cases we have studied. Are they true in general? The answer can be determined only by deductive procedures.

From the list of propositions that can be developed by induction, the following have been selected for inclusion in this text because of their generality. Each of these should be proved deductively. The proof of 9.08 is given because of its difficulty. Your proofs need not be limited to the propositions listed here. Prove or disprove any other conjectures you may have discovered by inductive reasoning.

Given: Parallelogram $ABCD$ with diagonal $\overline{BD}$

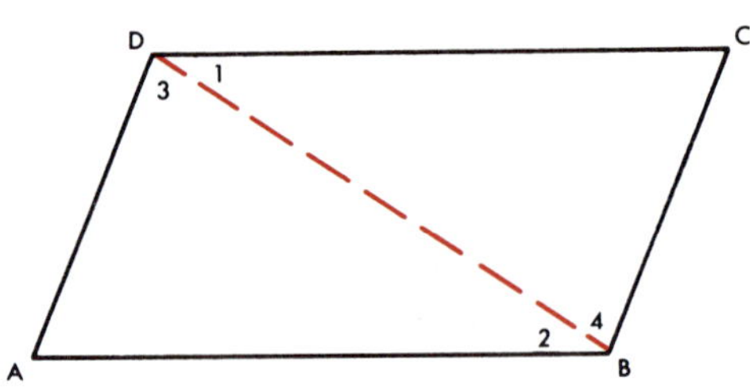

Figure 9–2

Conjecture: $\triangle ABD \cong \triangle CDB$

Proof: The proof is left to the student.

9.02 THEOREM

A diagonal of a parallelogram forms with the sides two congruent triangles.

9.03 THEOREM

The opposite sides, and the opposite angles, of a parallelogram are congruent. (See coordinate proof page 558.)

9.04 THEOREM

Two consecutive angles of a parallelogram are supplementary. (*Suggestion:* Use 5.26)

9.05 THEOREM

Segments of parallels included between parallels are congruent. (*Suggestion:* Use 9.03.)

9.06 THEOREM

Two parallel lines are everywhere equidistant.
(*Suggestion*: Construct perpendiculars to one parallel; then use 9.03. This demonstration proves assumption 2.44.)

We may also prove a proposition about the relationship of the two diagonals of a parallelogram.

Given: Parallelogram $ABCD$ and its two diagonals $\overline{DB}$ and $\overline{AC}$ intersecting at E

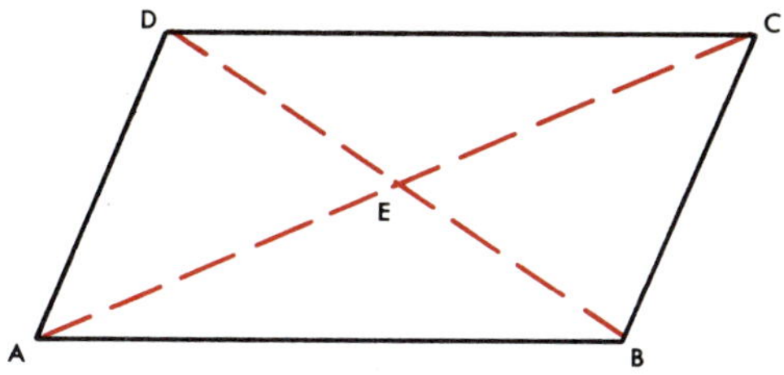

Figure 9–3

Conjecture: $\overline{DE} \cong \overline{BE}$ and $\overline{AE} \cong \overline{CE}$

Plan: Prove that $\triangle AEB \cong \triangle CED$.

Proof: The proof is left to the student.

9.07 THEOREM

The diagonals of a parallelogram bisect each other. (See problem 1, page 559.)

Exercises

1. Are the diagonals of every parallelogram congruent?
2. The measure of an angle of a parallelogram is $50°$. What is the measure of each of the other angles?
3. A diagonal of a parallelogram $ABCD$ divides angle A into two angles whose measures are $22°$ and $33°$. Give the measure of angles B, C, and D.
4. Do the diagonals of a parallelogram divide the figure into four congruent triangles? Explain.
5. What is the relationship of the bisectors of the opposite angles of a parallelogram? Prove your response.
6. The measures of two consecutive angles of a parallelogram are in the ratio 2:3. Find the measure of each angle of the parallelogram.
7. In any triangle ABC, draw a line through C parallel to AB, and a line through B parallel to AC. Name the intersection of these lines D. Prove that $ABDC$ is a parallelogram.
8. Prove that if the diagonals of a parallelogram are perpendicular to each other, the parallelogram is equilateral.
9. Prove that the intersections of the lines through the vertices of a quadrilateral parallel to the diagonals of the quadrilateral are the vertices of a parallelogram.
10. Prove that two parallel segments from two opposite vertices of a parallelogram to the diagonal joining the other two vertices are congruent.

QUADRILATERALS AND PARALLELOGRAMS

When is a quadrilateral a parallelogram? When it satisfies definition 9.00 is the obvious answer. Would this definition be satisfied by a quadrilateral whose opposite angles were congruent? By a quadrilateral whose diagonals bisect each other? It is now our purpose to examine the converses of the theorems of page 201 to see if they are true. If so, we will be in a position to recognize parallelograms when the conditions of definition 9.00 are not directly given.

Given: Case I. Quadrilateral $ABCD$, $\overline{AB} \cong \overline{CD}$ and $\overline{AD} \cong \overline{BC}$
Case II. $\angle A \cong \angle C$ and $\angle B \cong \angle D$

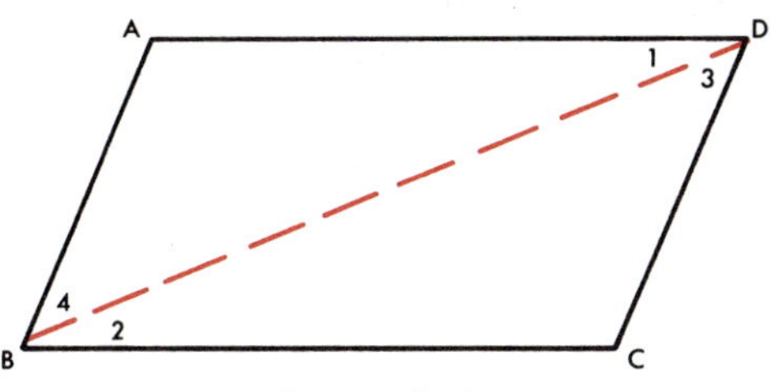

Figure 9–4

Conjecture: $ABCD$ is a parallelogram.

Plan: Use 9.00.

Proof:

	Statements	*Reasons*
Case I	1. Draw $\overline{BD}$	1. Why possible?
	2. $\overline{BD} \cong \overline{BD}$	2. Why?
	3. $\overline{AD} \cong \overline{BC}$ and $\overline{AB} \cong \overline{CD}$	3. Why?
	4. $\therefore \triangle ABD \cong \triangle CDB$	4. Why?
	5. $\angle 1 \cong \angle 2$ and $\angle 3 \cong \angle 4$	5. Why?
	6. $\overline{AD} \parallel \overline{BC}$ and $\overline{AB} \parallel \overline{CD}$	6. Why?
	7. $ABCD$ is a parallelogram.	7. Why?

Case II	1. $\angle A \cong \angle C$, $\angle B \cong \angle D$	1. Why?
	2. $m\angle A + m\angle B + m\angle C + m\angle D = 360°$	2. Why?
	3. $m\angle A + m\angle B + m\angle A + m\angle B = 360°$	3. Why?
	4. $m\angle A + m\angle B = 180°$	4. Why?
	5. $\overline{AD} \parallel \overline{BC}$	5. Why?
	6. $m\angle A + m\angle D + m\angle A + m\angle D = 360°$	6. Why?
	7. $m\angle A + m\angle D = 180°$	7. Why?
	8. $\overline{AD} \parallel \overline{CD}$	8. Why?
	9. $ABCD$ is a parallelogram.	9. Why?

9.08 THEOREM

If the opposite sides, or opposite angles, of a quadrilateral are congruent, the figure is a parallelogram.

Is any quadrilateral with *two* congruent sides a parallelogram? What conditions could be added so that the figure would necessarily be a parallelogram? Let us consider one possibility.

Given: $ABCD$ is a quadrilateral with $\overline{AD} \cong \overline{BC}$ and $\overline{AD} \parallel \overline{BC}$.

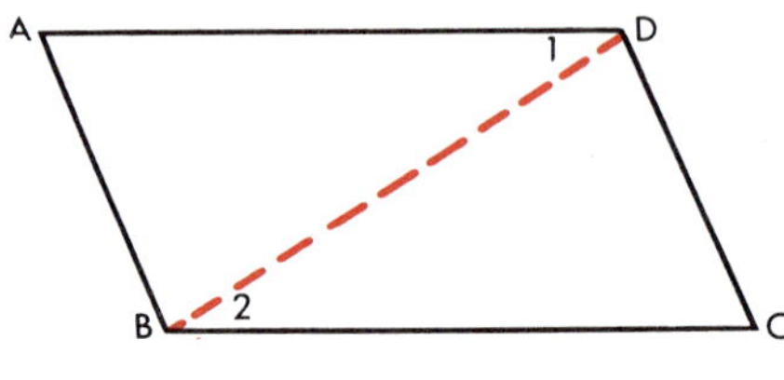

Figure 9–5

Conjecture: $ABCD$ is a parallelogram.

Plan: Prove $\triangle ADB \cong \triangle CBD$. Then use either 9.00 or 9.08.

Proof: The proof is left to the student.

9.09 THEOREM

If two sides of a quadrilateral are congruent and parallel, the figure is a parallelogram. (See problem 4, page 559.)

9.10 (Optional) THEOREM

If the diagonals of a quadrilateral bisect each other the figure is a parallelogram. (See problem 3, page 559.)

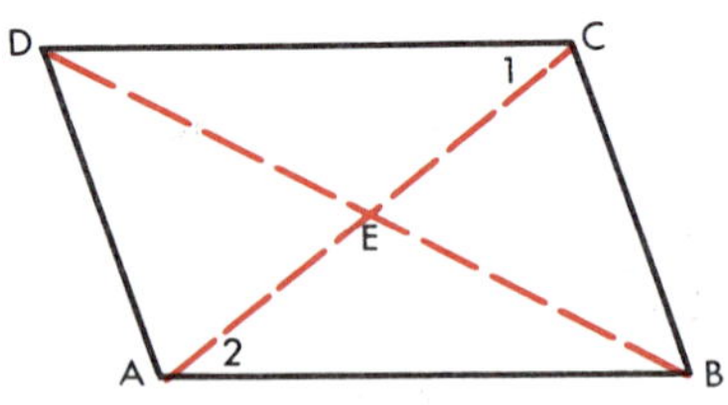

Figure 9–6

(*Suggestion*: Prove that $\triangle ABE \cong \triangle CDE$. Then apply 9.09.)

Exercises

1. If the measure of one angle of a parallelogram is 55°, what is the measure of the other angles?
2. The measure of one angle of a parallelogram is twice another. Find the measure of each angle of the parallelogram.
3. Are the bisectors of two consecutive angles of a parallelogram perpendicular? Prove your response.
4. If the midpoints of the four sides of a parallelogram are joined in order by segments to form a quadrilateral, is the quadrilateral a parallelogram? Prove your response.
5. Construct a parallelogram having two sides, 1 and 2 inches long, and a diagonal 2.5 inches long.
6. In quadrilateral $ABCD$, $\angle BAD \cong \angle DCB$ and $\angle DCA \cong \angle BAC$. Is $ABCD$ a parallelogram? Prove your response.
7. In plane x, two congruent isosceles triangles ABC and ABD have a common base $\overline{AB}$; C and D are on opposite sides of $\overline{AB}$. Is $ADBC$ a parallelogram? Prove your response.
8. Are two parallelograms formed by a segment joining the midpoints of two opposite sides of a parallelogram? Prove your response.
9. In parallelogram $WXYZ$, $\overline{XB} \perp \overline{WY}$ at B and $\overline{ZA} \perp \overline{WY}$ at A. Is $XAZB$ a parallelogram? Prove your response.
10. On diagonal $\overline{RT}$ of parallelogram $QRST$, $\overline{RE} \cong \overline{TF}$. Is $SFQE$ a parallelogram? Prove your response.

9.11 A rectangle is a parallelogram having one right angle. If a rectangle is a parallelogram, must the relationships of 9.00–9.10 be true for a rectangle? (Figure 9-7)

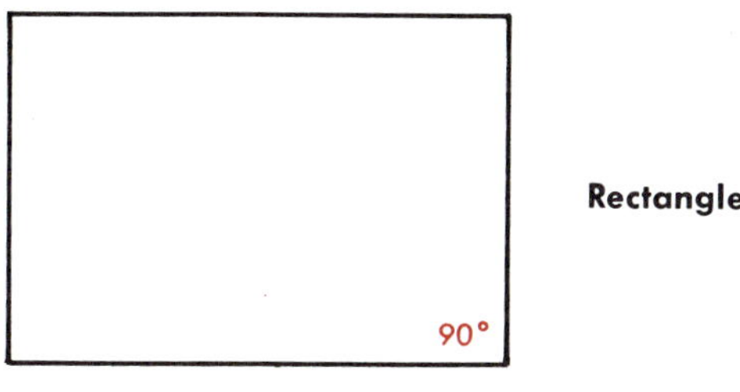

Rectangle

Figure 9–7

INDUCTIVE PROCEDURE WITH RECTANGLES

1. Construct models of several rectangles of different shapes and draw their diagonals.
2. Examine the rectangles for special properties not present in the general parallelogram.
3. State your conjectures and their converses.

The following propositions and their converses, selected from possible relationships developed by the inductive process, should be deductively proved by the student. This does not mean that the student cannot or should not attempt to prove other conjectures suggested by inductive procedures.

9.12 THEOREM

All angles of a rectangle are right angles. (Use 9.04.)

Will the diagonals of a rectangle have the same properties as the diagonals of a parallelogram? Do the diagonals of a rectangle have any special properties? Let us test a conjecture.

Given: $ABCD$ is a rectangle with diagonals $\overline{AC}$ and $\overline{BD}$.

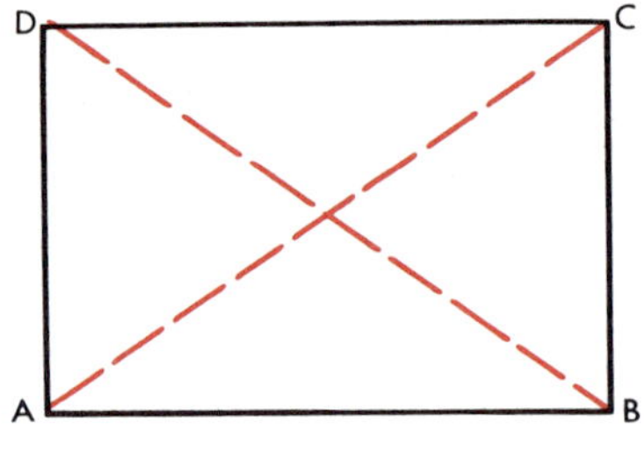

Figure 9–8

Conjecture: $\overline{AC} \cong \overline{BD}$

Plan: Prove that $\triangle ABC \cong \triangle BAD$.

Proof: The proof is left to the student.

9.13 THEOREM

The diagonals of a rectangle are congruent. (See the coordinate proof on page 558.)

9.14 (Optional) THEOREM

If the diagonals of a parallelogram are congruent, the figure is a rectangle. (See problem 2, page 559.)

(*Suggestion*: Use the figure for 9.13. Prove that $\triangle ABC \cong \triangle BAD$; then apply 5.26 and 9.11.)

Exercises

1. Are the diagonals of a rectangle congruent?
2. Are the diagonals of a rectangle perpendicular?
3. Do the diagonals of a rectangle bisect each other?
4. Do the diagonals of a rectangle bisect the angles?
5. Do the diagonals of a rectangle divide the rectangle into four congruent triangles?
6. If two consecutive angles of a parallelogram are congruent, is it a rectangle? Explain.
7. Construct a rectangle whose adjacent sides are 1 and 2 inches in length. Would two rectangles having these conditions necessarily be congruent?
8. Construct a rectangle, the length of whose diagonal is 3 inches. Would two rectangles having these conditions necessarily be congruent?
9. Are the four non-overlapping triangles, formed by the sides and diagonals of a rectangle, isosceles triangles? Prove your response.
10. In quadrilateral $ABCD$, $\overline{DA} \perp \overline{AB}$, $\overline{BC} \perp \overline{DC}$, and $\angle DCA \cong \angle BAC$. Is $ABCD$ a rectangle? Prove your response.
11. Construct a rectangle with a side $1\frac{1}{4}$ inches in length and a diagonal $2\frac{1}{4}$ inches in length. Would two rectangles drawn using these conditions necessarily be congruent?
12. Construct a rectangle whose diagonals are 4 inches in length and perpendicular to each other. Does this rectangle have any special properties?
13. If the diagonals of any quadrilateral are congruent, is the figure necessarily a rectangle?

9.15 A rhombus is a parallelogram with two adjacent sides congruent.

If a rhombus is a parallelogram, must the relationships of 9.00–9.10 be true for a rhombus?

INDUCTIVE PROCEDURE WITH RHOMBUSES

1. Construct several rhombuses of varied shapes and draw their diagonals.
2. Examine the rhombuses for special properties not present in the general parallelogram.
3. State any possible theorems and their converses.
4. Construct two segments, each being the perpendicular bisector of the other. Connect the ends of the segments. Find the lengths of the sides of the resulting quadrilateral. Does the figure seem to be a rhombus? Study the theorem of 8.13 in connection with the conditions imposed upon this figure.

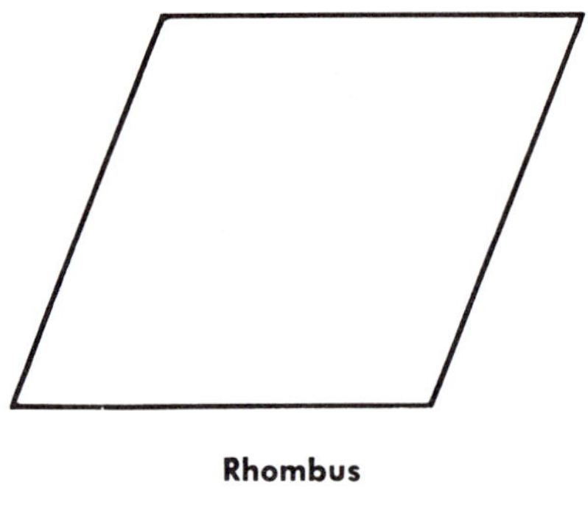

Rhombus

Figure 9–9

9.16 THEOREM

All sides of a rhombus are congruent. (Use 9.03.)

9.17 A square is a rectangle with two congruent adjacent sides.

The following proof is listed to show the use of a previous theorem to simplify our work.

Given: Rhombus $ABCD$ with diagonals $\overline{AC}$ and $\overline{BD}$

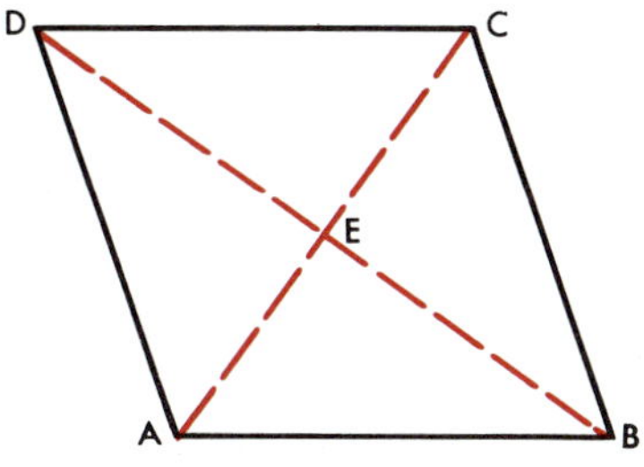

Figure 9–10

Conjecture: $\overline{AC} \perp \overline{BD}$ and $\overline{AC}$ bisects $\angle A$ and $\angle C$, $\overline{BD}$ bisects $\angle B$ and $\angle D$.

Plan: Use 8.12.

Proof: *Statements*	*Reasons*
1. $\overline{AB} \cong \overline{BC} \cong \overline{CD} \cong \overline{DA}$	1. Why?
2. $\overline{AC}$ and $\overline{BD}$ bisect each other.	2. Why?
3. $\overline{AC} \perp \overline{BD}$, $\overline{AC}$ bisects $\angle A$ and $\angle C$, and $\overline{BD}$ bisects $\angle B$ and $\angle D$.	3. Would it be best to use 8.12 or 8.14?

9.18 THEOREM

The diagonals of a rhombus are perpendicular to each other and bisect the angles of the rhombus. (See problem 7, page 559.)

Exercises

1. Is a square a rhombus? Explain.
2. Is a square a parallelogram? Explain.
3. Explain why all sides of a square must be congruent.
4. Name the properties of a square. Can you find ten properties of a square?
5. What kind of a quadrilateral is equilateral but not necessarily equiangular?

6. State the converse of the following statement and determine the truth of the converse: All theorems about a parallelogram are also theorems which apply to a rhombus.

7. Show by diagram the relationships of the set of quadrilaterals and the following subsets: Parallelograms, rectangles, squares, and rhombuses.

8. Construct a rhombus, the lengths of whose diagonals are $1\frac{1}{2}$ inches and 2 inches.

9. The perimeter of a rhombus is 6 inches. The measure of an angle of this rhombus is 45°. Use only compass and straight edge to construct this rhombus.

10. If the diagonals of a rectangle are perpendicular, is the rectangle a square? Prove your response.

11. If the midpoints of the four halves of the diagonals of a parallelogram are joined in order, is another parallelogram formed? Prove your response.

TRAPEZOIDS

Now we shall define a trapezoid and some of its related parts.

9.19 A trapezoid is a quadrilateral having only one pair of parallel sides called the *bases*. The nonparallel sides are sometimes called legs.

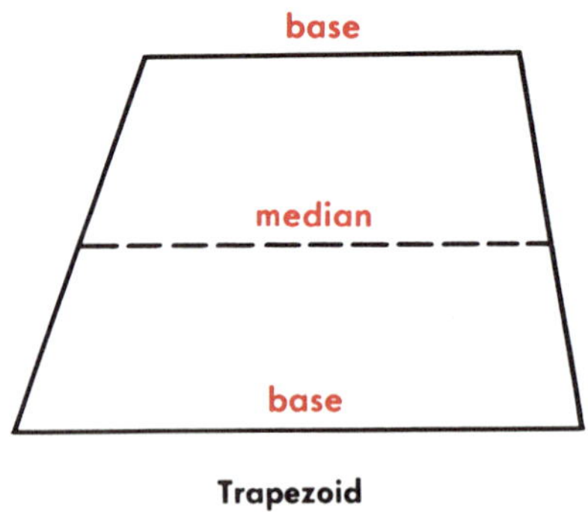

Trapezoid

Figure 9–11

9.20 The altitude of a trapezoid is a segment with an endpoint in each base and perpendicular to one base. The measure of the altitude is called the *height.*

9.21 The median of a trapezoid is the segment joining the midpoints of the nonparallel sides.

9.22 An isosceles trapezoid has congruent nonparallel sides.

INDUCTIVE PROCEDURE WITH TRAPEZOIDS

1. Construct several trapezoids, both isosceles and non-isosceles. Draw their diagonals and construct their medians.
2. Measure the angles, the diagonals, the medians, and the sides of the trapezoids.
3. Examine your data for possible recurring relationships and make a list of conjectures and their converses.
4. Determine the truth of your conjectures by deductive reasoning.

(*The following theorems were selected for proof in this text because of their general importance.*)

Given: Trapezoid $ABCD$, with $\overline{AD} \cong \overline{BC}$

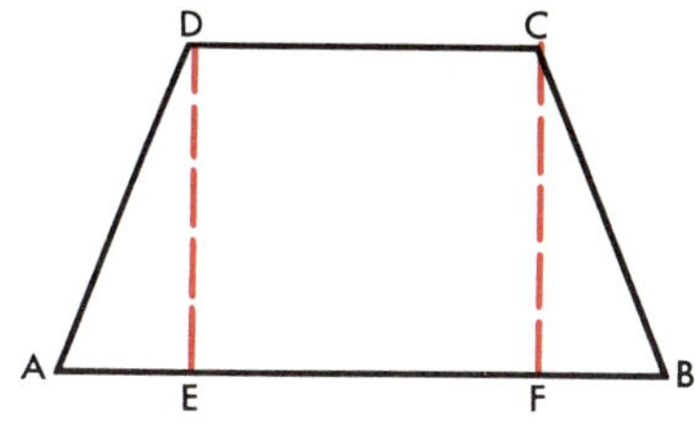

Figure 9–12

Conjecture: $\angle A \cong \angle B$ and $\angle D \cong \angle C$

Plan: Construct $\overline{DE}$ and $\overline{CF} \perp \overline{AB}$. Prove that $\triangle AED \cong \triangle BFC$.

Proof: The proof is left to the student.

9.23 THEOREM

The base angles of an isosceles trapezoid are congruent.

9.24 THEOREM

The diagonals of an isosceles trapezoid are congruent. (See problem 5, page 559.)

(The proof is left to the student.)

9.25 (Optional) THEOREM

If a trapezoid has congruent base angles or congruent diagonals, it is an isosceles trapezoid.

Exercises

1. Would it have been acceptable to have defined a trapezoid as "a quadrilateral with *at least one* pair of parallel sides"?
2. What is a definition?
3. If we had used the definition in Exercise 1, would this change any of our theorems?
4. Using the above definition, do all trapezoids become members of the set of parallelograms?
5. Determine the truth of the converse of your conjecture in Exercise 4.
6. Prove that a segment parallel to the base of an isosceles triangle and whose endpoints lie in the congruent sides form, with the base and sides, an isosceles trapezoid.
7. In quadrilateral $ABCD$, $\overline{AD} \cong \overline{CB}$, and diagonals $\overline{DB} \cong \overline{AC}$.
   (a) Is this quadrilateral necessarily a parallelogram? Explain.
   (b) Is this quadrilateral necessarily a trapezoid? Explain.
   (c) If the diagonals intersect in O, is $\overline{AO} \cong \overline{OB}$? Explain.
8. Prove that the median of an isosceles trapezoid divides it into two isosceles trapezoids.

PARALLEL LINES AND CONGRUENT SEGMENTS

Draw a scalene triangle. Find the midpoints of two sides of the triangle and connect these points with a segment. Compare the length of the segment to the length of the third side. What is the relationship? State your conjecture in "If-then" form.

Given: $\triangle ABC$, with D and E the midpoints of $\overline{AB}$ and $\overline{AC}$ respectively

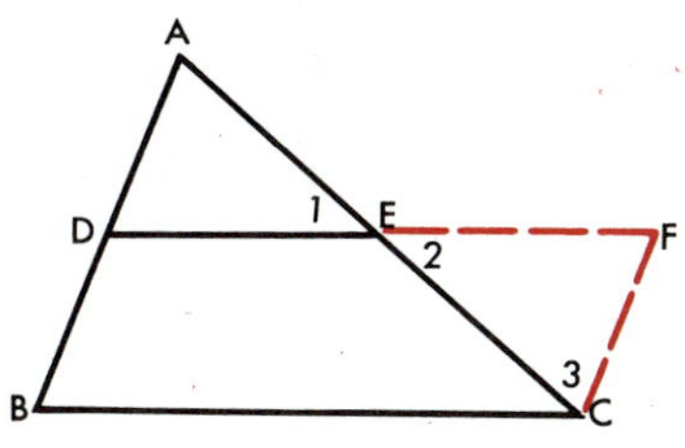

Figure 9–13

Conjecture: $\overline{DE} \parallel \overline{BC}$ and $m\overline{DE} = \frac{1}{2} m\overline{BC}$

Plan: Select F on $\overrightarrow{DE}$ so that $\overline{DE} \cong \overline{EF}$. Draw $\overline{CF}$. Prove that $DBCF$ is a parallelogram.

Proof:

Statements	*Reasons*
1. $\angle 1 \cong \angle 2$	1. Why?
2. $\overline{AE} \cong \overline{EC}$	2. Why?
3. $\overline{DE} \cong \overline{EF}$	3. Why?
4. $\therefore \triangle ADE \cong \triangle CFE$	4. Why?
5. $\angle 3 \cong \angle A$	5. Why?
6. $\overline{AB} \parallel \overline{CF}$	6. Why?
7. $\overline{AD} \cong \overline{CF}$	7. Why?
8. $\overline{AD} \cong \overline{DB}$	8. Why?
9. $\overline{DB} \cong \overline{CF}$	9. Why?
10. $\therefore DBCF$ is a parallelogram.	10. Why?
11. $\overline{DE} \parallel \overline{BC}$	11. Why?
12. $\overline{DF} \cong \overline{BC}$	12. Why?
13. $m\overline{DE} = \frac{1}{2} m\overline{DF}$	13. Why?
14. $m\overline{DE} = \frac{1}{2} m\overline{BC}$	14. Why?

Have we clearly proved the following theorem?

9.26 THEOREM

A segment joining the midpoints of two sides of a triangle is parallel to the third side and its length is one-half the length of the third side. **(See the coordinate proof, page 557.)**

Theorem 9.26 leads to the proof of some properties of the median of a trapezoid.

Given: Trapezoid $ABCD$, with median $\overline{EF}$

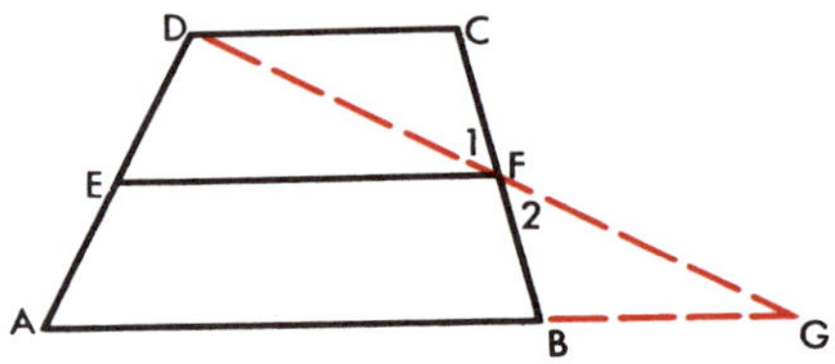

Figure 9–14

Conjecture: $\overline{EF} \parallel \overline{AB}$ and $\overline{CD}$. $m\overline{EF} = \frac{1}{2}(m\overline{AB} + m\overline{CD})$.

Plan: Draw $\overrightarrow{DF}$ to intersect $\overrightarrow{AB}$ at G. Prove that F is the midpoint of $\overline{GD}$. Then, by 9.26, $\overline{EF} \parallel \overline{AG}$ and $m\overline{EF} = \frac{1}{2} m\overline{AG} = \frac{1}{2}(m\overline{AB} + m\overline{BG})$. Prove that $m\overline{BG} = m\overline{DC}$.

Proof: The proof is left to the student.

9.27 THEOREM

The median of a trapezoid is parallel to the bases and its length is one-half the sum of their lengths.

Exercises

1. The lengths of the sides of a triangle are 4, 6, and 8 inches. If segments join the midpoints of the sides, find the measure of the sides of each of the four triangles thus formed. Are these four triangles congruent?
2. The lengths of the bases of a trapezoid are 8 and 12 inches. Find the length of the median.
3. The measure of the median and one base of a trapezoid are 7 and 5 centimeters respectively. Find the length of the other base.

4. Copy the chart below. Put check marks in appropriate spaces.

Relationships	*Opposite sides*		*All sides*	*Diagonals bisect*		*Opposite* $\angle$s	*Diagonals*	
	$\cong$	$\parallel$	$\cong$	*each other*	*the* $\angle$*s*	*congruent*	$\perp$	$\cong$
Parallelogram								
Rhombus								
Square	DO NOT MARK ON THIS BOOK							
Rectangle								
Trapezoid								
Isosceles Trapezoid								

5. Is a square necessarily a rectangle—a parallelogram—a rhombus?

6. Is a rectangle necessarily a square—a parallelogram—a rhombus?

7. Is a rhombus necessarily a square—a parallelogram—a rectangle?

8. List the members of the set of equiangular quadrilaterals. List the members of the set of equilateral quadrilaterals.

9. $ABCD$ is an isosceles trapezoid, $\overline{EF}$ is the median and $m\overline{CF} =$ 2 inches. Find $m\overline{AE}$. Explain. (Use Figure 9-15.)

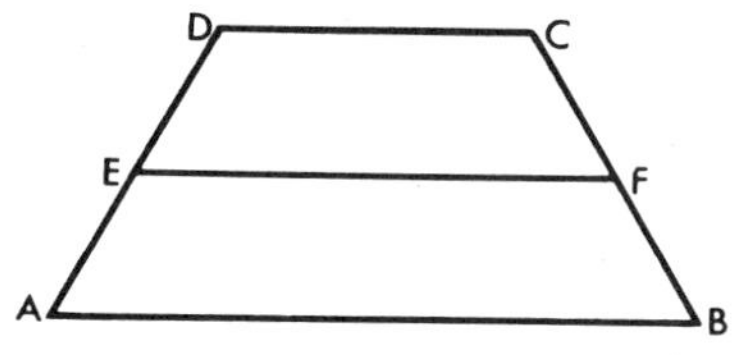

Figure 9–15

10. The length of the nonparallel sides of a trapezoid are 5 and 6 inches and the parallel sides are 10 and 14 inches. Find the length of the median.

11. Do the segments that connect the midpoints of the sides of an equilateral triangle form another equilateral triangle? If so, prove it.

12. Do the segments that connect the midpoints of the sides of a plane quadrilateral form a parallelogram? If so, prove it.

Draw a line and mark off three or more congruent segments on this line. At the ends of each segment construct perpendicular lines. It appears that the lines drawn perpendicular to the same line on our paper are parallel, and the distances between any two successive parallel lines are equal.

Draw several transversals of these parallel lines. Measure the segments of the transversals cut off by the parallel lines. Are the measures equal? Can you be sure your result will be the same every time you try this experiment?

Let us examine this idea deductively.

Given: The parallel lines l_1, l_2, l_3, and l_4, intersecting the equal segments $\overline{EF}$, $\overline{FG}$, and $\overline{GH}$ on transversal $\overleftrightarrow{AB}$, and also intersecting the transversal $\overleftrightarrow{CD}$ at points I, J, K and M.

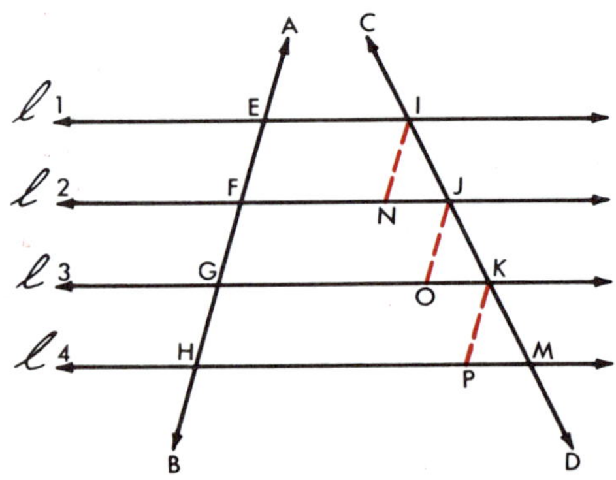

Figure 9–16

Conjecture: $\overline{IJ} \cong \overline{JK} \cong \overline{KM}$

Plan: Construct $\overline{IN}, \overline{JO}, \overline{KP}$ each $\parallel$ $\overleftrightarrow{AB}$. Show $\overline{IN} \cong \overline{EF}, \overline{JO} \cong \overline{FG}$, $\overline{KP} \cong \overline{GH}$ so that $\overline{IN} \cong \overline{JO} \cong \overline{KP}$. Then show $\triangle INJ \cong \triangle JOK \cong \triangle KPM$ so that $\overline{IJ} \cong \overline{JK} \cong \overline{KM}$.

Proof: The proof is left to the student. When the proof is completed is the following theorem justified?

9.28 THEOREM

If three or more parallel lines cut off congruent segments on one transversal, they cut off congruent segments on every transversal.

9.29 Construction problem: To divide a line segment into any number of congruent parts.

Given: Line segment $\overline{AB}$

Problem: To divide $\overline{AB}$ into any number of congruent parts (five for example).

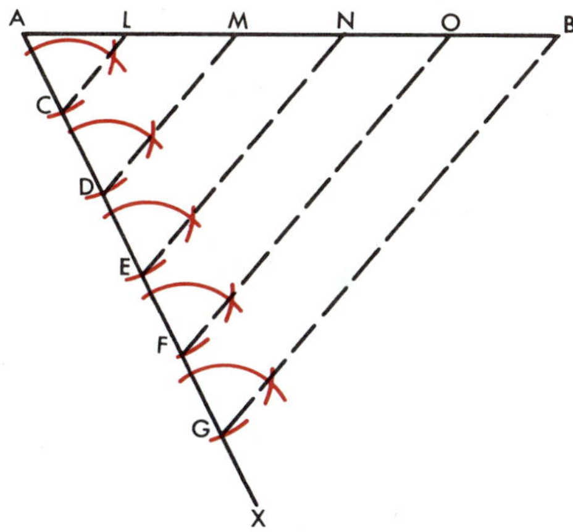

Figure 9–17

Construction:

1. Draw $\overrightarrow{AX}$ making a convenient angle with $\overrightarrow{AB}$.
2. From A, with any radius, mark off five congruent parts in succession on $\overrightarrow{AX}$ so that $\overline{AC} \cong \overline{CD} \cong \overline{DE} \cong \overline{EF} \cong \overline{FG}$.
3. Draw $\overline{BG}$ and through C, D, E, and F construct lines parallel to $\overline{BG}$. The intersections of $\overline{AB}$ and the parallels (L, M, N. O) are the required points of division.

Why *must* $\overline{AL}$, $\overline{LM}$, $\overline{MN}$, $\overline{NO}$, and $\overline{OB}$ be congruent if the instructions are followed precisely? Which theorem or corollary applies here?

Theorems 9.30 and 9.31 are easily established by using Theorem 9.28.

9.30 THEOREM

If a line is parallel to one base of a trapezoid and bisects one of the nonparallel sides, it bisects the other also.

9.31 THEOREM

If a line is parallel to one side of a triangle and bisects another side, it bisects the third side also.

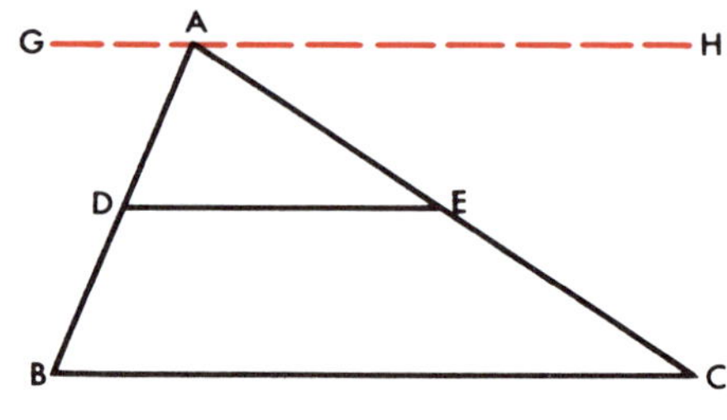

Figure 9–18

Hint: Draw $\overleftrightarrow{GH} \parallel \overline{BC}$ through A. Use 9.28.

Exercises

1. Complete the construction problem of 9.29 on your paper.
2. Divide a segment into three congruent parts using 9.29.
3. Draw a scalene triangle and trisect two sides. Connect the respective points of trisection with line segments and compare their lengths with the length of the third side. State a conjecture. Is your conjecture the result of inductive, or deductive reasoning?
4. Using the construction method of 9.29, divide a given segment into two parts whose measures have the ratio 2:3.
5. Four sides and one diagonal of a quadrilateral are given. Construct the quadrilateral.
6. Given: $\overline{CF} \cong \overline{DE}$, $\overline{DF} \cong \overline{CE}$. Is $\angle 8 \cong \angle 5$? Is $\angle 1 \cong \angle 4$? Is $\overline{CG} \cong \overline{EG}$? Is $\angle 7 \cong \angle 6$? Is $\overline{FG} \cong \overline{EG}$? Is $\overline{FE} \parallel \overline{CD}$?

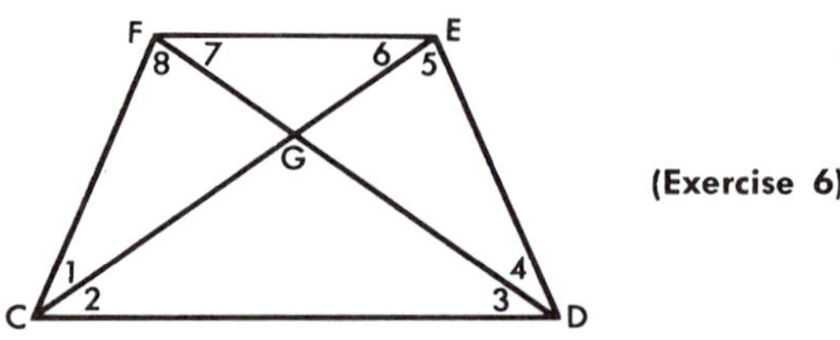

(Exercise 6)

Figure 9–19

7. Explain how you could use a sheet of lined notebook paper to divide certain segments into congruent parts.
8. Draw a segment across your paper. Construct a rhombus whose perimeter is equal to the length of this segment and which has an angle whose measure is 60°.
9. Given the positions of the midpoints of the three sides of a triangle, construct the triangle.
10. Explain why the lines joining two opposite vertices of a parallelogram to the midpoints of two parallel sides divide into three congruent segments the diagonal joining the other two vertices.

SYMMETRY (OPTIONAL)

Some of the most pleasing geometric forms, either man-made or formed by nature, are the symmetric figures. Symmetry is one of the basic laws of design and to most of us it means "balance."

9.32 In geometry, a figure is symmetric if it is balanced with respect to a point, line, or plane. Most animals are nearly symmetric; this is most evident in a butterfly. A snowflake is symmetric about its center point.

POINT SYMMETRY

9.33 If a point can be found such that all segments drawn through the point and terminating on the figure are bisected by the point, the figure has point symmetry. The figures below have point symmetry.

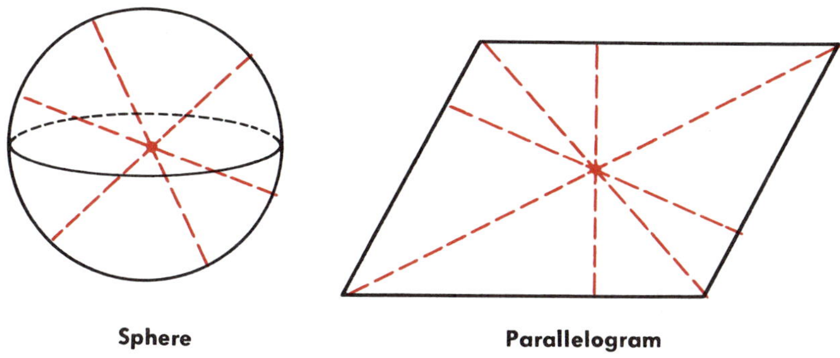

Figure 9–20

LINE SYMMETRY

9.34 A geometric figure is said to be symmetric with respect to a line as an axis if this line bisects all segments perpendicular to the line and terminated by the figure. This is sometimes called axial symmetry. A plane figure has line symmetry if there is a line along which it can be folded in such a manner that the corresponding parts will coincide. Lines *a, b,* and *c* in the figures below represent the axis or line of symmetry.

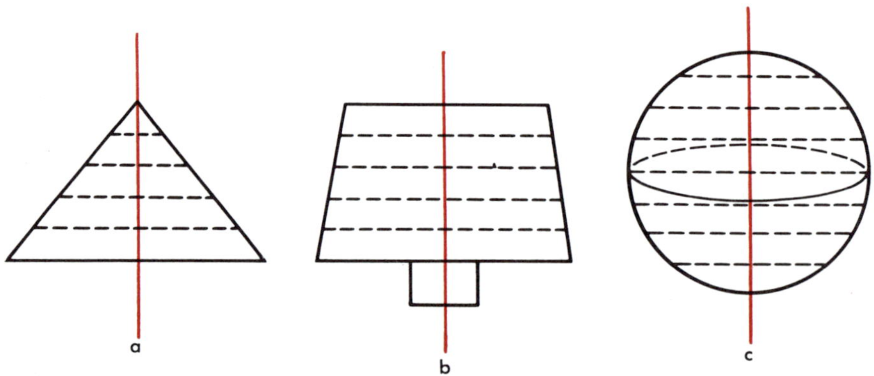

Figure 9–21

PLANE SYMMETRY

9.35 An object is said to be symmetric with respect to a plane if this plane bisects all segments perpendicular to the plane and terminated by the figure.

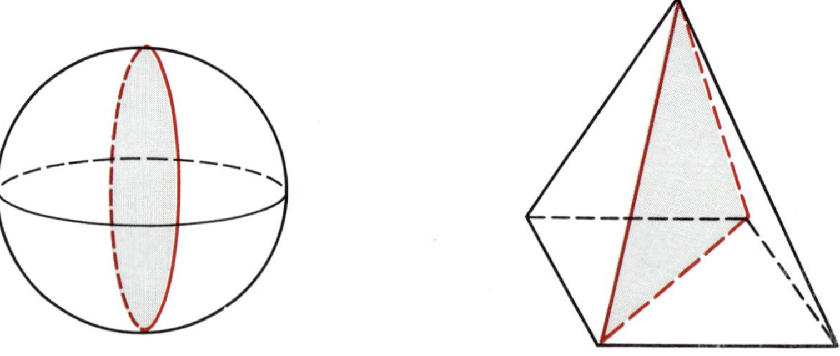

Figure 9–22

Courtesy of the AMERICAN MUSEUM OF NATURAL HISTORY

This monarch butterfly displays the symmetry found in many animals. What kind of symmetry is evident here?

Exercises

1. Where is the axis of symmetry of the letter B? Of the letter A? Of the letter D?
2. Name two letters of our alphabet which have point symmetry.
3. What type of symmetry is represented by the skeleton of the human body?
4. Name the types of symmetry represented by the following figures:

   (a) Isosceles triangle
   (b) Regular hexagon
   (c) Circle
   (d) Football

5. Prove or disprove the conjecture that "no triangle can have point symmetry."

Vocabulary List

parallelogram
rectangle
rhombus
square
median of a trapezoid
isosceles trapezoid
symmetry
axial
trapezoid

Chapter Review

1. List the methods that we may use to prove that a quadrilateral is a parallelogram.
2. Show by a Venn diagram the relationships of the set of quadrilaterals and the following subsets; parallelograms, rectangles, squares, rhombuses, and trapezoids.

Experimentally determine whether the following statements are true or false. Prove or disprove each statement.

3. The segments joining the midpoints of the sides of a rectangle form a rhombus.
4. The perpendiculars to the third side of a triangle from the midpoints of two sides are congruent.
5. The median of a trapezoid bisects each diagonal.
6. The diagonals of an isosceles trapezoid bisect the angles of the trapezoid.
7. If parallels to the congruent sides of an isosceles triangle are drawn from any point in the base, a parallelogram is formed whose perimeter is equal to the sum of the lengths of the congruent sides.

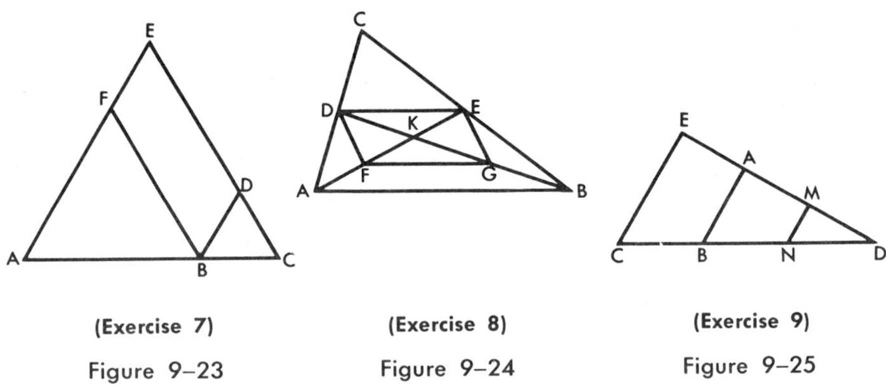

(Exercise 7) Figure 9–23

(Exercise 8) Figure 9–24

(Exercise 9) Figure 9–25

8. $\overline{AE}$ and $\overline{BD}$ are medians of $\triangle ABC$. Points F and G are the midpoints of $\overline{AK}$ and $\overline{BK}$ respectively. Therefore, $FGED$ is a parallelogram.
9. Given $\triangle CDE$ with $\overline{AB} \parallel \overline{MN} \parallel \overline{CE}$. Also $\overline{AB}$ and $\overline{MN}$ trisect $\overline{CD}$. Therefore, $m\overline{AB} + m\overline{MN} = m\overline{CE}$.

10. The length of the median of a trapezoid included between the diagonals is equal to one-half the difference between the lengths of the bases of the trapezoid.
11. The segments which connect the midpoints of two sides of a triangle to a point one-fourth the length of the third side from the nearest vertex are parallel.
12. The segments joining in order the midpoints of the sides of a trapezoid form a rhombus.
13. The line through the midpoints of the bases of an isosceles trapezoid is perpendicular to the bases.
14. If, from two opposite vertices of a parallelogram, non-intersecting line segments are drawn to the midpoints of two opposite sides, then these line segments are parallel.
15. In a right triangle the distance from the vertex of the right angle to the midpoint of the hypotenuse is equal to half the length of the hypotenuse.

Chapter 9 Test

Choose the word "always," "sometimes," or "never" to correctly complete statements 1–15.

1. The sides of a parallelogram are _?_ congruent.
2. A square is _?_ a rhombus.
3. The opposite angles of a parallelogram are _?_ congruent.
4. The diagonals of a parallelogram are _?_ congruent.
5. A square is _?_ a rectangle.
6. The diagonals of a rectangle are _?_ congruent.
7. The non-parallel sides of a trapezoid are _?_ congruent.
8. The diagonals of a trapezoid _?_ bisect each other.
9. The diagonals of a rhombus _?_ bisect the angles of the rhombus.
10. A parallelogram is _?_ a trapezoid.
11. If the diagonals of a quadrilateral are congruent it is _?_ a rectangle.

12. If a pair of opposite sides of a quadrilateral are congruent and parallel it is _?_ a parallelogram.

13. If two adjacent sides of a parallelogram are perpendicular it is _?_ a rectangle.

14. If the diagonals of a quadrilateral do not bisect each other it is _?_ a parallelogram.

15. If the diagonals of a quadrilateral are perpendicular it is _?_ a rhombus.

Beside the number of the following statements write the letter of the choice or choices that correctly completes the statement.

16. The median of a triangle always (**a**) bisects an angle (**b**) bisects a side (**c**) is perpendicular to a side of the triangle.

17. A segment joining the midpoints of two adjacent sides of any quadrilateral (**a**) is parallel to a diagonal of the quadrilateral (**b**) is equal in measure to one-half of the measure of a diagonal of the quadrilateral (**c**) is neither (a) nor (b).

18. The segments joining in order the midpoints of the sides of an isosceles trapezoid form a (**a**) square (**b**) rhombus (**c**) rectangle.

19. The median of a trapezoid (**a**) is parallel to the bases (**b**) is equal in measure to one-half the sum of the measures of the bases (**c**) bisects any line segment whose endpoints lie in the parallel bases.

20. The line joining the midpoints, A and B, of the diagonals of a trapezoid (**a**) bisects the non-parallel sides of the trapezoid (**b**) $m\overline{AB} = \frac{1}{2}$ the difference of the measures of the bases (**c**) neither (a) nor (b) is true.

10

Inequalities in Triangles And Indirect Proof

A large portion of our time is spent in dealing with the relations of equality and congruency for they are, indeed, important concepts in our world. Even more prevalent, however, is the relation of inequality.

As with equalities and congruences, we must agree upon a basic set of rules (assumptions) to deal with the study of inequalities. In the development of this set of assumptions, the following symbols will be used:

$>$ means "is greater than."
$<$ means "is less than."

For example, $a > b$ means "a is greater than b;" $c < d$ means "c is less than d." The symbol always points to the smaller term. If a and b represent real numbers, then $a > b$ means that $a = b + k$ where k is a positive real number.

10.00 Two inequalities are of the same order if in each case the symbols point in the same direction, and of the opposite order if they do not.

Same Order		*Opposite Order*
a. $5 < 10$ $3 < 8$	b. $10 > 5$ $8 > 3$	c. $10 > 5$ $3 < 8$

INDUCTIVE APPROACH TO INEQUALITIES

Addition. Let us experiment with inequalities. If we add equals to unequals are the results equal or unequal?

a. Since $5 < 8$ and $6 = 6$, is $5 + 6 < 8 + 6$?

b. Since $7 > 3$ and $4 = 4$, is $7 + 4 > 3 + 4$?

c. If $A > B$ and $C = C$, is $A + C > B + C$? (A, B, and C are real numbers.)

Are the sums unequal in the same order as the original inequalities, or in the opposite order? Would this always be correct even if negative numbers were used?

What is the result when we add unequals to unequals?

Test the conclusions that you draw from the following questions by writing other examples of adding inequalities.

a. Since $2 < 9$ and $5 < 7$, is $2 + 5 < 9 + 7$?

b. Since $9 > 3$ and $2 < 7$, is $9 + 2 < 3 + 7$?

c. Since $9 > 2$ and $5 < 7$, is $9 + 5 > 2 + 7$?

d. Since $7 > 3$ and $2 > 1$, is $7 + 2 > 3 + 1$?

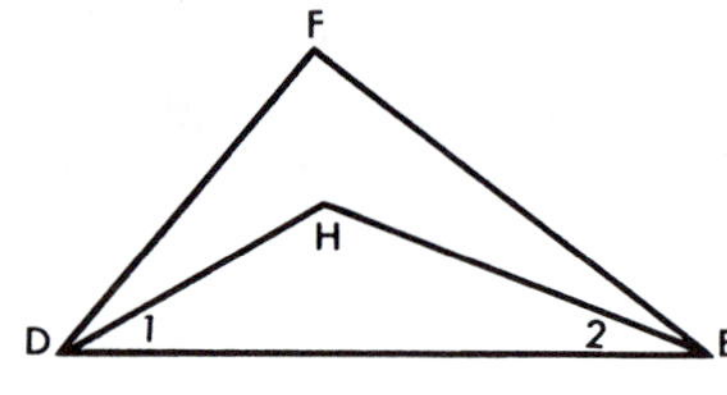

Figure 10–1

If $m\angle FDH > m\angle FEH$ and $m\angle 1 > m\angle 2$, is $m\angle FDE$ greater than, or less than $m\angle FED$? Why? See Figure 10-1.

Do you see a relationship that exists when we add inequalities to inequalities, or when equalities and inequalities are added? Does the order of the inequalities matter?

Subtraction. If we subtract equals from unequals are the results unequal in the same or in the opposite order? Try this with negative numbers.

a. Since $7 > 4$ and $5 = 5$, is $7 - 5 > 4 - 5$?
b. Since $5 < 10$ and $12 = 12$, is $5 - 12 < 10 - 12$?
c. Since $3 < 12$ and $2 = 2$, is $3 - 2 < 12 - 2$?
d. If $A > B$ and $C = C$, is $A - C > B - C$? (A, B, and C are real numbers.)

Is there a recurrent relationship evident when unequals are subtracted from equals? What is your conjecture from the following experiments?

a. Since $8 = 8$ and $5 > 2$, is $8 - 5 < 8 - 2$?

b. Since $14 = 14$ and $2 > -5$, is $14 - 2 < 14 - (-5)$?

c. Since $10 = 10$ and $7 < 10$, is $10 - 7 > 10 - 10$?

d. Since $9 = 9$ and $12 > 10$, is $9 - 12 < 9 - 10$?

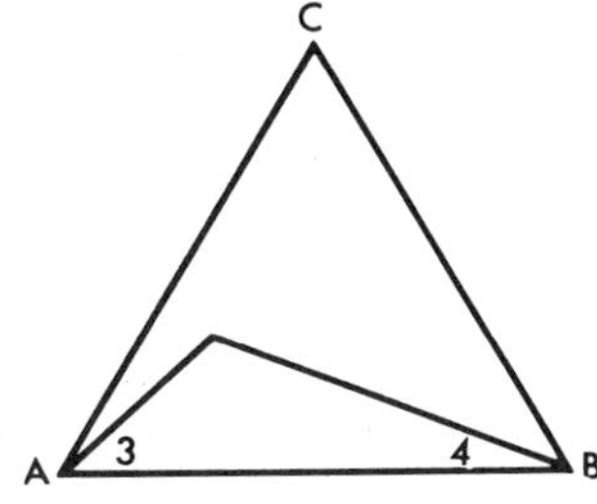

Figure 10-2

If $m\angle CAB = m\angle CBA$ and $m\angle 3 > m\angle 4$, is $m\angle CAB - m\angle 3$ greater than, or less than $m\angle CBA - m\angle 4$? See Figure 10-2.

State the various relationships that are evident to you from the preceding examples. Have we proved these relationships, or are they merely conjectures?

Multiplication and Division. Examine the following multiplication and division of equalities and inequalities:

Dividing unequals by equals: Is $\frac{9}{3} > \frac{6}{3}$? Is $\frac{4}{2} < \frac{8}{2}$?

If $A > B$, is $\frac{A}{C} > \frac{B}{C}$? ($A$, B, and C are real numbers, $C > 0$)

Multiplying: Is $(5)(2) > (3)(2)$? Is $(6)(9) < (8)(9)$?
Let A, B, and C be real numbers. If $A > B$ and $C = C$, is $AC > BC$? Can C be a negative number? Zero?

Is there a recurring relationship here? If so, state it. Is the relationship the same if negative numbers are used?

Multiplying inequalities by inequalities:

a. Since $5 > 3$ and $4 > 2$, is $(5)(4) > (3)(2)$?

b. Since $8 > 2$ and $6 > 2$, is $(8)(6) > (2)(2)$?

c. Since $5 > 3$ and $2 < 4$, is $(5)(2) < (3)(4)$?

d. Since $8 > 2$ and $2 < 6$, is $(8)(2) < (2)(6)$?

Let A, B, C, and D be real numbers. If $A > B$ and $C > D$, is AC greater than, or less than BD? Is AD greater than, or less than BC?

Dividing inequalities by inequalities:

a. Since $8 > 6$ and $4 > 2$, is $\frac{8}{4} > \frac{6}{2}$?

b. Since $8 > 6$ and $4 > 3$, is $\frac{8}{4} > \frac{6}{3}$?

c. Since $24 > 20$ and $4 < 5$, is $\frac{24}{4} > \frac{20}{5}$?

d. Since $24 > 20$ and $6 > 4$, is $\frac{24}{6} > \frac{20}{4}$?

Can we establish any relationships after examining the preceding examples? Try experimenting with negative numbers. Repeat the above experiments using various combinations of positive and negative numbers. Does this change your conjectures?

After close examination we find that some relationships seem to occur repeatedly while others depend upon the value of the quantities used and thus are not consistent. This thinking yields the following axioms. Can you add others?

AXIOMS OF INEQUALITY

Let a, b, and c denote real numbers. Remember that $a > b$ means that $a = b + k$ where k is a positive real number.

10.01 *Transitive Axiom: If $a > b$ and $b > c$, then $a > c$, or if $a < b$ and $b < c$, then $a < c$.*

If the first of three numbers is greater than the second and the second is greater than the third, then the first is greater than the third. You may replace "greater than" by "less than."

10.02 *Addition Axiom: If $a > b$, then $a + c > b + c$, or if $a < b$, then $a + c < b + c$. Also, if $a > b$ and $c > d$, then $a + c > b + d$.*

A. If equals are added to unequals, the results are unequal in the same order.

B. If unequals are added to unequals of the same order, the sums are unequal in the same order.

10.03 *Subtraction Axiom: If* $a > b$, *then* $a - c > b - c$ *and* $c - a < c - b$. *Also if* $a < b$, *then* $a - c < b - c$ *and* $c - a > c - b$.

A. If equals are subtracted from unequals, the results are unequal in the same order.

B. If unequals are subtracted from equals, the results are unequal in the opposite order.

10.04 *Multiplication Axiom: If* $a > b$ *and* $c > 0$, *then* $ac > bc$. *Also if* $a > b$ *and* $c < 0$, *then* $ac < bc$.

A. If unequals are multiplied by positive equals, the results are unequal in the same order.

B. If unequals are multiplied by negative equals, the results are unequal in the opposite order.

10.05 *Division Axiom: If* $a > b$ *and* $c > 0$, *then* $\frac{a}{c} > \frac{b}{c}$.

$$\left(\text{Also if } a > b > 0, \text{ then } \frac{c}{a} < \frac{c}{b}.\right)$$

If $a > b$ *and* $c < 0$, *then* $\frac{a}{c} < \frac{b}{c}$.

$$\left(\text{Also if } a > b > 0, \text{ then } \frac{c}{a} > \frac{c}{b}.\right)$$

A. If unequals are divided by positive non-zero equals, the results are unequal in the same order.

B. If unequals are divided by negative equals, the results are unequal in the opposite order.

10.06 *Whole-parts Axiom: If* $a \geq b + c$, *where* a, b, *and* c *are positive real numbers, then* $a > b$ *and* $a > c$.

10.07 *Comparison Axiom: For any two real numbers* a *and* b, *either* $a > b$, $a = b$, *or* $a < b$.

Of two real numbers, the first is greater than, equal to, or less than the second.

Why did we not include a reflexive axiom and an axiom of symmetry for inequality?

Exercises

Supply the missing symbol if the relationship can be determined; otherwise, place a question mark on your paper. Which axiom is your authority? Do not mark on the text! (a, b, c, d, x, and y represent real numbers.)

1. If $a > b$ and $c = d$, then $ac \underline{?} bd$.
2. If $a > b$ and $b > c$, then $a \underline{?} c$.
3. If $a = 2x$, $b = 2y$, and $x > y$, then $a \underline{?} b$.
4. If $a = b$ and $c < d$, then $a + c \underline{?} b + d$.
5. If $a = b$ and $c < d$, then $a - c \underline{?} b - d$.
6. If $a = \frac{1}{2}b$ and $c = \frac{1}{2}d$, then $a \underline{?} c$.
7. If $a = b$ and $c > d$, then $\frac{c}{a} \underline{?} \frac{d}{b}$.
8. If $a > b$ and $c > d$, then $a + c \underline{?} b + d$.
9. If $a < b$ and $c < d$, then $a + c \underline{?} b + d$.
10. If $a < b$ and $x < y$, then $a - b \underline{?} x - y$.

Use Figure 10-3 for Exercises 11–14.

11. If $m\angle A > m\angle C$ and $m\angle 1 = m\angle 3$, then $m\angle 2 \underline{?} m\angle 4$.
12. If $m\angle 1 < m\angle 3$ and $m\angle 2 < m\angle 4$, then $m\angle A \underline{?} m\angle C$.

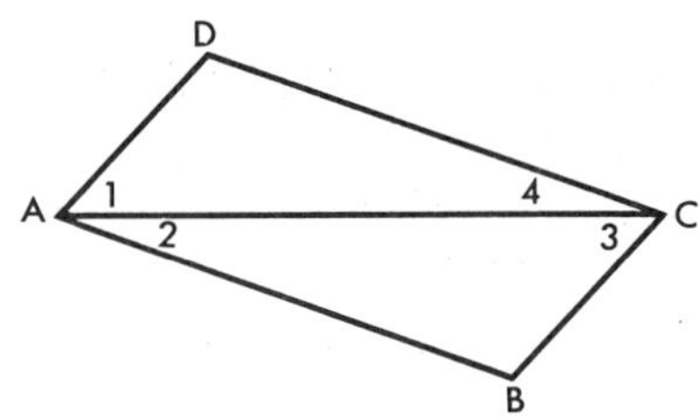

(Exercises 11-14)

Figure 10–3

13. If $m\angle A = m\angle C$ and $m\angle 1 > m\angle 3$, then $m\angle 2 \underline{?} m\angle 4$.
14. $m\angle A \underline{?} m\angle 1$.
15. Is the angle formed by the bisectors of two angles of a triangle obtuse? Prove your response.

INEQUALITY RELATIONSHIPS OF ANGLES

Is the sum of the measures of the exterior angles of a triangle greater than, less than, or equal to the sum of the measures of the angles of the triangle? Does this relationship hold true for quadrilateral? For a pentagon? Refer to Sections 6.19, 6.21, and 6.23 to review some of the relationships of angles and exterior angles of polygons.

Consider one exterior angle of a triangle. What is its relationship to the adjacent angle of the triangle? What is the relationship of this exterior angle to one nonadjacent angle of the triangle? To both nonadjacent angles of the triangle? Can you prove these conjectures? (Figure 10-4)

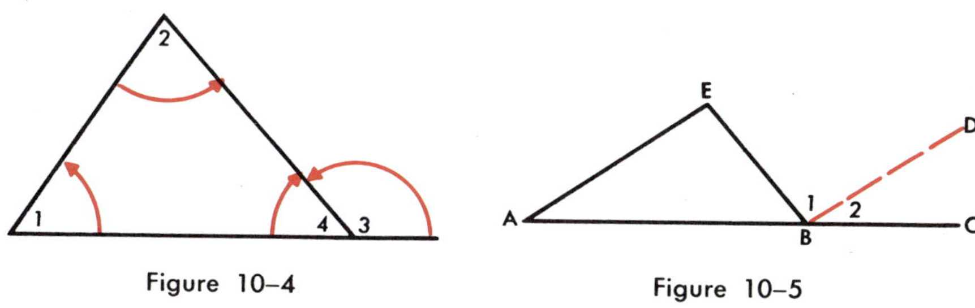

Figure 10–4

Figure 10–5

If you cannot think of a method of proof, the figures and accompanying suggestions may help. See Figure 10-5.

Given: $\triangle ABE$, with exterior $\angle CBE$

Construction: $\overline{DB} \parallel \overline{AE}$

Conjecture: $m\angle CBE = m\angle A + m\angle E$

Proof: The proof is left to the student.

Do you agree with the following general statement of the theorem?

10.08 THEOREM

The measure of an exterior angle of a triangle is equal to the sum of the measures of the nonadjacent angles of the triangle.

Theorem 10.08 states a relationship between the measure of the exterior angle of a triangle and the sum of the measures of the non-adjacent angles. How does the measure of the exterior angle compare to the measure of one nonadjacent angle? Test your conjecture in the following proof.

Given: $\triangle ABC$ with exterior $\angle ACD$

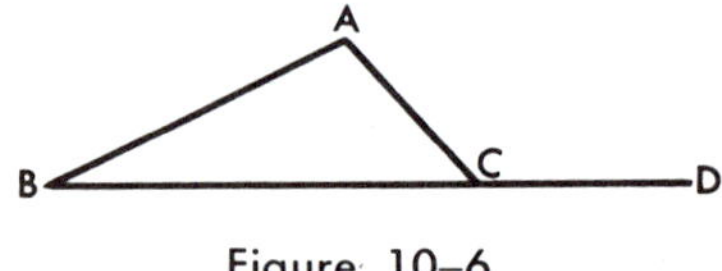

Figure 10–6

Conjecture: $m\angle ACD > m\angle B$; $m\angle ACD > m\angle A$

Plan: $m\angle ACD = m\angle A + m\angle B$. Why? Therefore, $m\angle ACD > m\angle A$ or $m\angle ACD > m\angle B$. Why?

Proof: The proof is left to the student.

10.09 THEOREM

The measure of an exterior angle of a triangle is greater than the measures of either of the nonadjacent angles of the triangle.

Exercises

1. The measure of an exterior angle of the vertex of an isosceles triangle is 100°. What is the sum of the measures of the base angles of the triangle?
2. The measure of an exterior angle at the base of an isosceles triangle is 150°. Find the measure of the vertex angle.
3. Given: $\overleftrightarrow{AOB}$ and $\overleftrightarrow{DCB}$ with $\overline{BC} \cong \overline{OC} \cong \overline{OD}$. Find $m\angle AOD$ if $m\angle B = 40°$. If $m\angle B = x°$. (Figure 10-7)

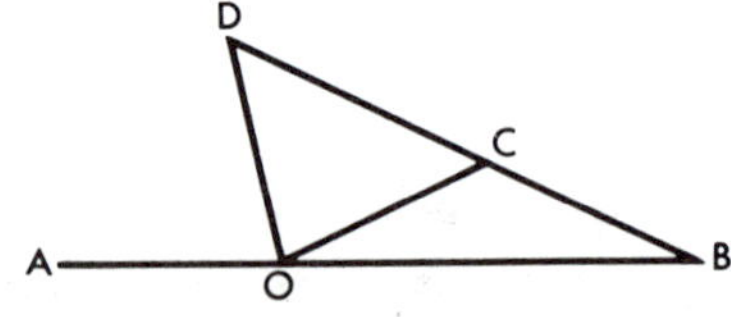

Figure 10–7

4. The measure of an exterior angle of a triangle is 110° and the measure of one nonadjacent angle is 31°. Find the measures of the other two angles of the triangle.

5. Is the bisector of the exterior angle at the vertex of an isosceles triangle parallel to the base? Prove your answer.

6. From any point in the interior of $\triangle ABC$ draw $\overline{DA}$ and $\overline{DB}$. Prove $m\angle ADB > m\angle ACB$. (*Hint:* Draw $\overrightarrow{AD}$ to intersect $\overline{BC}$ at E.)

INEQUALITIES IN A TRIANGLE

By definition, the measures of the sides of a scalene triangle are unequal. Must the measures of the angles also be unequal? In what order does the inequality seem to occur?

It appears that the measures of the angles of a scalene triangle are unequal in the same order as the measures of the opposite sides. Let us try to prove deductively that this is true.

Given: $\triangle ABC$ with $m\overline{AC} > m\overline{AB}$

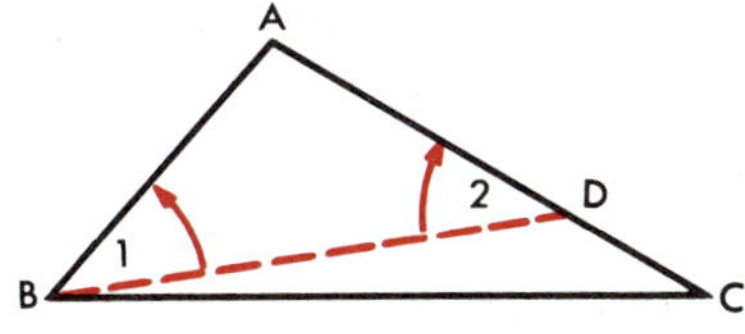

Figure 10–8

Conjecture: $m\angle B > m\angle C$

Plan: Draw $\overline{BD}$ so $\overline{AB} \cong \overline{AD}$. Show $m\angle 2 > m\angle C$ and $m\angle B > m\angle 2$.

Proof: The proof is left to the student and the general statement of the theorem follows:

10.10 THEOREM

If the measures of two sides of a triangle are unequal, the measures of the angles opposite those sides are unequal in the same order.

INDIRECT REASONING

Indirect reasoning is a useful method of proof, particularly when dealing with inequalities.

We often use indirect reasoning. For example, we say, "The basketball game must be over; the crowd is leaving the fieldhouse." We have not actually seen the game end, but since the crowd is leaving the fieldhouse, we are led to believe that the game is over by rejecting the other possible solutions. Other examples are "The neighbors must be home; their lights are on;" or "The car won't start. The trouble must be in the carburetor. The gas tank is full, the spark plugs are functioning properly, and the battery 'turns the motor' easily."

This type of reasoning is common in the fields of law and criminology. For example, a crime is committed and Mr. X is implicated. There are only two possibilities: (1) He did commit the crime, or (2) he did not. If Mr. X can show that the premise that he committed the crime is false, he has proved indirectly that the remaining possibility is true. He did not commit the crime charged against him.

Indirect reasoning is a variation of the deductive method. This consists in *examining all the possible solutions* of a given problem and in *showing that all but one are false* or impossible. The remaining solution, therefore, must be true—if a solution exists. Review "contradiction," page 91.

10.11 ***Assumption: If a conjecture leads to a contradiction of an assumption or a theorem, the conjecture must be false.***

EXAMPLE

Examine the following problem in indirect reasoning.

Three applicants for the position of assistant to a scientist gather in his office. As a test the scientist fastens a small disc of paper to each applicant's forehead. He then announces that the discs are either red or white and that as soon as the applicant sees a red disc he is to come to the door of the inner office. The first man who can prove whether his own disc is red or white, without removing it, will win the job. (No mirrors are available.)

All three men look at each other. Then each arises and goes to the door of the inner office and hesitates. Finally one man steps forward and says that the color of his disc is red.

He reasoned as follows:

1. Each one saw a red disc because they all went to the office door; therefore there must have been at least two red discs.
2. His disc is either white or red. If it had been white, one of the other two men would have realized his disc was red and claimed the job. Neither did, so his disc had to be red.

The steps involved in careful indirect reasoning are

1. State the various possibilities which can exist in the situation under consideration.
2. Assume that the possibility or possibilities other than the one you wish to prove are true; then show that this leads to a contradiction. For example, if we wish to show that two quantities are equal, we can assume that they are unequal and then show that this possibility is inconsistent with the known facts.
3. Then we may conclude that the remaining possibility is true. If the quantities cannot be unequal, then they must be equal.

The method is sometimes called *reductio ad absurdum*, which means "reduction to absurdity."

Exercises

1. A group of boys went on a picnic and, after eating, Tom, Dick, and Harry became ill. The table below shows what each boy ate. What was the probable cause of their illness?

	Milk	Candy Bar	"Hot Dogs"	Hamburger	Cheese	Water	Watermelon	"Pop"
Tom	x	x	x	x		x	x	
Bob	x		x	x	x	x		x
Bill	x	x	x			x		x
Dick		x	x	x		x	x	
John	x	x	x	x		x		x
Harry	x	x		x		x	x	

2. Three men are placed, one behind the other, facing forward. From a pile of three red hats and two black hats, a hat is placed on each man's head. Each man is unable to see the color of the hat he has or that of anyone behind him. The man in the rear is asked the color of his hat. He replies, "I don't know." The man in the middle says he doesn't know either. The front man tells the color of his hat. What color is it?

INDIRECT PROOF

We have had some exercises on indirect reasoning and two indirect proofs have been attempted earlier. (Exercise 18, page 67, and Exercise 15, page 121.) The basis for our indirect proof is assumption 10.11.

The best form for presenting such a proof is in three parts as follows:

a. State all possibilities including the one we desire to prove.

b. Show by direct proof that each of the other possibilities leads to some contradiction.

c. The remaining possibility is then proved.

As an example, we shall attempt to prove the converse of Theorem 10.10 by the indirect method. We must show that if the measures of two angles of a triangle are unequal, the measures of the sides opposite them are unequal in the same order. Here is the proof; study it carefully to help you understand this very useful method. It is sometimes called the *Method of Exclusion*.

Given: $\triangle ABC$ with $m\angle B > m\angle C$

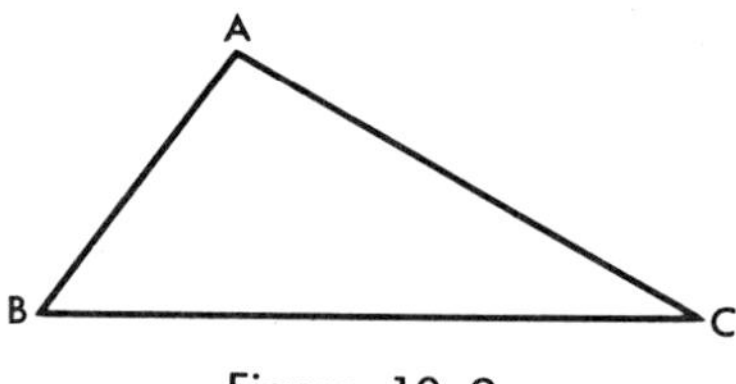

Figure 10–9

Conjecture: $m\overline{AB} < m\overline{AC}$

Plan: Use indirect proof. Show that of the three possibilities (a) $m\overline{AB} = m\overline{AC}$, (b) $m\overline{AB} > m\overline{AC}$, (c) $m\overline{AB} < m\overline{AC}$, that (a) and (b) lead to contradictions and are therefore false.

Proof: *Statements*	*Reasons*
1. There are 3 possibilities: (a) $m\overline{AB} = m\overline{AC}$ (b) $m\overline{AB} > m\overline{AC}$ (c) $m\overline{AB} < m\overline{AC}$	1. Section 10.07
2. Suppose (a) $m\overline{AB} = m\overline{AC}$	2. Statement 1(a)
3. then $m\angle B = m\angle C$	3. Why?
4. but $m\angle B > m\angle C$	4. Given
5. $\therefore m\overline{AB} = m\overline{AC}$ is false	5. Assumption 10.11
6. Suppose (b) $m\overline{AB} > m\overline{AC}$	6. Statement 1(b)
7. then $m\angle B < m\angle C$	7. Why?
8. but $m\angle B > m\angle C$	8. Why?
9. $\therefore m\overline{AB} > m\overline{AC}$ is false	9. Assumption 10.11
10. Therefore, $m\overline{AB} < m\overline{AC}$	10. If two or more possibilities exist for a given situation, and all except one can be proved false, then the remaining possibility is true. (Method of Exclusion)

Upon the basis of the preceding proof, the following theorem is now evident.

10.12 THEOREM

If the measures of two angles of a triangle are unequal, the measures of the sides opposite those angles are unequal in the same order.

Exercises

1. The measures of the sides of a triangle are 5, 7, and 11 inches. Which angle of the triangle is the largest? Which angle is the smallest?

2. The measures of two angles of a triangle are 60° and 70°. Opposite which angle of the triangle is the shortest side?

3. Using the angle measures indicated in Figure 10-10, which segment is shortest?

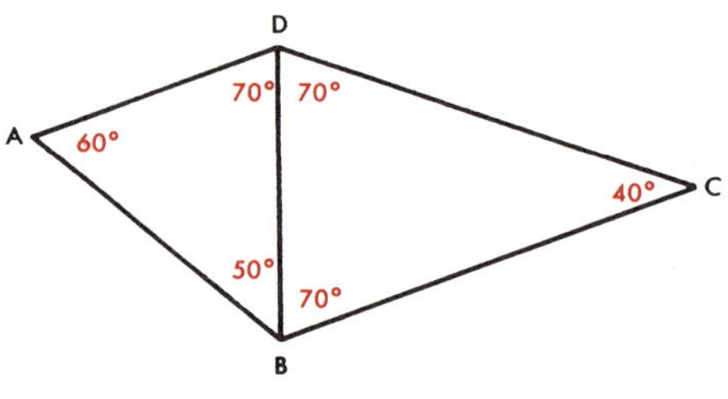

Figure 10–10

4. If the measures of two sides of a parallelogram are unequal, must the measures of the angles be unequal? Explain.

5. Which side of a right triangle is the longest side? Prove this.

6. Which subset of triangles has no congruent angles? Prove this.

7. What is the relationship of the measure of an altitude of a triangle to the measure of the sides adjacent to the altitude? Is the measure of the altitude equal to half the sum of the measures of these sides, more than half their sum, or less than half their sum? Prove your conjecture.

8. $\overline{AB}$ and $\overline{CD}$ intersect at O, $m\angle ACO > m\angle CAO$, and $m\angle BDO > m\angle DBO$. Prove $m\overline{AB} > m\overline{CD}$.

9. Using the angle measures indicated in Figure 10-11, which segment is the shortest? Explain.

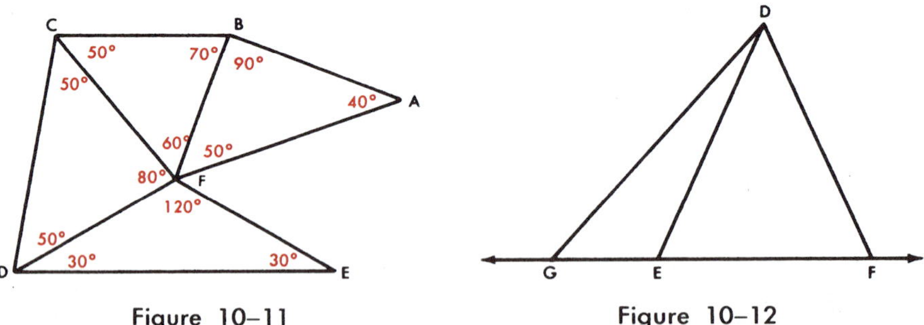

Figure 10–11

Figure 10–12

10. In $\triangle DEF$ of Figure 10-12, $\overline{DE} \simeq \overline{DF}$, G is any point of $\overleftrightarrow{EF}$ not in $\overline{EF}$. Prove $m\overline{DG} > m\overline{DF}$.

We agreed on the distance from a point to a line or a plane as the length of the perpendicular segment from that point to the line or plane (page 65). This implied that the length of the perpendicular is the shortest distance from the point to the line or plane. We may now prove this conjecture.

Given: $\overline{PA} \perp$ plane MN (or $\overleftrightarrow{MN}$). $\overline{PB}$ any other segment drawn to plane MN (or $\overleftrightarrow{MN}$).

Conjecture: $m\overline{PA} < m\overline{PB}$

Plan: Show that $m\overline{PB} > m\overline{PA}$, the perpendicular segment from the point to the plane or line.

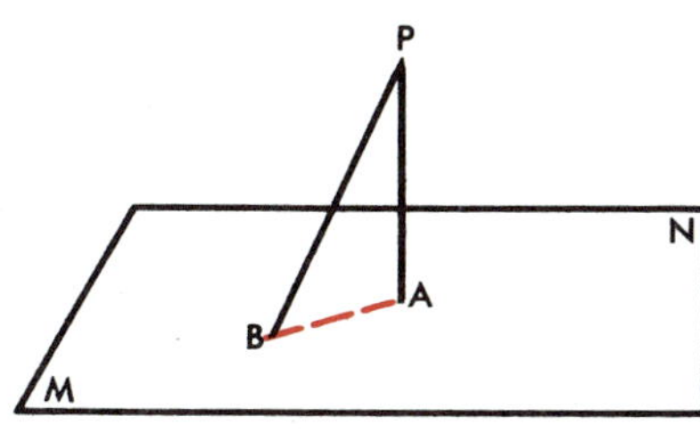

Figure 10–13

Proof: The proof is left to the student. When completed, it justifies the follow theorem.

10.13 THEOREM

The least distance from a point to a line or a plane is the measure of the segment from the point perpendicular to and terminated by the line or plane.

Draw a scalene triangle. Measure the length of two sides, and compare the sum with the measure of the remaining side. Do this with several triangles of different shapes. What is your conjecture? By using an auxiliary line to form right triangles, can you apply Exercise 5, page 238, to the proof of your conjecture?

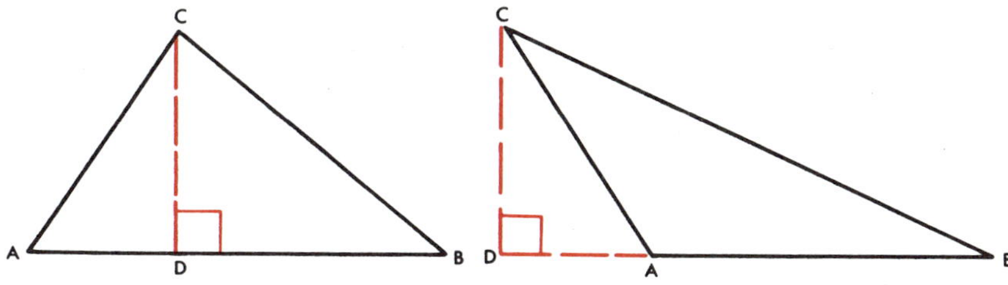

Figure 10–14

Has your proof established the following theorem?

10.14 THEOREM

The sum of the measures of two sides of a triangle is greater than the measure of the third side.

Using Theorem 10.14, it is possible to prove another proposition which is often accepted as a postulate.

10.15 THEOREM

The least distance between two points is the measure of the segment joining them.

Exercises

1. In quadrilateral $ABCD$, $\overline{AC} \perp \overline{BD}$ at E, so mAE $<$ _?_ and $<$ _?_. State the theorem which justifies your response.
2. In Figure 10-15, m$\overline{AB}$ + m$\overline{AD}$ $>$ _?_. State the theorem which justifies your response.

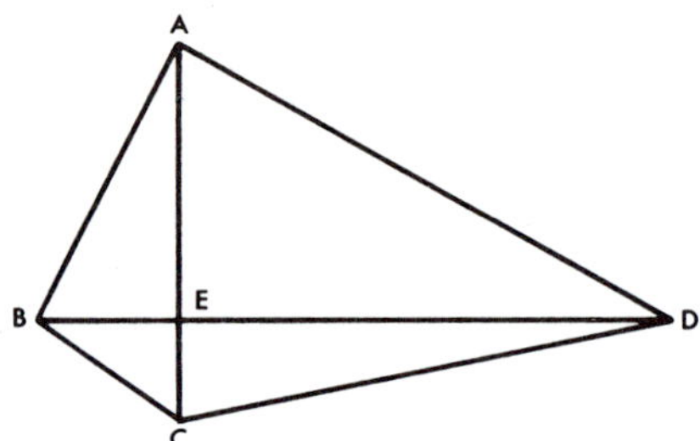

Figure 10–15

3. From any point D in the interior of $\triangle ABC$ draw $\overline{DA}$ and $\overline{DB}$. Prove m$\overline{AC}$ + m$\overline{BC}$ $>$ m$\overline{AD}$ + m$\overline{BD}$. (*Hint:* Draw $\overrightarrow{AD}$ to intersect $\overline{BC}$ at E.)
4. Is the length of any side of a triangle greater than the difference of the lengths of the other two sides? Prove your response.
5. Is the length of any side of a triangle less than one-half the perimeter of the triangle? Prove your response.

Harvard University News Office

The French architect, Le Corbusier, contrasted straight and curved walls, open and closed spaces, and heavy and light masses in the design for this art center at Harvard University. What other kinds of inequalities can you find here?

6. Given P is any point in the interior of $\triangle ABC$. Is $m\overline{PA} + m\overline{PB} + m\overline{PC} < \frac{1}{2}(m\overline{AB} + m\overline{BC} + m\overline{CA})$? Prove your response.
7. Is the sum of the measures of the diagonals of a convex quadrilateral less than the perimeter of the quadrilateral? Prove your response.
8. $\overline{AO}$ and $\overline{BO}$ are noncollinear segments in plane s such that $m\overline{AO} > m\overline{BO}$. $\overline{PO}$ is perpendicular to plane s. Develop a conjecture concerning $m\overline{AP}$ and $m\overline{BP}$ and prove it. (*Hint:* Make $\overline{B'O} \cong \overline{BO}$ with B' on $\overline{AO}$; then use 10.12.)
9. Point P is in the interior of a convex quadrilateral $ABCD$. Determine the position of P such that $m\overline{PA} + m\overline{PB} + m\overline{PC} + m\overline{PD}$ is the least possible number. Justify your decision.
10. If A and B are two points on the same side of $\overleftrightarrow{DE}$ in plane s, find the point C of $\overleftrightarrow{DE}$ for which $m\overline{AC} + m\overline{BC}$ is the least number. Justify your response. (Light from A, reflected from a mirror DE to point B, would follow path ACB.)

We have proved that there is a relationship between unequal angles and unequal sides within one triangle (10.12). If two sides of one triangle are respectively congruent to two sides of another triangle, what will influence the measure of the third sides?

Given: $\triangle ABC$ and $\triangle DEF$, in which $\overline{AC} \cong \overline{DF}$, $\overline{BC} \cong \overline{EF}$, and $m\angle C > m\angle F$

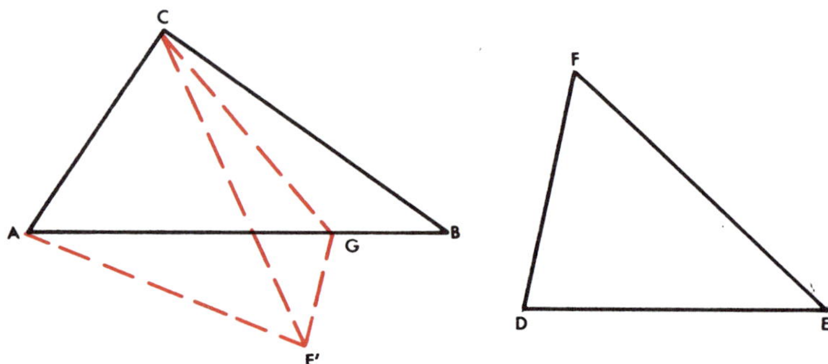

Figure 10–16

Conjecture: $m\overline{AB} > m\overline{DE}$

Plan: 1. In the plane of $\triangle ABC$ and on the same side of $\overleftrightarrow{AC}$ as B, draw auxiliary segment $\overline{CE'}$ congruent to $\overline{FE}$ such that $\angle ACE' \cong \angle DFE$. Since $\overline{CA} \cong \overline{FD}$, $\triangle CAE' \cong \triangle FDE$.

2. $\overline{CE'}$ will fall within $\angle ACB$, since $m\angle ACB > m\angle F$.
3. Bisect $\angle E'CB$ by $\overrightarrow{CG}$ intersecting $\overline{AB}$ at G.
4. Prove that $\triangle E'CG \cong \triangle BCG$.
5. $m\overline{AG} + m\overline{GE'} > m\overline{AE'}$. Why?
6. $m\overline{GB} = m\overline{GE'}$. Hence, $(m\overline{AG} + m\overline{GB}) > m\overline{AE'}$ or $m\overline{AB} > m\overline{DE}$. Why?

Proof: Optional

10.16 THEOREM

If two sides of one triangle are congruent to two sides of another triangle and the measure of the included angle of the first is greater than the measure of the included angle of the second, then the measure of the third side of the first is greater than the measure of the third side of the second.

Let us attempt to prove a converse of 10.16.

Given: $\triangle ABC$ and $\triangle DEF$, in which $\overline{AB} \cong \overline{DE}$, $\overline{AC} \cong \overline{DF}$, and $m\overline{BC} > m\overline{EF}$

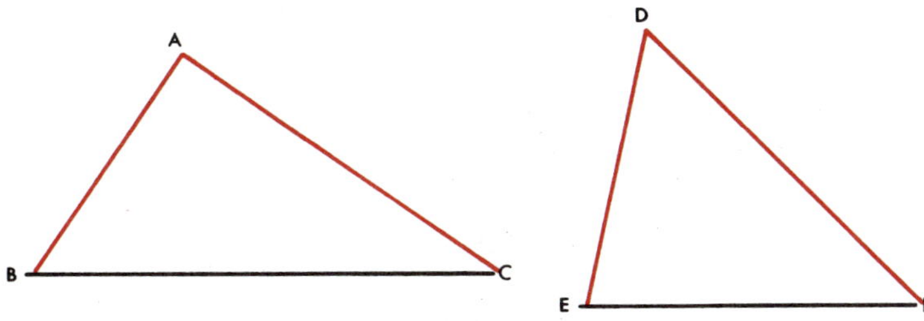

Figure 10–17

Conjecture: $m\angle A > m\angle D$

Plan: Use indirect proof.

Proof: Optional. The proof is left to the student. (Refer to the proof of 10.12 as a guide.)

10.17 THEOREM

If two sides of one triangle are congruent to two sides of a second triangle and the measure of the third side of the first triangle is greater than the measure of the third side of the second triangle, then the measure of the angle opposite the third side of the first is greater than the measure of the angle opposite the third side of the second.

Exercises

1. In $\triangle ABC$, $m\angle A < m\angle B$ and $m\angle B < m\angle C$. What is the relationship of $m\angle A$ to $m\angle C$? Why?
2. Why is $m\angle ABD > m\angle BCD$? If $\overline{BD} \perp \overline{AC}$ is $m\overline{CD} > m\overline{BD}$? Why? Why is $m\overline{BC} + m\overline{BD} > m\overline{CD}$?

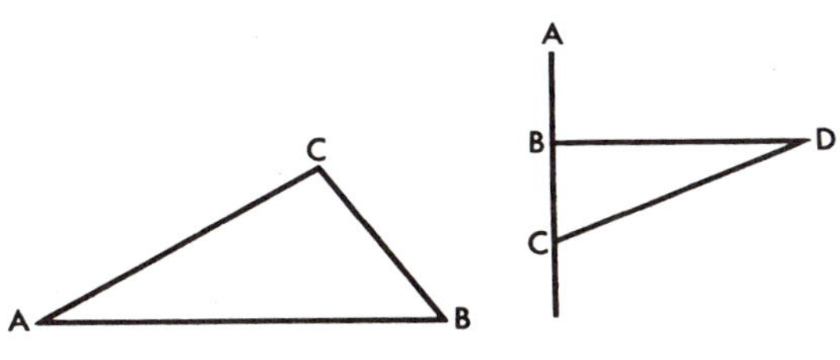

Figure 10–18

3. In $\triangle CDE$, $\overline{CD} \cong \overline{ED}$, B is a point of $\overline{CE}$ other than the midpoint. Prove that $\overrightarrow{DB}$ does not bisect $\angle CDE$.
4. Given $\triangle ABC$ with D in $\overline{AB}$ such that $\overline{AD} \cong \overline{BC}$. Prove $m\overline{AC} > m\overline{DB}$.
5. Given parallelogram $ABCD$ such that $m\angle BAD > m\angle CBA$. The diagonals intersect at E. Prove $m\overline{BE} > m\overline{AE}$.

VOCABULARY LIST

inequality	indirect reasoning
same order	contradiction
opposite order	method of exclusion
transitive	axioms

Chapter Review

Use Figure 10-19 for Exercises 1–3.

1. If $m\overline{BC} > m\overline{AB}$, $\overrightarrow{AD}$ bisects $\angle A$ and $\overrightarrow{CD}$ bisects $\angle C$, what is the relationship of $m\angle 1$ to $m\angle 2$? Why?
2. If $\overline{AD} \cong \overline{AB}$ and $m\angle 2 > m\angle 1$, what is the relationship of $m\overline{BC}$ to $m\overline{DC}$? Why?
3. $\triangle ABC$ is equilateral. ACD is a straight line. What is the relationship of $m\angle ABD$ to $m\angle D$? Why?

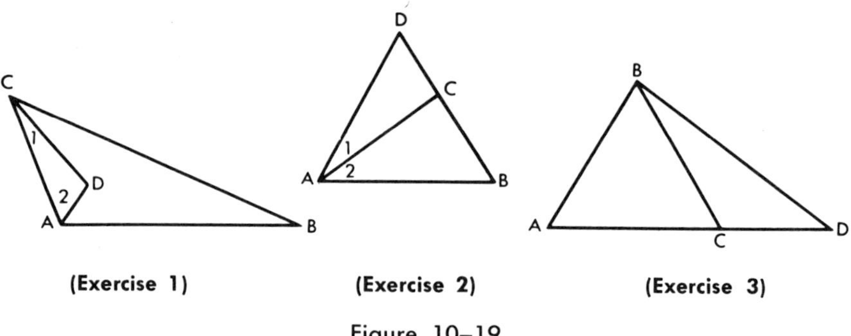

Figure 10–19

Use Figure 10-20 for Exercises 4–6.

4. $\overline{CD}$ is a median of $\triangle ABC$. $m\overline{BC} > m\overline{AC}$. What is the relationship of $m\angle 1$ to $m\angle 2$? Why?

5. $\overline{AC} \cong \overline{BC}$. $\overline{CD}$ is not an altitude. Is m$\angle 1 >$ m$\angle 2$? What do you know about m$\angle 3$ and m$\angle 4$?

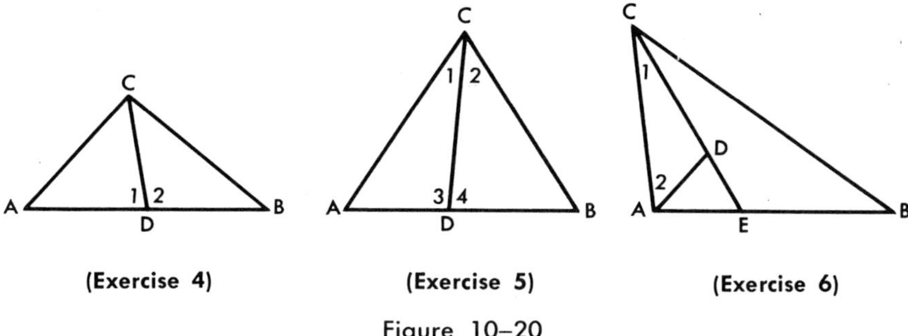

Figure 10–20

6. m$\overline{BC}$ > m$\overline{BA}$, $\overrightarrow{AD}$ bisects $\angle BAC$, and $\overrightarrow{EC}$ bisects $\angle BCA$. What is the relationship of m$\angle BAC$ to m$\angle BCA$? Of m$\angle 1$ to m$\angle 2$? Of m$\overline{AC}$ to m$\overline{EC}$?

7. In Figure 10-21, $\overline{DC} \parallel \overline{AB}$. m$\angle C >$ m$\angle ADC$. What is the relationship of m$\angle C$ to m$\angle ABD$? Why?

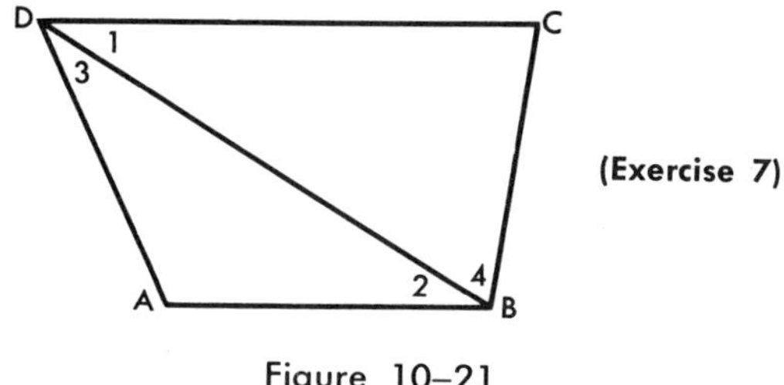

Figure 10–21

8. Does the relationship of inequality have reflexive, symmetric, and transitive properties? Explain.

9. If the universal set is the number line, does $x \geq 3$ represent a ray, a half-line, a segment, or a region?

10. If the universal set is the number line, does $0 \leq x \leq 2$ represent a ray, a half-line, a segment, or a region?

11. Give examples of the following axioms of inequality: Transitive, Addition, Multiplication.

12. Give an example of indirect reasoning you have used, other than in geometry.

Chapter 10 Test

Supply the missing symbol, word, or phrase that correctly completes the statement. (In exercises 1–5, x, y, and z represent real numbers.)

1. "If $x > y$, then $x + z > y + z$" represents the ? axiom of inequality.
2. If $x = y + z$, then $x > y$ is justified by the ? axiom of inequality.
3. "If $x < y$ and $y < z$, then $x < z$" represents the ? axiom of inequality.
4. If $x > y$, then $xz > yz$ if and only if ? .
5. If $x < y$ and $z < 0$, then $\frac{x}{z}$? $\frac{y}{z}$.

Use Figure 10-22 and the conditions listed in each exercise, 6–9.

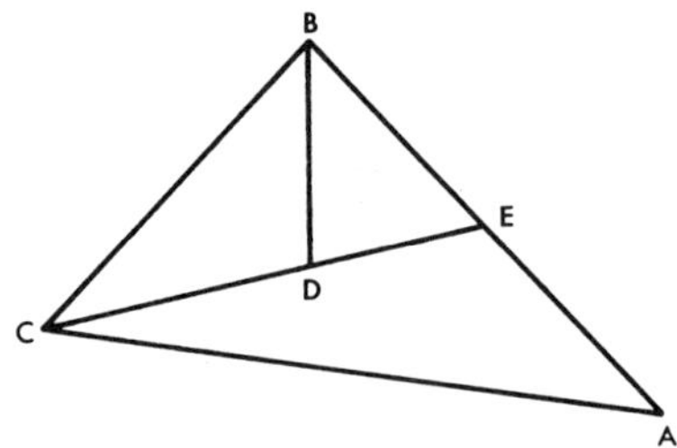

Figure 10–22

6. In $\triangle BDE$, if $BD \perp CE$, then $m\overline{BE}$? $m\overline{DE}$.
7. If $m\overline{AC} > m\overline{AB}$, then $m\angle BCA$? $m\angle CBA$.
8. If $m\angle BDE = 70°$, $\overrightarrow{BD}$ bisects $\angle CBA$ and $\overrightarrow{CD}$ bisects $\angle BCA$, then $m\angle A =$? .
9. If $\overline{CE}$ bisects $\overline{BA}$ and $m\angle CEA > m\angle CEB$, then ? .
10. Use indirect proof to show that $m\overline{AC} = m\overline{BC}$ in $\triangle ABC$ if $m\angle A = m\angle B$.

11

Three-Dimensional Concepts

Some three-dimensional concepts have been introduced in earlier chapters as an outgrowth of two-dimensional concepts. However, in preceding chapters we have been concerned primarily with plane figures. This emphasis was necessary to establish the basic concepts and the language of geometry, but it would be unrealistic to restrict our work in geometry to plane figures. We live in a three-dimensional world and need to consider the relationships within it. Some of the basic three-dimensional concepts which were established in Chapter 2 should be reviewed now.

We make use of lines and planes every day. The desk surface on which you write represents a plane; the floor and walls of your classroom resemble planes (often interrupted by projections). Does the water of a lake on a completely calm day represent a plane surface? Remember that a geometric plane is a surface with no thickness; any straight line connecting two points in the plane lies completely in the plane.

Carpenters build walls so that they are perpendicular to the floor; they construct roofs that generally are combinations of several planes. Artists sometimes make attractive forms in three dimensions by means of many small planes of metal or glass.

11.00 A 3 inch by 5 inch card can be used to represent a plane. Can you balance a card on a pin point? Even if you succeed, the card is not stable and will fall if moved even slightly. Can the card be balanced on two points at approximately the same level? Once again the card is unstable. Will the card be stable if three points are used? What relationship between the points might cause the card to be unstable? Suppose a fourth point is added; will the card be stable? Will it touch all four points if the last pin point is placed at random? How many points in space "determine a plane"? A unique plane is determined if the conditions permit one plane to exist but no other plane will satisfy the conditions.

The same principle may be illustrated by imagining a large piece of plate glass fitting our concept of a plane and three pointed posts filling the role of the pins. If all three posts are of varied heights, what is the influence on determining a plane?

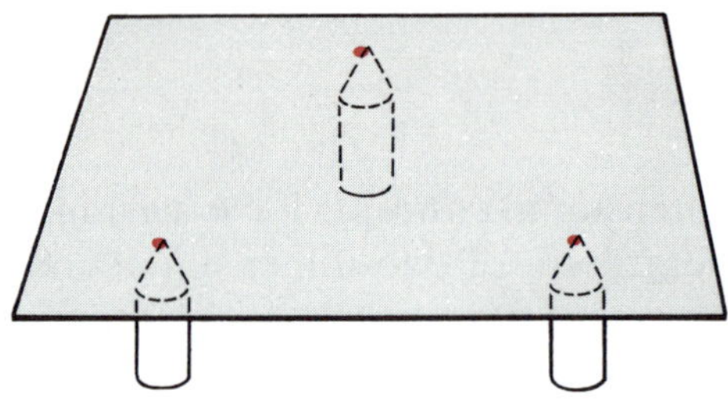

Figure 11–1

Are there other ways to determine a unique plane by means of lines and points? Will a card balance on a straight edge? If a pin point is placed outside the straight edge will the card be stable? Can two lines be arranged so that one and only one plane will contain them? How many such arrangements of lines will determine a plane? Are there any arrangements of two lines in space which do not determine a plane?

Examine the following diagrams to see if they might represent ways to determine a plane.

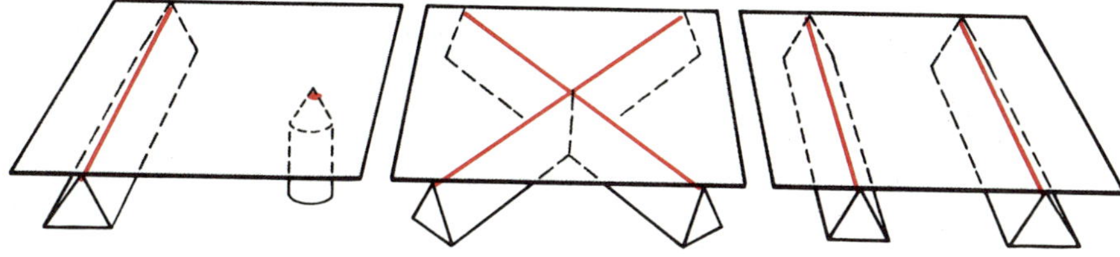

Figure 11–2

PLANES IN SPACE

You will remember that in 2.08 we decided that through any one point an infinite number of straight lines may be drawn. That is, one point does not determine or fix a straight line. Now we may ask, how many distinct planes may be passed through (and contain) one straight line? Would one straight line determine a plane? Explain.

11.01 *Assumption: An infinite number of planes may be passed through (contain) one straight line.*

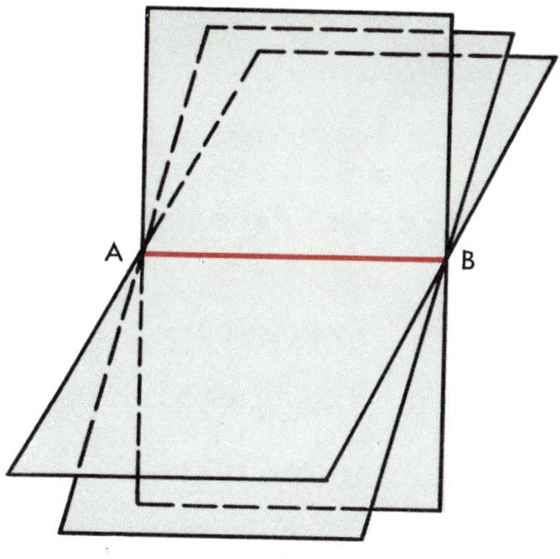

Figure 11–3

To understand the previous assumption take a piece of cardboard to represent a portion of a plane. Draw a line across the cardboard. Now rotate the cardboard about this line. As it rotates it assumes an infinite number of positions. Each such position represents a distinct plane containing the same line.

Let us stop this revolving plane at some point not on the line. Does this point and the line about which the plane rotates determine a distinct plane?

11.02 *Assumption: A straight line and a point outside the line determine one and only one plane.*

Would any two points which determine the line of 11.02 and the point outside the line comprise three points not in a straight line?

11.03 THEOREM

Three points not in a straight line determine one and only one plane.

Can the three points in 11.03 determine two intersecting lines?

11.04 THEOREM

Two intersecting lines determine one and only one plane.

If l_1 and l_2 are parallel lines, there is a plane containing them by definition(2.41). If P is any point of l_1, then 11.02 states that there is one and only one plane containing P and l_2. Therefore k is the only plane containing l_1 and l_2.

11.05 THEOREM

Two parallel lines determine one and only one plane.

Exercises

1. Is a triangle necessarily a plane figure? Explain.
2. Do two skew lines determine a plane?
3. Explain how a sidewalk is smoothed to become a plane surface. Which assumption or theorem is the basis of this method?
4. How many planes are determined by three points not in a straight line? By four non-coplanar points?

5. Why are telescopes, cameras, and surveying instruments mounted on tripods?
6. Why are some chairs and tables "wobbly"?
7. The intersecting diagonals of a quadrilateral are parallel to a given plane. Are the sides of the quadrilateral parallel to the plane? Explain.
8. If a line is parallel to a plane, do all the perpendiculars from the line to the plane lie in one plane?
9. How many different planes are determined by three parallel lines not all in the same plane?
10. If three lines are concurrent, must they always lie in one plane? How many planes may they determine?
11. If a line intersects two sides of a triangle, why must the line lie in the plane of the triangle?
12. If a line is perpendicular to each of two lines in a plane, is it perpendicular to the plane of the lines? Draw a figure to illustrate your answer.
13. Give the names of two classes of quadrilaterals that are necessarily plane figures.
14. Two planes have three non-collinear points in common. Do the planes coincide?
15. If three sides of a quadrilateral lie in a plane, does the quadrilateral lie in that plane?
16. If a side and one diagonal of a parallelogram are parallel to a given plane, is the plane of the parallelogram parallel to the plane?

DIHEDRAL ANGLES

Although we have limited our discussion of angles so far to angles formed by straight lines, it is obvious that when planes meet they form an angle. Angles between certain curved lines are also dealt with in geometry (13.16).

As discussed on pages 30 and 31, a plane is a set of points. A straight line in a plane separates the points of the plane into three distinct sets (2.22): the set of points of the line, the set of points on one side of the line, and the set of points on the other side of the line. The sets of points other than the line are called half-planes and the line is the edge of, but not a part of, either half-plane.

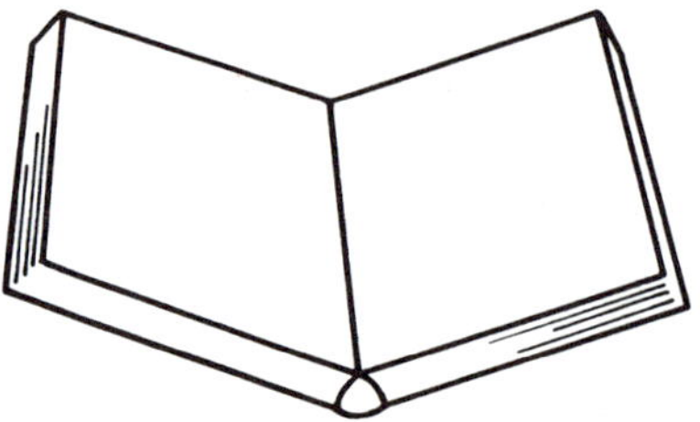

The pages of an open book represent a dihedral angle.

Figure 11–4

The arrows indicate the dihedral angles formed by the wings of the airplane.

Figure 11–5

Figure 11-3 (page 249) represents three intersecting planes—or six half-planes and their common edge. We consider two half-planes having a common edge to be an angle, as defined below.

11.06 A dihedral angle is the union of a line and two half-planes having the line as their common edge. The half-planes are called the *faces* of the angle and their common edge is called the *edge* of the angle.

A *dihedral angle* is denoted by naming a point in each face and two points of the edge. The two points on the edge are always the middle letters. In Figure 11-6 the dihedral angle is denoted by $\angle O\text{-}NM\text{-}L$. A dihedral angle may also be labeled by naming its edge, (MN in Figure 11-6), if this designation will not cause confusion.

Figure 11-7 shows two intersecting planes, AB and CD. The "upper right" angle should be labeled $\angle A\text{-}EF\text{-}D$, or $\angle D\text{-}FE\text{-}A$. The "lower left" angle should be labeled $\angle C\text{-}EF\text{-}B$ or $\angle B\text{-}FE\text{-}C$.

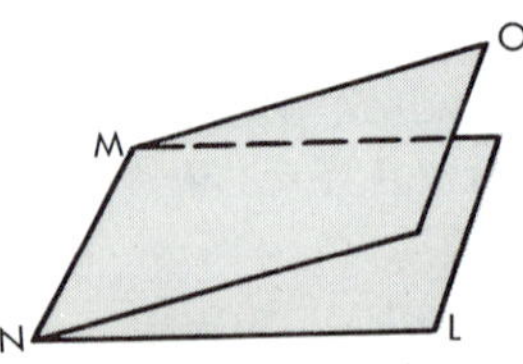

Figure 11–6

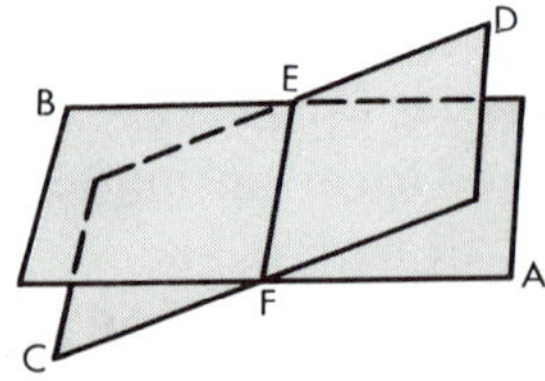

Figure 11–7

How can we find the measure of a dihedral angle? Let us fold a 3″ by 5″ card along a line to form a model of dihedral angle *O-NM-L*, similar to that in Figure 11-6. If $\overleftrightarrow{MN}$ is perpendicular to the edge of the card how would you measure the dihedral angle? If $\overleftrightarrow{MN}$ is not perpendicular to the edge, could the angle be measured in the same way?

Open your geometry book to represent a dihedral angle. How would you measure this dihedral angle? Can you form a dihedral angle whose measure is 90°? 180°? 30°?

How would you measure the dihedral angle if the intersecting planes forming the dihedral angle are represented by irregularly shaped pieces of cardboard as shown in Figure 11-8?

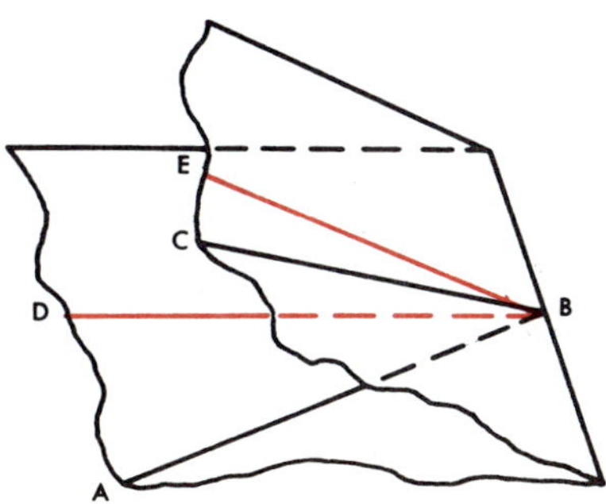

Figure 11–8

Some students respond to the question in the preceding paragraph by saying, "Draw lines in these irregular planes to represent the plane angle formed by the top or bottom edges of the book." Is this acceptable to you? If so, would you draw these lines oblique to the edge of the dihedral angle, or perpendicular to the edge? Why?

The plane angle formed by these lines is defined as follows:

11.07 A plane angle of a dihedral angle is formed by two rays, one in each face, *perpendicular to the edge at the same point.* This point is the endpoint of both rays. A dihedral angle is measured by its plane angle.

11.08 Measures and relationships of dihedral angles are classified in the same way as plane angles. The terms obtuse, acute, right, straight, adjacent, vertical, supplementary, and complementary, have the same meaning when applied to dihedral angles as when applied to plane angles. In Figure 11-9 $\angle DOC$ represents a plane angle of the dihedral angle.

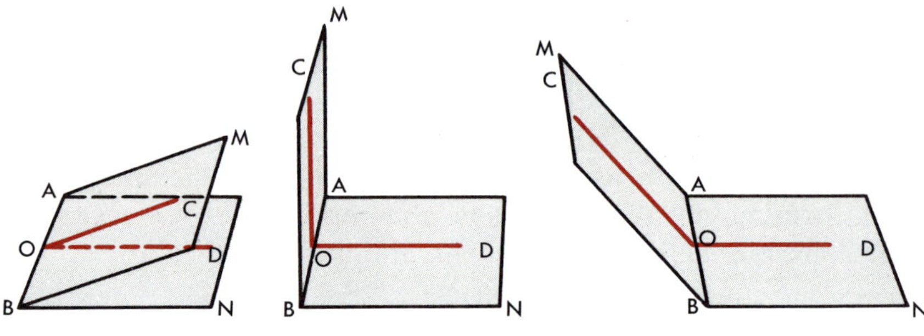

Figure 11–9

Exercises

In the figure below, two planes, RS and LM, are cut by a third plane, NP. The lines of intersection are $\overleftrightarrow{AB}$ *and* $\overleftrightarrow{CD}$.

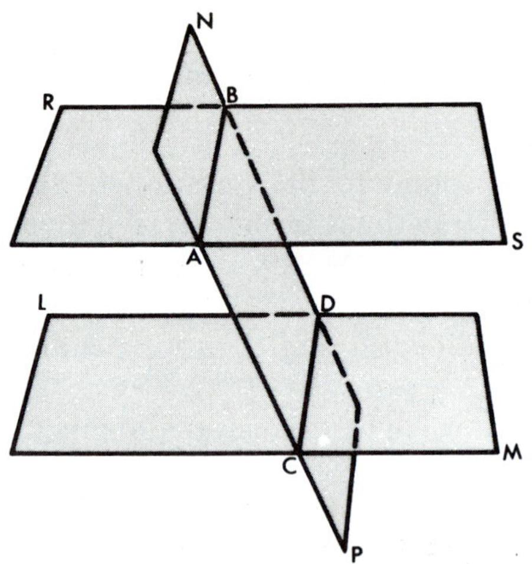

Figure 11–10

1. If planes RS and LM are parallel, are the alternate interior dihedral angles equal?

Apply the names alternate interior, alternate exterior, and corresponding angles to the dihedral angles named as follows:

2. $\angle R\text{-}AB\text{-}C$ and $\angle M\text{-}DC\text{-}B$
3. $\angle N\text{-}AB\text{-}S$ and $\angle L\text{-}CD\text{-}P$
4. $\angle N\text{-}AB\text{-}R$ and $\angle P\text{-}DC\text{-}M$
5. $\angle N\text{-}AB\text{-}S$ and $\angle B\text{-}CD\text{-}M$
6. $\angle S\text{-}BA\text{-}C$ and $\angle L\text{-}CD\text{-}A$
7. $\angle R\text{-}AB\text{-}C$ and $\angle P\text{-}DC\text{-}L$
8. Name two pairs of vertical dihedral angles.
9. Name an obtuse dihedral angle; an acute dihedral angle.
10. Name a pair of adjacent dihedral angles.

RELATIONSHIPS BETWEEN LINES AND PLANES

The congruence relationship is defined for dihedral angles as well as for angles in a plane. The dihedral angles determined by two or more intersecting planes are often used to classify the relationship of the planes. The following assumptions and definitions concern dihedral angles, planes, and lines in space.

11.09 *Assumption: Dihedral angles are congruent if their plane angles are congruent.*

11.10 *Assumption: If dihedral angles are congruent, their plane angles are congruent.*

11.11 *Assumption: Plane angles of the same dihedral angle are congruent. (Assumption 11.11 is given here so that it may be used in subsequent proofs. It may be proved as a theorem in 11.33.)*

11.12 The bisector of a dihedral angle is the half-plane whose edge is the edge of the angle and which contains the bisector of a plane angle of the dihedral angle.

11.13 Perpendicular planes are planes which intersect to form congruent adjacent dihedral angles. (See Figure 11-11, page 256; $\angle E\text{-}AB\text{-}D \cong \angle E\text{-}AB\text{-}C$.)

11.14 If two distinct planes meet at an angle other than a right angle they are called oblique planes.

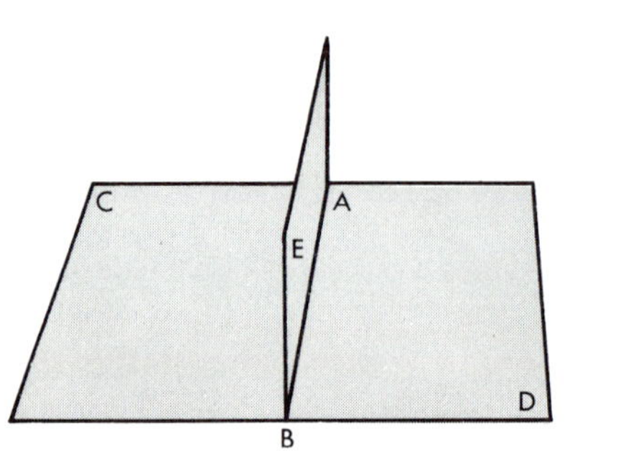

Figure 11–11

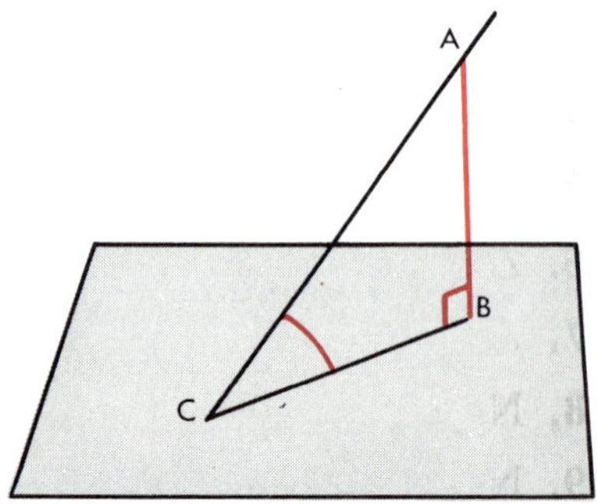

Figure 11–12

11.15 To measure the angle formed by a plane and a line oblique to the plane, construct a perpendicular to the plane from any point in the line ($\overleftrightarrow{AB}$ in Figure 11-12). Draw the segment ($\overline{BC}$) in the plane connecting the foot of the perpendicular (B) and the foot of the oblique line (C). Measure the angle formed by the oblique line and the line in the plane ($\angle ACB$). Is this the angle with the least measure formed by $\overleftrightarrow{AC}$ and any other line in the plane?

USING YOUR INTUITION

The following illustration shows two planes intersected by a third plane. We have assumed (2.24) that two planes intersect in a straight line. In the illustration suppose that the planes (m and n) are not parallel. Can the lines of intersection be parallel lines? Must the lines of intersection be parallel lines? Explain your answers by drawing pictures of two planes intersected by a third plane.

If the two planes are parallel, *must* the lines of intersection be parallel lines? Explain. (Since you are unfamiliar with three-dimensional proofs, the following proof is presented for your study.)

Proposition: If two parallel planes are intersected by a third plane, the lines of intersection are parallel.

Given: Parallel planes m and n. Plane h intersecting planes m and n, in $\overleftrightarrow{AB}$ and $\overleftrightarrow{CD}$.

Conjecture: The lines of intersection, $\overleftrightarrow{AB}$ and $\overleftrightarrow{CD}$, are parallel.

Plan: Prove $\overleftrightarrow{AB}$ and $\overleftrightarrow{CD}$ are in the same plane and never meet.

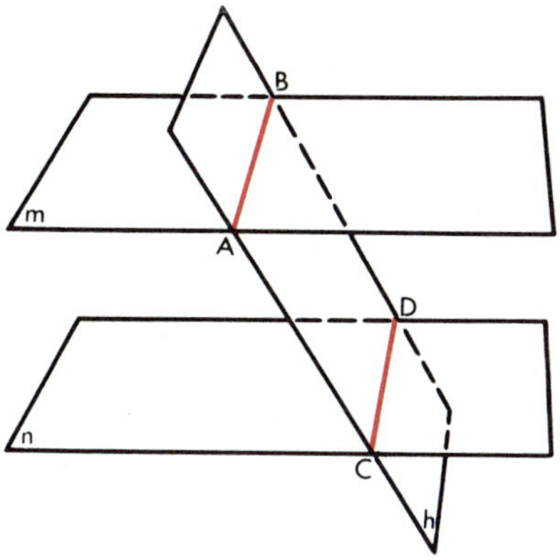

Figure 11–13

Proof:

Statements	*Reasons*
1. Plane h intersects parallel planes m and n, with $\overleftrightarrow{AB}$ and $\overleftrightarrow{CD}$ as the lines of intersection.	1. Given
2. $\overleftrightarrow{AB}$ and $\overleftrightarrow{CD}$ lie in the same plane, h.	2. Given
3. $\overleftrightarrow{AB}$ in plane m cannot meet $\overleftrightarrow{CD}$ in plane n.	3. Parallel planes never meet.
4. $\overleftrightarrow{AB} \parallel \overleftrightarrow{CD}$	4. Lines in the same plane which never meet are parallel.

We have now established the following:

11.16 THEOREM

If two parallel planes are intersected by a third plane, the lines of intersection are parallel.

Would the converse of this theorem necessarily be true? If you think the converse is true, try to prove it deductively.

PERPENDICULAR LINES AND PLANES

It is reasonable to accept the following three theorems without proof. Nevertheless, their proofs are not difficult and do provide an insight into demonstrations involving three-dimensional relationships. All students should become familiar with the theorems, though the proofs may be considered as optional. It is suggested that you study them.

The definition of 3.23 stated that a line is perpendicular to a plane if it is perpendicular to every line in the plane through its foot. To prove that a line is perpendicular to a plane would be impossible if we had to prove that every line in the plane through the foot of the line was perpendicular to this line. Is it possible to show that a line is perpendicular to a plane if it is perpendicular to a very few lines in the plane through its foot? How many is a few?

Our intuition should help us here. We fashion bases for Christmas trees, and for many other vertical objects from two crossed wood braces. Would two lines then be the minimum number? Let us examine this idea.

Given: $\overleftrightarrow{CB}$ and $\overleftrightarrow{BD}$ are distinct lines in plane s, intersecting at B. $\overleftrightarrow{AF} \perp \overleftrightarrow{CB}$ and $\overleftrightarrow{AF} \perp \overleftrightarrow{BD}$ at B in plane s.

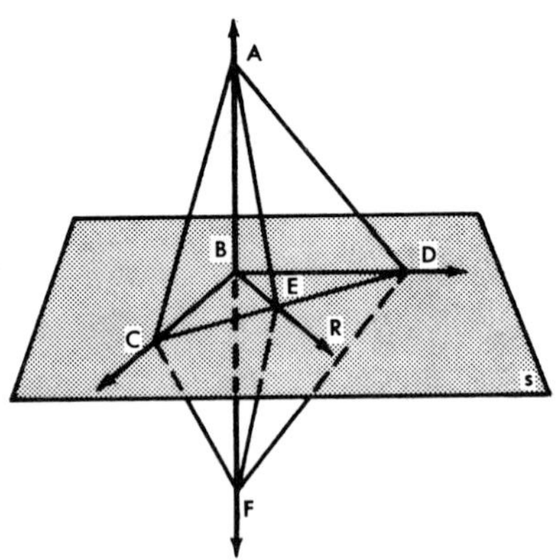

Figure 11–14

Conjecture: $\overleftrightarrow{AF}$ is perpendicular to all lines ($\overleftrightarrow{BR}$) in s through B.

Proof:

Statements	*Reasons*
1. $\overleftrightarrow{AF} \perp \overleftrightarrow{CB}$ and $\overleftrightarrow{AF} \perp \overleftrightarrow{BD}$ at B in plane s. $\overleftrightarrow{CB}$ and $\overleftrightarrow{BD}$ in plane s.	1. Given
2. Construct any $\overleftrightarrow{BR}$ in s.	2. 2.08
3. Construct $\overleftrightarrow{CD}$.	3. 2.10
4. $\overleftrightarrow{CD}$ intersects $\overleftrightarrow{BR}$ at E.	4. $\overleftrightarrow{BD}$ and $\overleftrightarrow{BC}$ contain points on each side of $\overleftrightarrow{BR}$ in s. Let C be on one side of $\overleftrightarrow{BR}$ and D on the other side, then use 2.28.
5. Construct $\overline{BA} \cong \overline{BF}$,	5. 2.38
6. then $\overline{AC} \cong \overline{CF}$ and $\overline{AD} \cong \overline{DF}$.	6. 8.13
7. $\overline{CD} \cong \overline{CD}$,	7. Reflexive Axiom.
8. then $\triangle ACD \cong \triangle FCD$.	8. *s.s.s*
9. So $\angle ACE \cong \angle FCE$.	9. *c.p.c.t.c.*
10. $\overline{CE} \cong \overline{CE}$,	10. Reflexive Axiom.
11. then $\triangle ACE \cong \triangle FCE$	11. *s.a.s.*
12. So $\overline{AE} \cong \overline{EF}$.	12. *c.p.c.t.c.*
13. $\overline{AB} \perp \overline{BR}$,	13. 8.14 (Points B and E are equidistant from A and F.)
14. $\therefore \overline{AF} \perp s$.	14. Since $\overleftrightarrow{BR}$ represents all lines in s through B, we apply Definition 3.23.

Since $\overleftrightarrow{AF}$ represents any line perpendicular to two distinct lines in a plane through the foot of $\overleftrightarrow{AF}$ and $\overleftrightarrow{BR}$ represents all other lines in the plane through the foot of $\overleftrightarrow{AF}$, then we have justified the following theorem.

11.17 THEOREM

If a line is perpendicular to two lines in a plane through its foot, then it is perpendicular to the plane.

Given: The planes r and s, each $\perp$ to $\overleftrightarrow{AB}$

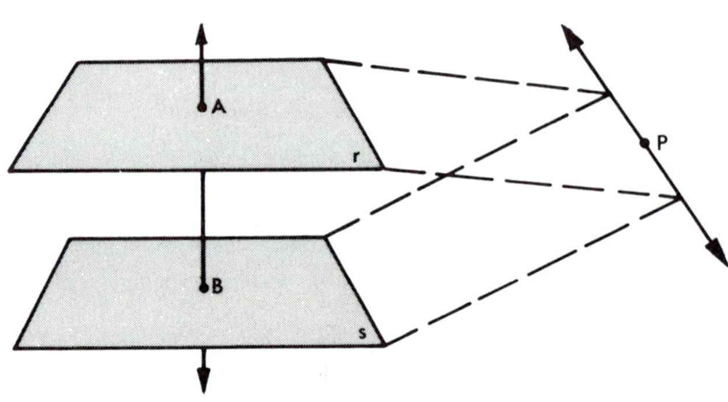

Figure 11–15

Conjecture: $r \parallel s$

Plan: Prove that planes r and s can never meet. (Indirect proof)

Proof:

Statements	*Reasons*
1. Either planes r and s are parallel or they intersect.	1. Assumption for Indirect Proof. (Exercise 4, page 41)
2. Suppose r and s intersect in a line containing point P.	2. 2.24 and an assumption for Indirect Proof.
3. Planes r and s are each perpendicular to $\overleftrightarrow{AB}$,	3. Given.
4. then there are two planes through $P \perp \overleftrightarrow{AB}$.	4. Statements 2 and 3.
5. But statement 4 is impossible.	5. There can be one and only one perpendicular from a point to a line. (3.21)
6. The assumption of statement 2 is false.	6. Assumption 10.11
7. Therefore, plane $r \parallel$ plane s.	7. Statements 1 and 6. (Method of Exclusion, page 236.)

11.18 THEOREM

Two planes perpendicular to the same line are parallel.

If a line is perpendicular to one of two parallel planes, what is its relationship to the other plane? Test your conjecture.

Given: Parallel planes r and s, and line $\overleftrightarrow{CD} \perp$ plane r

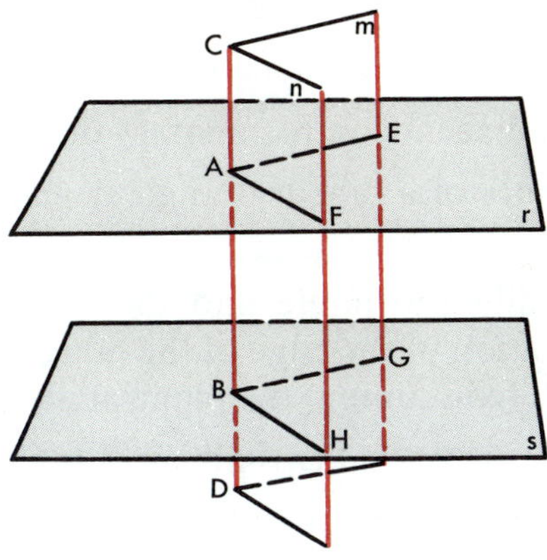

Figure 11–16

Conjecture: $\overleftrightarrow{CD} \perp s$.

Plan: Prove that $\overleftrightarrow{CD} \perp$ to two intersecting lines in plane s.

Proof:

Statements	*Reasons*
1. Through $\overleftrightarrow{CD}$ pass plane n, intersecting planes r and s in $\overleftrightarrow{AF}$ and $\overleftrightarrow{BH}$, respectively; also through $\overleftrightarrow{CD}$ pass plane m, intersecting planes r and s in $\overleftrightarrow{AE}$ and $\overleftrightarrow{BG}$, respectively.	1. Construction (11.01, 2.24)
2. $\overleftrightarrow{AF} \parallel \overleftrightarrow{BH}$ and $\overleftrightarrow{AE} \parallel \overleftrightarrow{BG}$	2. 11.16
3. $\overleftrightarrow{CD} \perp \overleftrightarrow{AE}$ and $\overleftrightarrow{AF}$	3. 3.23
4. $\therefore \overleftrightarrow{CD} \perp \overleftrightarrow{BH}$ and $\overleftrightarrow{BG}$	4. 5.24
5. So $\overleftrightarrow{CD} \perp s$	5. 3.23

11.19 THEOREM

A line perpendicular to one of two parallel planes is perpendicular to the other.

Exercises

1. How many dihedral angles are formed by the floor, walls, and ceiling of your classroom? What is the probable measure of these dihedral angles?
2. Can a line bisect a dihedral angle?
3. Illustrate two adjacent supplementary dihedral angles.
4. Illustrate two adjacent complementary dihedral angles.
5. Does the measure of a dihedral angle depend upon the size of its faces?
6. Draw a right dihedral angle and then draw its plane angle. What is the relation of the edge of the dihedral angle to a plane passed through (containing) the plane angle?
7. How could you prove that two dihedral angles are congruent?
8. Are the half-planes perpendicular which bisect a pair of supplementary adjacent dihedral angles? Explain.
9. If two planes are perpendicular, what kind of dihedral angles do they form?
10. Through a line perpendicular to a plane, how many planes perpendicular to the given plane may be drawn? (*Note:* "Through" means *containing* the line.)
11. Through a line oblique to a plane, how many planes perpendicular to the given plane can be drawn? Prove your response.
12. If two intersecting planes are each perpendicular to a third plane, is their intersection perpendicular to the third plane? Prove your response.

POLYHEDRAL ANGLES

We have agreed that the figure formed when two rays have a common endpoint is called a plane angle. Also, we agree that the union of two half-planes and their common edge forms a dihedral angle (11.06). Look at a corner of your classroom. If it is formed by plane surfaces, at least three such planes meet there in a common point. Could more than three planes meet to form such a "corner"?

Can we speak of the "angle" formed by these three or more planes? Can this figure properly be called an angle?

The answer to these questions is "Yes," of course. We can define an angle as we wish, for a definition is an agreement concerning the meaning of a term (page 21). The following definition is merely one of several that are widely accepted. See whether you can find or devise another definition that will describe the type of figure we have in mind and satisfy the conditions of page 21.

11.20 A polyhedral angle is the figure formed by the union of (1) a plane polygon, (2) a fixed point not in the plane, and (3) the set of rays which have the fixed point as origin and contain a point of the polygon. Each ray of the set contains a different point of the polygon, and for each point of the polygon there is a ray of the set.

The fixed point is called the *vertex* of the angle. The rays through the vertices of the polygon are called the *edges*. All other rays are called *elements*. The angles formed by consecutive edges are called *face angles* and the unions of the face angles and their interiors are called the *faces* of the polyhedral angle.

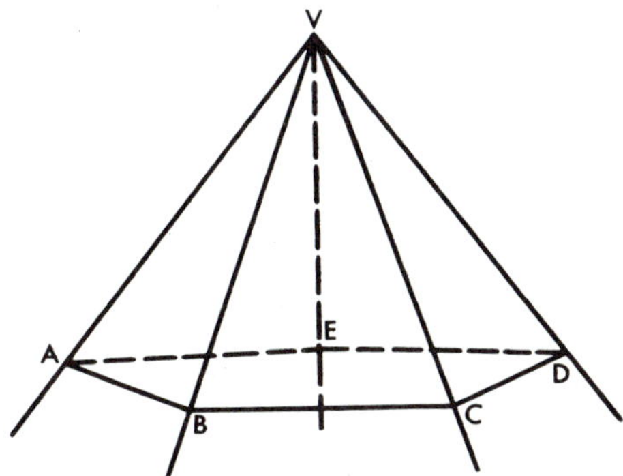

Figure 11–17

The figure above is a model of polyhedral angle $V\text{-}ABCDE$, where $ABCDE$ is the plane polygon and V is the vertex. It may also be read "polyhedral angle V," if that notation is not ambiguous. $\angle AVB$ is a face angle, as are $\angle BVC$ and the others. Why must each face be a plane figure?

11.21 A polyhedral angle is convex if the determining polygon is a convex polygon. We will consider only convex polyhedral angles.

11.22 Two polyhedral angles are vertical if the edges of one are the opposite rays of the edges of the other.

11.23 The measure of any polyhedral angle depends upon the measure of its face angles and on the order in which these angles are arranged.

11.24 *Assumption: Polyhedral angles are congruent if their corresponding face angles and dihedral angles are congruent and are arranged in the same order.*

11.25 Two polyhedral angles are symmetrical if the corresponding face angles and dihedral angles are congruent but are arranged in the opposite order.

11.26 Polyhedral angles are named according to the number of their faces.

Trihedral angle	3 faces	Pentahedral angle	5 faces
Tetrahedral angle	4 faces	Hexahedral angle	6 faces

(For further names refer to the prefixes used for polygons in 6.00.)

We see many examples of polyhedral angles about us. Two simple examples are shown below.

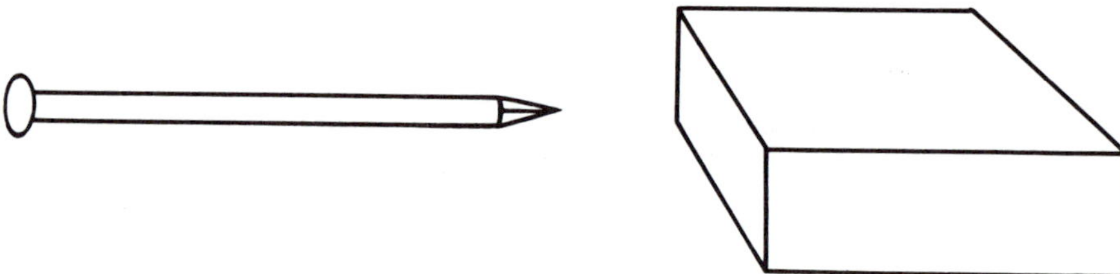

Figure 11–18

Exercises

1. How many trihedral angles are formed by the walls, floor, and ceiling of your classroom? What is the probable measure of each of their face angles?
2. How many dihedral angles does a tetrahedral angle have? A trihedral angle?

3. Construct a model of a polyhedral angle using models of five congruent equilateral triangular regions to form the faces.

4. Can you make a model polyhedral angle as in exercise 3 using six equilateral triangular regions? Explain.

5. About point O on your paper, draw successive angles of 70°, 50°, and 90° as shown. Cut out the figure $AODCBA$. Fold along OB and OC to represent a trihedral angle. Now draw the same angles but in the order 50°, 70°, 90°, and follow the instructions above. Are the two trihedral angles so obtained congruent? Are they symmetric?

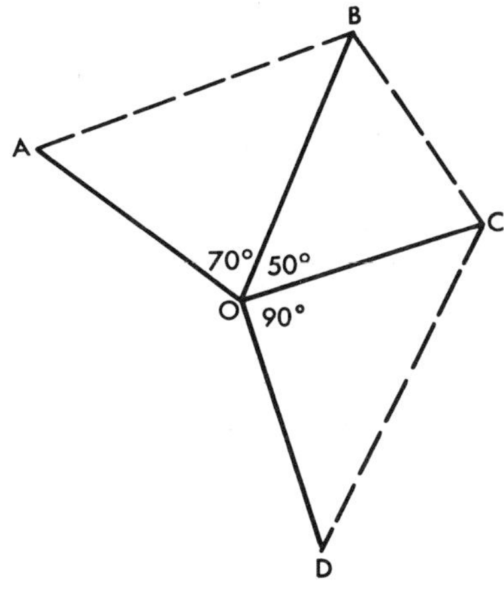

Figure 11–19

6. What is the sum of the measures of the face angles of a trihedral angle modeled by a figure whose faces are equilateral triangular regions? Whose faces are square regions?

7. Complete: The sum of the measures of two face angles of a trihedral angle is __?__ __?__ the third. What if the sum of the measures of two face angles equals the third?

8. Represent on paper two vertical trihedral angles. Are they congruent or symmetric?

9. Is it possible to have a trihedral angle, only one of whose dihedral angles is a right angle? With two right dihedral angles? With three right dihedral angles?

10. The sum of the measures of the face angles of a trihedral angle is 260°. What is the greatest measure that any one of the face angles can have?

11. If three distinct planes intersect in three parallel lines, what is the sum of the measures of the three interior dihedral angles?

12. Complete: The sum of the measures of the face angles of a convex polyhedral angle is ? ? ? .

(OPTIONAL) PROOFS OF PREVIOUS CONJECTURES

Many of the assumptions and theorems concerning the relations of points and lines in a plane may be extended to three-dimensional situations. Although the proofs of the theorems in this section are optional, the student should know and be able to apply the theorems themselves.

We accepted as true the assumption that, in a plane, there is one and only one line that is perpendicular to a given line at a given point of the line (3.20). In three-space, how many lines could be perpendicular to a given line at a given point of the line? How many planes could be perpendicular to a line at a given point of the line? (See 3.21)

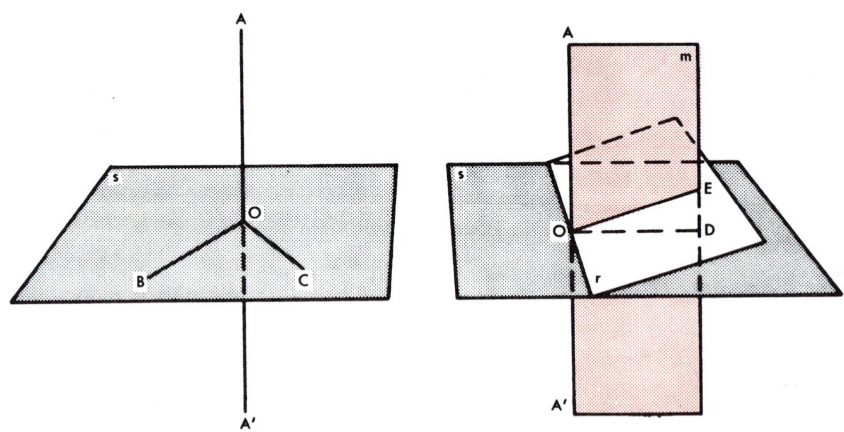

Figure 11–20

Outline of proof: Using Figure 11-20 left, let $\overleftrightarrow{AA'}$ be the given line and O the given point in line $\overleftrightarrow{AA'}$. Draw $\overleftrightarrow{OB}$ and $\overleftrightarrow{OC}$ each $\perp$

$\overleftrightarrow{AA'}$ at O. Then $\overleftrightarrow{OB}$ and $\overleftrightarrow{OC}$ determine plane s (11.04). Then plane $s \perp \overleftrightarrow{AA'}$ at O (11.17).

Using Figure 11-20 right, suppose another plane $r \perp \overleftrightarrow{AA'}$ at O. Pass a plane m through $\overleftrightarrow{AA'}$ and D, a point in s but not in r. Plane m intersects s in $\overleftrightarrow{OD}$ and r in $\overleftrightarrow{OE}$. In plane m, $\overleftrightarrow{OD}$ and $\overleftrightarrow{OE}$ are both $\perp$ to $\overleftrightarrow{AA'}$ at O. But this is impossible (3.20). Therefore s is the only plane $\perp \overleftrightarrow{AA'}$ at O.

11.27 THEOREM

Through a given point in a given line there can be one plane, and only one, perpendicular to the line.

If the given point is *not* in the given line, how many planes can contain the point and be perpendicular to the line? Can you prove that there is one, and only one?

Outline of proof: Using Figure 11-20 left, let $\overleftrightarrow{AA'}$ be the given line and B the given external point. Draw $\overleftrightarrow{OB}$ and $\overleftrightarrow{OC} \perp \overleftrightarrow{AA'}$. $\overleftrightarrow{OB}$ and $\overleftrightarrow{OC}$ determine plane $s \perp \overleftrightarrow{AA'}$ at O. Why?

Any plane $s \perp \overleftrightarrow{AA'}$ and passing through O contains a $\perp$ from B to $\overleftrightarrow{AO}$ (3.23). Since there is only one $\perp$ from B to $\overleftrightarrow{AA'}$ (3.20), all planes through $B \perp \overleftrightarrow{AA'}$ must intersect $\overleftrightarrow{AA'}$ at O. Therefore s is the only plane $\perp \overleftrightarrow{AA'}$ through B. (11.27)

11.28 THEOREM

Through a given external point there can be one plane, and only one, perpendicular to a given line.

Using Theorem 11.27 and Theorem 11.28, Theorem 11.29 follows directly.

11.29 THEOREM

All the perpendiculars to a line at a point in that line lie in a plane which is perpendicular to the line at that point.

Theorem 11.18 stated that two planes perpendicular to the same line are parallel. Are two lines perpendicular to the same plane parallel?

Given: $\overleftrightarrow{AB} \perp$ to plane p and $\overleftrightarrow{CD} \perp$ to plane p

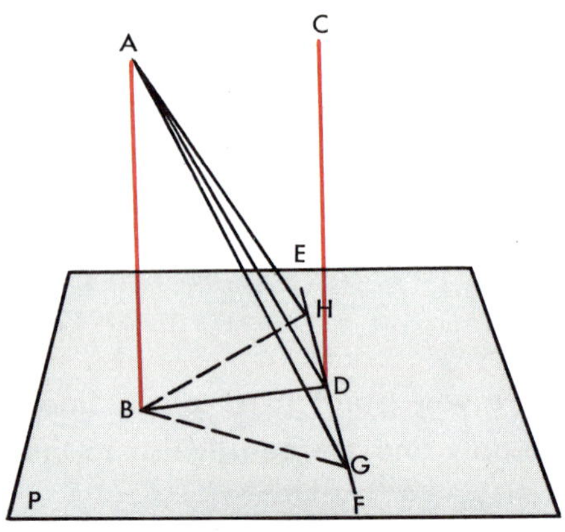

Figure 11–21

Conjecture: $\overleftrightarrow{AB} \parallel \overleftrightarrow{CD}$

Plan: To prove $\overleftrightarrow{AB}$ and $\overleftrightarrow{CD}$ are in the same plane and $\perp$ to the same line $\overleftrightarrow{BD}$

Outline of proof: Draw $\overline{BD}$ and $\overline{AD}$. In plane p, draw $\overleftrightarrow{EF} \perp \overleftrightarrow{BD}$ and construct $\overline{HD} \cong \overline{DG}$. Draw $\overline{AH}$, $\overline{AG}$, $\overline{BH}$, and $\overline{BG}$. Prove $\overline{BH} \cong \overline{BG}$ and $\overline{AH} \cong \overline{AG}$. Then, $\overleftrightarrow{AD} \perp \overleftrightarrow{HG}$ and $\overleftrightarrow{CD} \perp \overleftrightarrow{EF}$. (8.12). $\therefore \overline{BD}$, $\overline{AD}$, and $\overleftrightarrow{CD}$ lie in the same plane (11.29). $\overleftrightarrow{AB}$ is also in this plane, so $\overleftrightarrow{AB} \perp \overleftrightarrow{BD}$ and $\overleftrightarrow{CD} \perp \overleftrightarrow{BD}$; $\therefore \overleftrightarrow{AB} \parallel \overleftrightarrow{CD}$.

11.30 THEOREM

Two lines perpendicular to the same plane are parallel.

How can you form a converse of Theorem 11.30? How many conditions are implied in the antecedent? In the consequent? A converse is formed by exchanging a number of conditions in the antecedent with an equal number of conditions in the consequent. The following demonstration outlines a proof of a converse of 11.30.

Given: $\overleftrightarrow{AB} \perp$ plane m and $\overleftrightarrow{AB} \parallel \overleftrightarrow{CD}$

Conjecture: $\overleftrightarrow{CD} \perp m$

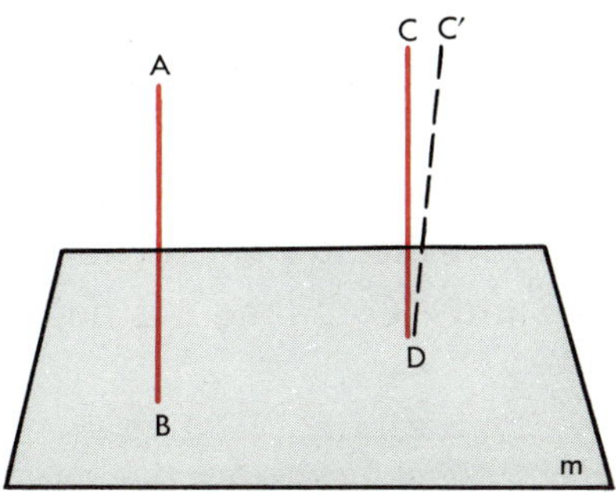

Figure 11–22

Plan: Draw $\overleftrightarrow{C'D} \perp$ plane m at D. Then it follows that $\overleftrightarrow{C'D} \parallel \overleftrightarrow{AB}$ and therefore coincides with $\overleftrightarrow{CD}$, since only one line can be drawn $\perp$ to a plane at a given point in the plane.

11.31 THEOREM

If one of two parallel lines is perpendicular to a plane, the other is also perpendicular to the plane.

You have already proved that if two lines in a plane are parallel to a third line in the same plane, then the two lines are parallel to each other (5.32). Is this proposition true even if it is not required that all three lines be in the same plane?

Given: $\overleftrightarrow{AB} \parallel \overleftrightarrow{EF}$ and $\overleftrightarrow{CD} \parallel \overleftrightarrow{EF}$

Conjecture: $\overleftrightarrow{AB} \parallel \overleftrightarrow{CD}$

Plan: Draw plane $n \perp \overleftrightarrow{EF}$; therefore $\overleftrightarrow{CD}$ and $\overleftrightarrow{AB} \perp$ plane n (11.31), so $\overleftrightarrow{AB} \parallel \overleftrightarrow{CD}$ (11.30).

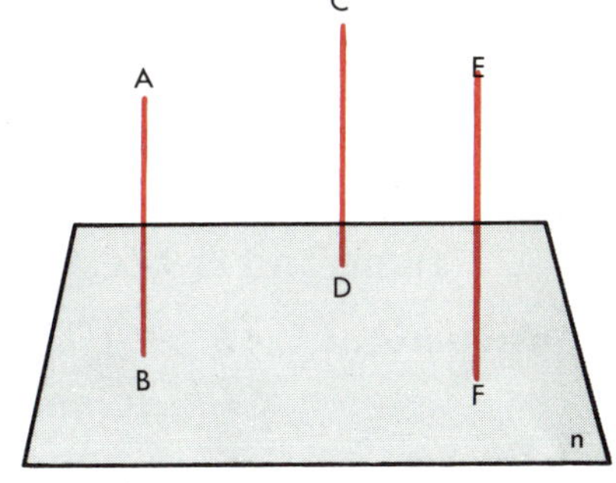

Figure 11–23

11.32 THEOREM

If two lines are parallel to a third line, they are parallel to each other.

Another theorem limited to lines in one plane stated that if two angles in the same plane have their sides respectively parallel, extending in the same direction, they are congruent (5.29). Is this proposition true if we require that the two angles are not in the same plane?

Given: The sides of $\angle BAC$ and $\angle B'A'C'$ respectively parallel and extending in the same direction. Plane m is determined by $\overleftrightarrow{AB}$ and $\overleftrightarrow{AC}$. Plane n is determined by $\overleftrightarrow{A'B'}$ and $\overleftrightarrow{A'C'}$.

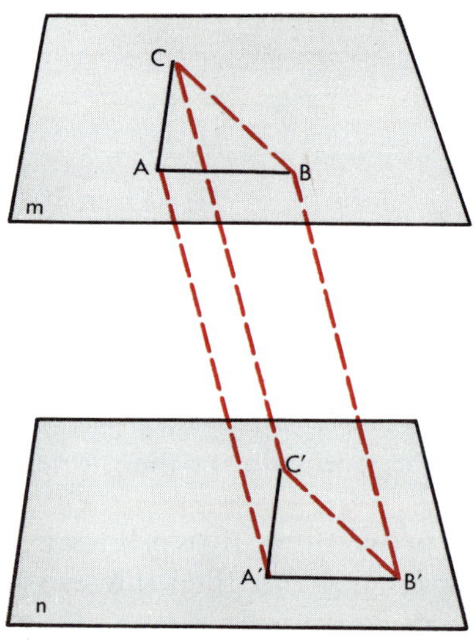

Figure 11–24

Conjecture: $\angle BAC \cong \angle B'A'C'$

Plan: Prove that $ABB'A'$, $ACC'A'$, and $BCC'B'$ are parallelograms and $\triangle BAC \cong \triangle B'A'C'$.

Construction: $\overline{AB} \cong \overline{A'B'}$ and $\overline{AC} \cong \overline{A'C'}$. Draw $\overline{AA'}$, $\overline{BB'}$, $\overline{CC'}$, $\overline{BC}$ and $\overline{B'C'}$.

Proof: The proof is left to the student. Note that this proves Assumption 11.11.

11.33 THEOREM

If two angles, not in the same plane have their sides respectively parallel and extending in the same direction, they are congruent.

Vocabulary List

dihedral
polyhedral
trihedral
tetrahedral
hexahedral
coplanar points
plane angle
face angle
pentahedral

Chapter Review

Answer 1–5 by using "sometimes," "always," or "never."

1. If a segment does not intersect a plane, it is parallel to the plane.
2. A plane perpendicular to one of two skew lines is perpendicular to the other.
3. A plane parallel to one of two skew lines is parallel to the other.
4. A plane perpendicular to one of two intersecting planes must intersect the other.
5. If two planes are perpendicular, a line perpendicular to one of them at any point of their intersection lies in the other.

Correctly complete the statements of 6-15.

6. The measure of a face angle of a polyhedral angle must be less than _?_.
7. The measures of two face angles of a trihedral angle are 80° and 110°. The measure of the third face angle must be less than _?_.
8. Four non-coplanar points determine _?_ planes.
9. The number of planes determined by five points, if no three are collinear and no four are coplanar, is _?_.
10. Through a given point there _?_ plane(s) perpendicular to a given plane.
11. Through a given point there _?_ plane(s) parallel to a given plane.
12. The faces of a dihedral angle are _?_.
13. The plane angles of the dihedral angles determined by oblique planes are _?_.

14. A ? angle of a ? angle is formed by two ?, one in each face, perpendicular to the edge at the same point.

15. The four methods of determining a plane are ?.

Some of your answers to the following questions will be conjectures you may wish to prove.

16. What theorem or corollary implies that the spokes of a wheel, which are perpendicular to the axle, lie in one plane?

17. If a plane parallel to the edge of a dihedral angle intersects the faces of the angle, are the intersections parallel lines?

18. Are two dihedral angles congruent if their faces are parallel? Are they supplementary?

19. Can a plane be passed through any point within a dihedral angle so that it is perpendicular to each face of the angle?

20. If a plane is perpendicular to one of two parallel planes, is it perpendicular to the other?

21. Are two planes parallel to each other if they are perpendicular to the same plane?

22. Are a plane, and a line not in the plane, parallel if they are both perpendicular to the same line?

23. Are the segments of parallel lines cut off by two parallel planes congruent? Prove your answer.

24. Do the segments which connect in succession the midpoints of the sides of a skew quadrilateral form a parallelogram? (A skew quadrilateral is one in which the vertices are not all in the same plane.) Prove your answer.

25. Is the sum of the measures of the angles of a skew quadrilateral equal to, less than, or more than 360°? Explain.

26. If $ACFD$ and $BCFE$ are parallelograms, prove that $ABED$ is a parallelogram. Is $\angle ACB \cong \angle DFE$? Explain. (Use Figure 11-25.)

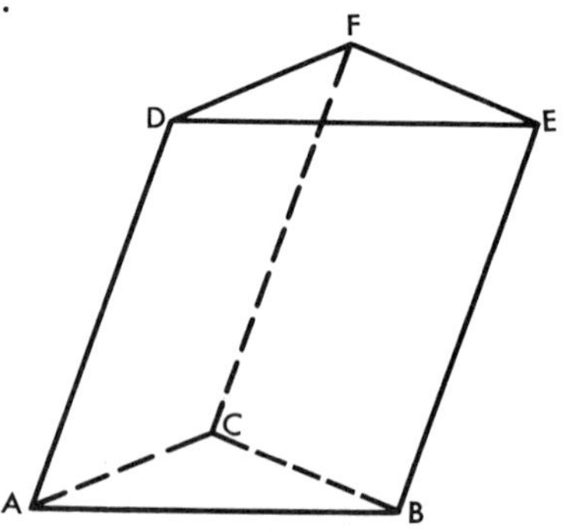

Figure 11–25

Chapter 11 Test

Place the letter of the correct answer opposite the number of the question on your paper.

1. A plane is *not* determined by (a) two intersecting lines; (b) two skew lines; (c) two parallel lines.
2. Two planes perpendicular to the same plane (a) are always parallel; (b) are never parallel; (c) may or may not be parallel.
3. Two planes perpendicular to the same line (a) are always parallel; (b) are never parallel; (c) may or may not be parallel.
4. A line cannot be perpendicular to a line in a plane if it (a) is oblique to the plane; (b) is parallel to the plane; (c) lies in the plane.
5. A line cannot be drawn perpendicular to each of two (a) skew lines; (b) intersecting lines; (c) intersecting planes.
6. When two planes intersect, the figure(s) formed (a) is a plane angle; (b) are dihedral angles; (c) is a polyhedral angle.
7. The sum of the measures of the face angles of a convex polyhedral angle (a) is 180°; (b) is less than 360°; (c) can be any value.
8. The point of intersection of a line and a plane is called the (a) foot; (b) vertex; (c) end, of the line.
9. A hexahedral angle has (a) 4; (b) 6; (c) 7, face angles.
10. To measure the size of a dihedral angle, we must (a) measure the size of its faces; (b) measure its face angle; (c) measure its plane angle.
11. It is not possible to construct a trihedral angle the measures of whose face angles are respectively 80°, 20°, and (a) 50°; (b) 70°; (c) 100°.
12. The number of lines that can be drawn perpendicular to a given line from a point not on that line is (a) none; (b) one; (c) an unlimited number.
13. The number of lines that can be drawn perpendicular to a given line from a point on that line is (a) none; (b) one; (c) an unlimited number.
14. The intersection of two faces of a polyhedral angle is called (a) an edge; (b) a vertex; (c) an element of the angle.

15. The point of concurrency of the lines of intersection of the faces of a polyhedral angle is called the (**a**) edge; (**b**) vertex; (**c**) axis, of the angle.

Complete the following statement by placing on your paper the words which correctly fill the numbered blanks.

16–19. A(n) (16) angle of a dihedral angle is formed by two rays, one in each (17), drawn (18) to the (19) at the same point.

20–25. Prove the theorem: If two parallel planes are cut by a third plane, the lines of intersection are parallel.

12

Circles, Spheres, Tangents, Secants, and Chords

Two of the most interesting, pleasing, and useful geometric figures are the circle and the sphere. They appear frequently in art, design, engineering, and construction. Without the circle many of our advances in modern living would not have occurred. Our civilization literally moves on, and sometimes in, circles on our "spherical" Earth.

Circle, radius, diameter, and arc of a circle have previously been defined in Sections 2.47 and 2.48. We suggest that you review these sections at this time. We will use the word radius to denote the *length of the radius* and the *segment*. This ambiguity should not be confusing since the meaning will be clear from the context in which it is used. As in previous chapters, it will be necessary to establish additional definitions and assumptions about the geometric figures to be considered.

12.00 A sphere is the set of points in space each of whose points is a distance r from a given point S, called the *center* of the sphere. The distance r is a positive real number. If P is a point of the sphere, then $\overline{SP}$ is called a *radius* (plural: *radii*). If R and T are points of the sphere and $\overline{RT}$ contains S, then $\overline{RT}$ is called a *diameter*. (How does this compare with the definition of a circle in Section 2.47?)

Circles or spheres are commonly denoted by the letter which names the center of the circle or sphere.

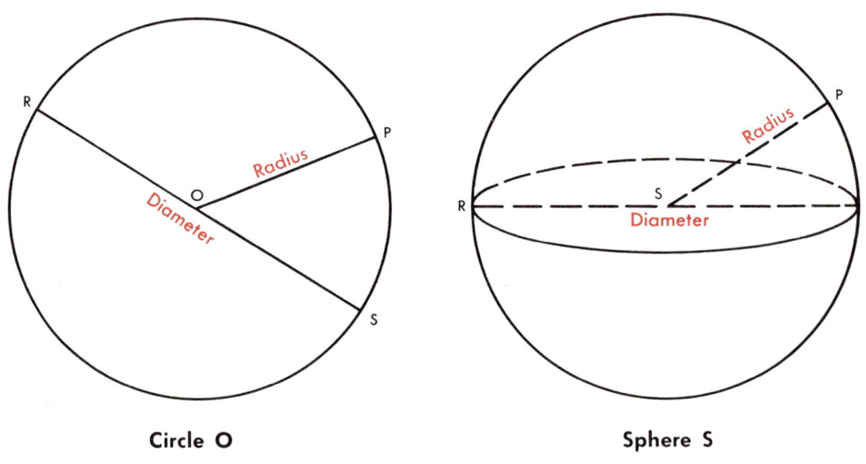

Figure 12–1

12.01 The circumference of a circle is the length of the circle.

We have occasions to work with "halves" of circles and spheres. We cannot just say that half of a circle is a semicircle, for someone might point out that the total of the lengths of several separate arcs might be half the length of the circle. Thus the following definitions are necessary even though the concept is simple.

12.02 A semicircle is the union of the endpoints of a diameter and the set of points of the circle on one side of the diameter. The *length of a semicircle* is considered to be one-half the length of the circle.

12.03 A hemisphere is the union of the intersection of a sphere and a plane containing the center of the sphere, and the set of points of the sphere on one side of the plane.

We have not stated the character of the figure formed by the intersection of the plane and the sphere. That must wait until we can prove it. What is your guess?

12.04 Concentric circles are coplanar circles having the same center and non-congruent radii. Spheres are also concentric if they have the same center and non-congruent radii.

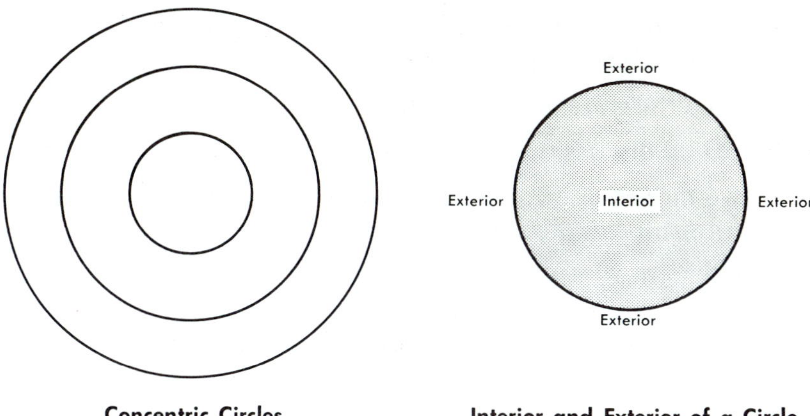

Concentric Circles **Interior and Exterior of a Circle**

Figure 12–2

The definition of a circle implies that it is a "line," not a "disc." The definition of a sphere classifies it as a "shell," not a "solid." Since we use the regions enclosed by circles and spheres, we need to define these ideas.

12.05 The interior of a circle is the union of its center and the set of points in the plane of the circle whose distances from the center are less than the measure of the radius.

12.06 The interior of a sphere is the union of its center and the set of points in space whose distances from the center are less than the measure of the radius.

12.07 A circular region is the union of a circle and its interior.

12.08 A spherical region is the union of a sphere and its interior.

12.09 The exterior of a circle includes all points in the plane of the circle not contained in the circular region. (Distances from the center are greater than the measure of the radius.)

Exercises

1. Define "exterior of a sphere."
2. Having given a circle, name the union of its circular region and the exterior of the circle.
3. Name the intersection of two semicircles of a circle determined by the same diameter.
4. Discuss whether the length of a semicircle, as defined, is actually one-half the length of the circle.
5. Discuss the statement "A diameter bisects a circle."
6. State and discuss the converse of the statement of Exercise 5.
7. If two circles or two spheres have congruent radii, what is the relationship of the circles or spheres? Can you prove your response?
8. If two circles or two spheres are congruent, what is the relationship of their radii? Can you prove your response?
9. Define exterior of a circle in terms of the distances of points from the center of the circle.
10. Complete: The intersection of a line and a circle or a sphere may be ___?___.
11. Find or devise a name for the region of a plane between two concentric circles, including the points of the circles.
12. Find or devise a name for the region between two concentric spheres, including the points of the spheres.

BASIC ASSUMPTIONS

The responses to several of the questions in the preceding exercise group were assumptions. Some of these will now be formalized so that they can be used in future proofs.

12.10 *Assumption: Circles having congruent radii are congruent.*

12.11 *Assumption: Congruent circles have congruent radii.*

12.12 *Assumption: Spheres having congruent radii are congruent.*

12.13 *Assumption: Congruent spheres have congruent radii.*

12.14 *Assumption: A point is in the exterior of a circle or a sphere if its distance from the center is greater than the measure of the radius.*

12.15 *Assumption: If a line contains a point of the interior of a sphere or circle, and lies in the plane of the circle, then it intersects the sphere or circle in exactly two points.*

CHORDS, SECANTS, TANGENTS

To continue our development of the structure of geometry, we must introduce some new terminology. The words *chord*, *secant*, and *tangent* have their derivation in antiquity. *Tangent* came from the Latin word *tangere* meaning "to touch;" *chord* came from the Greek word for "string," and *secant* from the Latin word *secare* meaning "to cut."

12.16 A chord is a segment whose endpoints are two distinct points of a circle or of a sphere.

12.17 A secant is a line, segment, or ray which contains two distinct points of a circle or a sphere and points of its exterior.

12.18 A line is tangent to a sphere or a circle if it intersects the circle in exactly one point and is coplanar with the circle. A ray or segment which is a portion of the tangent line and also contains the point of tangency is said to be tangent to the sphere or circle.

12.19 A plane is tangent to a sphere if it intersects the sphere in exactly one point.

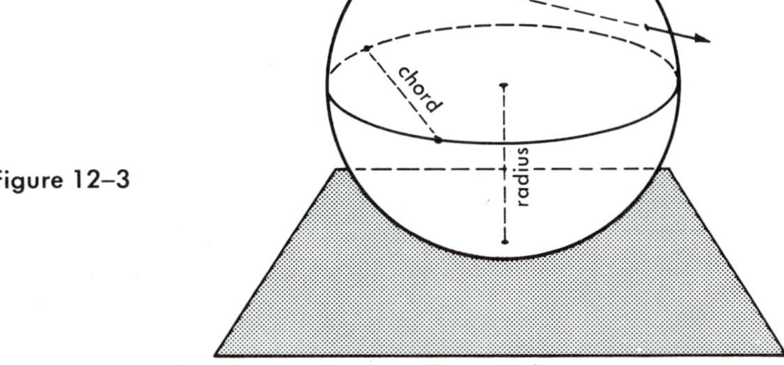

Figure 12–3

12.20 The line of centers of two coplanar circles or of two spheres is the straight line determined by the centers of the circles or the spheres.

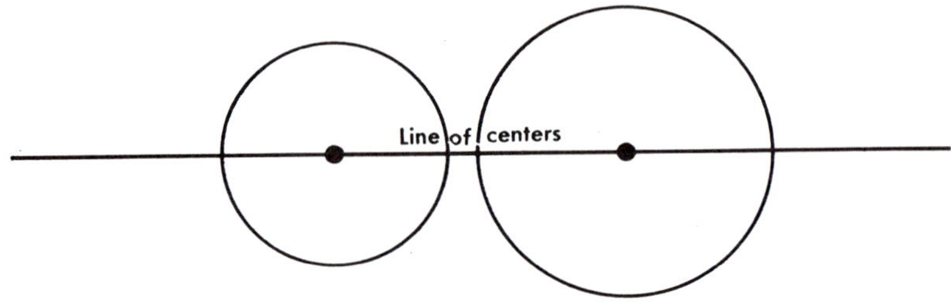

Figure 12–4

ARCS AND DEGREE MEASURE

We will use the symbol $\widehat{AB}$ to mean arc AB, where A and B are the endpoints of the arc. The measure of the length of arc AB will be denoted by m$\widehat{AB}$. "The degree measure of $\widehat{AB}$ is n" will be written "m$\widehat{AB} = n°$."

Although "arc" was defined in Section 2.48, there are other possible non-conflicting definitions of arc, one of which we pose below. First we need to define "central angle."

12.21 A central angle of a circle or of a sphere is an angle whose vertex is the center of the circle or the sphere.

12.22 A minor arc of a circle is the arc on and in the interior of a central angle of the circle. We shall say that the arc subtends the central angle and conversely.

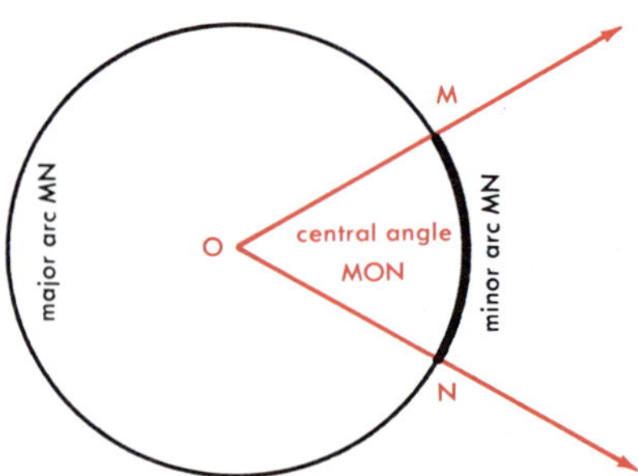

Figure 12–5

12.23 A major arc of a circle is the arc on and in the exterior of a central angle of the circle.

12.24 The degree measure of a minor arc of a circle is equal to the degree measure of the central angle it subtends.

Exercises

1. Does a circle have point symmetry? Line symmetry?
2. How many planes can be tangent to a sphere at a given point of the sphere? How many lines?
3. How many tangents may be drawn to a circle from a point in the exterior of the circle? To a sphere from a point in the the exterior of the sphere? Can you prove your response?
4. Is a diameter of a circle also a chord of the circle?
5. Is a chord of a circle also a diameter of the circle?
6. A segment intersects a sphere in exactly one point. Is it tangent to the sphere? Explain.
7. A line intersects a circle in exactly one point. Is it tangent to the circle? Explain.
8. May a secant of a circle also be tangent to this circle? Explain.
9. In a plane, how many distinct circles of the same radius may be tangent to a line at a given point.
10. A chord of a circle is drawn whose measure equals the measure of a radius of the circle. Radii are drawn to the ends of this chord. Name the measure of the central angle determined by the two radii. Explain your response.
11. If the measure of a minor arc of a circle is 80°, what is the measure of the major arc of this circle having the same endpoints? Prove your response.
12. A triangle is formed by two radii and a chord joining the ends of the radii. The measure of an angle formed by the chord and one of the radii is 30°. What is the measure of the minor arc determined by the two radii?
13. Is the length of the chord joining the endpoints of two radii determining an angle whose measure is 100° twice the length of the chord joining the endpoints of two radii of the same circle which determine an angle whose measure is 50°? Explain.

ARCS, CHORDS, AND ANGLES

Could two arcs having the same degree measure possibly have different lengths? See Figure 12-6. Do $\widehat{EF}$ and $\widehat{AB}$ have the same degree measure? The same length? Could two arcs having the same length have different degree measures? Would two arcs of the same or congruent circles having the same length have the same degree measure? Can you prove it? This leads to the following assumption.

12.25 *Assumption: In the same circle or in congruent circles, congruent central angles subtend congruent arcs and congruent arcs subtend congruent central angles.*

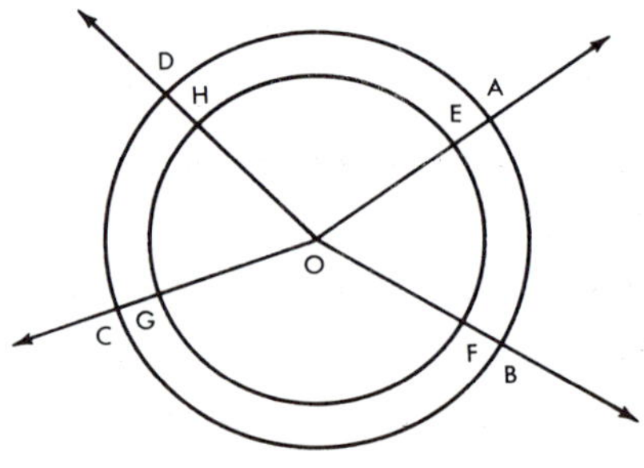

Figure 12–6

EXAMPLE: Using Figure 12-6, if $\angle AOB \cong \angle COD$, then $\widehat{AB} \cong \widehat{CD}$. Also if $\widehat{HG} \cong \widehat{FE}$, then $\angle HOG \cong \angle FOE$.

Using Sections 12.24 and 12.25, the following theorems are evident.

12.26 THEOREM

Congruent arcs of the same circle or of congruent circles have the same degree measure.

12.27 THEOREM

In the same circle or in congruent circles, the longer of two minor arcs subtends the central angle whose measure is greater.

12.28 THEOREM

Of two central angles in the same circle or in congruent circles, the one with the greater measure subtends the longer arc.

In the same manner that central angles and arcs subtend each other, the minor arc, the chord, and the central angle determined by the endpoints of the arc are said to subtend each other. What is the relationship of arcs subtending congruent chords of the same or congruent circles? Can you prove your conjecture using Figure 12-7?

Given: Congruent circles O and O'. Chord $\overline{AB} \cong \overline{A'B'}$.

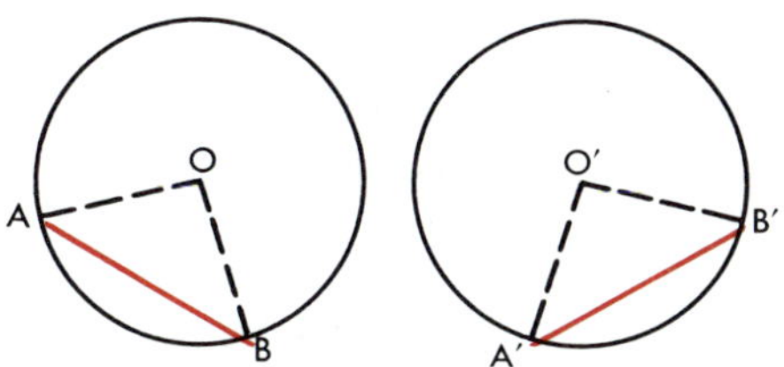

Figure 12–7

Conjecture: $\widehat{AB} \cong \widehat{A'B'}$

Plan: Prove the triangles congruent, then use 12.25.

Proof: The proof is left to the student.

12.29 THEOREM

In the same circle or in congruent circles, congruent chords subtend congruent arcs.

The converse of Theorem 12.29 may also be proved and is stated as follows. (The proof is left to the student.)

12.30 THEOREM

In the same circle or in congruent circles, congruent arcs subtend congruent chords. (The proof is left to the student.)

INEQUALITIES (Optional)

If the measure of one chord is greater than the measure of another chord of the same circle, what is the relationship of the arcs and central angles subtended by these chords? Can you prove your answer? Perhaps you can if you review the theorems of Sections 10.16 and 10.17. The use of indirect proof (pages 234–237) may also be helpful.

12.31 THEOREM

In the same circle or in congruent circles, the longer of two minor arcs subtends the longer chord.

(*Suggestion:* Use 12.27 and 10.16)

12.32 THEOREM

In the same circle or in congruent circles, the longer of two chords subtends the longer minor arc.

INSCRIBED AND CIRCUMSCRIBED POLYGONS

It is easy to see that chords, joining in order three or more distinct points of a circle, form polygons. Would the intersections of tangents to the circle at these points be the vertices of a polygon of the same number of sides as the polygon formed by the chords?

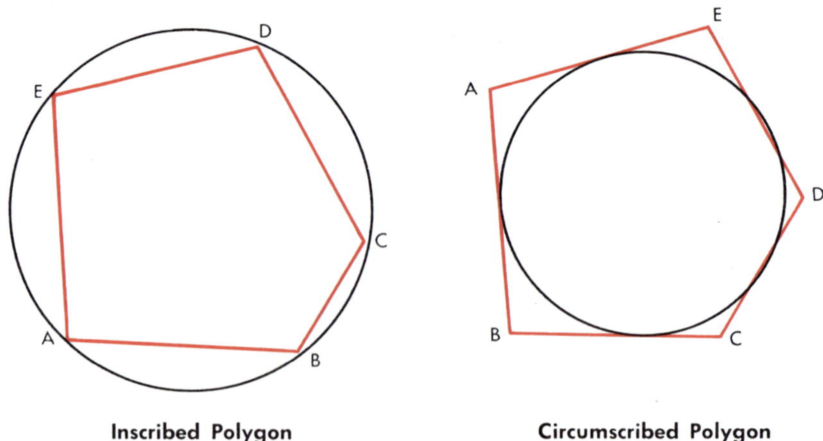

Figure 12–8

12.33 A polygon is inscribed in a circle if its vertices lie on the circle. The circle is then said to be *circumscribed* about the polygon.

12.34 A polygon is circumscribed about a circle if each side of the polygon is tangent to the circle. The circle is then said to be *inscribed* in the polygon.

Exercises

1. How can you prove that two chords of a circle are congruent?
2. How can you prove that two arcs of a circle are congruent?
3. Construct an equilateral hexagon with its vertices on a given circle. Is this an inscribed or a circumscribed polygon? Is this a regular hexagon? Explain.
4. In a circle, draw a diameter and construct a diameter perpendicular to it. Explain why the circle *must* be divided into four congruent arcs by the ends of the diameters.
5. Can any three points of a circle be the points of tangency of a circumscribed triangle? Explain the limitations of the position of the points.
6. State the authority for saying that "In the same or congruent circles, the longer of two chords subtends the central angle with the greater measure."
7. Theorem 10.13 states that the least distance from a point to a line is the measure of the segment from the point perpendicular to the line. What is the shortest chord that can be drawn through a given point in the interior of a circle? Prove it.

Perform the following three experiments.

8. Construct a circle, draw a chord, and construct a radius perpendicular to the chord.
   (a) Measure the two parts of the chord.
   (b) Measure the arcs intercepted by chord and radius.
   (c) Measure the angles formed by the constructed radius and the radii drawn to the ends of the chord.
9. Construct a circle and construct the perpendicular bisector of a chord of the circle.
   (a) Does this perpendicular bisector pass through the center of the circle?
   (b) Does it bisect the arc of the chord?

10. Construct a circle and draw a radius of the circle. Construct a perpendicular to this radius at its endpoint on the circle.

(a) Can this line contain the endpoint of any other radius of this circle?

(b) Is this perpendicular line tangent to the circle?

(c) If a line contains the endpoint of a radius on the circle and is not perpendicular to the radius, *must* this line contain another point of the circle?

(d) Are your previous answers consistent with each other?

(e) Do your answers represent the process of inductive, or deductive reasoning?

There are many experiments one could perform with circles to determine possible relationships of chords, tangents, secants, and radii. The preceding list is meant to be a start, not a limit, for your inductive thinking.

DEDUCTIVE REASONING

The following propositions have been selected for deductive proof from the list of relationships developed by induction.

Given: Circle O with $\overleftrightarrow{OB} \perp \overline{AC}$ at D (Figure 12-9)

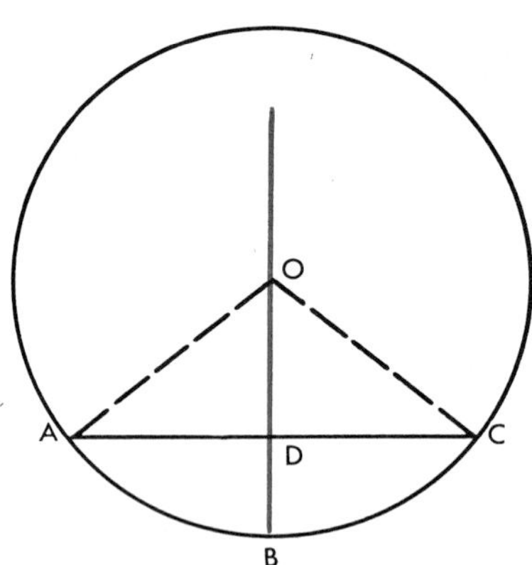

Figure 12–9

Conjecture: $\widehat{AB} \cong \widehat{CB}$, and $\overline{AD} \cong \overline{CD}$

Plan: Prove the triangles are congruent.

Proof: The proof is left to the student.

12.35 THEOREM

If a line through the center of a circle is perpendicular to a chord, it bisects the chord and its arc.

12.36 THEOREM

A line through the center of a circle that bisects a chord (which is not a diameter) is perpendicular to it.

The following proof considers another converse of 12.35.

Given: Circle O with $\overleftrightarrow{BD} \perp$ bisector of chord $\overline{AC}$ (Figure 12-10)

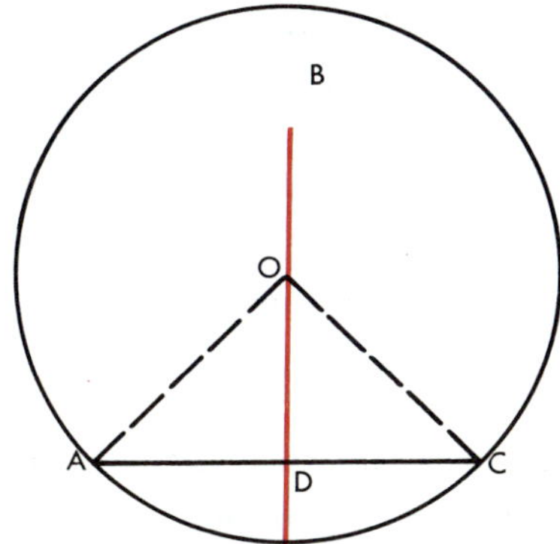

Figure 12–10

Conjecture: $\overleftrightarrow{BD}$ passes through O.

Plan: Every point on the perpendicular bisector of chord $\overline{AC}$ is equidistant from the ends of $\overline{AC}$. Why? Now prove that any point equidistant from the ends of a line is on the perpendicular bisector of the line. This can be done in $\triangle AOC$ (since $\overline{AO} \cong \overline{CO}$) by constructing $\overleftrightarrow{OD}$ as the bisector of $\overline{AC}$, and then proving $\overleftrightarrow{OD} \perp \overline{AC}$. Therefore O, the center of the circle, lies on the perpendicular bisector of the chord (3.20).

12.37 THEOREM

The perpendicular bisector of a chord of a circle passes through the center of the circle.

We now consider the relationship of congruent chords of a circle.

Given: Circle O with $\overline{AB} \cong \overline{CD}$, $\overline{OE} \perp \overline{CD}$, and $\overline{OF} \perp \overline{AB}$ (Figure 12-11)

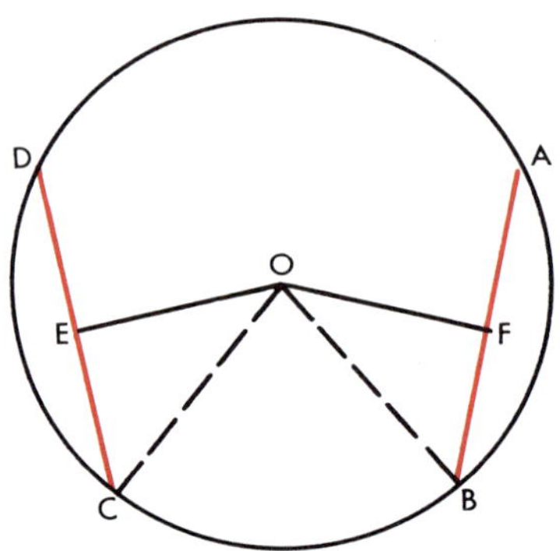

Figure 12–11

Conjecture: $\overline{OE} \cong \overline{OF}$

Proof: The proof is left to the student.

12.38 THEOREM

In the same circle or in congruent circles, congruent chords are equidistant from the center.

The converse of 12.38 is considered below.

Given: Circle O with $\overline{OE} \perp \overline{CD}$ and $\overline{OF} \perp \overline{AB}$, and $\overline{OE} \cong \overline{OF}$ (Figure 12-11)

Conjecture: $\overline{AB} \cong \overline{CD}$

Plan: Prove that $\overline{CE} \cong \overline{BF}$. Since $m\overline{CD} = 2m\overline{CE}$ and $m\overline{AB} = 2m\overline{BF}$, then $m\overline{CD} = m\overline{AB}$.

12.39 THEOREM

In the same circle or in congruent circles, chords equidistant from the center are congruent.

If two chords in a circle are not the same distance from the center, could they be congruent? Is there a pattern of inequality between the two chords?

Given: Circle O, in which $m\overline{AB} > m\overline{CD}$, $\overline{OG} \perp \overline{AB}$, and $\overline{OH} \perp \overline{CD}$ (Figure 12-12)

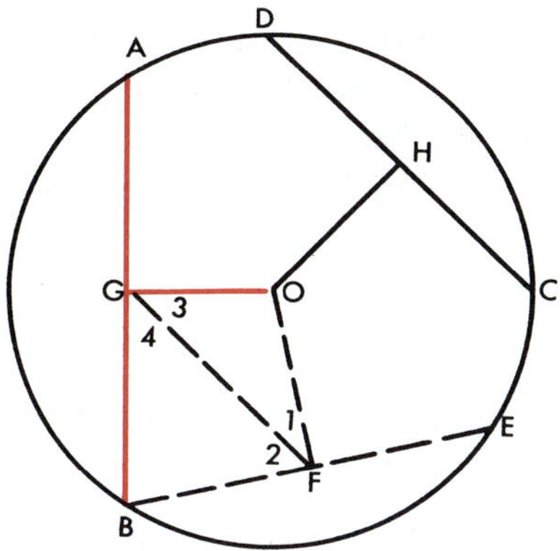

Figure 12–12

Conjecture: $m\overline{OH} > m\overline{OG}$

Plan: (1) From B, construct chord $\overline{BE} \cong \overline{CD}$, (2) make $\overline{OF} \perp \overline{BE}$, and (3) draw $\overline{FG}$. (4) $m\overline{AB} > m\overline{CD}$, (5) $m\overline{BE} = m\overline{CD}$; (6) hence, $m\overline{AB} > m\overline{BE}$. (7) $\overline{OF} \perp \overline{BE}$, $\overline{OG} \perp \overline{AB}$; (8) hence, $m\overline{BF} = \frac{1}{2}m\overline{BE}$, $m\overline{BG} = \frac{1}{2}m\overline{AB}$, and (9) $m\overline{BG} > m\overline{BF}$. (10) $\angle BFO$ and $\angle BGO$ are congruent right angles. (11) $m\angle 2 > m\angle 4$, $m\angle 3 > m\angle 1$; (12) hence, $m\overline{OF} > m\overline{OG}$. (13) $\overline{OH} \perp \overline{CD}$, (14) $m\overline{OH} = m\overline{OF}$; (15) hence, $m\overline{OH} > m\overline{OG}$.

Proof: The proper form and reasons are left to the student. When the proof is complete, the following theorem is established.

12.40 THEOREM

In the same circle or in congruent circles, the longer of two non-congruent chords is nearer the center of the circle.

12.41 THEOREM

In the same circle or in congruent circles, chords unequally distant from the center are unequal in length, the nearer being the longer.

SPHERES

If we draw a "line" on a sphere, must it be a "curved" line? If the line is closed must it be a circle? Is it possible to draw a line on a sphere that would be a circle?

Is a circle necessarily a plane figure? If a plane intersects a sphere in more than one point, would the intersection be a closed curved line? Would it necessarily be a circle?

To prove a closed curved line is a circle we must show that it satisfies the definition of a circle: (1) that it is in a plane, and (2) that every point of the curve is equidistant from a given point in the plane. Let us examine this idea in formal fashion.

Given: Sphere O intersected by plane MN, forming the intersection labeled ABC.

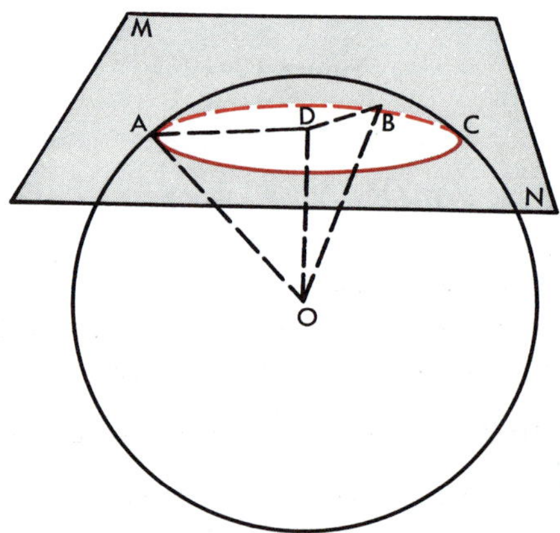

Figure 12–13

Conjecture: ABC is a circle.

Plan: Draw $\overline{OD} \perp$ plane MN. Draw $\overline{DA}$ and $\overline{DB}$. Prove that $\triangle OAD \cong \triangle OBD$. Since A and B are any points on the intersection of the plane and the sphere, and are equidistant from D, the section is a circle with D as center (2.47).

Proof: The formal proof is left to the student. Upon its completion the following theorem is established.

12.42 THEOREM

If a plane intersects a sphere in more than one point, the intersection is a circle.

By using a plan similar to the plan of proof for Theorem 12.42, the following proposition can be proved.

12.43 THEOREM

A radius of a sphere, perpendicular to the plane of a circle of the sphere, passes through the center of the circle.

12.44 The axis of a circle of a sphere is the diameter of the sphere perpendicular to the plane of the circle. (XY in Figure 12-14)

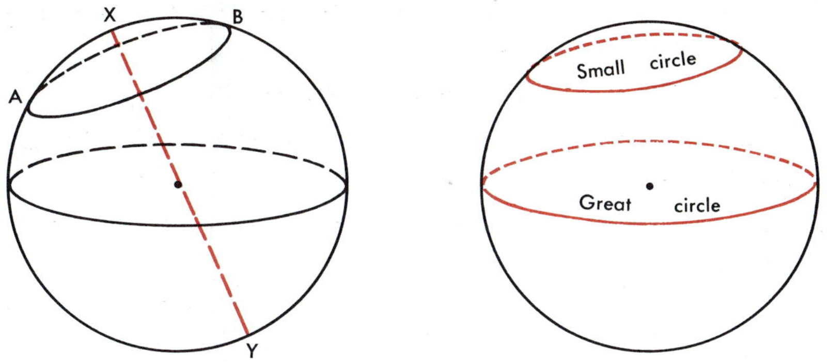

Figure 12–14

12.45 A great circle of a sphere is any circle on the sphere whose center is the center of the sphere. All other circles described on the sphere are called *small circles* of the sphere.

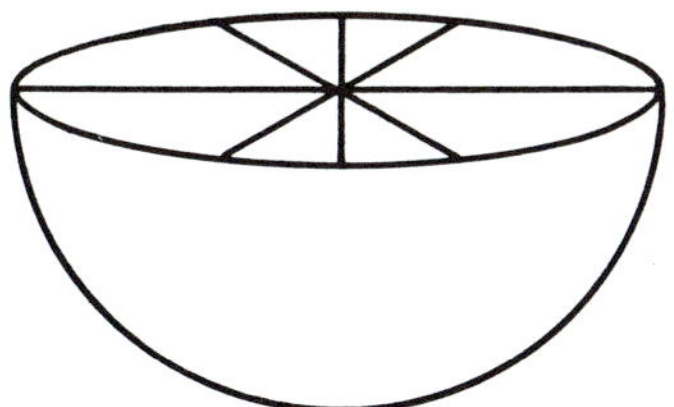

Figure 12–15

As we sit at our breakfast table with half a grapefruit on our plate, we are faced with the geometry of the sphere. The portion on our plate resembles a hemisphere. As the whole grapefruit is cut, the intersection of knife and grapefruit approximates a circle. The cross-sections of the interior of the grapefruit resemble radii and central angles. The core of the fruit corresponds to an axis of the sphere.

Exercises

1. How many small circles of a sphere can be passed through a point of the sphere? How many great circles?
2. How many small circles of a sphere can be passed through two points of the sphere? How many great circles? Explain.
3. If the earth were a sphere would the equator represent a great circle, or a small circle?
4. What is the axis of the equator of the earth?
5. Are all great circles of the same or congruent spheres congruent? Explain.
6. Are all small circles of the same or congruent spheres congruent? Explain.
7. In how many points can two great circles of a sphere intersect?
8. Are two chords of a sphere congruent if they are equidistant from the center?
9. Inscribe an equilateral quadrilateral in a given circle. Inscribe an equilateral octagon in a given circle. Are these figures necessarily regular polygons? Explain.
10. Complete: A circle is determined (located) on a plane if its ? and its ? are known. Would these conditions determine a particular sphere in space?
11. Are the perpendicular bisectors of the sides of an inscribed polygon concurrent? Explain.
12. Are the sides of an inscribed square equidistant from the center? Why?
13. Would any three distinct points on a sphere determine a circle? Explain.
14. The intersection of a sphere and a plane through its center is a (great, small) circle of the sphere.
15. Given a circle and a point in its interior, construct a chord that will have the point as its midpoint.

16. Explain how you could find the diameter of a broken circular disc if you have only a portion of the disc to work with. Describe the centering device used by industry—or devise your own version.

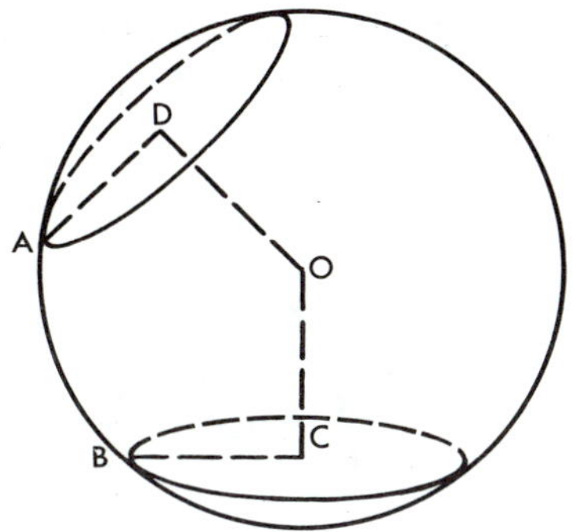

Figure 12–16

17. If the radii of two small circles of a sphere are congruent, are their planes equidistant from the center of the sphere? Prove your answer.

TANGENT CIRCLES AND SPHERES

Have you ever watched the sparks flying from a grinding wheel when it is grinding a piece of metal? These sparks travel in a path that is tangent to the wheel. Muddy car tires turning rapidly will throw off particles of mud in a path tangent to the outer edge of the tire. When a stone is whirled rapidly in a sling and suddenly released, its path will be tangent to its previous path in the sling. These and many similar instances of tangent paths, lines, and surfaces are necessarily part of the basic knowledge required of scientists and engineers. We will study the basic principles of tangency. Review 12.18 and 12.19.

12.46 Two circles are tangent to each other if they are in the same plane and are both tangent to the same line at the same point. Two spheres are tangent to each other if they are both tangent to the same plane at the same point.

Two spheres are not necessarily tangent to each other if they are each tangent to the same line at the same point. They must be tangent to a plane, at the same point, in order to be tangent to each other. Why are two spheres that are tangent to the same line at the same point not necessarily tangent to each other?

12.47 Spheres or coplanar circles may be placed so that one lies in the interior of the other and are tangent to the same line at the same point. These are called internally tangent circles or spheres. Circles O' and O'' are internally tangent. (Figure 12-17)

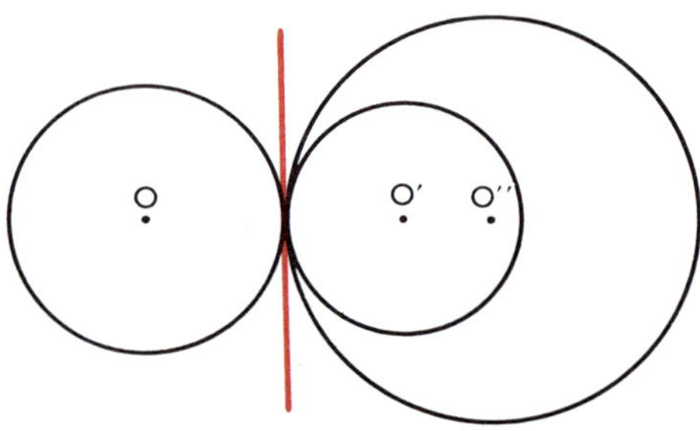

Figure 12–17

If spheres or coplanar circles lie outside each other and are tangent to the same line at the same point, they are externally tangent Circles O and O' and O and O'' are externally tangent.

12.48 Coplanar circles may have a line, ray, or segment tangent to both of them. If this common tangent intersects the segment whose endpoints are the centers of the circles, it is called a common internal tangent. ($\overleftrightarrow{AB}$ or $\overleftrightarrow{CD}$, Figure 12-18)

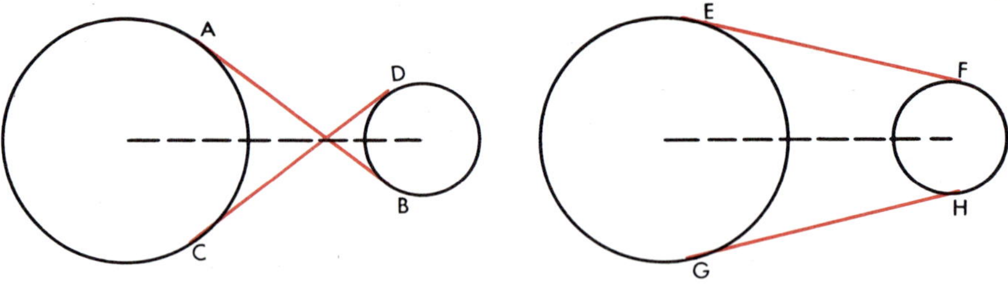

Figure 12–18

If the common tangent does not intersect the segment joining the centers, it is called a common external tangent. ($\overleftrightarrow{EF}$ or $\overleftrightarrow{GH}$, Figure 12-18)

TANGENTS IN OUR ENVIRONMENT

Although not strictly tangent, the bicycle sprocket wheels and chain, or the motor with belt and pulley are typical examples of tangents in everyday use. Tangent circles are suggested by the interlocking gear wheels of a watch or of various machines in industry.

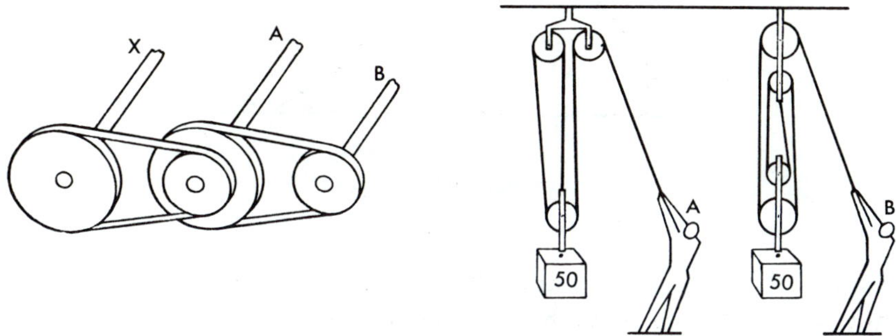

Figure 12–19

An interesting application of internally and externally tangent circles is that of compound curves.

12.49 A compound curve is composed of joined parts of the arcs of tangent circles. These arcs are joined at the point of intersection of the tangent circles. (Figure 12-20)

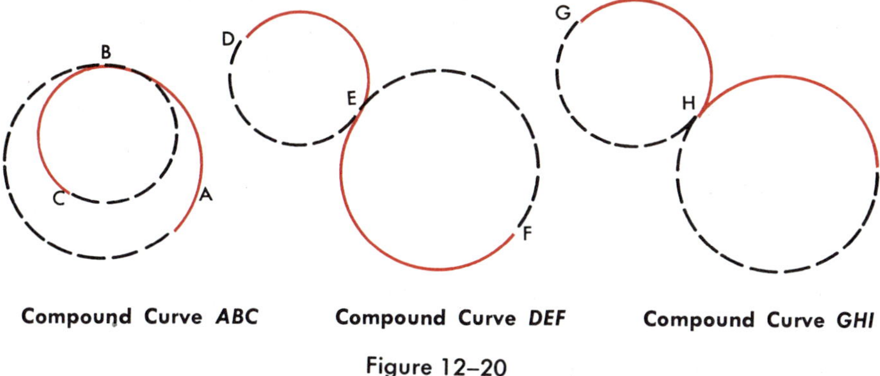

Compound Curve *ABC* **Compound Curve *DEF*** **Compound Curve *GHI***

Figure 12–20

Compound curves can be easily recognized in the design of the clover-leaf intersections of highways and the switching yards of a railroad. They are also much used in the field of art and design. Our digits 6, 8, 9 are common examples of compound curves.

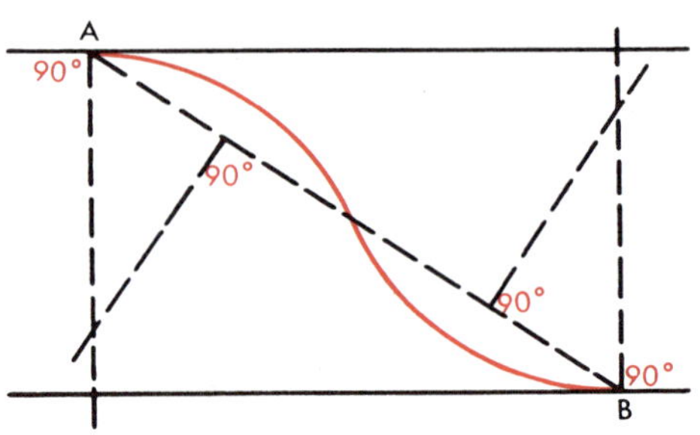

Figure 12–21

Figure 12-21 shows a method by which we could construct a connecting (compound) curve between two roads.

First draw a segment to join the desired points, A and B. Bisect that segment. Then find the center of the circles having those half-segments as chords and the roads as tangents.

Exercises

1. How many common internal tangents do the following have? (a) Two concentric circles. (b) Two internally tangent circles. (c) Two externally tangent circles. (d) Two coplanar circles with two points in common. (e) Two coplanar non-intersecting circles, the circles and their interiors having no common points.
2. Repeat Exercise 1, considering the common external tangents.
3. How many tangents can be drawn to a circle through a point in the interior of the circle? Through a point of the circle? Through a point in the exterior of the circle but coplanar with the circle?
4. How many tangent lines can be drawn to a sphere through a point in the interior of the sphere? Through a point of the sphere? Through a point in the exterior of the sphere?
5. How many tangent planes may be drawn to a sphere through a point of the sphere? Through a point in the exterior of the sphere?

TANGENTS AND RADII

Refer again to the definition of a line tangent to a circle. If we draw a radius to the point of tangency, what probably is the relationship of the radius and the tangent? Let us examine this premise in a formal, deductive manner.

Given: $\overleftrightarrow{AD}$ tangent to circle O at point C. $\overline{OC}$ is a radius.

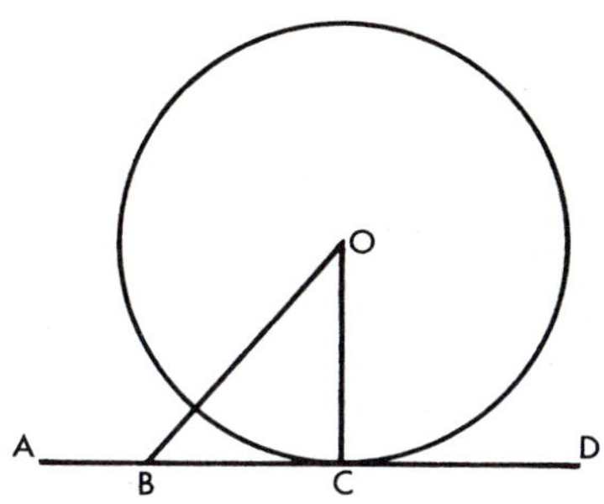

Figure 12–22

Conjecture: $\overleftrightarrow{AD} \perp \overline{OC}$

Plan: Draw $\overline{OB}$ to any point of $\overleftrightarrow{AD}$ except C. Prove $m\overline{OB} > m\overline{OC}$, then use 10.13.

Proof:

Statements	*Reasons*
1. From any point B of $\overleftrightarrow{AD}$, except C, draw $\overline{BO}$.	1. Why possible?
2. $\overleftrightarrow{AD}$ tangent to circle O at C	2. Why?
3. B is in the exterior of circle O.	3. Def. of a tangent
4. $\therefore m\overline{OB} > m\overline{OC}$	4. 12.09
5. so $\overline{OC} \perp \overleftrightarrow{AD}$ or $\overleftrightarrow{AD} \perp \overline{OC}$	5. Why?

Upon the basis of the preceding proof the following proposition is necessarily true (within the framework of geometry as we have developed it).

12.50 THEOREM

If a line is tangent to a circle, it is perpendicular to the radius drawn to the point of tangency.

The analogous proposition in three dimensions is outlined below.

Given: Plane MN tangent to sphere O at C. $\overline{OC}$ is a radius.

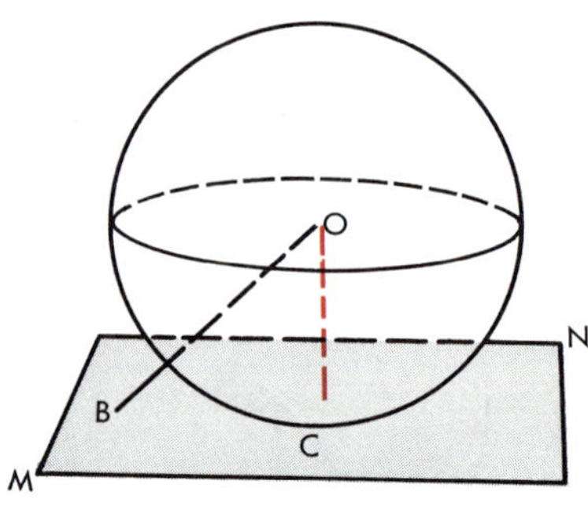

Figure 12–23

Conjecture: Plane $MN \perp \overleftrightarrow{OC}$

Proof: The proof is left to the student.

12.51 THEOREM

If a plane is tangent to a sphere, it is perpendicular to the radius drawn to the point of tangency.

Attempt to prove the following theorems.

12.52 THEOREM

If a line, coplanar with a circle, is perpendicular to a tangent to the circle at the point of tangency, it passes through the center of the circle. (*Suggestion*: Draw a radius to the point of tangency and show that it is contained in the given perpendicular.)

12.53 THEOREM

If two coplanar circles are tangent to the same line at the same point, the line of centers contains the point of tangency. (*Suggestion:* Use 12.52 to show that a perpendicular to the tangent at the point of tangency passes through the center of both circles.)

12.54 THEOREM

If two spheres are tangent to the same plane at the same point, the line of centers contains the point of tangency. (*Suggestion*: Use 12.53 and draw a great circle of each sphere through the point of tangency and in the same plane.)

12.55 THEOREM

If a line is perpendicular to a radius at its point of a sphere or of a circle, and the line is in the plane of the circle, the line is tangent to the sphere or circle. (*Suggestion*: Use Figure 12-22. Prove that $m\overline{OB} > m\overline{OC}$. Since B is any point in $\overleftrightarrow{AD}$ except C, this proves that every point in $\overleftrightarrow{AD}$, except C, lies in the exterior of the circle.)

INDUCTIVE PROCEDURE

The possibilities for experimentation with circles and spheres are almost limitless. By letting coins represent circles and marbles represent spheres, we can discover many relationships between lines, planes, circles, and spheres. Try the following experiments and formulate conjectures which seem to describe the conditions you believe are true. Compare these with theorems in this chapter.

1. Place a coin flat on a desk and arrange two pencils or wires so that they are tangent to the coin. If the wires intersect, what can be said of the segment joining the point of intersection to the center of the coin?

2. How can you represent the common internal tangents to two non-congruent circles? How can you represent two common external tangents? What conjectures can now be stated?

3. Make a wedge-shaped figure by folding a piece of cardboard along a line. Place two non-congruent marbles in the dihedral angle formed. What conjectures can be stated describing this situation?

4. Represent concentric circles by placing a dime on top of a quarter with their centers as nearly coincident as possible. What can be said of chords of the quarter which are tangent to the dime? Do congruent central angles intercept congruent

arcs? State as many conjectures that relate to concentric circles as you can. Are there corresponding conjectures about concentric spheres?

5. Make several other arrangements of lines, planes, circles, and spheres and formulate as many conjectures about their relationships as you can.

Earlier in this chapter you have proved some of the conjectures suggested by the inductive procedures above. Through the method of deduction we shall now test the two additional conjectures that are stated below as theorems. The theorems are followed by a list of exercises. In the list of exercises, you will find some of the ideas you discovered in your experimentation with coins and marbles. Can you propose other theorems to challenge your classmates?

DEDUCTIVE PROOFS

What relation did you find between the segments tangent to a circle from an exterior point and the segment joining the point and the center of the circle? How are the two tangent segments related?

Given: $\overline{AB}$ and $\overline{CB}$ tangents to circle O at points A and C respectively

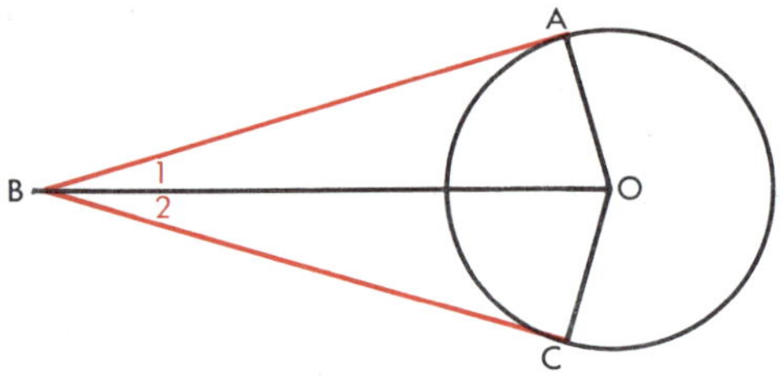

Figure 12–24

Conjecture: $\overline{AB} \cong \overline{CB}$ and $\angle 1 \cong \angle 2$

Proof: The proof is left to the student.

12.56 THEOREM

The tangent segments to a circle from a point in the exterior of the circle are congruent and make congruent angles with a segment joining the point to the center of the circle.

If two circles in a plane intersect in two points, they have a common chord joining the two points. Does the line of centers intersect this chord? In what way?

Given: Coplanar circles O and O' intersecting at A and B

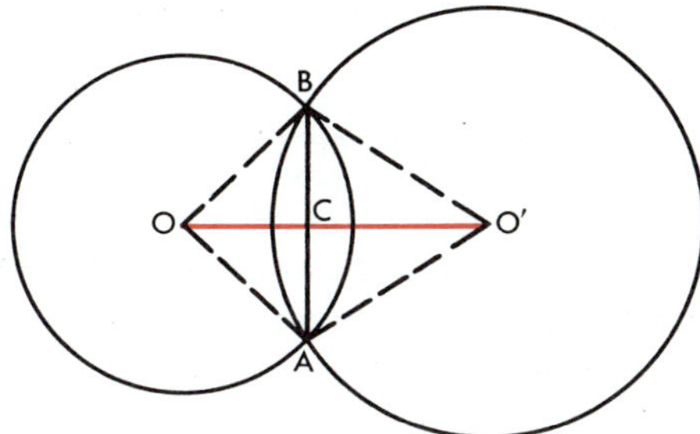

Figure 12–25

Conjecture: $\overline{OO'} \perp \overline{AB}$ and $\overline{OO'}$ bisects $\overline{AB}$

Proof: The proof is left to the student.

12.57 THEOREM

If two circles in a plane intersect in two points, the segment joining their centers is the perpendicular bisector of the common chord.

Vocabulary List

sphere
circumference
semicircle
hemisphere
concentric
chord
tangent
line of centers
central angle
minor arc
major arc
subtend
secant
inscribed
circumscribed
axis of a circle
internal tangent
external tangent
compound curve

Chapter Review

1. Construct a tangent to a circle at a given point of the circle.
2. Construct a line that will be tangent to a given circle and parallel to a given line.
3. Circumscribe a square about a given circle.

4. If a line is tangent to a sphere, is it tangent to any of the great circles of the sphere containing the point of tangency? Explain.
5. If we were to rotate the figure for 12.57 about OO', what figure would be generated? (Cut the figure out of stiff paper to examine it better.)
6. What figure would be formed at the intersection of two spheres?
7. How many spheres can intersect in three given non-collinear points?
8. If two small circles of a sphere are not congruent, the plane of the greater is (nearer to, farther from) the center of the sphere than the plane of the smaller. Explain.
9. Suppose we have a circle but don't know its center. Explain how 12.56 can be used to make a device with which we could find the center of the circle.
10. Explain why a line tangent to a sphere is perpendicular to a radius drawn to the point of tangency.

Within the framework of geometry as we have developed it, the following propositions are either true or false. Prove deductively those you believe to be true.

11. If a plane is tangent to a sphere, a line perpendicular to this plane at the point of tangency passes through the center of the sphere.
12. If two circles are concentric, the chords of the larger which are tangent to the smaller are congruent.
13. $\overline{AB}$ is the diameter of the larger of two concentric circles. $\overline{CD}$ is a diameter of the smaller circle. Figure $ADBC$ is a rhombus.
14. The tangents from a point in the exterior of a circle form congruent angles with the chord joining the points of tangency.
15. Two non-congruent circles are tangent externally at D. A common external tangent is $\overline{AB}$. A right triangle is determined by points A, B, and D.
16. The tangents to a circle at the ends of a diameter are parallel.
17. If two circles are coplanar and their interiors have no common points, the segments of the internal tangents between the points of tangency are congruent.

18. Circle A is internally tangent to circle B at point O. A line through point O intersects circle A at point C and circle B at point D. m$\overline{AC}$ equals one-half m$\overline{BD}$.

19. If one circle lies entirely in the exterior of another in the same plane, the segments of their common external tangents between the points of tangency are congruent.

20. If two circles are coplanar and externally tangent, their common internal tangent bisects the segment of their common external tangent between the points of tangency.

21. If two circles O and O' are tangent externally at B and $\overline{AC}$ is drawn through the point of tangency and terminated by the circles, $\overline{OA}$ is parallel to $\overline{O'C}$. (Figure 12-26)

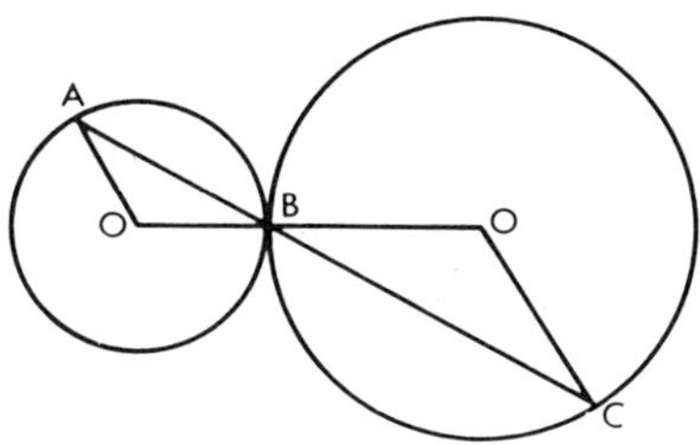

Figure 12–26

22. The sums of the lengths of the pairs of opposite sides of a circumscribed quadrilateral are equal.

23. If the common internal tangents of two non-congruent circles are drawn and the points of tangency are joined to form a quadrilateral, the figure is a trapezoid.

Chapter 12 Test

True-False: If the statement is not always true, then it is false.

1. The center of a circle is not part of the circle.

2. A circular region does not include the circle.

3. A secant of a circle can not be a chord of the circle.

4. A chord of a circle can not be a secant of the circle.

5. A radius of a sphere is a chord of the sphere.

6. A diameter of a circle is a chord of the circle.
7. Concentric circles are congruent circles.
8. A diameter of a circle separates it into two semicircles having no points in common.
9. A hemisphere contains the center of the sphere.
10. There is one and only one line tangent to a circle at a given point of the circle.
11. There is one and only one line tangent to a sphere at a given point of the sphere.
12. If two spheres are tangent to the same plane at the same point the spheres are tangent to each other.
13. If two circles are tangent to the same line at the same point the circles are tangent to each other.
14. The measure of the segment joining the centers of two intersecting circles is less than the measure of the radius of either.
15. The point of concurrency of the common external tangent and the common internal tangent of two coplanar externally tangent circles is equidistant from the three points of tangency.

Completion: On your paper opposite the number of the question place the word or phrase that best completes the statement.

16. The measure of a central angle of a circle is always less than or equal to __?__.
17. A circle is divided into five distinct congruent arcs. The measure of a central angle which subtends one of these arcs is __?__.
18. If the sides of a polygon circumscribed about circle A are chords of circle B, where A and B are concentric circles, the polygon is __?__.
19. In the plane of a circle the point of intersection of the perpendicular bisectors of two chords of the circle is the __?__.
20. A diameter of a sphere containing the center of a small circle of the sphere is called the __?__ of the sphere.
21. The intersection of a sphere and a plane containing the center of the sphere is called a __?__.
22. Two chords of a sphere are not congruent. The __?__ chord is nearest the center of the sphere.

23. The planes of two small circles of a sphere are equidistant from the center of the sphere. An equilateral triangle is inscribed in each small circle. The triangles (*are*, *are not*) necessarily congruent.

24. The center of a sphere, a point on the sphere, and a point in the exterior of the sphere determine a triangle. It is a(n) __?__ triangle if the point on the sphere and in the exterior determine a tangent to the sphere.

25. The common __?__ tangent of two coplanar circles intersects the line of centers of the circles.

13

Angles Determined by Tangents, Secants, and Chords of Circles

The definition of section 12.24 tells us that the degree measure of a minor arc of a circle is equal to the degree measure of the central angle it subtends. In Chapter 3 we established the method of measuring an angle. The proof of Theorem 3.19 assured us that the sum of the measures of all the consecutive adjacent angles about a point in a plane is 360°.

Consider a given circle in which a minor and a major arc have the same endpoints. What is the union of the minor and major arcs? Can you prove this? Review Exercise 11, page 281. This allows us to make the following statement.

13.00 THEOREM

The degree measure of a circle is 360.

We are now prepared to attack the problems involving the measures of angles, other than central angles, intercepting arcs of circles.

13.01 The sides of an angle intercept an arc of a circle if all points of the arc except the endpoints lie in the interior of the angle and if each side of the angle contains at least one endpoint of the arc.

INDUCTIVE PROCEDURE

Our previous comments should lead some of you to ask "What angles can be drawn so that their sides intercept an arc or arcs of circles and what are the relationships of the measures of the angles and the measures of the intercepted arcs?"

We suggest you do some experimenting and develop some conjectures. If you are at a loss as to how to proceed, the following picture and questions may help.

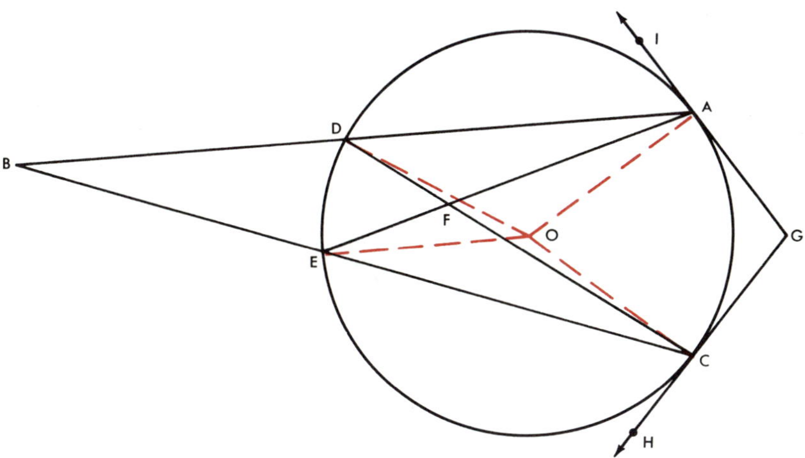

Figure 13–1

1. Find the measure of:

(a) $\widehat{DA}$	(f) $\angle ABC$	(k) $\angle ECH$
(b) $\widehat{AC}$	(g) $\angle AEC$	(l) $\angle ADC$
(c) $\widehat{DE}$	(h) $\angle AFC$	(m) $\angle OCG$
(d) $\widehat{EC}$	(i) $\angle DCG$	(n) $\angle HGI$
(e) $\widehat{DC}$	(j) $\angle DCE$	(o) $\angle DAI$

2. Do any pairs of angles in Figure 13-1 have the same measure? Is one measure a multiple of another?
3. Are there relationships between arc measures, or sums and differences of arc measures, and angle measures?
4. State any conjectures you can reasonably derive from your experiments.
5. Do not stop with these experiments. Devise more of your own and state the conjectures derived therefrom.

INSCRIBED ANGLES

It is impractical to proceed further without naming some of the types of angles indicated in Figure 13-1.

13.02 An inscribed angle is an angle determined by two chords of a circle or of a sphere which intersect on the circle or on the sphere.

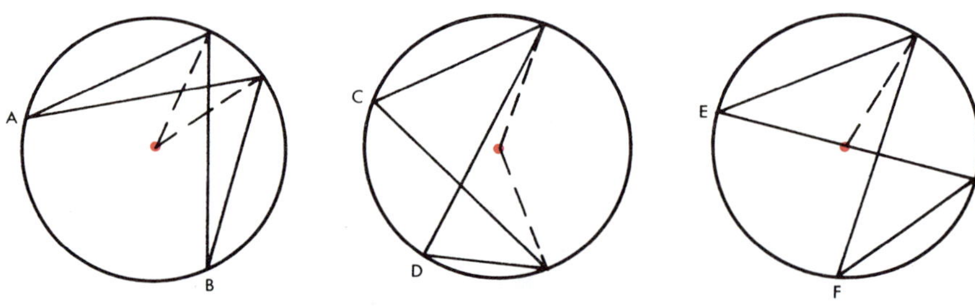

Figure 13–2

Experiment: Measure each of the inscribed angles and compare their measure with the measure of the central angle intercepting the same arc. What conjecture can you state from this experiment?

DEDUCTIVE PROCEDURE

Your conjectures are anticipated—in part at least. One involving inscribed angles is considered in the following deductive approach to the problem.

Given: $\angle BAC$ inscribed in circle O. $\overline{AD}$ a diameter. Let $\widehat{BC} = x°$.

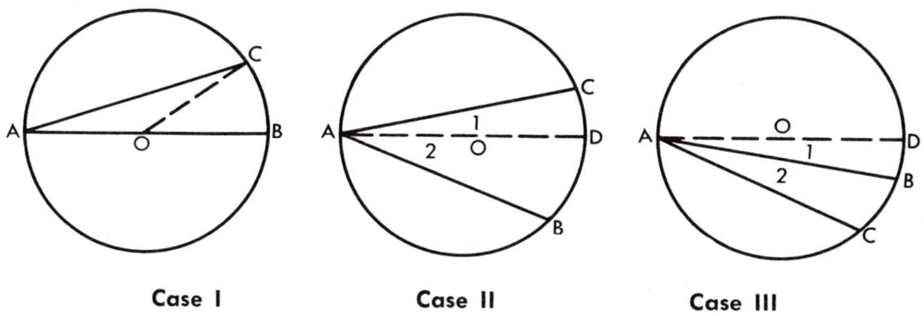

Figure 13–3

Conjecture: $m\angle BAC = \frac{1}{2}x°$.

Plan: Use 10.08 and 12.24.

Case I. The diameter determines one side of the angle. Prove that $m\angle A = \frac{1}{2} m\angle BOC = \frac{1}{2}x°$.

Case II. The center of the circle lies in the interior of the angle. Apply Case I to $\angle 1$ and $\angle 2$. Use addition axiom.

Case III. The center of the circle lies in the exterior of the angle. Apply Case I to $\angle 1$ and $\angle 2$. Use subtraction axiom.

Proof: The proof is left to the student.

13.03 THEOREM

The measure of an inscribed angle is half the degree measure of its intercepted arc.

NOTE: The theorem as stated applies to all three cases. It is unnecessary to state a particular case.

The following five theorems develop directly from Theorem 13.03 with little proof.

13.04 THEOREM A

Inscribed angles which intercept the same arc are congruent.

13.05 THEOREM B

Congruent inscribed angles intercept congruent arcs in the same or congruent circles.

If an angle is inscribed in a semicircle, what kind of arc does it intercept? Does this suggest what the measure of the inscribed arc is?

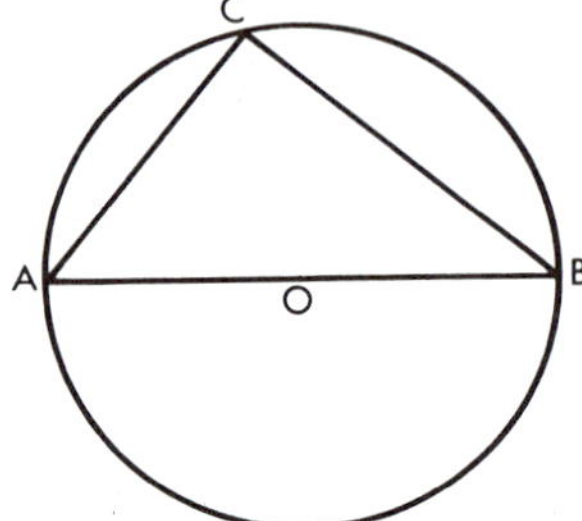

Figure 13–4

Given: $\angle C$ inscribed in semicircle $\widehat{ACB}$

Conjecture: $m\angle C = 90°$

13.06 THEOREM C

An angle inscribed in a semicircle is a right angle.

A common and reasonable conjecture is that parallel lines intercept congruent arcs of a circle. This can be proved with the help of 13.05.

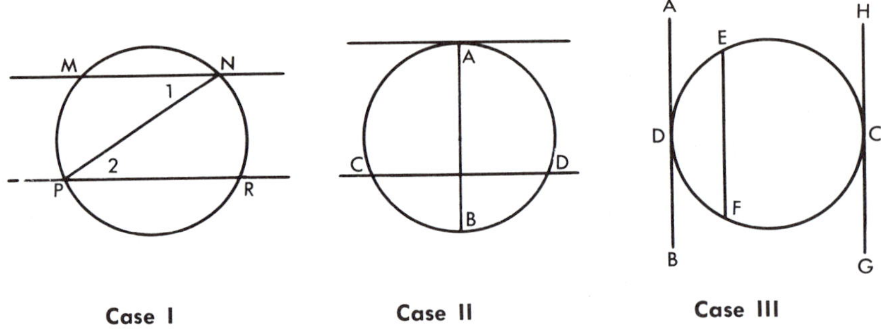

Case I Case II Case III

Figure 13–5

Proof: Case I. $\angle 1 \cong \angle 2$. Why? Therefore the arcs are congruent. Why?

Case II. Draw diameter $\overline{AB}$. Use 12.35 and the Subtraction Axiom.

Case III. Draw $\overleftrightarrow{EF} \parallel \overleftrightarrow{AB}$. Use Case II and the Addition Axiom.

This justifies the following theorem.

13.07 THEOREM D

Parallel lines intercept equal arcs on a circle.

13.08 THEOREM E

If a quadrilateral is inscribed in a circle, the opposite angles are supplementary. (See Figure 13-6)

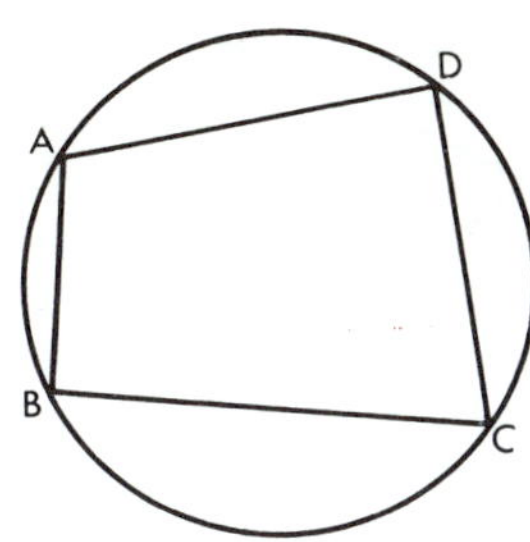

Figure 13–6

EXERCISES AND INDUCTIVE REASONING WITH ARCS AND ANGLES

1. Name the central and inscribed angles in the figures for Exercises 2–8 and with each angle name its intercepted arc.
2. O is the center of the circle. $\widehat{AB} = 30°$, $\widehat{BC} = 120°$, and $\widehat{CD} = 70°$. What is the measure of each of the angles?

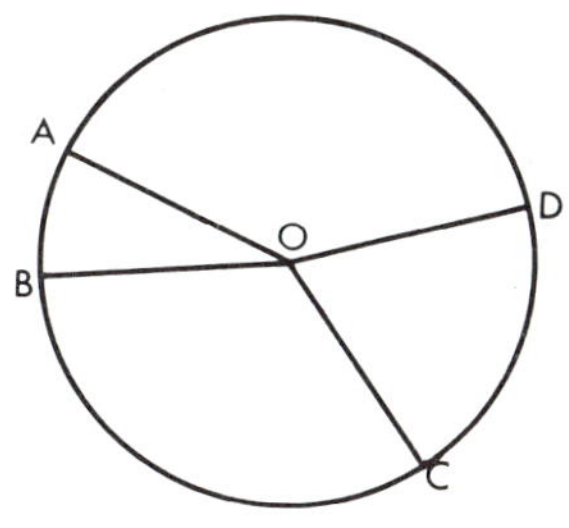

(Exercise 2)

Figure 13–7

3. $\overline{AB}$ and $\overline{CD}$ are two chords intersecting at E. Find m$\angle AEC$ if $\widehat{AC} = 30°$ and $\widehat{BD} = 140°$? If $\widehat{AC} = 60°$ and $\widehat{BD} = 100°$? If $\widehat{AC} = x°$ and $\widehat{BD} = y°$?

4. $\overleftrightarrow{ABC}$ and $\overleftrightarrow{EDC}$ are secants. Find m$\angle C$ if $\widehat{AE} = 60°$ and $\widehat{BD} = 40°$? If $\widehat{AE} = 100°$ and $\widehat{BD} = 60°$? If $\widehat{AE} = x°$ and $\widehat{BD} = y°$?

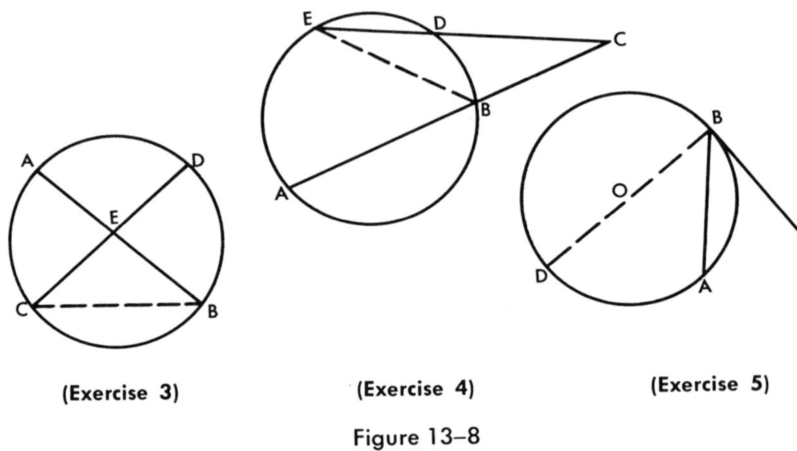

(Exercise 3) (Exercise 4) (Exercise 5)

Figure 13–8

5. $\overline{BOD}$ is the diameter, $\overline{AB}$ a chord, and $\overrightarrow{BC}$ a tangent. Find m$\angle ABC$ if $\widehat{AB} = 40°$? $80°$? $x°$?

6. State problems 3, 4, and 5 as theorems.

7. Name the inscribed angles in the figures for Exercises 8-14 that are congruent because they intercept the same arc.

8. If $\overline{AC}$ and $\overline{BD}$ are two chords that intersect at E, prove the angles of $\triangle AEB$ are congruent to the angles of $\triangle DEC$.

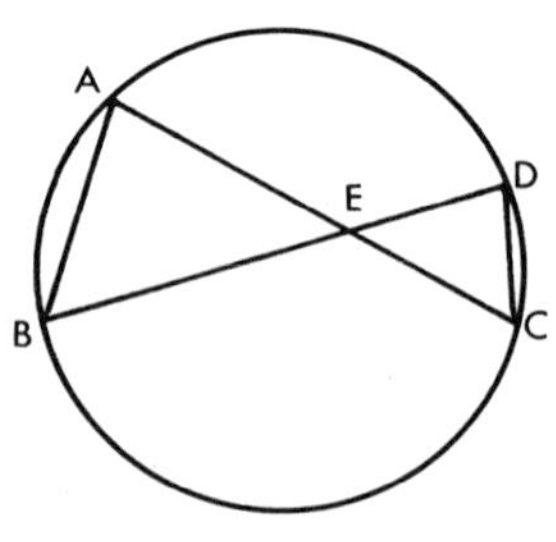

(Exercise 8)

Figure 13–9

9. $\overline{BD}$ is a diameter and $\overline{DE} \perp \overline{AC}$. Prove the three angles of $\triangle ADE$ are congruent to the three angles of $\triangle BCD$.

10. Using the conditions of Exercise 9, what is the measure of each angle of the figure if $\widehat{CD} = 120°$ and $\widehat{AB} = 100°$?

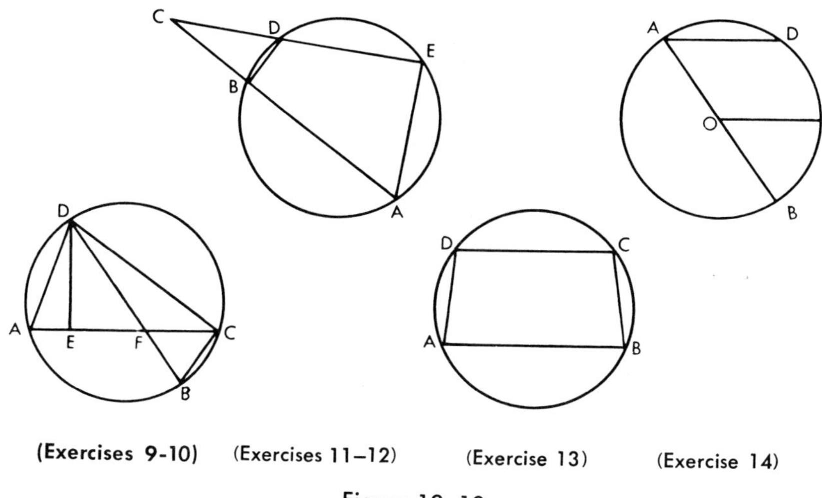

(Exercises 9-10) (Exercises 11–12) (Exercise 13) (Exercise 14)

Figure 13–10

11. If $\widehat{AE} = 90°$, $\widehat{AB} = 100°$, and $\widehat{BD} = 30°$, what is the measure of each angle of the inscribed quadrilateral?

12. $\overleftrightarrow{AC}$ and $\overleftrightarrow{EC}$ are secants. Prove the three angles of $\triangle ACE$ are congruent to the three angles of $\triangle BDC$.

13. $\overline{DC} \parallel \overline{AB}$. Prove $\overline{AD} \simeq \overline{BC}$.

14. $\overline{AB}$ is a diameter, $\overline{OC}$ is a radius, and $\widehat{DC} \stackrel{\circ}{=} \widehat{CB}$. Prove that $\overline{AD}$ is parallel to $\overline{OC}$. (See Figure 13-10.)

15. Explain how you could find the center of a circle, using a piece of paper whose corner is a right angle. (See Figure 13-11.)

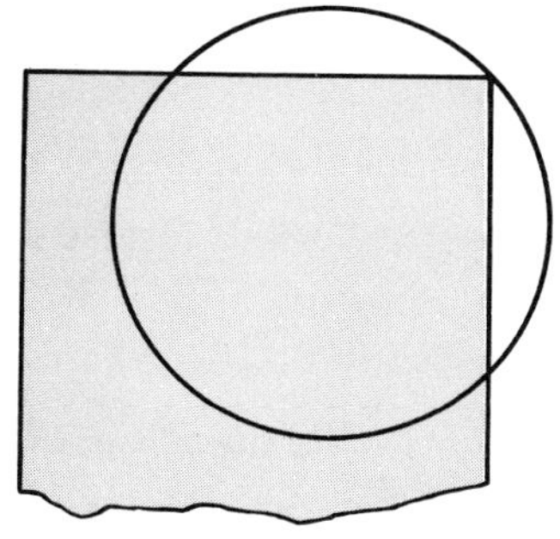

Figure 13–11

Review of Algebra

Solve for x and check

1. $\frac{x + 40}{2} = 8$
2. $\frac{x + 12}{3} = 10$
3. $\frac{x}{3} + 8 = 5$
4. $5 - \frac{x}{2} = 6$
5. $\frac{x - 10}{2} = 30$
6. $\frac{4x - 8}{5} = 12$
7. $\frac{50 - x}{2} = 20$
8. $\frac{18 - 2x}{3} = 12$
9. $\frac{7 + x}{3} = \frac{10}{8}$
10. $\frac{x + 60}{2} = 57\frac{1}{2}$
11. $\frac{3}{2}x + 8 = x$
12. $\frac{2}{3}x - 5 = \frac{x}{2}$
13. $\frac{40 - x}{2} = 2x$
14. $\frac{17 - 2x}{3} = 5$
15. $\frac{12}{x - 2} = 3$
16. $\frac{x + a}{2x} = 1$
17. $\frac{x + 3a}{2} = b$
18. $\frac{3x - 2a}{3} = 2b$

An angle may intercept arcs on a circle without being inscribed in the circle. What are some of the possible ways? In some of these cases, the measure of the angle may be determined by the measure of the intercepted arcs. The following two demonstrations consider two cases.

Given: Circle O in which $\angle ABD$ is determined by tangent $\overleftrightarrow{DBC}$ and chord $\overline{AB}$. Let $\widehat{AB} = a°$.

Conjecture: $m\angle ABD = \frac{1}{2}a°$.

Proof: (See Exercise 5, page 312.)

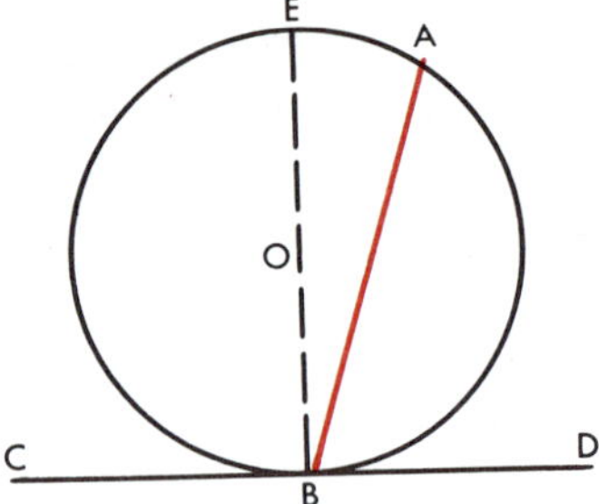

Figure 13–12

13.09 THEOREM

The measure of an angle determined by a tangent and a chord intersecting at the point of tangency to a circle is half the degree measure of the intercepted arc.

How many angles are determined by two intersecting chords, when the intersection is not on the circle? Let us consider how the measure of such angles may be determined.

Given: A circle with two chords, $\overline{AB}$ and $\overline{CD}$ intersecting at E. Let $\widehat{AC} = a°$ and $\widehat{BD} = b°$.

Conjecture: $m\angle AEC = \dfrac{a° + b°}{2}$

Proof: (See Exercise 3, page 312.)

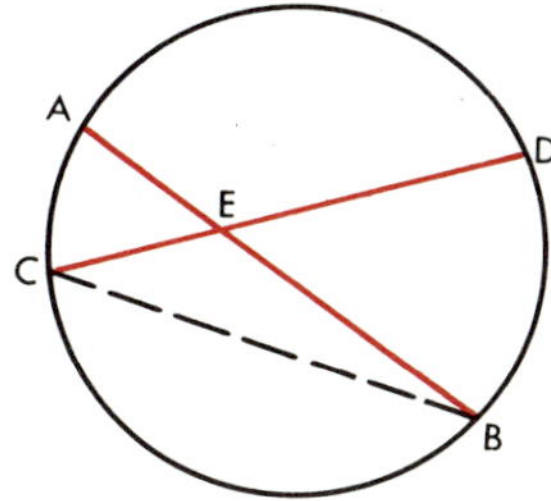

Figure 13-13

13.10 THEOREM

The measure of an angle determined by two chords intersecting in the interior of a circle is half the sum of the degree measure of the two arcs intercepted by it and its vertical angle.

Exercises

Use Figure 13-13 for Exercises 1–6.

1. If $\widehat{AC} = 50°$ and $\widehat{BD} = 100°$, find $m\angle AEC$.
2. If $\widehat{AC} = 15°$ and $\widehat{BD} = 70°$, find $m\angle DEB$.
3. If $\widehat{AD} = 110°$ and $\widehat{BC} = 120°$, find $m\angle AEC$.
4. If $\overline{AB}$ is a diameter, $\widehat{AC} = 40°$ and $\widehat{AD} = 100°$, find $m\angle BED$.
5. If $m\angle AEC = 80°$ and $\widehat{AC} = 60°$, find the degree measure of $\widehat{BD}$. (*Hint:* Let $\widehat{BD} = x°$ and develop an equation using 13.10.)
6. If $m\angle BED = 60°$ and $\widehat{AC} = 30°$, find the degree measure of $\widehat{BD}$.

Use Figure 13-12 for Exercises 7–10.

7. If $\widehat{AB} = 50°$, find $m\angle ABD$.
8. If $\widehat{AB} = 75°$, find $m\angle ABD$.
9. If major $\widehat{AB} = 200°$, find $m\angle ABD$.
10. If $\widehat{EA} = 30°$, find $m\angle CBA$.

If the vertex of an angle is in the exterior of a circle and the sides of the angle intersect the circle as tangents or secants, can the measure of the angle be determined by the intercepted arcs? How do the intercepted arcs differ from those in the cases already discussed?

Given: $\overleftrightarrow{AB}$ and $\overleftrightarrow{AC}$ intercepting the two arcs whose measures are x° and y° on circle O.

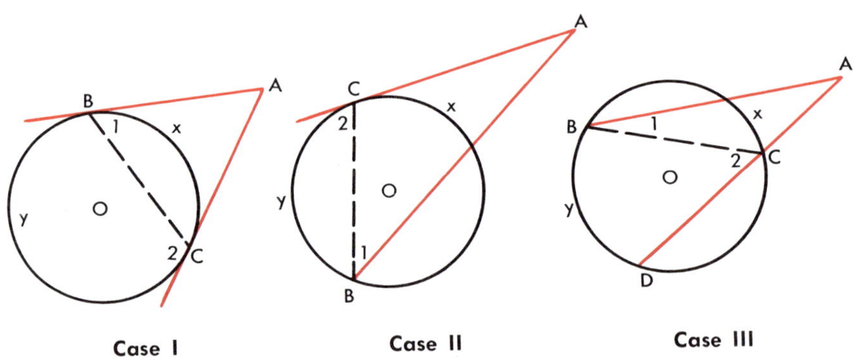

Figure 13–14

Conjecture: $m\angle A = \frac{1}{2}(y^\circ - x^\circ)$

Proof: (See Exercise 4, page 312.)

13.11 THEOREM

The measure of an angle determined by two tangents, or a tangent and a secant, or two secants intersecting in the exterior of a circle, is equal to half the difference of the degree measures of the intercepted arcs.

Exercises

1. In Figure 13-14, Case II, find $m\angle A$ if $\widehat{x} = 50^\circ$ and $\widehat{y} = 80^\circ$. Find $m\angle A$ if $\widehat{x} = 57^\circ$ and $\widehat{y} = 84^\circ$.
2. In Case II, if $m\angle A = 20^\circ$ and $\widehat{y} = 100^\circ$, what is the degree measure of $\widehat{x}$?
3. In the figure for Case I, find $m\angle A$ if $\widehat{x} = 120^\circ$. If $\widehat{y} = 200^\circ$.
4. In Case III, find $m\angle A$ if $\widehat{y} = 90^\circ$ and $\widehat{x} = 20^\circ$.

5. In Case III, what is the degree measure of $\overset{\frown}{x}$ if $m\angle A = 35°$ and $\overset{\frown}{y} = 120°$?
6. In Case I, is there an upper limit for $m\angle A$, depending upon the positions of points B and C? Explain.
7. In Case I, is there a minimum limit for $m\angle A$? Explain.
8. Would Cases II and III have the same limits for $m\angle A$ as Case I?

By defining the degree measure of an arc that is concave toward the vertex of a given angle as positive and one that is convex as negative, we may combine the theorems of 13.03, 13.09, 13.10, 13.11 into one theorem.

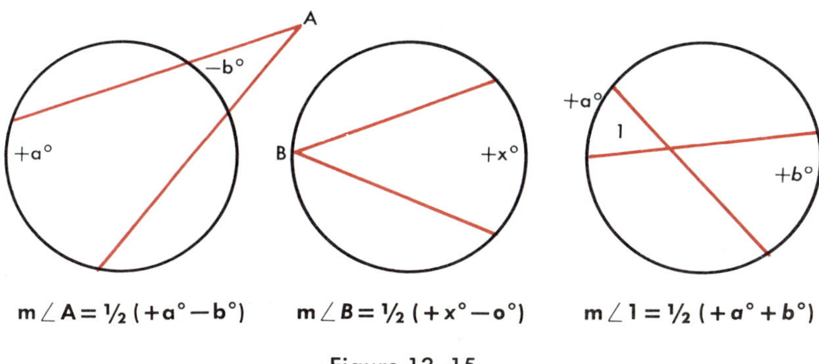

Figure 13–15

13.12 THEOREM (Optional)

The measure of any angle determined by two straight lines both intersecting a circle and lying in the plane of the circle is half the algebraic sum of the degree measures of the two intercepted arcs.

Does a Regular Polygon Have a Center Point?

(1) Can we divide the degree measure of a circle (360°) into any desired number of equal parts? For example, in Figure 13-16, page 318, can we make $\overset{\frown}{AB} \overset{\circ}{=} \overset{\frown}{BC} \overset{\circ}{=} \overset{\frown}{CD} \overset{\circ}{=} \overset{\frown}{DE} \overset{\circ}{=} \overset{\frown}{EA}$?

(2) Do congruent central angles subtend congruent arcs and congruent chords? In Figure 13-16 is $\overline{AB} \cong \overline{BC} \cong \overline{CD} \cong \overline{DE} \cong \overline{EA}$ if the central angles are congruent? Justify your answers by referring to previous assumptions or theorems.

(3) Is $\widehat{ABCD} \cong \widehat{BCDE} \cong \widehat{CDEA} \cong \widehat{DEAB} \cong \widehat{EABC}$? Explain.

(4) Is $\angle A \cong \angle B \cong \angle C \cong \angle D \cong \angle E$? Explain.

(5) Is $ABCDE$ a regular polygon? Explain.

(6) Is point O, the center of the circle, equidistant from the vertices and from the sides of the polygon? Explain.

After studying these questions, we make the following assumption.

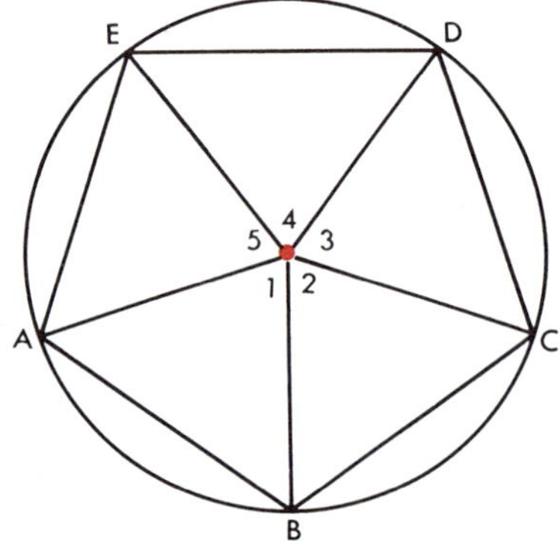

Figure 13–16

13.13 *Assumption: There is a point in the plane of a regular polygon, called the center, that is equidistant from the vertices and from the sides of the polygon.*

13.14 To Construct a Tangent to a Circle From a Given External Point

Given: Circle O and a point P outside the circle

To Construct: A tangent from point P to circle O

Construction: Draw $\overline{OP}$. Bisect $\overline{OP}$ and call the midpoint A.

With A as center draw the semicircle with radius $\overline{AO}$, intersecting the circle at B. Draw $\overline{BP}$. Then $\overline{BP} \perp \overline{OB}$. Why? $\therefore$ $\overline{BP}$ is tangent to circle O. Why?

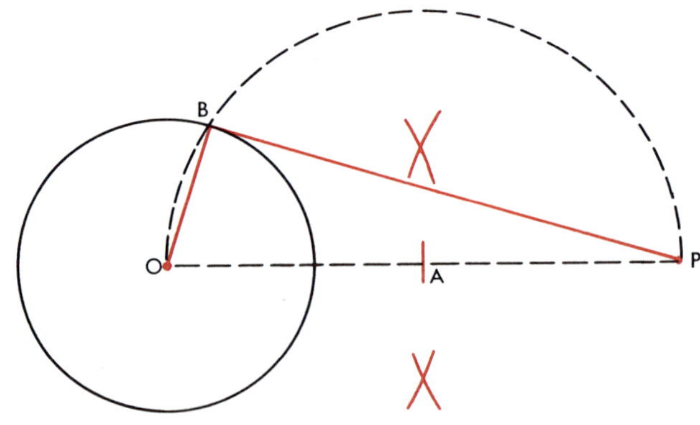

Figure 13–17

SPHERICAL ANGLES (Optional)

13.15 When the arcs of two great circles intersect on a sphere, the figure formed is called a spherical angle.

The measure of a spherical angle (ACB in Figure 13-18) is obtained by drawing the tangents ($\overleftrightarrow{A'C}$ and $\overleftrightarrow{B'C}$) to the great circles at the vertex of the angle and measuring the plane angle thus formed.

A spherical angle may also be measured by the arc of a great circle ($\widehat{AB}$) included between the sides of the spherical angle and drawn so that it is perpendicular to the intersection of the planes of the great circles forming the angle.

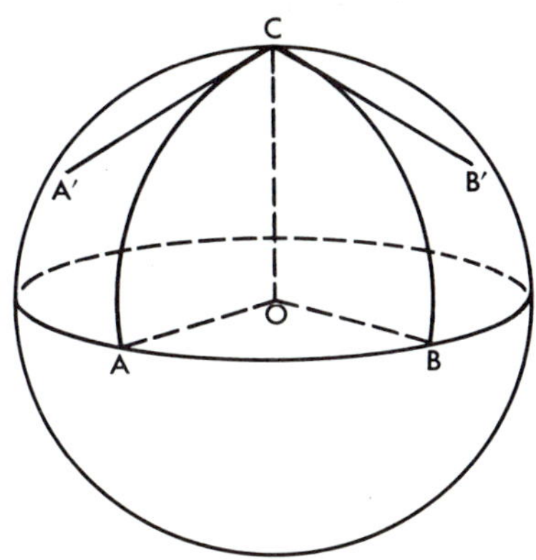

Figure 13–18

Proof: $\overleftrightarrow{A'C} \parallel \overline{AO}$ and $\overleftrightarrow{B'C} \parallel \overline{BO}$. Why? $\therefore \angle A'CB' \cong \angle AOB$ (11.33). $\angle AOB$ is measured by $\widehat{AB}$. $\therefore$ $\angle A'CB'$ could be measured by $\widehat{AB}$.

SPHERICAL POLYGONS (Optional)

13.16 Generally speaking a spherical polygon is a closed figure on a sphere composed of three or more arcs of great circles. The *sides* of the polygon are the arcs of the circles, and their intersections are the *vertices* of the polygon. The *angles* of the polygon are the spherical angles formed by the sides of the polygon. A more exact definition of a spherical polygon would require substituting "arcs of great circles" for "segments" in a definition similar to that of section 2.49.

Connect the vertices of spherical triangle *ABC* (Figure 13-19) to the center of the sphere. Angle *O-ABC* is a trihedral angle. The degree measures of the sides of the spherical triangle correspond to the measures of the face angles of the trihedral angle, and the measures of the angles of the spherical triangle are equal to the measures of the dihedral angles of the polyhedral angle.

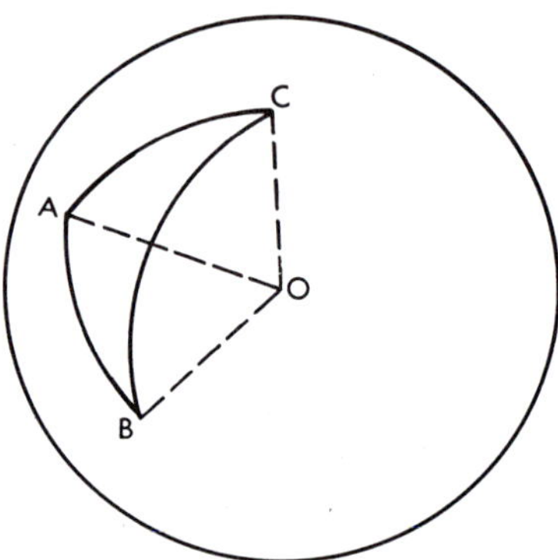

Figure 13–19

It can be shown that the sum of the degree measures of the sides of a spherical polygon is less than 360° (What if they totaled 360°?) and that the sum of the measures of the angles of a spherical triangle is greater than 180° and less than 540°.

LATITUDE AND LONGITUDE (Optional)

The earth is not a sphere, but for practical purposes we may consider it as such. Actually the diameter of the earth at the poles is about 7899 miles and the diameter at the equator is about 7925 miles.

Measurements by earth satellites indicate that the earth is slightly pear-shaped. The deviation of the shape of the earth from a true sphere is so slight relative to its size, however, that if the earth were reduced to the size of an apple, high precision measurements would be needed to detect the variations.

13.17 To locate positions on the earth's surface a "grid" is formed by drawing two sets of circles as shown in Figure 13-20. One set of great circles is drawn through the poles. These circles are called meridians or *lines of longitude.*

The equator is a great circle whose plane is perpendicular to the diameter (axis of the earth) connecting the North Pole and the South Pole. Small circles are drawn with their planes parallel to the plane of the equator. These are called the parallels of latitude. The zero circles (coordinates) are the equator and the meridian through Greenwich, England, called the *Prime Meridian.*

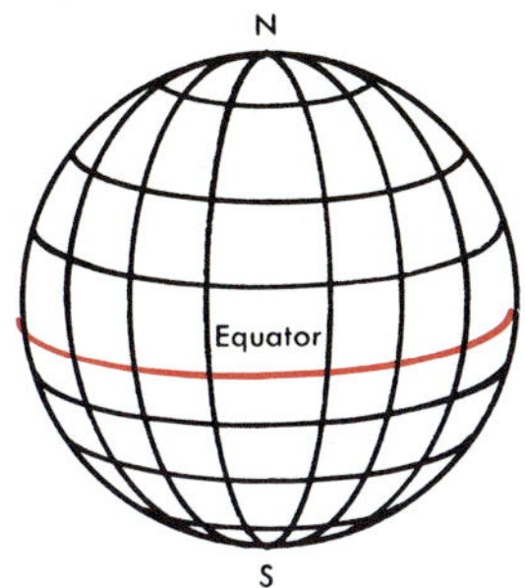

Figure 13–20

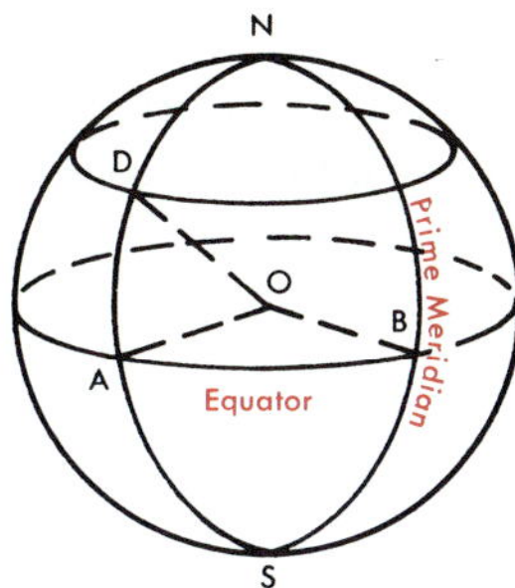

Figure 13–21

The International Date Line follows roughly the 180th Meridian.

A location may be determined by the intersection of a meridian with a parallel of latitude. For example, Des Moines, Iowa (Point D Figure 13-21), lies on parallel 41.32° North and the meridian 93.32° West. This means that arc AB is 93.32° and arc AD is 41.32°, where O is the center of the earth.

Exercises

1. At what angle does a meridian intersect the equator?
2. The measure of an angle formed by two meridians on the earth's surface is 30°. What is the degree measure of the arc of the equator included between the meridians?
3. What is the latitude of the North Pole? Of the South Pole?
4. What is the greatest latitude of any place on the earth? What is the greatest longitude on the earth?
5. Which great circle of the earth corresponds to the x-axis on a graph? To the y-axis.

6. What position on a map of the earth corresponds to a point whose coordinate on a graph is (0,0)?

7. Figure 13-22 represents the earth. $\overline{NS}$ is the axis of the earth connecting the North and South Poles. $\overline{OA}$ lies in the plane of the equator and is $\perp \overline{NS}$. $\overrightarrow{OD}$ is a vertical ray and $\overleftrightarrow{BC}$ a horizontal line $\perp \overrightarrow{OD}$ at E on the earth's surface. $\overrightarrow{EP}$ represents a ray of light from Polaris, the North Star. Because of the great distance to Polaris, we may consider $\overrightarrow{EP} \parallel \overline{NS}$. Arc $\widehat{AE}$, measured along a great circle (line of longitude) through the North and South Poles, is our latitude North.

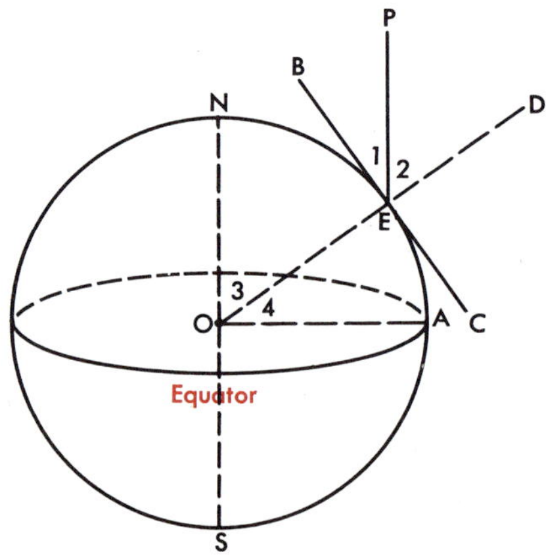

Figure 13–22

(a) Prove that the measure of the angle of elevation of the North Star ($\angle 1$) equals the measure of the latitude of point E.

(b) Find the approximate latitude of your home by measuring the angle of elevation of the North Star. Check with the nearest weather bureau.

GREAT CIRCLE ROUTES (Optional)

We proved in section 10.15 that the shortest distance between two points is the length of the segment joining them. This is true in a plane or in 3-space. However, if we restrict ourselves to the surface of a sphere, we cannot follow a straight line. For instance, when we travel on the earth's spherical surface the shortest distance between two points lies along the arc of a great circle. How would you find the shortest distance between two points on a sphere? If a rubber band or a string is stretched so that it passes through the given points on a sphere, will the shortest path on the surface be determined? On a globe, find the shortest path between New York and London. Find the shortest path between two points on the equator. Find the shortest path between two points on the same meridian. What can you conclude about the shortest path between any two points on the surface of the earth?

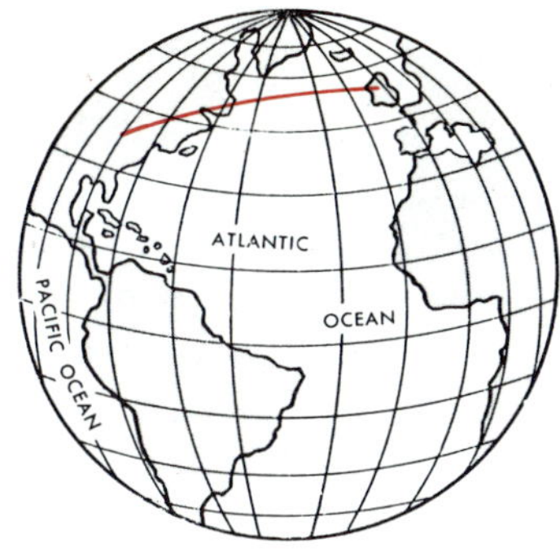

Figure 13–23

We shall make the following assumption based on these experiments:

13.18 *Assumption: The shortest distance between two points on the surface of a sphere is the length of the minor arc of the great circle passing through the two points.*

If two points on the surface of a sphere are the end points of a diameter (as the North and South Poles), what would be the shortest distance between them?

Distance along a great circle of the earth is measured in degrees, minutes, and seconds. The distance may be converted into *nautical* miles (on water), *statute* miles (on land), or meters.

A minute of a great circle on the earth may be considered as equal to a nautical mile. As there are sixty minutes in each degree, and 360 degrees in a complete circle, can you figure the earth's circumference in nautical miles? A nautical mile is approximately 6,080 feet, while a statute mile is 5,280 feet. What is the circumference of the earth in statute miles? Are your answers approximately the same as the accepted values?

TIME ZONES (Optional)

13.19 Before 1883 we had no standard system of time measurement. Each community set its clocks by the sun, and neighboring communities to the east or west had differing times. Railroad timetables were difficult, if not impossible, to interpret.

The earth completes one rotation on its axis in approximately 24 hours (actually 23 hours 56 minutes 4.091 seconds). It follows, then, that the earth completes 1/24th of a rotation in one hour, and 1/24th of 360° (a complete rotation) equals 15°, so each 15° represents 1 hour of time. Starting from the Prime Meridian at Greenwich, time zones of 15° each were marked off. The zero meridian is in the middle of its zone. The 15° meridian East and the 15° meridian West are in the middle of their respective time zones, and so forth.

The Standard Time kept everywhere within each zone is exactly the same. As one moves from one zone to another, going east, the time advances an hour. If one goes west it moves back. Actually the time belts do not, like the meridians, follow perfectly straight lines. Sometimes the boundary of a belt may be shifted to make the time used in a community the same as that used in its nearby economic and commercial area. In some parts of the world this zone system has not been adopted and there is no legal time. The time zone meridians are located approximately in the center of each zone. Those used in the continental United States are the 75th (Eastern), 90th (Central), 105th (Mountain), and 120th (Pacific). Use a globe to find cities which are near the center of each of these time zones.

Since a person traveling westward will lose one hour of time for each 15° of travel, a complete circle of the earth would lose one day. If he had made the circuit eastward, he would have gained a day. For practical reasons it is necessary to add a day as we travel westward or subtract a day as we travel eastward. By international agreement, the change-of-date line is near the 180th meridian.

For simplicity in figuring time in different parts of the world, the 24-hour method is often used. In this method noon is 1200 hours and midnight 2400 hours. One minute after midnight is written as 0001, one o'clock in the afternoon is written 1300 and thirty-seven minutes after nine in the evening is written 2137.

Exercises

1. How long does the earth require to turn one degree in longitude? Five degrees?
2. How many hours west of Greenwich is the International Date Line?
3. If it is noon in Greenwich, what time is it 180° to the west of Greenwich?
4. How many hours of time must be added to Greenwich time to give the local time in the following longitudes? (a) 45° W? (b) 165° E? (c) 105° W? (d) 60°E?
5. If it is 0815 in Des Moines, Iowa, what is the Greenwich time?
6. New York is in longitude 74° W. If local standard time in New York is 1325, what is the standard time in London?
7. If San Francisco's longitude is about 122°W, what will be the local standard time there when the standard time in New York is 1545?
8. Find local time if it is 0420 Greenwich time on Tuesday in the following longitudes: (a) 28°E. (b) 175° W. (c) 93° E.
9. Is 1 minute of longitude at 40° latitude equal to, less than, or more than a nautical mile? Explain.
10. The difference in latitude between two places is 4° 18′. What is the distance between them measured on a meridian in nautical miles?
11. The distance between two places on the equator is 250 nautical miles. What is the difference in their longitudes?
12. The latitude of Seattle, Washington, is 47° 36′ North. What is the distance from Seattle to the geographic North Pole, measured on a meridian in nautical miles?
13. The distance from New York to Glasgow, Scotland, is 3086 nautical miles. What is this distance in statute miles?

TEMPERATE ZONES (Optional)

13.20 Certain circles of latitude divide the earth into zones (Figure 13-24). The line of latitude (23° 28′ N) marking the southern boundary of the North Temperate Zone is called the Tropic of Cancer. The line (23° 28′ S) marking the northern boundary of the South Temperate Zone is called the Tropic of Capricorn.

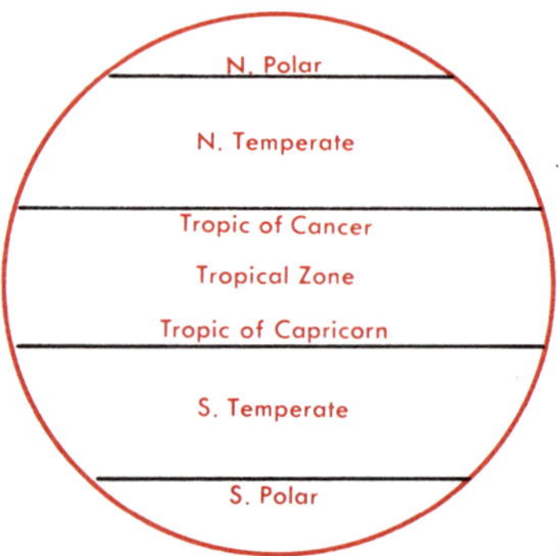

Figure 13–24

Because the earth's axis tilts at an angle to the plane of its rotation about the sun, the Tropic of Cancer and the Tropic of Capricorn mark the northernmost and southernmost points at which the sun's rays are normal (at right angles to a tangent plane) to the earth's surface. (See Figure 13-25.)

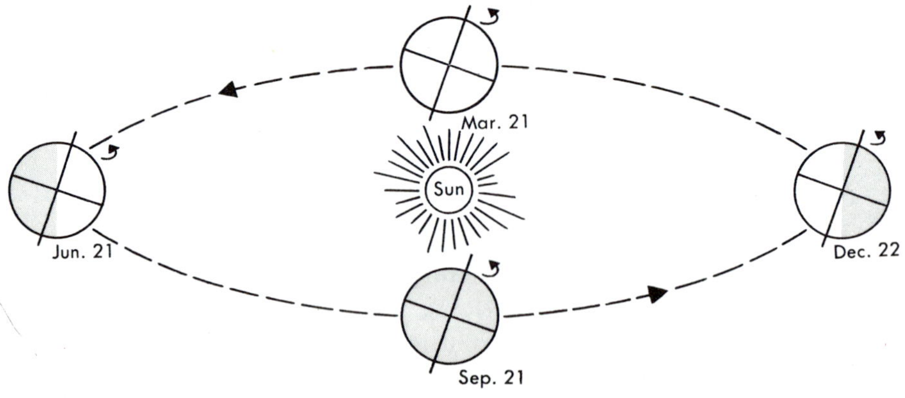

Figure 13–25

The orbit of the earth is elliptical, but only slightly so. If reduced to a diameter of three inches, the variation from a circle could hardly be detected except with precision instruments. The paths of some planets (notably Pluto) are much more elliptical than that of the earth.

Figure 13-25 is an exaggerated representation of the elliptical plane of the earth's rotation about the sun. The earth is nearer the sun in winter than in summer (that is, in the Northern Hemisphere). Investigate the angle of the sun as measured from the plane of the equator at noon on March 21? June 21? September 21? December 22?

Vocabulary List

intercept	nautical mile
spherical angle	statute mile
spherical polygon	temperate zone
meridian	Cancer
longitude	Capricorn
latitude	elliptical

Chapter Review

1. Construct a circle whose radius is 1 inch. Place a point $2\frac{1}{2}$ inches from the center and construct two tangents to the circle from that point.
2. The center of a circle is O, $\overline{GC}$ is a diameter, $\overline{EG}$ and $\overline{BE}$ are secants, $\overline{AG}$ and $\overline{AB}$ are tangents, $\widehat{BG} = 110°$, $\widehat{GF} = 68°$, and $\widehat{FD} = 74°$. Find, in order, the measure of all the numbered angles.

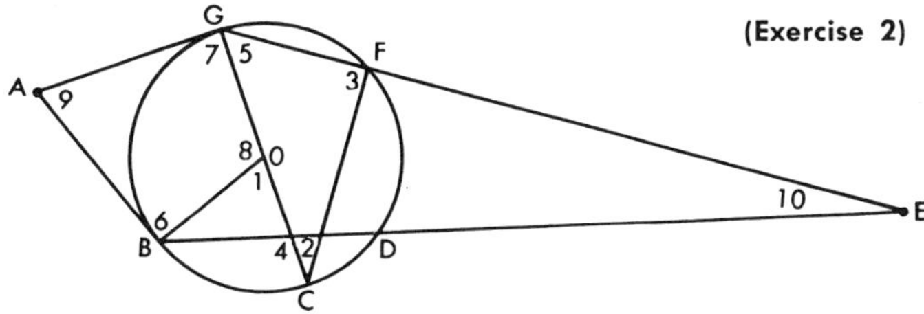

Figure 13–26

3. A great circle and a small circle of a sphere pass through the same two points on the sphere. Which of the two arcs is shorter, the arc of the great circle, or of the small circle?

4. Inscribe a regular hexagon in a circle whose radius is 3 cm. How long is each side of the hexagon? (*Hint:* What is the measure of each angle of a triangle formed by two radii and a side of the hexagon?)

5. Inscribe an equilateral triangle in a circle whose radius is 1 inch.

6. An angle formed by two secants intersecting outside a circle intercepts two arcs on the circle. An angle inscribed on the circle intercepts the greater of these two arcs. Which angle has the greater measure? Prove your answer.

7. A ship sailing out of a bay into the ocean avoids marked shoals, keeping the measure of its "danger angle" (greater than, less than) the measure of the inscribed angle of the "safety arc." A and C represent lighthouses. Arc $\widehat{ABC}$ = 250°. Find $m\angle ABC$.

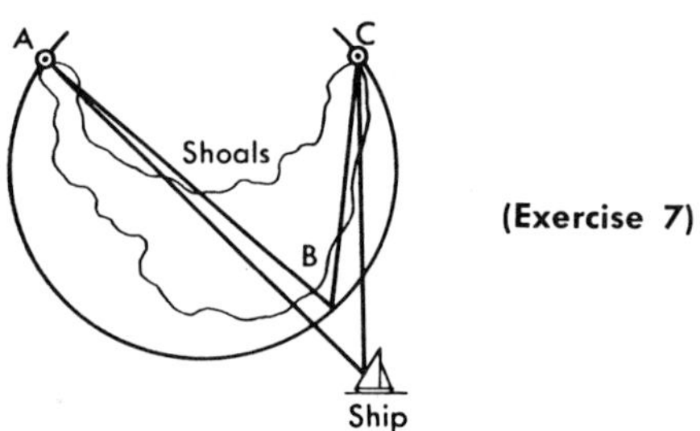

Figure 13–27

8. An isosceles trapezoid $ABCD$ is inscribed in a circle. $\overline{AD}$ is a diameter 24 inches in length. If $\widehat{AB} = 60°$, find the length of chord $\overline{BC}$.

9. Is the midpoint of the arc intercepted by the angle formed by a tangent and a chord equidistant from the tangent and the chord? Prove your answer.

10. If one leg of a right triangle is a diameter of a circle, must the tangent at the point where the circle cuts the hypotenuse bisect the other leg? Prove your answer.

11. Construct a common external tangent to two coplanar, nonintersecting circles. (*Hint:* Draw a circle whose radius is the difference between the radii of the two circles and whose center is the center of the larger circle.)

Chapter 13 Test

1. In circle O, $\widehat{AB} = 70°$. Name $m\angle AOB$.
2. Triangle ABC is inscribed in circle O. If $\widehat{AB} = 110°$ and $\widehat{AC} = 100°$, find $m\angle A$.
3. If the measures of two adjacent angles of an inscribed quadrilateral are 50° and 100°, name the measures of the remaining angles of the quadrilateral.
4. Tangents are drawn to a circle from an external point. If the measure of the smaller arc intercepted by the tangents is 110°, find the measure of the angle determined by the tangents.
5. A and C are points of circle O. If $\overline{SC}$ is tangent to the circle and $\widehat{CA} = 80°$, find $m\angle SCA$.
6. Quadrilateral $ABCD$, with $\overline{AB} \parallel \overline{CD}$, is inscribed in circle O where $\overline{AB}$ is a diameter. If $\widehat{AD} = 40°$, find $m\angle A$.
7. In circle O, chords $\overline{AB}$ and $\overline{CD}$ intersect at E. If $\widehat{AC} = 30°$ and $\widehat{BD} = 120°$, find $m\angle AEC$.
8. Two rays from P in the exterior of a circle each contain two points of the circle. If the measures of the arcs in the interior of $\angle P$ are 20° and 120°, find $m\angle P$.
9. Circle O is drawn with diameter $\overline{BC}$. Circle B (B is the center) is drawn intersecting circle O at A and D. Show that $\overleftrightarrow{AC}$ is tangent to circle B.
10. Equilateral pentagon $ABCDE$ is inscribed in circle O. $\overleftrightarrow{AE}$ and $\overleftrightarrow{CD}$ intersect at S. The tangents at B and C intersect at P. Prove $m\angle BPC = 3m\angle ESD$.

14

Geometric Solids

As you now realize, our geometric definitions and assumptions are patterned after the objects in our environment. This seems contradictory when you remember that most of the geometric figures we have studied thus far have been two-dimensional. It is necessary to study two-dimensional figures first, however, because many of our three-dimensional geometric figures will be formed from certain combinations of two-dimensional figures.

Our senses of sight and touch perceive objects in our environment. We think of these objects as physical solids. Geometric solids and closed surfaces will now be defined.

14.00 A closed geometric surface divides space into three distinct sets of points, the surface, the interior points enclosed by the surface, and the exterior points. A geometric solid is the union of a geometric surface and its interior.

14.01 A polyhedron is a closed geometric surface formed by the union of polygonal regions (6.04) so arranged that the intersections of the regions is the null set, a side, or a vertex of each of them. The regions are called faces, the intersection of two sides is called an edge, and the intersection of three or more vertices is called a vertex of the polyhedron. The name of a face of a polyhedron is the name of the geometric figure forming the boundary of the face. Faces may be triangular, quadrangular, etc.

A cube is a polyhedron, as are all geometric surfaces with rectangular faces. If a rectangular block of wood is run through a saw a number of times, is the result a polyhedron if care is taken that all surfaces are planes?

14.02 A diagonal of a polyhedron is a segment whose endpoints are any two vertices not in the same face. In Figure 14-1, $\overline{AH}$ is a diagonal of the polyhedron.

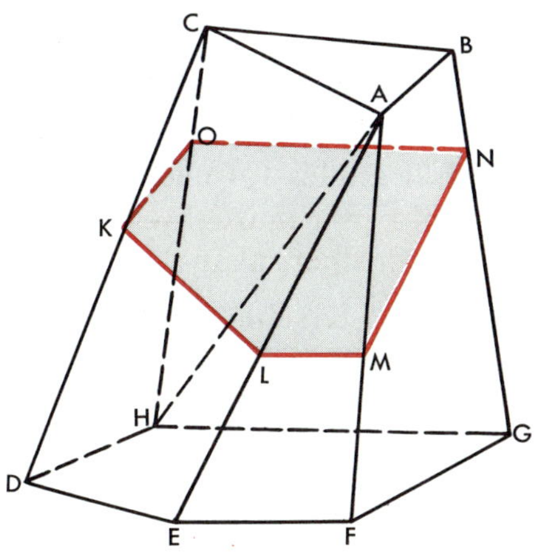

Figure 14–1

14.03 A section of a geometric surface is the plane figure formed by the intersection of a plane and the surface.

In Figure 14-1, plane KMN intersects the polyhedron to form section $KLMNO$. The section of a polyhedron is necessarily a polygon. Why? (2.23 and 2.48)

14.04 A polyhedron is convex if every section of it is a convex polygon. Only convex polyhedrons will be considered here.

14.05 Polyhedrons may be classified according to the number of faces.

tetrahedron—four faces
pentahedron—five faces
hexahedron—six faces
heptahedron—seven faces
octahedron—eight faces
nonahedron—nine faces
decahedron—ten faces
dodecahedron—twelve faces
icosahedron—twenty faces

Exercises

1. Is a sphere a closed geometric surface?
2. What is a section of a sphere?
3. Name the polyhedron in Figure 14-1 by the number of its faces.
4. Name the type of polyhedron represented by your classroom.
5. What is the least number of faces a polyhedron may have? The greatest number?
6. In the polyhedron in Figure 14-1, how many faces are triangular? How many faces are quadrangular?
7. How many dihedral angles formed by the faces of the polyhedron can you find in Figure 14-1? How many trihedral angles? How many tetrahedral angles?
8. List the trihedral and tetrahedral angles of Figure 14-1 by the vertex letter.
9. In Figure 14-1, name the plane section (type of polygon) formed by passing a plane through D, A, and G. By passing a plane through D, B, and H.
10. Represent a rectangular-shaped room and a section of it made by a plane intersecting only three of its faces.

REGULAR POLYHEDRONS

14.06 A regular polyhedron is a polyhedron whose faces are congruent regular polygonal regions and whose polyhedral angles are congruent.

Regular physical polyhedrons are formed when certain liquids are solidified. The solidification produces crystals with definite shapes. Salt crystals represent regular hexahedrons—cubes. Diamonds and certain compounds of copper are in the form of regular octahedrons. Most minerals of the earth occur in crystalline form.

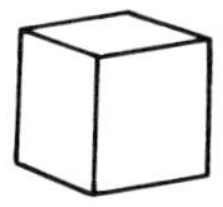

Tetrahedron Octahedron Icosahedron Hexahedron Dodecahedron

Figure 14–2

Exercises

Because each polyhedron contains several polyhedral angles, we need to review and extend our understanding of polyhedral angles.

1. Construct a model of a trihedral angle the measure of whose face angles are 100°, 60°, and 80°. Attempt to construct another trihedral angle, using the same size face angles, which is not congruent to the first. Is this possible? Is it possible for the measures of two tetrahedral angles to be equal (sums of the measures of their face angles are equal) although the angles are not congruent? Explain.
2. What is the least number of faces that a polyhedral angle can have?
3. Must the sum of the measures of two face angles of a trihedral angle be greater than the measure of the third face angle? Explain. (*Hint*: Consider the case where the sum of the measures of two face angles is less than the third; equal to the third. Make paper models to illustrate the three possibilities.)
4. If the measure of one face angle of a trihedral angle is 130°, the sum of the measures of the other two face angles is (less than, equal to, greater than) 130°. Select the correct phrase.
5. The sum of the measures of the face angles of a trihedral angle is 150°. What is the largest value one face angle may have?
6. Is it possible to have a trihedral angle the measures of whose face angles are 80°, 30°, and 40°? Explain.
7. **(a)** Attempt to construct models of convex polyhedral angles having three, four, five, and six congruent equilateral triangular faces. **(b)** What is the sum of the measures of the

"face angles" of the polyhedral angle having six faces? What type of figure is represented? (c) What type of figure would result if you formed a polyhedral angle using seven or more equilateral triangular faces?

8. (a) Attempt to construct polyhedral angles having respectively three and four congruent square regions as faces. (b) What type of figure is represented when four congruent square faces form the polyhedral angle? What is the sum of the measures of the "face angles"?

9. Attempt to construct models of convex polyhedral angles using (a) regular pentagonal regions and (b) regular hexagonal regions to form the face angles. (c) Is this possible? Discuss the results.

10. Is the sum of the measures of the face angles of convex polyhedral angles always the same or does the sum vary with various polyhedral angles? If the sum varies, does it have an upper limit? Explain your answer.

It is interesting to note that only a limited number of distinctly different convex regular polyhedrons may be constructed. The following questions were designed to direct your thinking toward the reasons for such limitations.

11. Must the faces of regular polyhedrons be regular polygonal regions? Must these polygonal regions be congruent? What is the relationship of the polyhedral angles of a regular polyhedron?

12. Based on the results of Exercise 7, how many distinctly different types of convex polyhedral angles can be formed using congruent equilateral triangular faces? Since the polyhedral angles of a regular polyhedron must be congruent (14.06), what is the greatest number of non-similar regular polyhedrons that could be formed using equilateral triangular faces?

13. How many non-similar regular polyhedrons might be formed using square regions as faces? (Re-examine Exercise 8.) Using pentagonal faces? (See Exercise 9.) Using hexagonal faces?

14. How may different shaped convex regular polyhedrons can be formed?

15. Have we proved that the number of regular polyhedrons in Exercise 14 actually exist?

It is possible to prove the existence of five regular convex polyhedrons and to construct them. The proof is too difficult for this course, however. Models of the regular polyhedrons can easily be made by drawing on stiff paper the following diagrams. Cut along the solid lines and fold along the dotted lines.

Suggestion: Leave a flange for gluing the faces together as shown in the first figure.

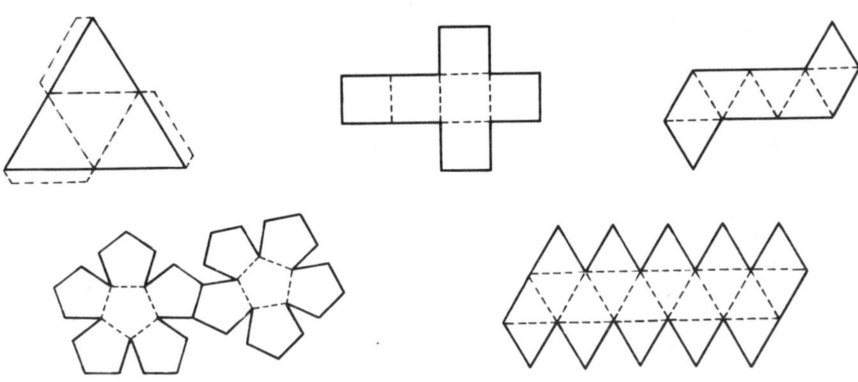

Figure 14–3

Exercises

1. Make a table similar to the following and fill in the blanks for the indicated regular polyhedrons. (Do not write in this book!)

Name	*No. of faces*	*No. of edges*	*No. of vertices*	*Form of faces*	*No. of diagonals*
Tetrahedron					
Hexahedron					
Octahedron					
Dodecahedron					
Icosahedron					

2. From the preceding table, verify the formula $F + V = E + 2$, where F is the number of faces of a regular polyhedron, V the number of vertices, and E the number of edges.

3. How many faces does a regular polyhedron have if its faces are equilateral triangular regions, three of which meet at each vertex of the polyhedron? Name this polyhedron.
4. How many faces does a regular polyhedron have if its faces are equilateral triangular regions, four of which meet at each vertex of the polyhedron? Name this polyhedron.
5. How many faces does a regular polyhedron have if its faces are equilateral triangular regions, five of which meet at each vertex of the polyhedron? Name this polyhedron.
6. How many faces does a regular polyhedron have if its faces are square regions? Name this polyhedron.
7. How many faces does a regular polyhedron have if its faces are regular pentagonal regions, three of which meet at each vertex of the polyhedron? Name this polyhedron.
8. If we place two congruent regular tetrahedrons together to form a polyhedron, is it a regular polyhedron? Explain.

PRISMS

The *prism* forms a special class of polyhedra in which two faces are congruent parallel polygonal regions. Prisms are probably the most commonly used three-dimensional figures; nearly all boxes are prismatic in shape, as are most buildings and packages. A prism might be thought of as the solid generated by moving a polygonal region parallel to itself from one position to another. It is easier to define a prism if we first investigate a special surface called a closed prismatic surface.

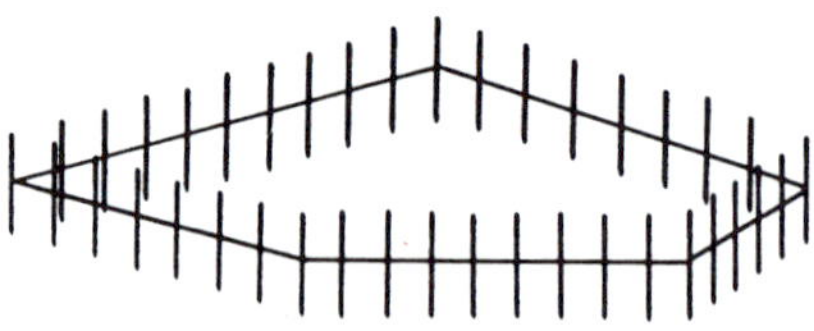

Figure 14–4

14.07 A closed prismatic surface (Figure 14-4) is *generated* by a moving straight line which always intersects and moves completely about a fixed polygon, always remaining parallel to a fixed straight line not in the plane of the polygon.

14.08 This moving line (*generatrix*) in any one of its positions is called an element of the surface. The fixed polygon is called the *directrix*.

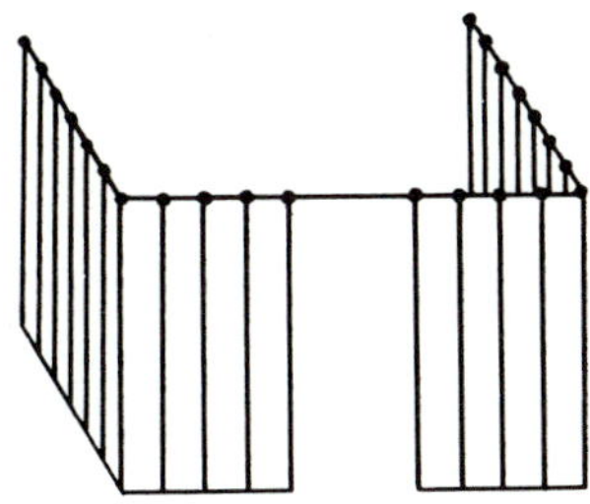

Figure 14–5

In Figure 14-5 above, the closing of a shower curtain may be compared to the prismatic surface generated by a straight line as it moves about a polygon. The leading edge of the curtain may be compared to the generatrix. The rod the curtain hangs on is similar to the directrix.

14.09 A prism (Figure 14-6) is the union of the part of the closed prismatic surface between two parallel planes intersecting all the elements and the polygonal regions determined by the plane sections.

The polygonal regions of the sections formed by the two parallel planes are the bases of the prism ($LMNOP$ and $L'M'N'O'P'$). The remaining faces of the prism are called lateral faces, and the intersections of the lateral faces are called the lateral edges. The segments of the elements of the prismatic surface between the bases will be called elements of the prism.

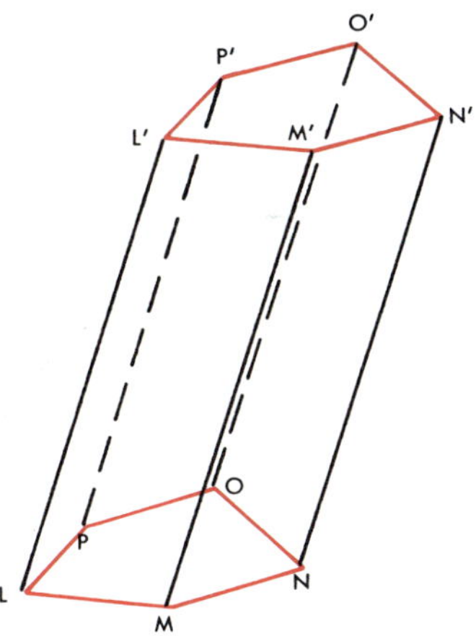

Figure 14–6

A glass prism in which the bases are equilateral triangular regions is often used in experiments with light to separate white light into the colors of the spectrum. If you can borrow such a prism from the science laboratory of your school, you can see how this is accomplished. To understand many important principles of science, you must have a good background in mathematics.

Exercises

1. If the elements of a prism (14.08) are always parallel to some fixed line in space (14.07), must they be parallel to each other? Explain.
2. If the elements of a prism are cut by two parallel planes (14.09), must the segments be the same length? Explain.
3. Must the lateral faces of a prism be quadrangular?
4. Does it follow that the lateral faces of a prism are always parallelogram regions? Explain.
5. Will the bases of a prism always be quadrangular? Will the bases always be congruent figures? Explain.

Some of the conjectures made in these exercises may be proved as theorems. Proof of the following propositions is left to the student.

14.10 THEOREM

The lateral edges of a prism are congruent and parallel.

(Refer to Exercise 23, page 272).

14.11 THEOREM

The lateral faces of a prism are parallelogram regions.

How are the bases of a prism formed? Will each base have the same number of sides? Will they be congruent?

Given: Any prism $ABCD\text{-}A'B'C'D'$

Figure: 14–7

Conjecture: $ABCD \cong A'B'C'D'$

Plan: Use 9.09 and 11.33.

Proof: The proof is left to the student.

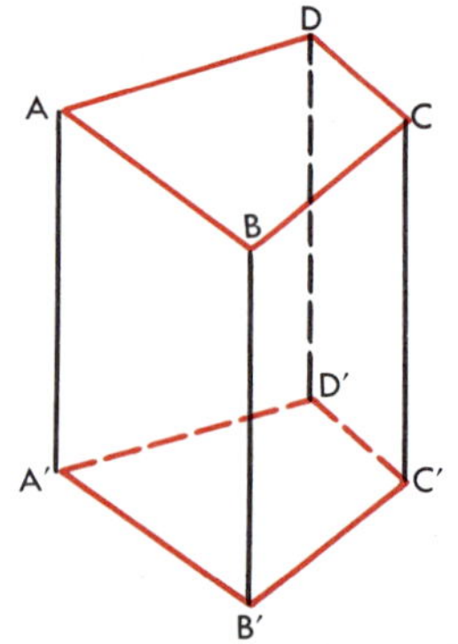

Figure 14–7

14.12 THEOREM

The bases of a prism are congruent polygonal regions.

14.13 The altitude of a prism is a segment perpendicular to and terminated by the planes of the bases. The measure of the altitude is called the *height*.

14.14 A right section of a prism (*ABCDE*, Figure 14-8) is a section made by a plane perpendicular to one and intersecting all the lateral edges. An oblique section is made by a plane oblique to one and intersecting all the lateral edges.

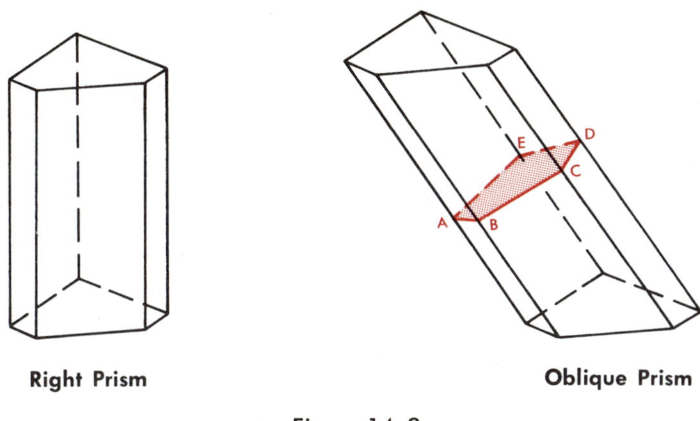

Right Prism **Oblique Prism**

Figure 14–8

14.15 A right prism is a prism whose lateral edges are perpendicular to the bases. An oblique prism is a prism whose lateral edges are not perpendicular to the bases.

14.16 THEOREM

The lateral faces of a right prism are rectangular regions.

The proof is left for the student. (*Hint*: See 14.11.)

14.17 A regular prism is a right prism whose bases are regular polygonal regions. The axis of a prism whose bases are regular polygonal regions is the line segment connecting the centers of the bases.

14.18 Prisms are triangular, quadrangular, pentagonal, etc., according to whether their bases are triangular, quadrangular, pentagonal, etc.

14.19 A truncated prism is the union of the part of the closed prismatic surface between two non-parallel planes intersecting all the elements and the polygonal regions determined by the plane sections.

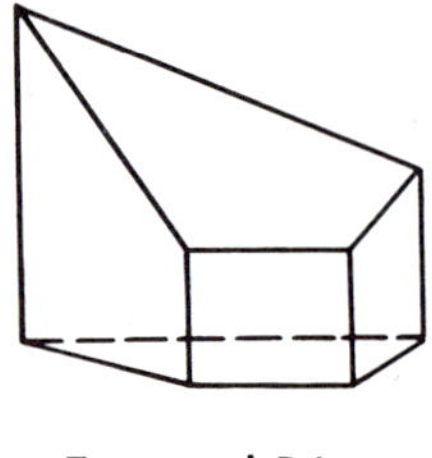

Truncated Prism

Figure 14–9

14.20 A parallelepiped is a prism whose bases are parallelogram regions.

A right parallelepiped is a parallelepiped that is a right prism.

A rectangular parallelepiped is a right parallelepiped whose bases are rectangular. A rectangular parallelepiped solid is commonly called a rectangular solid.

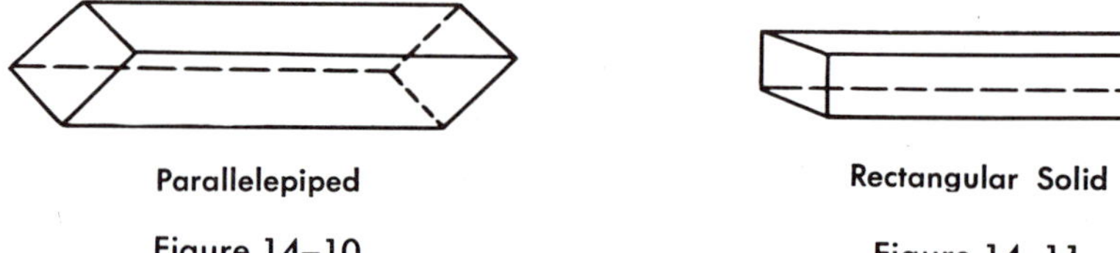

Parallelepiped

Figure 14–10

Rectangular Solid

Figure 14–11

14.21 A cube (Figure 14.12) is a rectangular parallelepiped whose edges are all congruent.

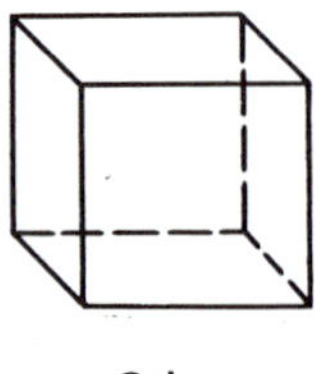

Cube

Figure 14–12

14.22 A prismatoid (Figure 14-13) is a polyhedron all of whose vertices are in two parallel planes.

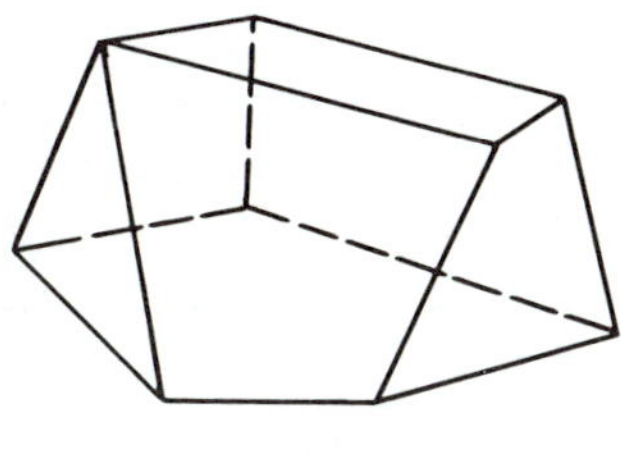

Prismatoid

Figure 14–13

Exercises

1. Is a parallelepiped a regular prism? A prismatoid?
2. Is a cube a rectangular parallelepiped? A prism? A prismatoid?
3. How many faces does a parallelepiped have? How many diagonals? Are the diagonals congruent?
4. Are all the faces of a rectangular solid rectangular? Explain.
5. Are all the faces of a parallelepiped parallelogram regions? Explain.
6. What type of figures are the faces of a cube? Explain.
7. What is the section of a parallelepiped made by a plane containing two nonadjacent edges? Why?
8. If a parallelepiped has a rectangular base is it necessarily a right parallelepiped?
9. Represent a regular pentagonal prism.
10. Represent a right triangular prism and an oblique triangular prism with congruent bases and congruent altitudes.
11. Under what conditions is an altitude of a prism perpendicular to a right section of the prism?
12. Could a quadrangular prism which is not regular be a right prism? Could its bases be squares? Could it have both of these properties together?

14.23 A diagonal section of a prism is a section of the prism containing two edges not in the same face.

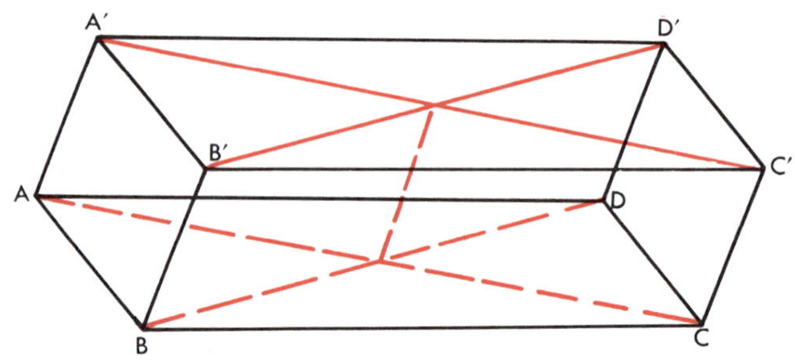

Figure 14–14

1. Is $B'BDD'$ a diagonal section of parallelepiped $ABCD$–$A'B'C'D'$? How many diagonal sections can be drawn in Figure 14-14?
2. Are these diagonal sections parallelograms or are they skew quadrilaterals? (See Exercise 24, page 272.) Explain.
3. How many diagonals (14.02) can you draw in Figure 14-14? Would they all lie in the regions of the two diagonal sections shown? Would they all be congruent? Are the diagonals necessarily concurrent?
4. Would the diagonals bisect each other? (*Hint*: The diagonals would first have to be proved concurrent.)

14.24 THEOREM

A diagonal section of a parallelepiped is a parallelogram.

Let us explore the conjecture made in exercise 4.

Given: Parallelepiped $ABCD$–$A'B'C'D'$

Conjecture: $\overline{AC'}$, $\overline{A'C}$, $\overline{B'D}$, $\overline{BD'}$ bisect each other.

Plan: Use 9.07.

Proof: The proof is left to the student.

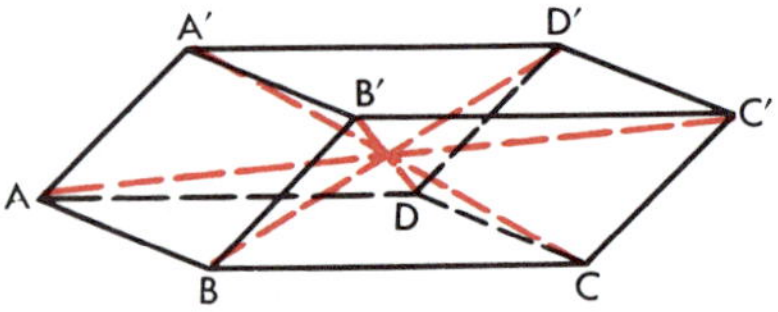

Figure 14–15

14.25 THEOREM

The diagonals of a parallelepiped are concurrent and bisect each other.

14.26 THEOREM

The diagonals of a rectangular parallelepiped are congruent.

(The proof is left to the student.)

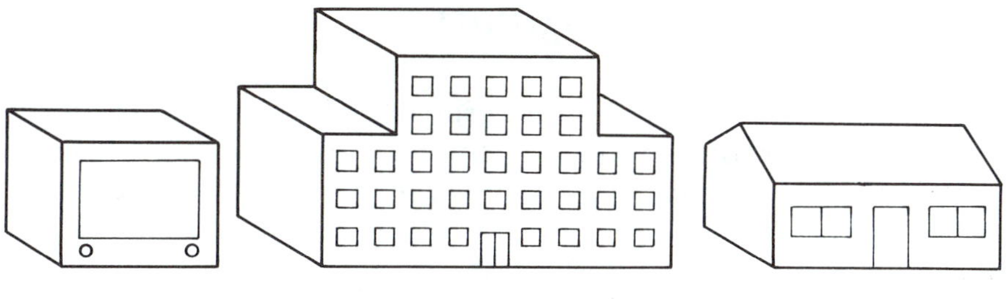

Prisms in Our Environment

Figure 14–16

Miscellaneous Exercises

1. What is the least number of faces a prism can have? The greatest number?
2. How many faces has a triangular prism? A pentagonal prism?
3. How many edges has a triangular prism? A pentagonal prism?
4. Can a section of a prism be a trapezoid? Explain.
5. What type of figures are the lateral faces of a prism?
6. Are all edges of a prism congruent?
7. Is the axis of a regular prism congruent to the altitude?
8. Are the lateral edges of a right prism congruent to the altitude?
9. What type of figures are the lateral faces of a right prism?
10. What is the measure of a dihedral angle formed by two adjacent lateral faces of a regular pentagonal prism?
11. Find the sum of the measures of the face angles of a polyhedral angle of a regular hexagonal prism.

12. What is the name of the prism whose lateral faces are rectangular?

13. Is the right section of a regular quadrangular prism a square? Is the oblique section a square?

14. If an oblique prism has a square base region, is a right section of it a square?

15. Name the various sections of a cube formed by the planes containing the midpoints, and only the midpoints, of the edges.

CYLINDERS

Most students in your class are probably able to give a fairly accurate definition of a cylinder. However, it is doubtful if many members of your class have compared the properties of a cylinder with those of a prism. Can you think of any properties of a cylinder that a prism would also have?

14.27 As you may recall from 14.07, a *prismatic surface* is generated by a moving straight line (parallel to a fixed straight line *not* in the plane of the polygon) that intersects and travels about a fixed polygon. Similarly, a cylindrical surface is generated when the *directrix* is a closed curve rather than a polygon. You can see, then, that a *cylindrical surface* is similar to a *prismatic surface*, except that its directrix is a closed curve rather than a polygon.

Figure 14–17

14.28 A cylinder is the union of the parts of a closed cylindrical surface between two parallel planes intersecting all of the elements (14.08), and the curved regions determined by the plane sections. The segments of the elements of the cylindrical surface between the parallel planes will be called elements of the cylinder. (Fig. 14-18)

The curved regions are the bases of the cylinder, and the lateral surface is the portion of the cylindrical surface between the parallel planes.

What similarities can you now see in the bases, directrix, elements, and surface of a cylinder as compared with those of a prism?

14.29 THEOREM

The elements of a cylinder are parallel and congruent.

14.30 THEOREM

The bases of a cylinder are congruent.

14.31 The altitude of a cylinder is a segment perpendicular to and terminated by the planes of the bases. The measure of the altitude is called the *height*.

14.32 A right section of a cylinder is a section made by a plane perpendicular to and intersecting all elements.

14.33 A right cylinder is a cylinder in which the planes of the bases are perpendicular to one of the elements. Are the planes perpendicular to all the elements? Why?

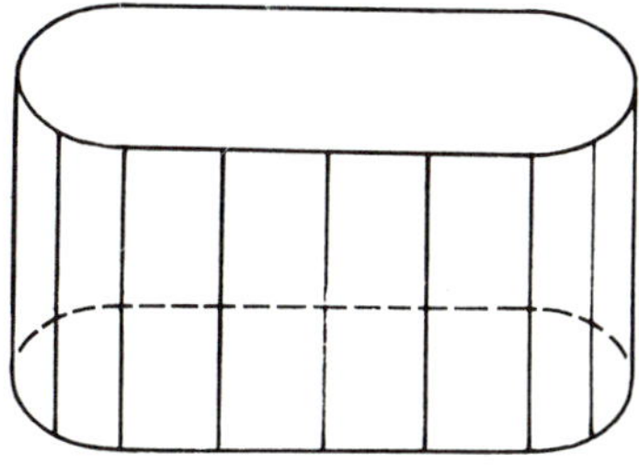

Figure 14–18

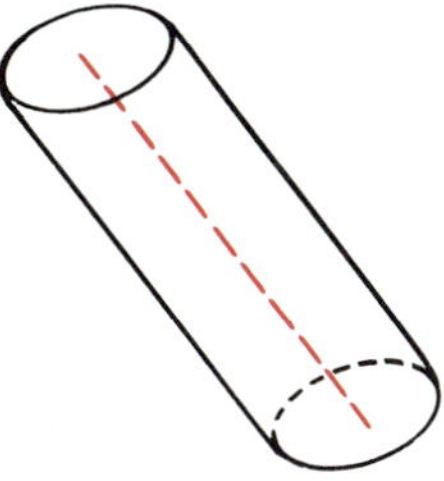

Figure 14–19

14.34 An oblique cylinder is a cylinder in which the planes of the bases are oblique to one of the elements. Are they then oblique to all the elements?

14.35 A circular cylinder (Figure 14-19) is a cylinder whose bases are circular regions (12.07).

We can see by the definitions above that, just as a cylinder, itself, is similar to a prism, the properties of a cylinder are similar to the properties of a prism. From this relationship, state a definition for the *axis of a circular cylinder*. If necessary, refer to 14.17.

Because a right circular cylinder may be generated by revolving a rectangle about one of its sides as an axis, it is sometimes called a cylinder of revolution (Figure 14-20).

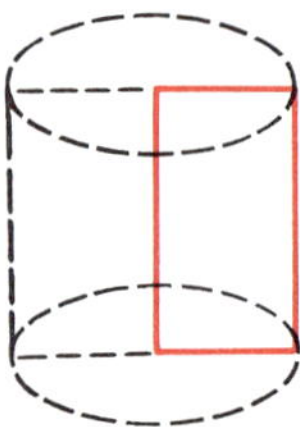

Figure 14–20

Exercises

1. Are all circular cylinders necessarily cylinders of revolution? Explain.
2. Are all cylinders of revolution necessarily circular cylinders? Explain.
3. Name the section of a cylinder made by a plane containing two elements of the cylinder.
4. In what type of cylinder would all sections containing two elements be a rectangle?
5. Is it possible that a section of an oblique circular cylinder would be a rectangle? Explain.
6. What is the name of a right section of a right circular cylinder? Can an oblique section of a right circular cylinder be a circle?
7. Is it possible for a section of a non-circular cylinder to be a circle? Explain.
8. If we cut a right circular cylinder along one element, discard the bases, and make it a plane surface, what type of figure results?
9. Repeat Exercise 8 with an oblique circular cylinder. Illustrate the result with a drawing.
10. A section of a circular cylinder containing the axis would form what type of polygon?

PYRAMIDS

You have all seen pictures of the ancient pyramids of Egypt which were built as tombs for rulers. In modern times, some camping tents are also in the shape of pyramids. From these examples and others that you can imagine, we see that two characteristics of a pyramid are that it has only one base and that it comes to a point. How does a pyramid compare with a prism? Could the same base serve as one base of a prism and the base of a pyramid?

Perform the following experiment: Tie several elastic strings (or broken rubber bands) into a knot at one end. Place as many thumbtacks as you have strings (four or five are satisfactory) at vertices of a polygon on a bulletin board. Then tie the loose ends of the elastic strings to the thumbtacks and pull the knot so the strings are taut. The strings and the polygon form the edges of a pyramid. Can the knot be moved to different positions? Can the polygon have a greater or fewer number of vertices?

Before we consider pyramids, let us define a pyramidal (pĭ răm′ ĭ dal) surface.

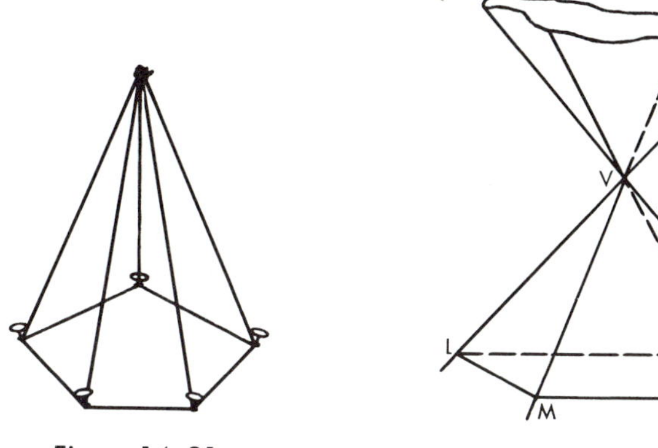

Figure 14–21

Figure 14–22

14.36 A pyramidal surface (Figure 14-22) is *generated* by a line which always intersects and moves around a given fixed polygon ($LMNP$ in Figure 14-22) and passes through a fixed point (V) not in the plane of the polygon.

The moving line in any one of its positions is an element of the surface. The fixed point (V) is the vertex of the surface.

14.37 The moving line (generatrix), moving along one side of the polygon (directrix), generates a plane determined by the vertex and the side. Why?

14.38 As shown in Figure 14-22, a pyramidal surface consists of two parts separated by the vertex. Each of these parts is called a nappe (năp). Each nappe contains the vertex.

14.39 A pyramid is the union of one nappe of a pyramidal surface and the section and its interior formed by a plane intersecting all its elements (Figure 14-23). The section and its interior (polygonal region) is called the base of the pyramid ($ABCD$ Figure 14-23). The segments of elements of the pyramidal surface between the base and the vertex will be called elements of the pyramid.

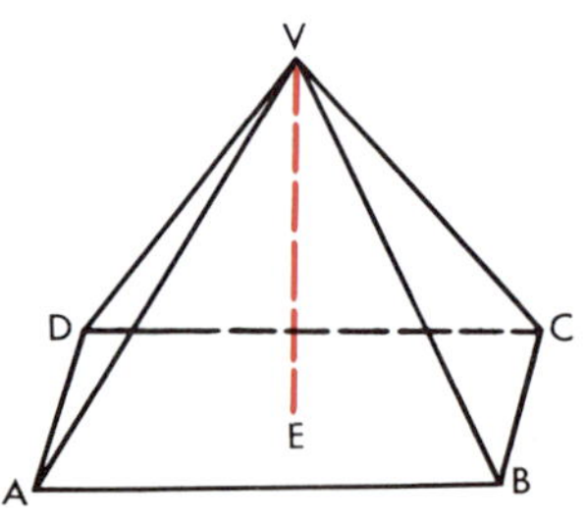

Figure 14–23

14.40 The triangular regions of a pyramid are called the lateral faces (AVB, DVC, BVC, AVD), and the intersections of the lateral faces are the lateral edges ($\overline{AV}$, $\overline{BV}$, $\overline{CV}$, $\overline{DV}$).

14.41 The vertex of the pyramid is the vertex of the pyramidal surface.

14.42 The altitude of the pyramid ($\overline{VE}$) is the segment from the vertex perpendicular to and terminated by the plane of the base. Is E necessarily the center of the base? The measure of the altitude is called the *height*.

14.43 A pyramid is triangular, quadrangular, or pentagonal, as the base is triangular, quadrangular, or pentagonal.

14.44 A regular pyramid is a pyramid whose base is a regular polygonal region and whose altitude contains the center of the base. (See 13.13.)

14.45 The axis of a pyramid whose base is a regular polygonal region is the segment connecting the vertex with the center of the base.

14.46 THEOREM

The lateral edges of a regular pyramid are congruent.

14.47 THEOREM

The lateral faces of a regular pyramid are congruent isosceles triangular regions.

14.48 The measure of the altitude of a lateral face of a regular pyramid is the slant height of the pyramid.

Exercises

1. What type of polyhedral angle is determined by the union of the lateral faces of a quadrangular pyramid?
2. Are the polyhedral angles at the base of a pyramid always trihedral angles?
3. How many dihedral angles are there in a quadrangular pyramid? How many polyhedral angles?
4. How many edges does a hexagonal pyramid have?
5. Could a regular octahedron be separated into two regular quadrangular pyramids? Explain.
6. What is the section of a pyramid formed by a plane containing two lateral edges not in the same face?
7. If segments connect the center of a cube with each point of each edge, how many pyramids are formed?
8. Does every pyramid have a slant height? Explain.
9. In what type of pyramids is the axis perpendicular to the base?
10. Is a regular tetrahedron a regular triangular pyramid?
11. Is a regular triangular pyramid always a regular tetrahedron?

12. Are the altitudes of the triangular faces of a regular pyramid congruent? Explain.

13. $V\text{-}CDE$ is a regular tetrahedron. A and B are the midpoints of $\overline{CD}$ and $\overline{VE}$ respectively. Prove $\overline{AB}$ is perpendicular to $\overline{CD}$ and $\overline{VE}$. (Figure 14-24)

Are $\overleftrightarrow{CD}$ and $\overleftrightarrow{VE}$ skew lines? Do you believe two skew lines have a common perpendicular? More than one common perpendicular?

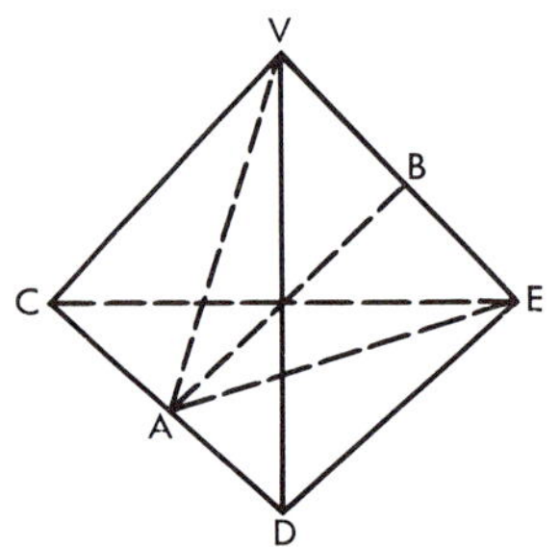

Figure 14–24

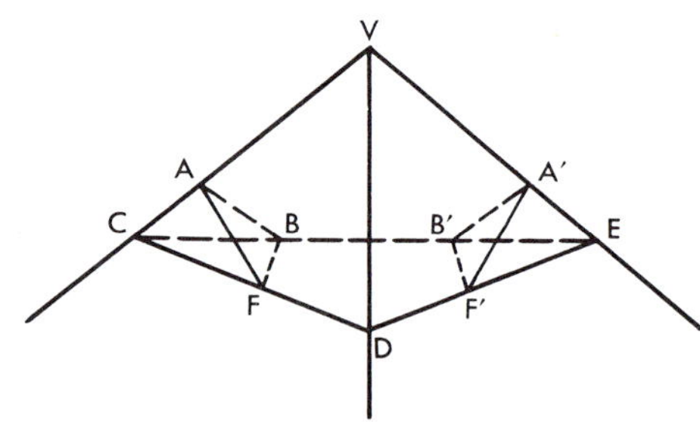

Figure 14–25

14. In trihedral angle $V\text{-}CDE$, $\overline{VC} \simeq \overline{VD} \simeq \overline{VE}$, and $\angle CVD \simeq \angle EVD$. $\overline{CA} \simeq \overline{EA'}$. $\overline{FA}$, in face CVD is $\perp \overline{VC}$. $\overline{BA}$, in face CVE, is $\perp \overline{VC}$. $\overline{F'A'}$, in face DVE, is $\perp \overline{VE}$. $\overline{B'A'}$ in face CVE, is $\perp \overline{VE}$. Prove dihedral angles VC and VE are congruent. (Figure 14-25)

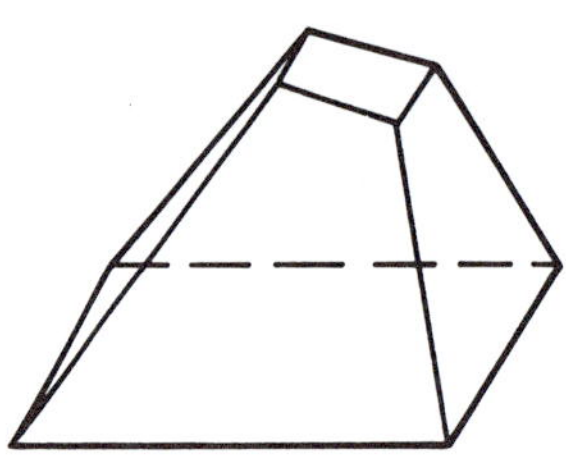

Figure 14–26

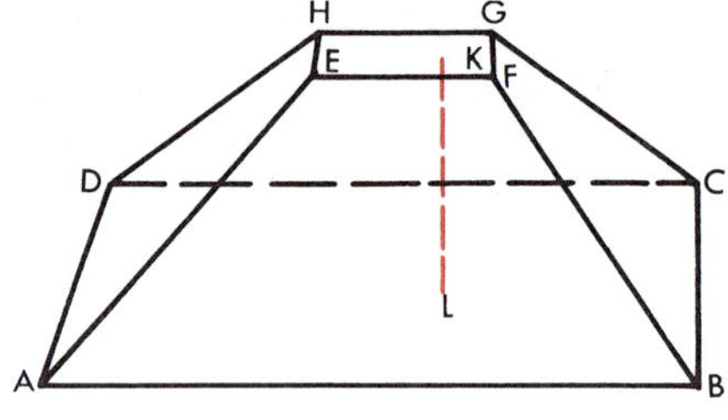

Figure 14–27

14.49 A truncated pyramid (Figure 14-26) is the union of the part of a pyramid containing the base and the section and its interior formed by a plane intersecting all the elements of the pyramid. The section and its interior is called a base of the truncated pyramid. If the two bases are parallel, the figure is called a frustum of a pyramid.

14.50 The altitude of a frustum of a pyramid is a segment ($\overline{KL}$) perpendicular to and terminated by the planes of the bases. (Figure 14-27) The measure of the altitude is called the *height*.

14.51 The slant height of a frustum of a regular pyramid is the measure of the altitude of one of its lateral faces.

14.52 THEOREM

The lateral faces of a frustum of a pyramid are trapezoid regions. (9.19)

14.53 THEOREM

The lateral faces of a frustum of a regular pyramid are congruent isosceles trapezoid regions. (9.22)

Pyramids and Frustums of Pyramids in Our Environment

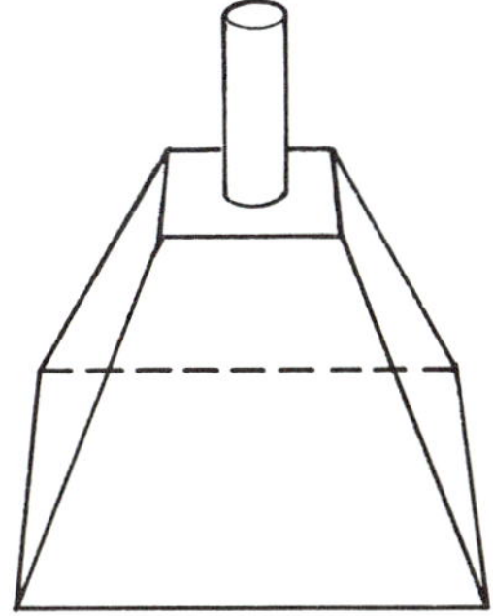

Footings for Construction

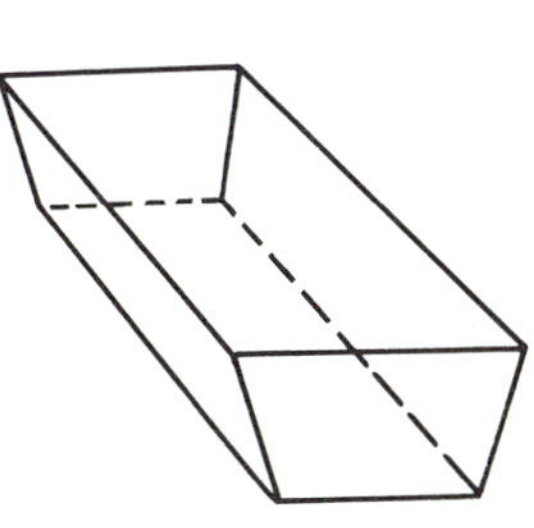

Flower Pots, Feed Troughs, Dump Truck Boxes

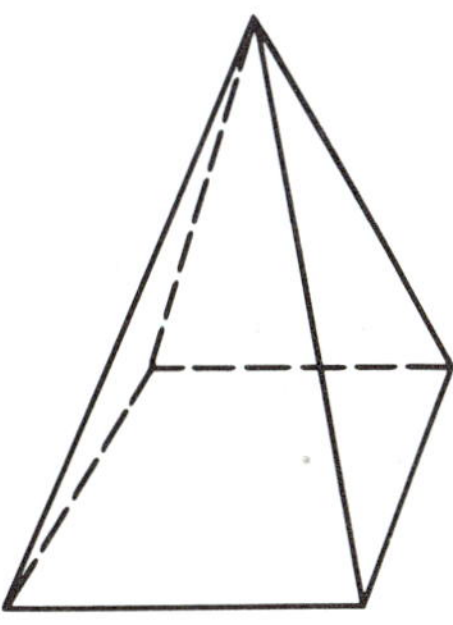

Tents, Spires, Pyramids of Egypt

Figure 14–28

Exercises

1. Is a frustum of a pyramid a truncated pyramid?
2. Under what conditions would the lateral faces of a truncated pyramid become trapezoid regions?
3. What distinguishes frustums of pyramids from the general set of truncated pyramids?

4. Are the lateral edges of a frustum of a regular pyramid congruent?
5. Name the type of frustums of pyramids where the name "slant height" applies. Does this name apply to frustums of pyramids in general?
6. Distinguish between the terms "altitude" and "slant height" of a frustum of a pyramid.
7. Is a truncated pyramid a member of the set of polyhedrons?
8. Is it possible that a pyramid could be a regular polyhedron? Explain.
9. Is it possible that a truncated pyramid could be a regular polyhedron? Explain.
10. Is it possible that a prism could be a regular polyhedron? Explain.
11. Is it possible that a prismatoid could be a regular polyhedron? Explain.

CONES

14.54 As a cylinder is related to a prism, a cone is related to a pyramid. Given the conditions of 14.36, but with the directrix a closed curve rather than a polygon, a conical surface will be generated (Figure 14-29).

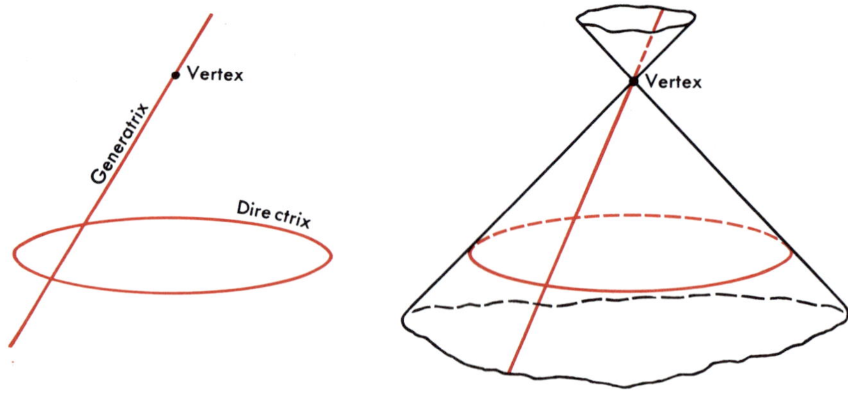

Figure 14–29

Courtesy of Mexican Government Tourism Department

About a thousand years ago, in what is now Guatemala and Yucatan, the Mayas built temples like this one. What geometric solids do you see? How does this differ from an Egyptian pyramid?

14.55 A cone (Figure 14-30) is the union of one nappe of a closed conical surface and the section and its interior formed by a plane intersecting all the elements. The section and its interior is called the base of the cone. The segments of the elements of the conical surface between the vertex and the base will be called elements of the cone.

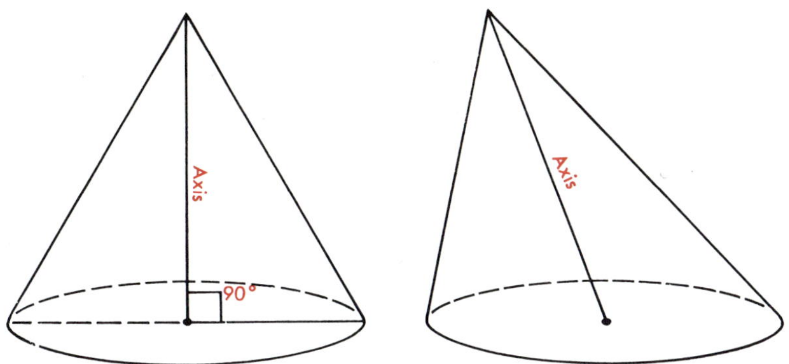

Figure 14–30

The lateral surface is the union of the elements of the cone. The altitude of the cone is the segment from the vertex perpendicular to and terminated by the plane of the base. The measure of the altitude is called the *height*.

14.56 A circular cone is a cone whose base is a circular region. The axis of a circular cone is the segment from the vertex to the center of the base.

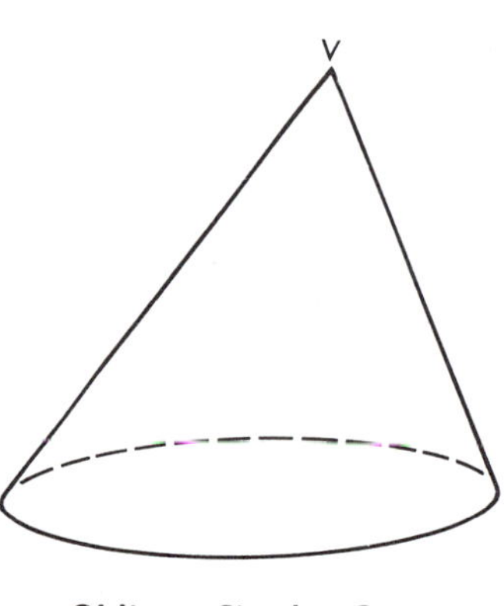

Oblique Circular Cone

Figure 14–31

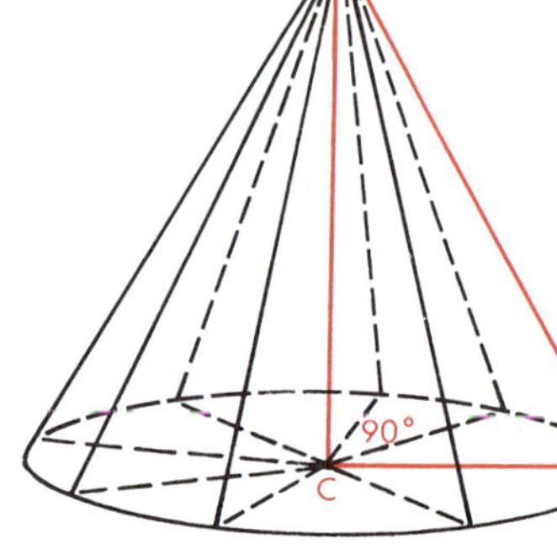

Right Circular Cone

Figure 14–32

14.57 A right circular cone is a cone whose axis is perpendicular to the plane of the base. Rotation of a right triangle about one of its legs as an axis *generates a right circular cone*, which is sometimes called a *cone of revolution.*

A cone whose axis is oblique to the plane of the base is called an oblique cone.

14.58 THEOREM

The altitude and axis of a right circular cone are the same segment.

(The proof is left to the student.)

14.59 THEOREM

The elements of a right circular cone are congruent.

(The proof is left to the student.)

14.60 A truncated cone is the union of the base and all elements of the cone between the base and the region determined by a section intersecting all elements of the cone. The section and its interior is called a *base* of the truncated cone. If the two bases are parallel, the figure is called a frustum of a cone (Figure 14-33).

14.61 The altitude of the frustum of a cone is a segment perpendicular to and terminated by the planes of the bases. The measure of the altitude is called the *height*.

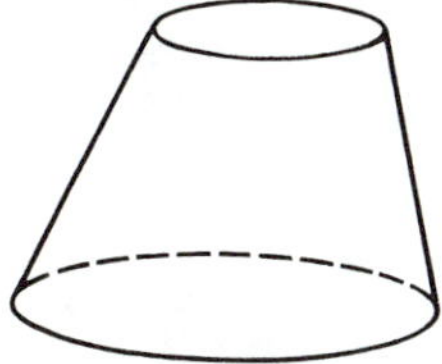

Figure 14–33

14.62 The slant height of a right circular cone is the length of an element of the cone.

The slant height of a frustum of a right circular cone is the length of an element of the frustum.

Exercises

1. Are all cones circular cones?
2. What is the name of the segment connecting the vertex and the center of the base of a cone of revolution?
3. To what types of cone does the term *slant height* apply?
4. What is the section of a cone made by a plane containing the axis?
5. What is the section of a right circular cone made by a plane perpendicular to the axis?
6. Is the axis of a cone perpendicular to the base?
7. When a right triangle is moved to generate a right circular cone, the hypotenuse and one leg moves. Which side of the triangle is the altitude of the cone?
8. In a cone of revolution, which side of the right triangle would we measure to find the slant height of the cone?
9. Name the section made by a plane containing the axis of a right circular cone.
10. List at least one use of each of the following in your home or in your community: (a) prism, (b) cylinder, (c) pyramid, (d) frustum of a pyramid, (e) cone, (f) frustum of a cone.

Vocabulary List

polyhedron	icosahedron	cube
section	prism	prismatoid
tetrahedron	prismatic	cylindrical
pentahedron	generatrix	cylinder
hexahedron	directrix	pyramidal
heptahedron	generates	pyramid
octahedron	element	slant height
nonahedron	lateral	frustum
decahedron	truncated	conical
dodecahedron	parallelepiped	cone
	circular	

Chapter Review

1. Name the set of regular convex polyhedrons.
2. Name the greatest measure of a polyhedral angle of any regular convex polyhedron and name the polyhedron.
3. Name the number of diagonals of a tetrahedron. Of a cube.
4. Is a prism a polyhedron?
5. Name the least number of faces a prism may have.
6. What is the intersection of two lateral faces of a prism called?
7. What are the properties of a regular prism?
8. What is the name of a prism whose lateral edges are not perpendicular to the base?
9. Cite the difference between a prism and a prismatoid.
10. Is a parallelepiped a prism? Explain.
11. What other name may be applied to a rectangular prism?
12. What distinguishes a truncated prism from other prisms?
13. Name the type of prism which has congruent diagonals.
14. Are the diagonals of regular prisms concurrent? Explain.
15. A cylinder compares closely with what other geometric surface?
16. Are all cylinders circular?
17. Describe the axis of a circular cylinder.

18. What is another name for a cylinder of revolution?
19. Do all pyramids have square base regions?
20. Name the properties of a regular pyramid.
21. What is the altitude of a pyramid?
22. What is the slant height of a pyramid?
23. Do all pyramids have "slant height?" Explain.
24. What is a frustum of a pyramid?
25. Name the polygonal regions forming the faces of a frustum of a pyramid.
26. A cone compares closely with what other geometric surface?
27. Are all cones circular cones?
28. What is the relationship of the elements of a right circular cone?
29. In what type cone is the axis perpendicular to the base?
30. Explain "slant height" of a cone.

Chapter 14 Test

True-False

1. A section of a pyramid intersecting all the lateral edges of a quadrangular pyramid is necessarily a quadrilateral.
2. If the lateral faces of a pyramid are congruent isosceles triangular regions it is a regular pyramid.
3. A section of a right cylinder containing an element of the cylinder is a rectangle.
4. The lateral faces of a frustum of a pyramid are necessarily trapezoid regions.
5. A section of a circular cone may be an ellipse.
6. A right section of an oblique circular cylinder is a circle.
7. A section of a frustum of a cone may be a trapezoid.
8. The measure of the axis of a regular pyramid is equal to the measure of its altitude.
9. If a diagonal section of a prism is a rectangle the prism is a right prism.

10. If a right section of a prism is a square the prism is a regular prism.

Completion

11. The least number of faces a polyhedron may have is _?_.

12. The greatest number of faces a regular polyhedron may have is _?_.

13. The prism whose base is a parallelogram is called a _?_.

14. The number of regular polyhedrons having hexagonal regions as faces is _?_.

15. The number of distinctly different regular polyhedrons having triangular regions as faces is _?_.

16. The measure of an _?_ of a right circular cone is called its slant height.

17. Another name for a cylinder of revolution is _?_.

18. A triangular prism has _?_ diagonals.

19. A diagonal section of a _?_ is necessarily a rectangle.

20. The diagonals of a _?_ polyhedron are concurrent but are not necessarily congruent.

21. Name the prism which is also a regular polyhedron.

22–23. Name the conditions necessary for a prism to be a regular prism.

24–25. Name the two regular polyhedrons that are also prismatoids.

26–27. Name the conditions necessary for a pyramid to be a regular pyramid.

28. Name the members of the set of regular polyhedrons whose faces are pentagonal regions.

29. Name the sum of the measures of the face angles of a polyhedral angle of a regular pentagonal prism.

30. Name the measure of a dihedral angle determined by two adjacent lateral faces of a regular hexagonal prism.

15

Locus Relationships

15.00 Locus (plural *loci*, pronounced lō′ sī) is the Latin word for *place*, or location.

The problem of locating an object or determining its path is ever-present in our daily life. We locate our position in a city by naming an address such as 100 Main Street, or our position in the United States by stating that 100 Main Street is in Home Town, Ohio, for example. We locate our position at sea by stating our latitude and longitude. We locate the shortest path from New York to London by finding the arc of a great circle of the earth through those points, (13.18). We open a door and the path of its outer edge must clear all obstacles. The fisherman must calculate the path of his hook and line as he casts into the water of a tree-shrouded stream.

The problem becomes more complicated as we consider the interaction of two or more objects, as when finding the point where the paths of two airplanes will cross, or the force and direction necessary for the quarterback to use when passing to the left end, who must elude the opponent's safety man.

The job of the postal service and the map-making industry presents many locus problems. Even the act of eating represents a locus problem. It becomes clear that every movement is a locus problem.

We *described* several loci in the previous section. To study the subject of locus we need to *define* the term carefully.

15.01 A locus is the set of points such that (1) *all points of the set satisfy the given conditions*, and (2) *all points satisfying the given conditions are members of the set* (or, such that any point not a member of the set does not satisfy the conditions).

EXAMPLES OF LOCUS PROBLEMS

EXAMPLE 1

What is the locus of the boundary of the surface grazed by a horse tied to a post by a rope?

Answer: The locus of the boundary of the area grazed is a circle. The center of the circle is the stake and the radius is the length of the rope (assuming that the rope does not wind around the stake).

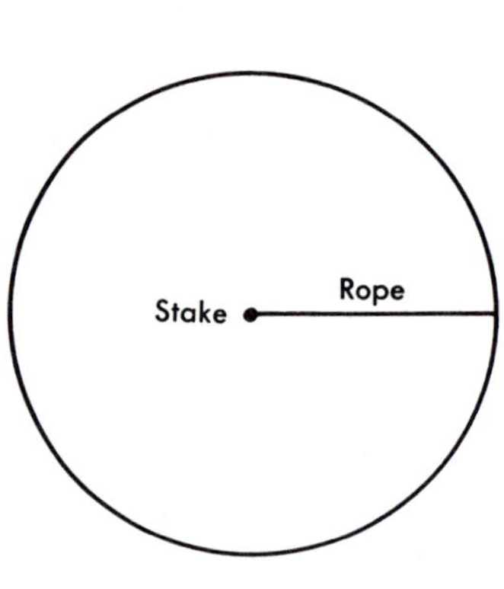

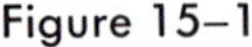

Figure 15–1

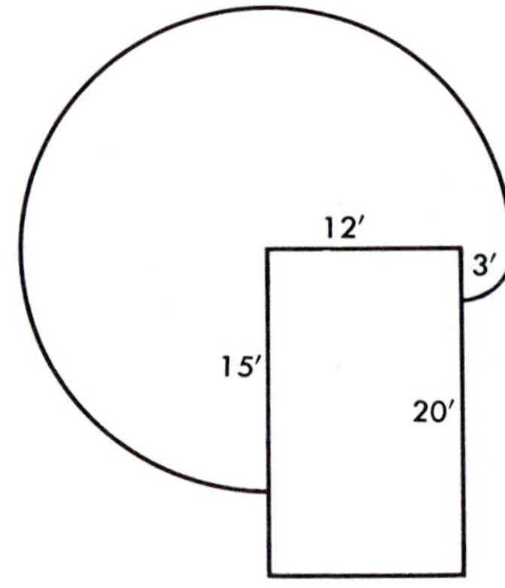

Figure 15–2

EXAMPLE 2

The horse mentioned above is tied with a 15′ rope to the corner of a shed 12′ × 20′. Find the locus of the boundary of its grazing area.

Answer: The locus is a compound curve composed of $\frac{3}{4}$ of a circle with radius 15′ and center the corner of the building, plus $\frac{1}{4}$ of the circle with radius 3′ and center the adjacent corner of the building on the 12′ side.

EXAMPLE 3

What is the locus of points *in space* equidistant from the points of a given circle?

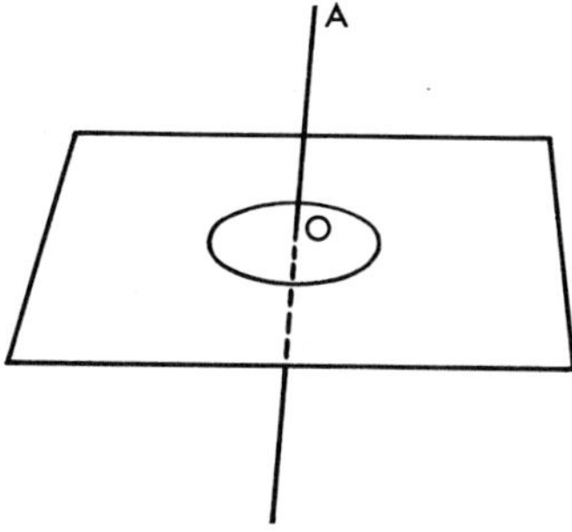

Figure 15–3

Answer: The locus is a line perpendicular to the plane of the circle at its center.
(Can you prove any point A on the line is equidistant from any two points on the circle?)

Oral Exercises

Discuss the following loci.

1. The tip of the hour hand of a clock.
2. The center of the hub of a wheel as it rolls down a level street.
3. A point on the tire of a wheel as it rolls along a level street.
4. All points equidistant from the corners of this page.
5. All points equidistant from the floor and ceiling of your classroom.
6. All points d distance from the plane of the top of the teacher's desk.

Exercises

Complete the following exercises about loci on your paper.

1. What is the locus in a plane of all points 5 inches from a given point in the plane? In space?
2. What is the locus in a plane of all points 5 inches from a given line in the plane? In space?
3. What is the locus of all points 5 inches from a given plane?
4. What is the locus in a plane of all points equidistant from two parallel lines in the plane? In space?
5. What is the locus of all points equally distant from two parallel planes?
6. What is the locus of all points equidistant from the sides of a plane angle that lie in the plane of the angle? In space?
7. What is the locus of all points equidistant from the sides of a dihedral angle?
8. What is the locus of all points in your classroom 10 feet from the front wall and three feet from a side wall?
9. In the definition of locus on page 360, is the statement in parentheses the converse, inverse, or contrapositive of (1)?

LOCUS THEOREMS

15.02 As we pursue our study of locus we shall see that a small number of basic locus theorems can be used, together or separately, to help solve most locus problems. Some of these theorems develop readily from previous definitions, and formal proof is not required. Some of the theorems are proved on following pages to show the proper procedure in the proof of locus theorems. The basic theorems, with both their two-dimensional and three-dimensional applications, are listed below without proof.

You should make enough drawings of special cases to satisfy yourself that the theorems are reasonable and intuitively true.

Locus in a Plane

1. The locus of points at a given distance from a given point is a circle, with the point as center and the given distance as radius. (Figure 15-4 left)

2. The locus of points at a given distance from a given line is a pair of lines, one on each side of the given line, parallel to it at the given distance from it. (Figure 15-4 right)

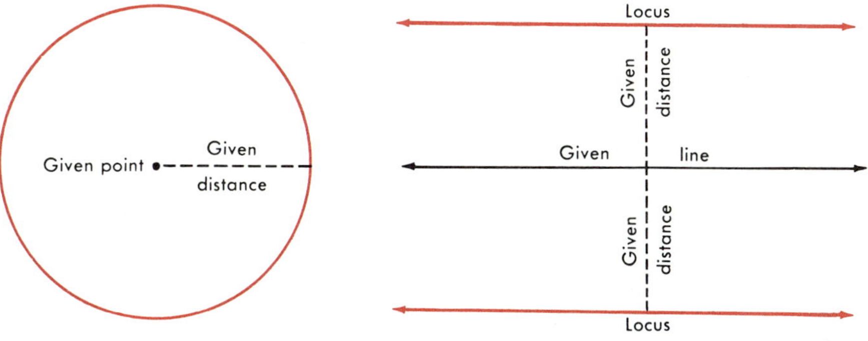

Figure 15–4

3. The locus of points equidistant from two parallel lines is a line parallel to them and midway between them. (Figure 15-5 left)

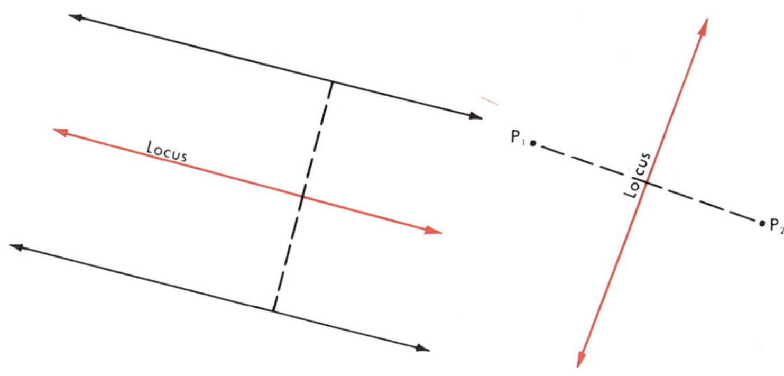

Figure 15–5

4. The locus of points equidistant from two points is the line which is the perpendicular bisector of the line segment joining the two points. (Figure 15-5 right)
5. The locus of points equidistant from the sides of an angle is the half-line determined by the ray which bisects the angle. (See 15.04.)

6. The locus of the vertex of the right angle of a right triangle with a given hypotenuse is the circle whose diameter is the hypotenuse. (See 15.05.)

Locus in Space

7. The locus of points at a given distance from a given point is a sphere with the point as center and the given distance as radius. (Figure 15-6 left)

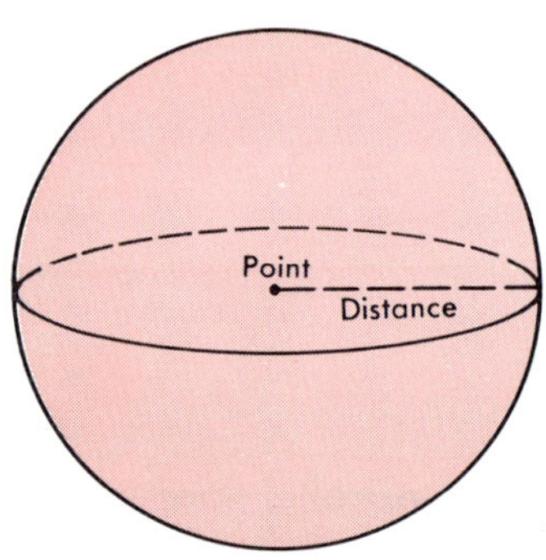

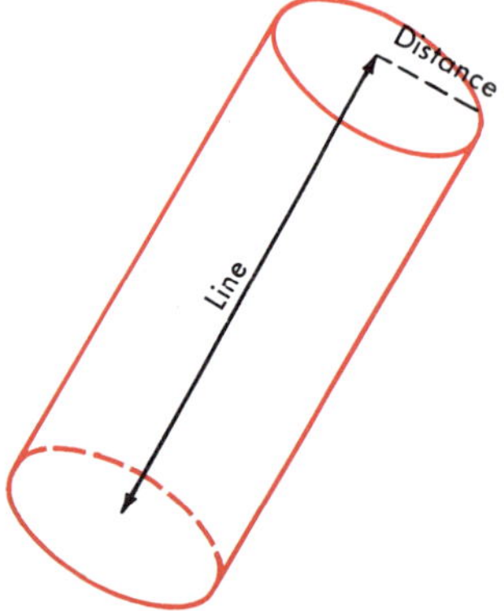

Figure 15–6

8. a. The locus of points at a given distance from a given line is a circular cylindrical surface whose axis is the given line and whose radius is the given distance. (Figure 15-6 right)
   b. The locus of points at a given distance from a given plane is a pair of parallel planes, one on each side of the given plane, parallel to it, and at the given distance from it. (Figure 15-7 left)

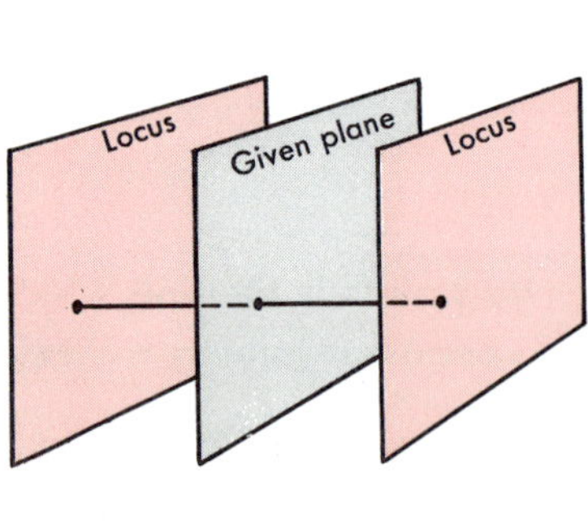

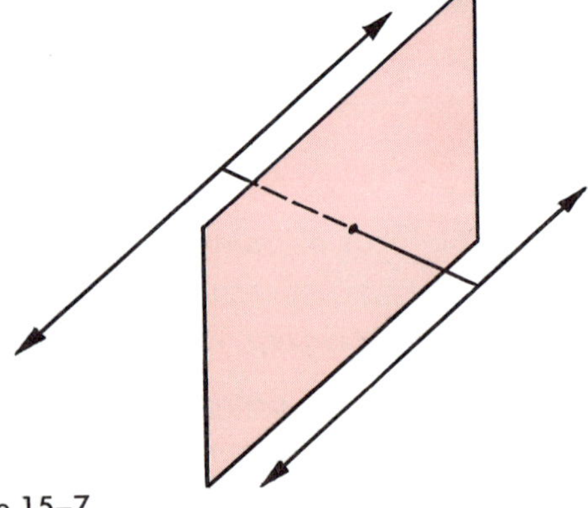

Figure 15–7

9. a. The locus of points equidistant from two parallel lines is a plane which is the perpendicular bisector of the common perpendicular joining the parallel lines. (Figure 15-7 right)
   b. The locus of points equidistant from two parallel planes is a plane parallel to them and midway between them. (Figure 15-8 left)

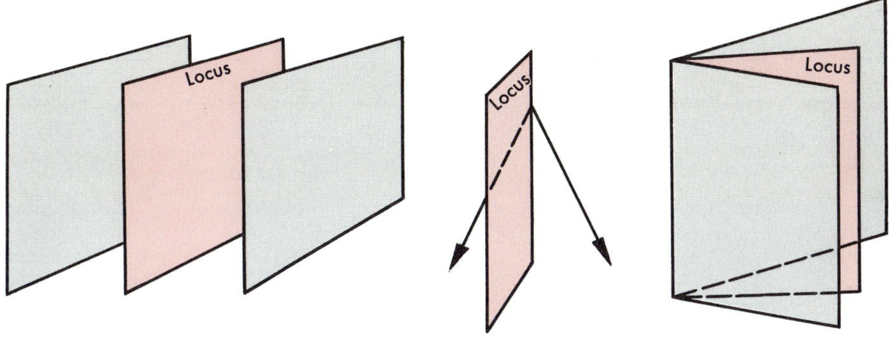

Figure 15–8

10. The locus of points equidistant from two points is a plane which is the perpendicular bisector of the line segment joining the two points. (See 15.03B.)
11. The locus of points equidistant from the sides of a plane or dihedral angle is the half-plane which bisects the angle, and, in a plane angle, is perpendicular to the plane of the angle. (Figure 15-8)
12. The locus of the vertex of the right angle of a right triangle with a given hypotenuse is the sphere whose diameter is the hypotenuse. (See 15.05.)

PROOF OF LOCUS THEOREMS (OPTIONAL)

The following proofs of selected locus theorems are presented to show you how both conditions of 15.01 must be satisfied before the proof is complete. That is, all points of the set must satisfy the conditions of the locus and all points satisfying the conditions must be included in the set.

Examine the proofs of 15.03A and 15.03B to see the comparison between two-dimensional and three-dimensional proofs of the same principle.

To begin the proof of the first theorem, we will show that any point satisfying the conditions is a member of a certain set.

Case I. Prove that any point (C) equidistant from two points (A and B) is on the line which is the perpendicular bisector of the segment joining the two points.

Given: $\overline{AC} \cong \overline{BC}$

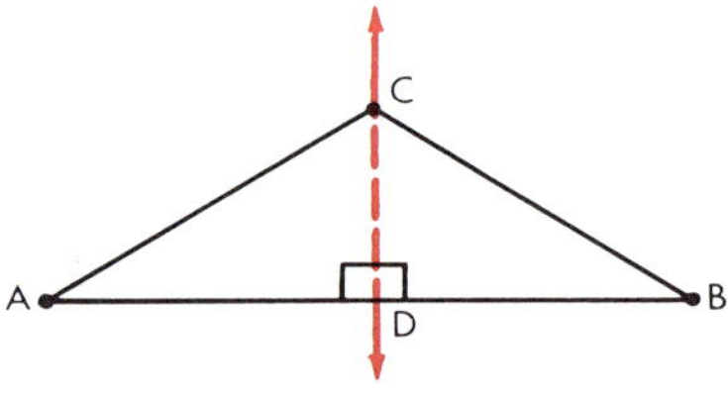

Figure 15–9

Conjecture: C is on the perpendicular bisector of $\overline{AB}$.

Construction: Through C draw $\overleftrightarrow{CD} \perp \overline{AB}$.

Proof:

Statements	Reasons
1. $\overline{CA} \cong \overline{CB}$, $\overleftrightarrow{CD} \perp \overline{AB}$	1. Given
2. $\overline{CD} \cong \overline{CD}$	2. Reflexive Axiom
3. Angles ADC and BDC are right angles	3. $\perp$ lines form right angles.
4. $\triangle ADC \cong \triangle BDC$	4. *h.s.*
5. $\overline{AD} \cong \overline{DB}$	5. If two $\triangle$s are $\cong$, their corresponding parts are congruent.
6. $\therefore$ any point C, equidistant from A and B, lies on the line which is the perpendicular bisector of $\overline{AB}$.	6. Steps 1 and 5, and 3.20.

Case II. (The Converse) Prove any point on the line, which is the perpendicular bisector of the line segment joining two points, is equidistant from the two points.

Given: Points A and B, and $\overleftrightarrow{CD} \perp$ bisector of $\overline{AB}$

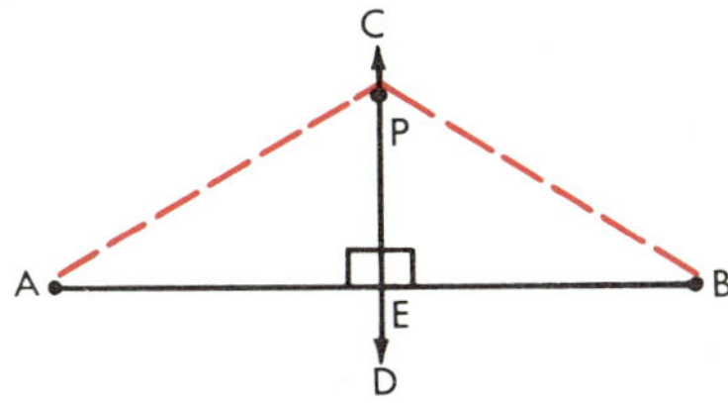

Figure 15–10

Conjecture: P is equidistant from A and B.

Proof:

Statements	Reasons
1. $\overleftrightarrow{CD} \perp$ bisector of $\overline{AB}$	1. Given
2. $\angle AEC \cong \angle BEC$	2. $\perp$ lines form congruent adjacent $\angle$s.
3. $\overline{AE} \cong \overline{BE}$	3. If a segment is bisected, it is divided into two congruent parts.
4. $\overline{PE} \cong \overline{PE}$	4. Reflexive Axiom
5. $\triangle AEP \cong \triangle BEP$	5. *s.a.s.*
6. $\therefore \overline{AP} \cong \overline{BP}$	6. If two $\triangle$s are $\cong$, their corresponding parts are congruent.
7. Any point (P) of $\overleftrightarrow{CD}$ is equidistant from A and B.	7. Statement 6.

15.03 A THEOREM

The locus of points in a plane equidistant from two points is the line which is the perpendicular bisector of the line segment joining the two points.

To prove the corresponding three-dimensional locus theorem, we again begin by attempting to prove that any point satisfying the conditions is in a particular set of points.

Case I. Any point equidistant from two points is on the perpendicular bisector of the segment joining the two points.

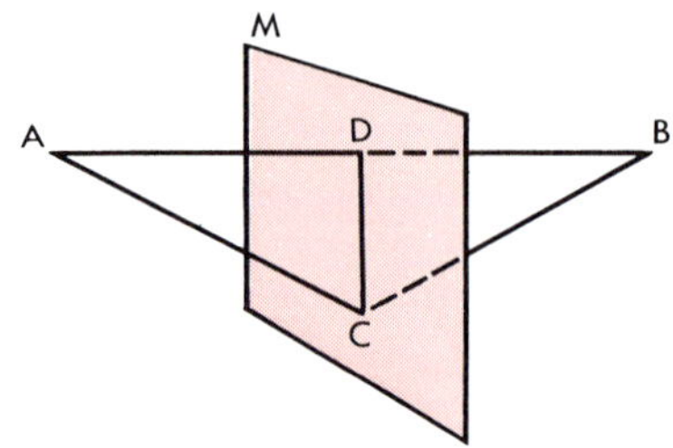

Figure 15–11

Given: Points A, B, and C.
$\overline{AC} \simeq \overline{BC}$.

Conjecture: C is on the perpendicular bisector of $\overline{AB}$.

Construction: Through C draw plane m perpendicular to $\overline{AB}$.

Proof:

Statements	*Reasons*
1. $\overline{CA} \simeq \overline{CB}$, plane $m \perp \overline{AB}$	1. Why?
2. $\overline{CD} \simeq \overline{CD}$	2. Why?
3. ADC and BDC are right angles.	3. Why?
4. $\triangle ADC \simeq \triangle BDC$	4. Why?
5. $\overline{AD} \simeq \overline{DB}$	5. Why?
6. $\therefore$ any point C, equidistant from A and B, lies on the plane which is the perpendicular bisector of $\overline{AB}$.	6. C is in plane m, which is the perpendicular bisector of $\overline{AB}$.

Case II. Any point on the perpendicular bisector of the line segment joining two points is equidistant from the two points.

Proof: This was proved in 8.13.

Since we have proved I and its converse II, which satisfies both conditions of the definition, we have proved the following theorem.

15.03 B THEOREM

The locus of points equidistant from two points is the plane which is the perpendicular bisector of the segment joining the two points.

The next theorem concerns the locus of all points equidistant from the sides of a plane angle.

Case I. Any point equidistant from the sides of a plane angle and in its plane is on the bisector of the angle.

Given: $\angle ABC$ and E located so that $\overline{DE}$ and $\overline{FE}$ are congruent perpendiculars to the sides of the $\angle$

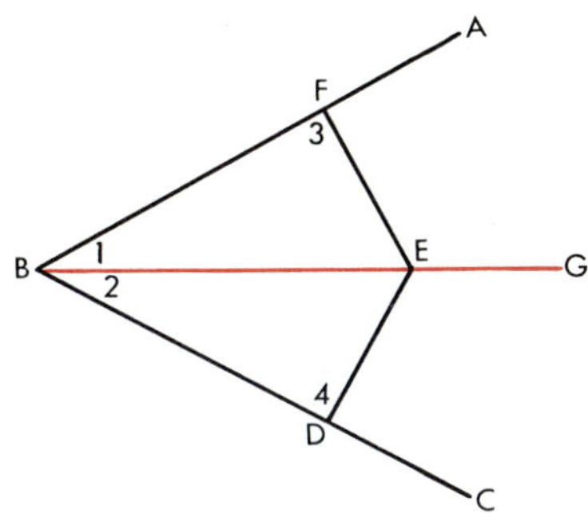

Figure 15–12

Conjecture: E is on the bisector of $\angle ABC$.

Plan: Draw $\overrightarrow{BG}$ through E. Prove $\angle 1 \cong \angle 2$; therefore, $\overrightarrow{BG}$ bisects $\angle ABC$, so E is on the bisector of $\angle ABC$.

Case II. Any point on the half-line determined by the bisector of a plane angle is equidistant from the sides of the angle.

Given: $\angle ABC$ with bisector $\overrightarrow{BG}$ and point E on $\overrightarrow{BG}$. $\overline{ED} \perp \overline{BC}$ and $\overline{EF} \perp \overline{AB}$.

Conjecture: $\overline{DE} \cong \overline{FE}$

Proof: The proof is left to the student. We have now satisfied both conditions of 15.01.

15.04 THEOREM

The locus of points equidistant from the sides of a plane angle and in its plane is the half-line determined by the ray which bisects the angle.

Which condition of the definition of locus is satisfied first in the following development?

Case I. If a circle is drawn with the hypotenuse of a right triangle as a diameter, the vertex of the right angle is on the circle.

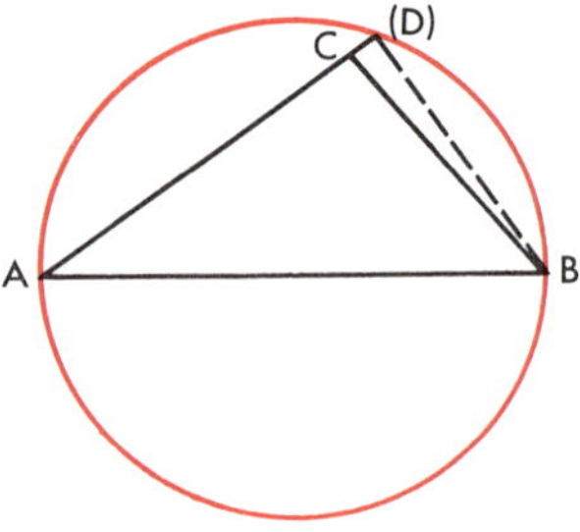

Given: Right triangle ACB, with hypotenuse $\overline{AB}$, and a circle with $\overline{AB}$ as a diameter.

Conjecture: C is on the circle.

Figure 15–13

Proof:

Statements	*Reasons*
1. $\overrightarrow{AC}$ must meet the circle at some point we shall call D.	1. Why?
2. Draw $\overline{DB}$.	2. Why possible?
3. Then $\angle ADB$ is a right angle.	3. 13.06
4. $\overline{BD} \perp \overline{AC}$	4. Why?
5. $\overline{BC} \perp \overline{AC}$	5. Why?
6. $\therefore \overline{BD}$ and $\overline{BC}$ are the same segment.	6. 3.20
7. Since D is on the circle, C is on the circle.	7. D and C coincide. (2.15)

Case II. If any point on a circle is joined to the extremities of a diameter, a right triangle is formed and the point is the vertex of the right angle.

Proof: This has been proved in 13.06.

Since we have now satisfied both conditions of 15.01, we may now state a theorem.

15.05 THEOREM

The locus of the vertex of the right angle of a right triangle with a given hypotenuse is a circle with the hypotenuse as diameter.

COMPOUND LOCI

In practical applications of the locus principles we are usually confronted with the problem of satisfying two or more sets of conditions imposed upon the set of points that are under consideration. Consider the following problems and their solution.

EXAMPLE 1

Mr. and Mrs. Neal are attempting to buy a home which is equidistant from his employer's home office in Chicago and from the branch manufacturing plant in Kansas City, both of which he visits regularly in addition to supervising the sales territory in the Midwest. The family also wishes to establish their home near the Mississippi River so that they may enjoy boating.

Mr. Neal's Solution:

By using the accompanying map, Mr. Neal drew the perpendicular bisector of the line segment (15.03) joining the points on the map representing Chicago and Kansas City. Where this line intersected the Mississippi River they began their search for their new home. Could the line intersect the river in more than one place?

Figure 15–14

EXAMPLE 2

Find the locus of all points in a plane 1 inch from a line in the plane and $1\frac{1}{2}$ inches from a point in the plane. (Figure 15-15)

Solution:

The locus is the intersection (L_1 and L_2) of the circle, whose radius is $1\frac{1}{2}$ inches and center at the given point, and the lines parallel to and one inch from the given line. The locus may be one, two, three, or four points, or the null set.

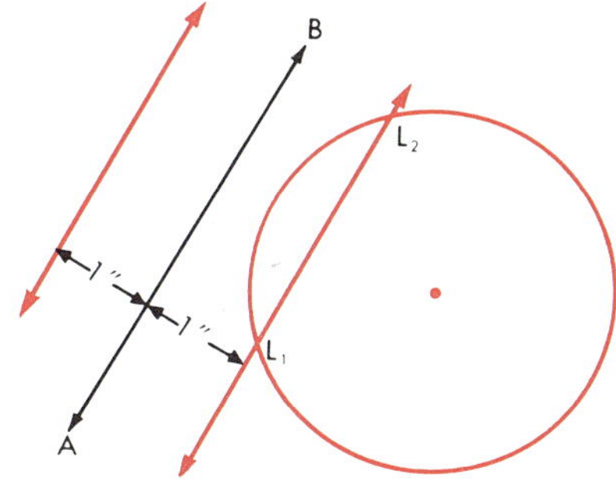

Figure 15–15

Exercises

Develop pictures and statements to describe the locus in Exercises 1–8.

1. A pirate buried his treasure 30 paces from the base of the cliff in Figure 15-16 (whose base represents a straight line) and equidistant from the foot of the only two trees near the cliff. Explain how you could find the spot to dig for the treasure. Is there only one reasonable place to dig? Explain.

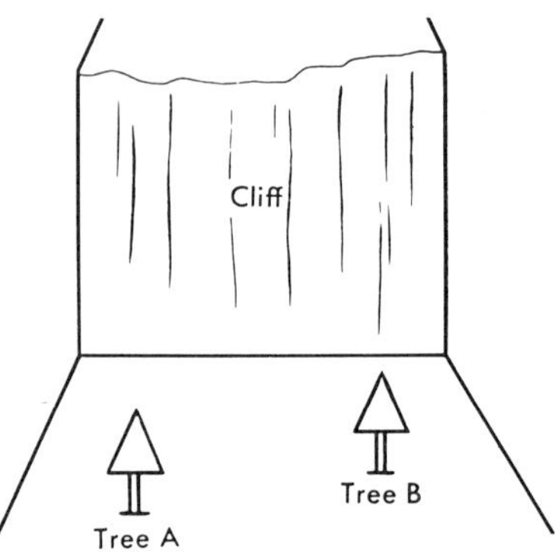

Figure 15–16

2. If the trees in Figure 15-16 above are 15 paces apart and another treasure lies buried 20 paces from tree A and 10 paces from tree B, how many places can you find that would satisfy these conditions? Explain how you determined your answer.

3. A house faces east toward a north-south street. One-hundred feet from the southeast corner of the house, under the center of the street, is the city water connection. Draw diagrams of the various possibilities for representing the location of the water connection.

4. What is the locus of the intersection of the blade of a butcher's slicing machine and a cylinder of cold meat?

5. In $\triangle ABC$, $m\overline{AB} = 1\frac{1}{2}$ inches, $m\overline{BC} = 2$ inches, and $m\overline{AC} = 3$ inches. Find all points equidistant from A and B that lie on the triangle.

6. Using the triangle in Exercise 5, find all points 3 inches from A and 2 inches from B which lie in the plane of the triangle.

7. Points C and D are in plane m, 2 inches apart. Find all points 3 inches from C and 2 inches from D in plane m. What is the locus of these points in space?
8. What is the locus of points equidistant from three collinear points?
9. Construct the following sets of segments, each in a different scalene triangle: (**a**) medians, (**b**) angle bisectors, (**c**) altitudes, (**d**) perpendicular bisectors of the sides. Is each set of these segments concurrent? Do your construction procedure and resultant conjecture represent deductive, or inductive, reasoning?
10. Could you use any of the construction procedures in Exercise 9 to find a single point equidistant from the three vertices of a triangle? Equidistant from the sides of the triangle? Explain your answers.

EXAMPLES OF COMPOUND LOCI

Most individuals who hold responsible jobs must eventually prepare a report on some phase of their work. Some of you will want to explain an original idea or procedure you may develop. The examples and subsequent exercises presented in this section are designed to provide a small measure of practice in the ability to express yourself in writing, as well as practice in compound locus problems—a real test of your communication skills.

A compound locus is a set of points which satisfies two or more conditions. A compound locus is usually the intersection of two or more simple loci. If there are no points in the intersection, we say that the locus is the *null set*, or empty set.

It is difficult to describe a compound locus in one sentence, and you may prefer to make two or more statements about the locus. This is especially true if there are several possible figures. For example, the compound locus in a certain problem may involve the possible intersection of a line and a plane as simple loci. There are three possibilities: the line may be parallel to the plane, it may intersect the plane in only one point, or it may lie entirely in the plane.

Observe the procedure shown in the following examples. First, make a neat drawing to represent the locus clearly and with reasonable accuracy. Then, prepare a statement or series of statements which describe the compound locus so that all points of the figure satisfy the conditions and that no other points do.

It is understood that *we are allowed three dimensions unless restricted by the statement of the conditions of the locus to a plane.*

EXAMPLE 1

What is the locus of points 5 inches from point A and equidistant from points A and B?

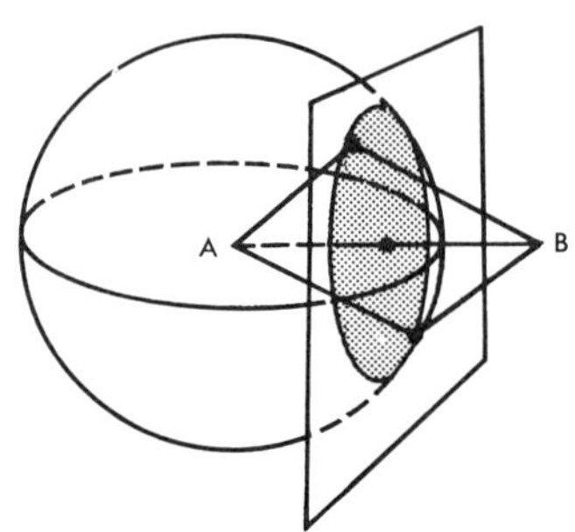

Figure 15–17

Solution:

First we find that the locus of points 5 inches from point A in space is a sphere with radius of 5 inches and with its center at A. This is true because every point of the sphere satisfies the condition that it is 5 inches from the center, and no other points do so.

Then the locus of points in space equidistant from points A and B is the plane which is the perpendicular bisector of $\overline{AB}$. This is true because every point of the plane (and no other point) satisfies the condition that it is equidistant from both A and B.

The compound locus is the set of all points satisfying both given conditions. Therefore, the compound locus is the intersection of two simple loci, the sphere and the plane. What is the intersection of the sphere and plane if the distance between A and B is exactly 10 inches? Less than 10 inches? Greater than 10 inches?

The compound locus is a circle, a point, or the null set according to whether $m\overline{AB} < 10''$, $m\overline{AB} = 10''$, or $m\overline{AB} > 10''$, respectively.

Answer:

As a single statement, we could say, "The locus of points 5 inches from point A and equidistant from points A and B is the circle, point, or null set formed by the intersection of the sphere, whose center is at A and whose radius is 5 inches, with the plane which is the perpendicular bisector of $\overline{AB}$."

EXAMPLE 2

Assume that a room is oriented so that one pair of walls faces north and south and the other pair faces east and west. What is the locus of points equidistant from the north and west walls of the room and also common to a line passing through the room?

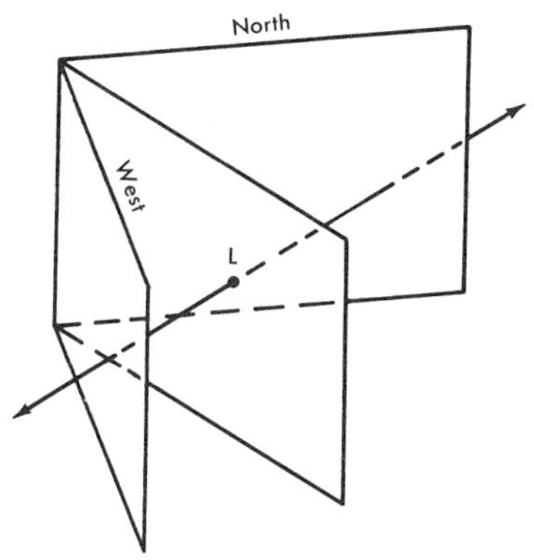

Figure 15–18

Solution:

The locus of points equidistant from the north and west walls is a half-plane through the northwest corner (edge) bisecting the dihedral angle formed by the walls. The given line may lie anywhere in the room.

Answer:

The compound locus of points equidistant from the north and west walls of a room and also common to a line passing through the room is a line, point, or null set formed by the intersection of the given line and the half-plane which is the bisector of the dihedral angle formed by the walls. When is the compound locus a line? When is it a point? When is it the null set?

Exercises

What is the locus of all points:

1. Equidistant from three points not in a straight line?
2. Equidistant from the vertices of a cube? Of a square?
3. Formed by the rotation of a semicircle about its diameter?
4. On a given line and at a given distance from a given point?

5. On line m and also at a given distance from line s?
6. At a given distance from a given plane c and also on another given plane d?
7. Equidistant from two given points and also equidistant from two given parallel planes?
8. Equidistant from the north and west walls of a room and from the floor?
9. Equidistant from four given points not in the same plane?
10. That are a given distance d from a fixed line and another given distance r from a point on the line?
11. Determined by the center of a sphere that contains two given points?
12. Determined by the center of a small circle of a sphere if the radius of the circle is constant? (The sphere does not move.)

15.06 Each problem in graphing an algebraic equation is a locus problem. Here we find the set of all points which satisfy the conditions of the equation, equations, or inequalities.

EXAMPLES

1. What is the locus (set) of all points satisfying the equation $y = 3x + 4$?

Answer:

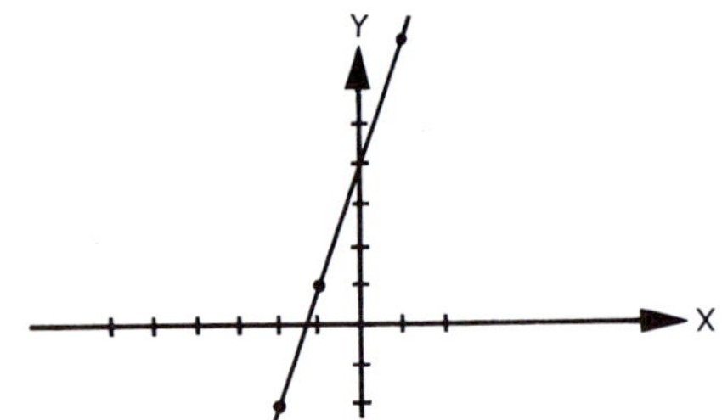

The line graph is the locus of $y = 3x + 4$.

2. What is the set of all points satisfying both the equations $2y + 3x = 6$ and $x - 2y = 2$?

Answer:

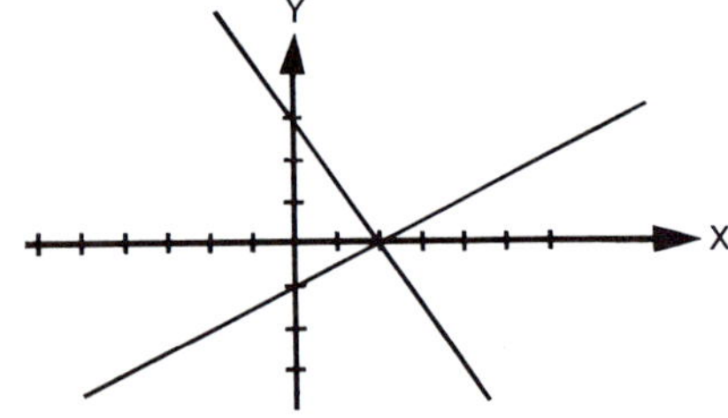

Figure 15–19

The point whose coordinates are (2,0) is the locus of points satisfying both $2y + 3x = 6$ and $x - 2y = 2$.

Exercises

Find the locus of the equations and inequalities in Exercises 1–12 on a rectangular coordinate system, where x and y are real numbers.

1. $x - 2y = 4$
2. $-2x - y = 3$
3. $x = 3$
4. $y = 2x$
5. $x + 2y = 3$
   $3x - y = 2$
6. $3 + x = y$
   $x - y = 1$
7. $y - x = \frac{1}{2}$
   $2y - 10x = 5$
8. $2y - x = 1$
   $4y - 2x = 2$
9. $x < 3$ and $y < 5$
10. $x + y < 5;$
    $x > 0; y > 0.$
11. $x > 4$ and $y \leq 3$
12. $3 \leq x \leq 7;$
    $0 \leq y \leq 4.$
13. Does $x > 3$ represent a ray, a segment, a line, a half-line, a half-plane with its edge (closed half-plane), a half-plane without its edge (open half-plane), or a plane?

CONIC SECTIONS (Optional)

15.07 Certain interesting curves are formed by the intersection of a plane and a circular conical surface. These curves were discovered by Menaechmus, who lived in the fourth century B.C., and were later thoroughly studied by Apollonius in the third century B.C. This was shortly after the period when Euclid compiled his famous book, *The Elements*, the basis of our present geometry.

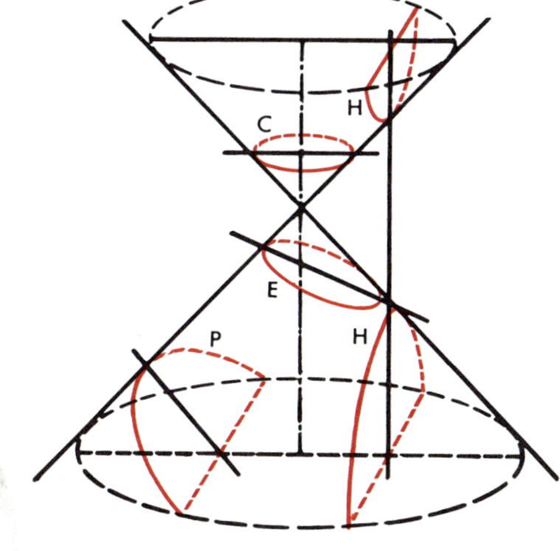

Figure 15–20

More than 1,000 years passed before practical use was made of these curves. Analytic geometry, developed by Descartes (about 1632), considers in detail certain of their important properties. Analytic geometry applies the tools of algebra and graphing to the study of geometric figures.

1. Circle

If a plane is passed perpendicular to the axis of a circular conical surface, the section is a circle (C, Figure 15-20). Such a section is the locus, in the plane, of all points at a given distance from a given point.

The properties of circles and their use in our environment are familiar to most of us.

2. Ellipse

If a plane cuts a circular conical surface obliquely to the axis and intersects all elements of a nappe, the section is an ellipse (E, Figure 15-20). Such a section is the locus of the points the sum of whose distances from two fixed points is constant. Is a circle a member of the set of ellipses?

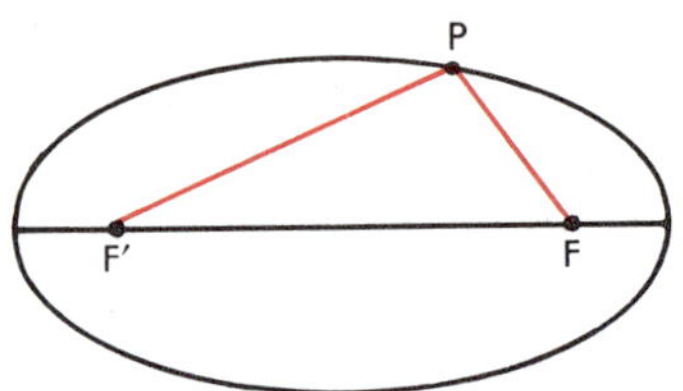

Figure 15–21

In Figure 15-21, F and F' are the foci, and P is any point on the ellipse. Then $m\overline{PF'} + m\overline{PF} = k$ (a constant) by definition.

Ellipses are commonly used in design. Arches in bridges and buildings are familiar examples. The astronomer Johannes Kepler (1571–1630) discovered that planets move in elliptical paths.

3. Parabola

If the intersecting plane is parallel to an element of a circular conical surface, the section is a parabola (P, Figure 15-20). Such a section is the locus of points whose distances from a fixed straight line and a fixed point are equal. The parabola extends indefinitely in one direction.

In the figure, F is the focus, AB is the directrix, and P is any point on the parabola. By definition, $m\overline{PF} = m\overline{PD}$.

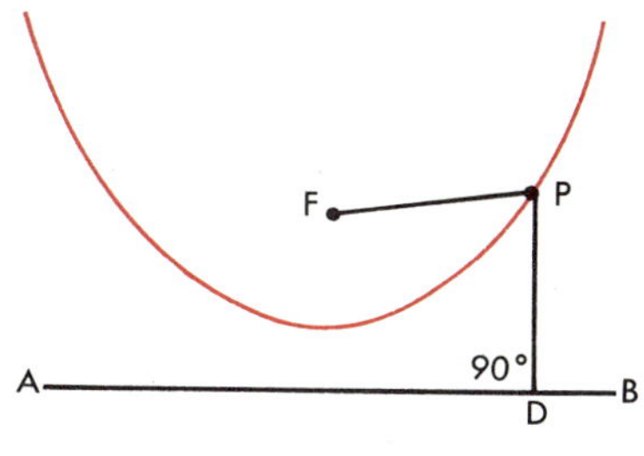

Figure 15–22

A baseball in flight follows a parabolic path, somewhat distorted by air friction, as does a stream of water projected upward. The steel cables of a suspension bridge form a parabolic curve. The reflectors in the headlights of automobiles are parabolic.

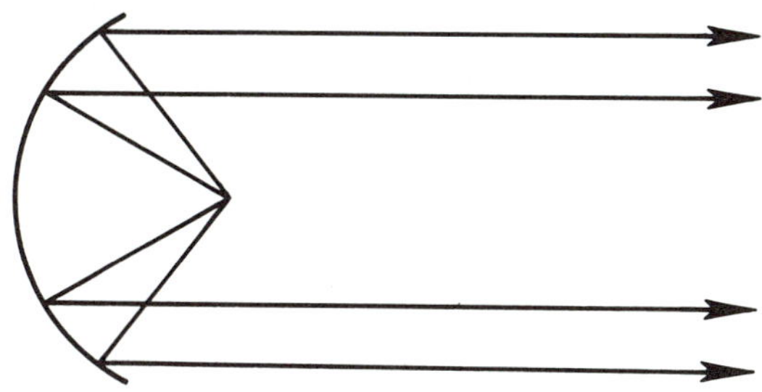

Parabolic Mirror
The light source placed at the focal point is reflected in parallel paths from all points of the reflector.

Figure 15–23

4. Hyperbola

If a plane intersects both nappes of a circular conical surface, the section is a hyperbola (H, Figure 15-20). Such a section is the locus of points the difference of whose distances from two fixed points is constant. The hyperbola consists of two symmetrical branches which extend indefinitely in opposite directions.

The hyperbola is used in electronics and in sound-detection devices. It is applied militarily in the location of hidden guns. In the figure, F and F' are the foci, and P is any point on either branch of the hyperbola. Then $m\overline{PF'} - m\overline{PF} = k$ by definition.

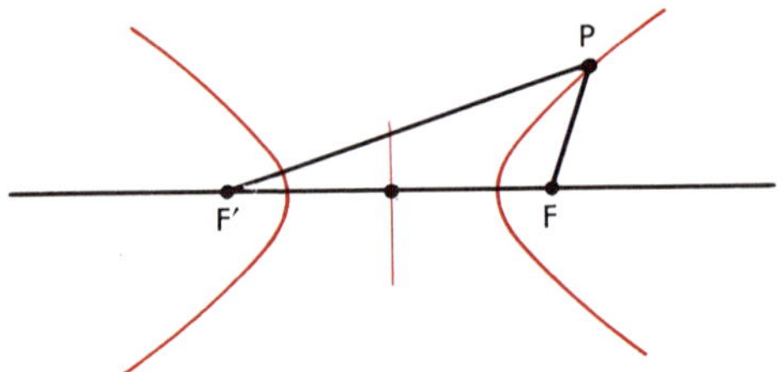

Figure 15–24

Exercises

1. Find the locus of all points the sum of whose distances from two fixed points is constant. (*Hint:* Fasten a string of length $a + b$ to the given points. Move a pencil within the limits allowed by the string.) What is the name of this curve?
2. Find the locus of points such that the difference between the distances from any one point to two fixed points (foci) is a constant. (*Hint:* Use the foci as the centers of circles with radii whose difference is the constant.) What is the name of this curve?
3. Find the locus of points which will be equidistant from a fixed line and fixed point. (*Hint:* Place the fixed point near the line.) What is the name of this curve?

CONCURRENT LINES— IMPORTANT LOCUS THEOREMS

An important and interesting aspect of geometry is the study of concurrent lines and their properties. In the exercises on page 373 we constructed the altitudes, angle bisectors, medians, and perpendicular bisectors of the sides of triangles. Most of you probably arrived at the tentative conclusion that the lines in each set of the constructed lines are *concurrent.* Is it possible that one or all of these points of concurrency are equidistant from the vertices of the triangle—or from the sides of the triangle?

We will now examine and attempt to generalize our inductive conjectures by means of deductive procedure.

Given: $\triangle ABC$ with $\overleftrightarrow{DE}$, $\overleftrightarrow{FG}$, and $\overleftrightarrow{HK}$ perpendicular bisectors of $\overline{AB}$, $\overline{BC}$, and $\overline{CA}$ respectively.

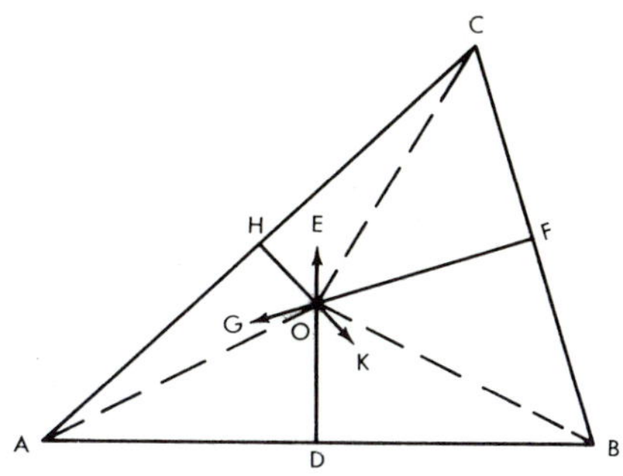

Figure 15–25

Conjecture: $\overleftrightarrow{DE}$, $\overleftrightarrow{FG}$, and $\overleftrightarrow{HK}$ meet at a point O, and $\overline{OA} \cong \overline{OB} \cong \overline{OC}$.

Proof: *Statements*	*Reasons*
1. $\overleftrightarrow{HK}$ will meet $\overleftrightarrow{DE}$ at some point O.	1. If they do not meet, they will be parallel and $\overline{AB}$ and $\overline{AC}$ will either be parallel or coincide. This is impossible since $\overline{AB}$ and $\overline{AC}$ are the sides of a triangle.
2. $\overline{OA} \cong \overline{OC}$, $\overline{OA} \cong \overline{OB}$	2. 15.03A
3. $\overline{OC} \cong \overline{OB}$	3. Transitive Axiom
4. O is on $\overleftrightarrow{FG}$	4. 15.03A
5. $\overline{OA} \cong \overline{OB} \cong \overline{OC}$	5. Statements 2 and 3. Substitution Axiom.

15.08 THEOREM

The perpendicular bisectors of the sides of a triangle are concurrent in a point equidistant from the vertices.

(See Exercise 8, page 562, for the coordinate method.)

15.09 Construction problem: To circumscribe a circle about a triangle.

Given: $\triangle ABC$

Required: To circumscribe a circle about $\triangle ABC$.

Procedure: We must find a point O equally distant from A, B, and C to use as a center of the circle. This can be done by using 15.08. The construction is left to the student.

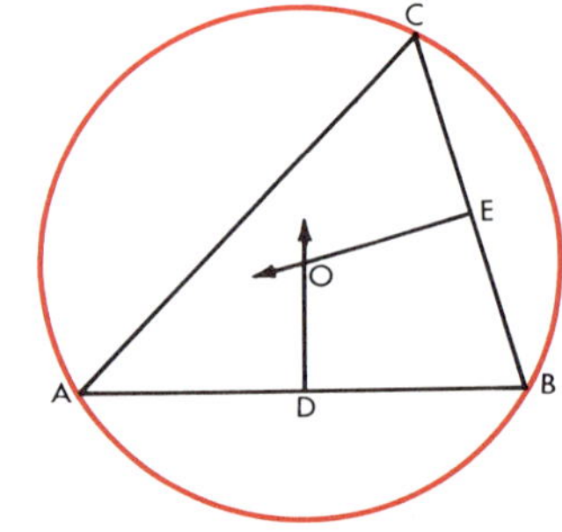

Figure 15–26

15.10 The point where the perpendicular bisectors of the sides of a triangle are concurrent is called the circumcenter of the triangle.

We have proved that the perpendicular bisectors of the sides of any triangle are concurrent. Let us now consider the bisectors of the angles of any triangle.

Given: $\triangle ABC$, with $\overrightarrow{AD}$, $\overrightarrow{BE}$, and $\overrightarrow{CF}$ bisectors of $\angle A$, $\angle B$, and $\angle C$ respectively.

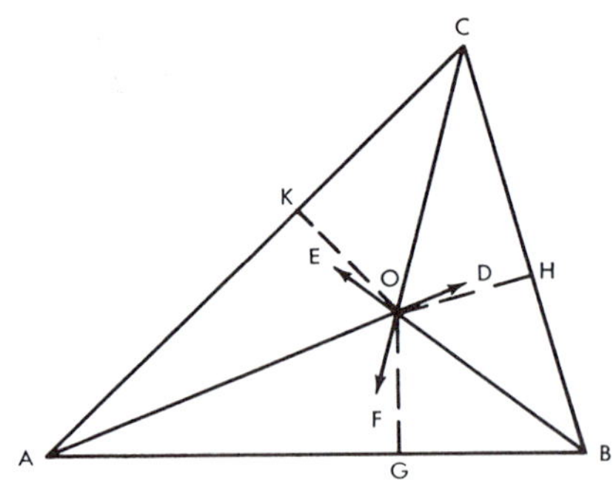

Figure 15–27

Conjecture: $\overrightarrow{AD}$, $\overrightarrow{BE}$, and $\overrightarrow{CF}$ meet at a point O, and O is equally distant from $\overline{AB}$, $\overline{BC}$, and $\overline{CA}$.

Plan: $\overrightarrow{AD}$ and $\overrightarrow{BE}$ will meet at some point O. (If they do not meet, they will be parallel, and $\angle OAB$ and $\angle OBA$ will be supplementary. This is impossible, for they are halves of two angles of a triangle). $\overline{OH} \perp \overline{BC}$, $\overline{OG} \perp \overline{AB}$, $\overline{OK} \perp \overline{AC}$. $\overline{OH} \simeq \overline{OG} \simeq \overline{OK}$ by 15.04. Show that O lies on $\overrightarrow{CF}$.

Proof: The formal proof is left to the student.

15.11 THEOREM

The bisectors of the angles of a triangle are concurrent in a point equidistant from the sides.

15.12 Construction problem: To inscribe a circle in a triangle.

Given: $\triangle ABC$.

Required: To inscribe a circle in $\triangle ABC$.

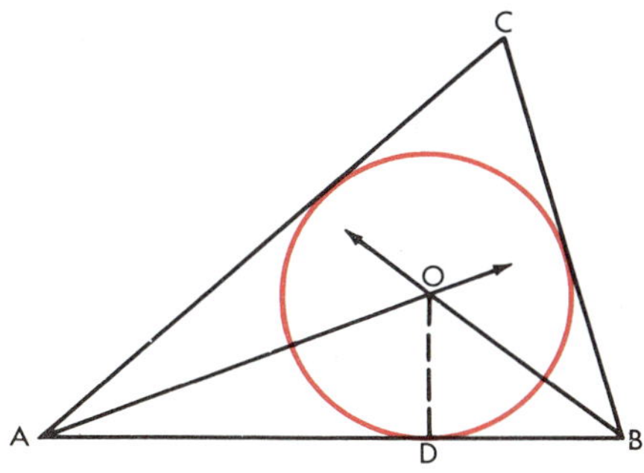

Figure 15–28

Procedure: We must find a point equidistant from the three sides of the triangle. This can be done by using 15.11, where O is the intersection of the angle bisectors and $\overline{OD}$ ($\overline{OD} \perp \overline{AB}$) is the radius of the circle.

15.13 The point where the bisectors of the angles of a triangle are concurrent is called the incenter of the triangle.

Exercises

1. Circumscribe a circle about an acute triangle; a right triangle; an obtuse triangle.
2. Construct an equilateral triangle whose sides are 3 inches long. Circumscribe a circle about the triangle and inscribe a circle within the triangle.

3. Does the center of an inscribed circle always fall within the triangle?
4. Name the type of triangle whose circumcenter falls (**a**) within the triangle, (**b**) outside the triangle, (**c**) on the triangle. (**d**) Describe the position of the point in (**c**).
5. Under what conditions are the incenter and circumcenter of a triangle the same point?
6. How many triangles may be circumscribed about a circle? Inscribed within a circle?
7. Construct a circle through three points not in a straight line.
8. Do three noncollinear points determine a circle? Explain.
9. Inscribe a circle in a given square. Circumscribe a circle about the square.
10. Can a circle be drawn through any four points? Explain.
11. Can a circle be drawn through the four vertices of a rectangle? A rhombus? An isosceles trapezoid? Explain each answer.
12. Are the bisectors of the congruent angles of an isosceles triangle concurrent with the altitude from the vertex angle? Explain.

Are the altitudes of a triangle concurrent? A deductive examination of this question follows.

Given: $\triangle ABC$, with altitudes $\overline{AD}$, $\overline{BE}$, and $\overline{CF}$.

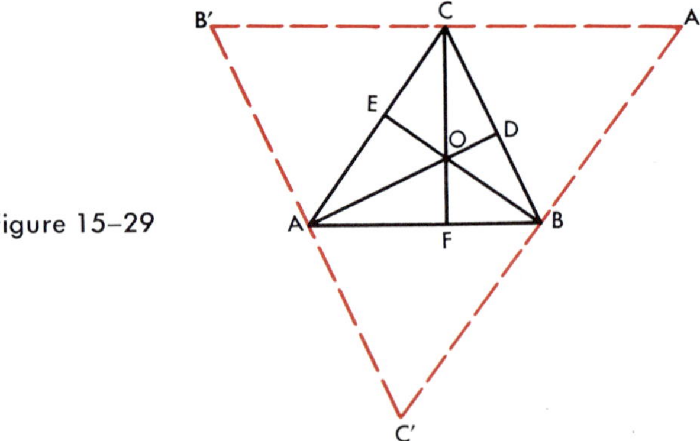

Figure 15–29

Conjecture: $\overleftrightarrow{AD}$, $\overleftrightarrow{BE}$, and $\overleftrightarrow{CF}$ meet at a point.

Plan: Show that the three lines containing the altitudes are the perpendicular bisectors of the sides of the larger triangle formed by drawing lines through the vertices of $\triangle ABC$ parallel to the opposite sides. Prove that $ABCB'$ and $ABA'C$ are parallelograms, so $\overline{B'C} \cong \overline{CA'}$, etc. Then use 15.08.

Proof: The formal proof is left to the student.

15.14 THEOREM

The altitudes or the lines determined by the altitudes of a triangle are concurrent. (See Exercise 7, page 562, for the coordinate proof.)

15.15 The point of intersection of the lines containing the altitudes of a triangle is called the orthocenter of the triangle. (Orthocenter relates to orthogonal, meaning right angle in Greek.)

Are the medians of a triangle concurrent? A deductive examination of thie question follows.

Given: $\triangle ABC$, with the medians $\overline{AE}$ and $\overline{BF}$ intersecting at O.

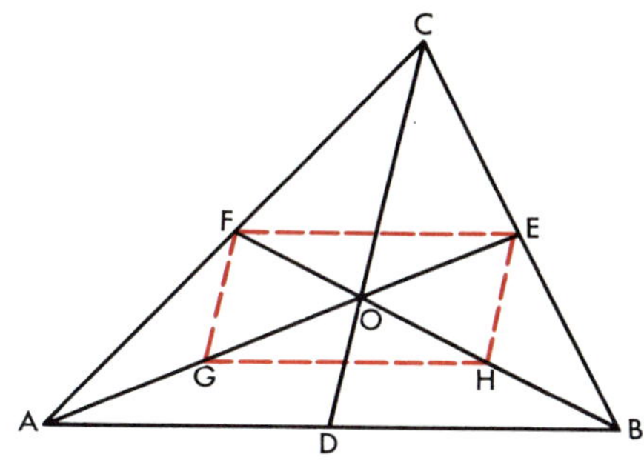

Figure 15–30

Conjecture: The median from C passes through point O, $m\overline{AO} = \frac{2}{3}\, m\overline{AE}$, $m\overline{BO} = \frac{2}{3}\, m\overline{BF}$, and $m\overline{CO} = \frac{2}{3}\, m\overline{CD}$.

Plan: Take G and H as the midpoints of $\overline{AO}$ and $\overline{BO}$. Show (1) that $EFGH$ is a ▱, (2) that $\overline{AE}$ and $\overline{BF}$ are trisected, and (3) that $\overline{CD}$ passes through O.

The formal proof follows on page 386.

Proof: *Statements*	*Reasons*
1. $\overrightarrow{AE}$ and $\overrightarrow{BF}$ will meet in some point O.	1. If they don't meet they are parallel but this is impossible. Why?
2. G and H are the midpoints of $\overline{AO}$ and $\overline{BO}$.	2. Every segment has a midpoint. (Exercise 11, page 40.)
3. In $\triangle AOB$, $\overline{GH} \parallel \overline{AB}$, $m\overline{GH} = \frac{1}{2} m\overline{AB}$	3. Why?
4. In $\triangle ABC$, $\overline{FE} \parallel \overline{AB}$, $m\overline{FE} = \frac{1}{2} m\overline{AB}$	4. Why?
5. $\therefore \overline{GH} \cong \overline{FE}$ and $\overline{GH} \parallel \overline{FE}$	5. Why?
6. $\therefore GHEF$ is a parallelogram.	6. Why?
7. $\therefore \overline{GO} \cong \overline{OE}$ and $\overline{HO} \cong \overline{OF}$	7. Why?
8. $\therefore \overline{AG} \cong \overline{GO} \cong \overline{OE}$ and $\overline{BH} \cong \overline{HO} \cong \overline{OF}$	8. Why?
9. $\therefore m\overline{AO} = \frac{2}{3} m\overline{AE}$ and $m\overline{BO} = \frac{2}{3} m\overline{BF}$	9. Why?

Take K as the midpoint of $\overline{CO}$. Draw $HKFD$. Show that it is a parallelogram. Follow the procedure of Steps 7 through 9. Show why $\overline{CD}$ must pass through O.

15.16 THEOREM

The medians of a triangle are concurrent in a point which lies two-thirds of the distance from each vertex to the midpoint of the opposite side. (See Exercise 12, page 559.)

15.17 The point where the medians of a triangle are concurrent is called the **centroid** (or medial point) of the triangle. It is the center of mass or of gravity of a triangular piece of material of uniform thickness and density. If this piece of material is suspended from any vertex, the centroid of the triangle will lie on a plumb line through the vertex.

Cut a "triangle" from a piece of cardboard, find its centroid, and attempt to balance it on the head of a pin. If you are careful it will balance.

Exercises

1. In what type of triangle would the orthocenter fall within the triangle? Outside the triangle? On the triangle? Describe the location of the point on the triangle.
2. Is it possible for the three medians of a triangle to coincide with the three corresponding altitudes and the three corresponding angle bisectors of the triangle? Explain.
3. Is it possible for a segment to be both a median and an altitude, but not an angle bisector of the triangle? Explain.
4. In Figure 15-30, if $m\overline{CD} = 6$ inches, find $m\overline{CO}$ and $m\overline{DO}$. If $m\overline{BO} = 8$ inches, find $m\overline{BF}$.
5. In Figure 15-30, if $m\overline{AE} = 9$ inches, find $m\overline{AO}$ and $m\overline{OE}$. If $m\overline{OE} = 4$ inches, find $m\overline{AO}$.
6. In Figure 15-30, if $m\overline{CD} = 5\frac{1}{2}$ inches, find $m\overline{CO}$ and $m\overline{OD}$.
7. The measure of the altitude of an equilateral triangle is 12 inches. What is the measure of the distance from each vertex of the triangle to the intersection of the altitudes?
8. If the measure of the altitude of an equilateral triangle is 9 inches, find the measure of the radius of the inscribed circle of the triangle.
9. If the measure of a median of an equilateral triangle is 6 inches, find the measure of the radius of the circumscribed circle of the triangle.
10. The measure of the altitude from the vertex angle of an isosceles triangle is 16 inches. What is the measure of the segment from the vertex of the isosceles triangle to the point of intersection of the angle bisectors of the triangle?
11. Does the point of concurrency of the bisectors of the congruent angles of an isosceles triangle and the altitude from the vertex angle move toward the base as the vertex angle increases in measure? (See Exercise 12, page 384.)

POLYHEDRONS

It is natural to attempt to extend two-dimensional concepts to three-dimensional figures. In terms of concurrent lines this leads us first to consider the polyhedrons. The simplest of the polyhedrons is the tetrahedron, which we will now examine.

Is it reasonable, for example, to surmise that the altitudes of a tetrahedron are concurrent? Should we, instead of considering lines, consider the concurrency of planes since we are dealing with three dimensions? Can three or more planes be concurrent? Study several tetrahedrons, and any other polyhedrons, with the preceding questions in mind.

Our next step is to examine deductively (15.19 and 15.21) several of our inductive conjectures.

15.18 Inscribed Sphere

A sphere is inscribed in a polyhedron if it is tangent to all the faces of the polyhedron. If the sphere is inscribed in the polyhedron, the polyhedron is *circumscribed* about the sphere.

Does this definition establish that a sphere can be inscribed in a polyhedron? We shall now examine this question in terms of a tetrahedron.

Given: Tetrahedron $VABC$

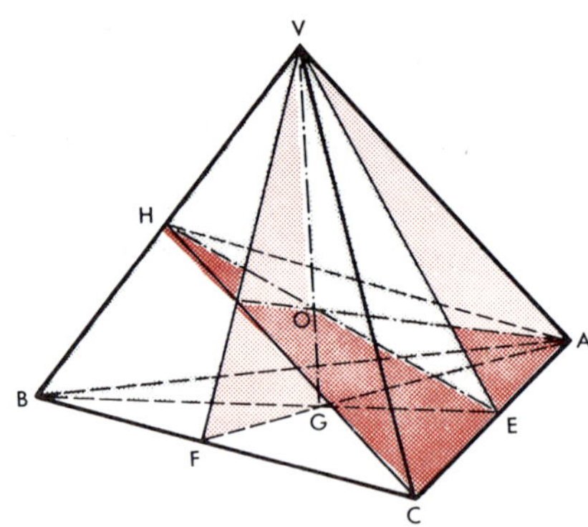

Figure 15–31

Conjecture: A sphere can be inscribed in tetrahedron $VABC$.

Plan: To prove point O is equally distant from the faces of the tetrahedron.

Proof: *Statements*	*Reasons*
1. Let half-plane VBE bisect dihedral angle VB, and half-plane VAF bisect dihedral angle VA.	1. 15.02, Theorem 11
2. Then VBE intersects VAF in $\overleftrightarrow{VG}$.	2. If two planes have a point in common (V), they intersect in a line. (2.24)
3. Any point in $\overleftrightarrow{VG}$ is equidistant from planes VBC and VBA, and equidistant from planes VCA and VBA.	3. 15.02, Theorem 11
4. $\therefore$ any point in $\overleftrightarrow{VG}$ is equidistant from planes VBC, VBA, and VCA.	4. Transitive Axiom of Equality
5. Plane ACH, which bisects dihedral angle AC, intersects $\overleftrightarrow{VG}$ in point O.	5. Why?
6. The point O is equidistant from planes BCA and VCA.	6. 15.02, Theorem 11
7. From (4) and (6), O is equidistant from all the faces of $VABC$.	7. Transitive Axiom of Equality
8. Therefore, the sphere with center O and radius congruent to the perpendicular from O to each face is tangent to each face of $VABC$.	8. 12.00, 12.51
9. $\therefore$ a sphere can be inscribed in $VABC$.	9. 15.18

15.19 THEOREM

A sphere can be inscribed in a tetrahedron.

15.20 Circumscribed Sphere

A sphere is circumscribed about a polyhedron if all the vertices of the polyhedron are points of the sphere. If the sphere is circumscribed about the polyhedron, it can be said that the polyhedron is *inscribed* in the sphere.

Does this definition establish that a sphere can be circumscribed about any polyhedron? Let us see if a sphere can be circumscribed about a tetrahedron.

Given: Tetrahedron $ABCD$

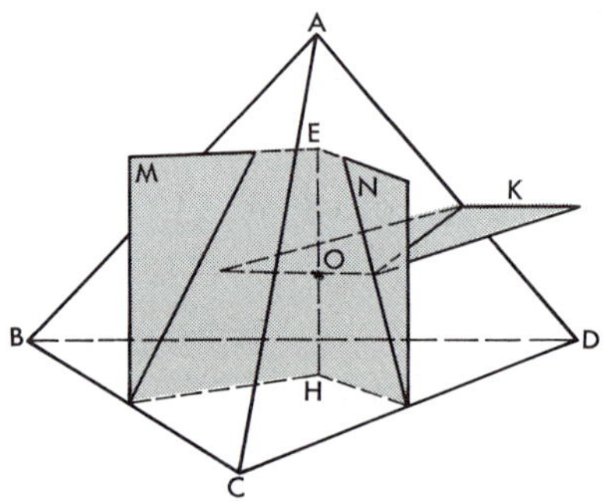

Figure 15–32

Conjecture: A sphere can be circumscribed about $ABCD$.

Plan: Prove point O equidistant from points A, B, C, and D.

Proof:

Statements	*Reasons*
1. Let planes m and n be the perpendicular bisectors of $\overline{BC}$ and $\overline{CD}$ respectively.	1. Why possible?
2. Planes m and n intersect in $\overleftrightarrow{EH}$.	2. If they were parallel, BCD would be a straight line, but BCD is not. Consequently, the planes must intersect in a line or coincide, but the planes do not coincide.
3. Any point in $\overleftrightarrow{EH}$ is equidistant from B, C, and D.	3. 15.02, Theorem 10

4. Let plane k be the perpendicular bisector of $\overline{AD}$.	4. Why possible?
5. Plane k intersects $\overleftrightarrow{EH}$ in O.	5. Why?
6. Every point of plane k is equidistant from A and D.	6. 15.02, Theorem 10
7. $\therefore O$ is equidistant from A, B, C, and D.	7. Transitive Axiom of Equality
8. $\therefore O$ is the center of a sphere which contains points A, B, C, and D.	8. 12.00

We have now established the following theorem.

15.21 THEOREM

A sphere can be circumscribed about a tetrahedron.

Exercises

1. Do Theorems 15.19 and 15.21 apply to all tetrahedrons, or to regular tetrahedrons only?
2. Do four non-coplanar points determine a sphere? Do three non-collinear points determine a sphere? Explain.
3. Can a cube be inscribed within a sphere? Explain.
4. Can a cube be circumscribed about a sphere? Explain.
5. Can a rectangular solid be inscribed within a sphere? Explain.
6. Can a parallelepiped be inscribed within a sphere? Explain.
7. Can a rectangular parallelepiped be circumscribed about a sphere? Explain.
8. Can a parallelepiped be circumscribed about a sphere? Explain.
9. Establish deductively the locus of points equidistant from three non-collinear points.
10. Prove or disprove the conjecture that the altitudes of a regular tetrahedron are concurrent.

11. If the corresponding sides of two triangles (ABC and $A'B'C'$) are not parallel and if lines drawn through the corresponding vertices are concurrent ($\overrightarrow{OA}$, $\overrightarrow{OB}$, $\overrightarrow{OC}$), then the points of intersection (R, S, T) of lines determined by the corresponding sides are collinear. (Desargue's Theorem.)

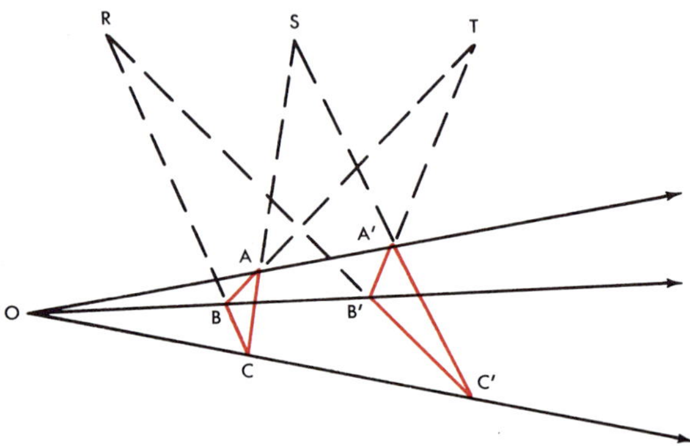

Figure 15–33

Vocabulary List

locus	parabola	incenter
loci	hyperbola	orthocenter
ellipse	circumcenter	centroid

Chapter Review

1. Describe how you would find the center and radius of a circle inscribed in a triangle.
2. Describe how you would find the center and radius of a circle circumscribed about a triangle.
3. Describe how you would find the centroid of a triangle.
4. Describe how you would find the orthocenter of a triangle.
5. Describe how you would find the center of a sphere inscribed in a tetrahedron.
6. Describe the locus of points in a plane whose distance from point P of the plane is exactly 2 inches.

7. Describe the locus of points in space whose distance from point P is exactly 2 inches.

8. Describe the locus of points in a plane whose distance from point P of the plane is less than 2 inches.

9. Describe the locus of points in a plane whose distance from point P of the plane is greater than 2 inches.

10. Describe the locus of points in the plane of and equidistant from two concentric circles.

11. Describe the locus of points equidistant from two concentric circles.

12. Describe the locus of points in space equidistant from two parallel lines.

13. Describe the locus of points equidistant from two distinct points.

14. Describe the locus of points equidistant from the vertices of a rectangular parallelepiped.

15. Describe the locus of points equidistant from the vertices of a tetrahedron.

16. Describe the locus of points equidistant from the faces of a dihedral angle and 6 inches from the edge of the angle.

17. Describe the locus of points exactly 3 inches from two points 2 inches apart.

18. Describe the locus of points equidistant from parallel planes x and y and 6 inches from point O in x, where x and y are 6 inches apart.

19. Describe the locus of the vertex of the right angle of a right triangle whose hypotenuse is given.

20. The statement of a locus must (1) _?_ all points that satisfy the given conditions, and (2) _?_ all points that do not satisfy the given conditions.

21. Complete the last half of the following locus statement to fulfill the conditions of Exercise 20. "In a plane, if a point is the distance r from the point O, it is on the circle whose center is O and whose radius is $r \cdots$."

22. Is the last half of the statement of Exercise 21 the converse, inverse, or contrapositive of the first half?

23. Write the contrapositive of the statement of Exercise 21.

24. Complete the last half of the statement of Exercise 23 to fulfill the conditions of Exercise 20.

25. Are the statements of Exercise 21 and Exercise 23 equivalent statements? Justify your response.

Chapter 15 Test

Matching: Following the number of the phrase in Column I place the letter of the phrase in Column II that best completes the statement. Phrases from Column II may be used more than once.

Column I	Column II
1. The centroid of a triangle is determined by...	A. concurrent altitudes.
2. The incenter of a triangle is determined by...	B. concurrent medians.
3. The circumcenter of a triangle is determined by...	C. concurrent angle bisectors.
4. The orthocenter of a triangle is determined by...	D. concurrent perpendicular bisectors of the sides.
5. The circumcenter of a tetrahedron is determined by...	

Multiple choice: Following the number of the problem place the letter of the phrase that makes it a true statement.

6. To completely prove a locus theorem we need to establish **(a)** that all points of the locus set satisfy the given conditions **(b)** that all points satisfying the given conditions are members of the locus set **(c)** both (a) and (b) **(d)** none of these.

7. The locus of points in a plane equidistant from the sides of an angle in the plane is **(a)** a half-line **(b)** a line **(c)** a segment **(d)** a ray.

8. The locus of points at a given distance from a given line is **(a)** a line **(b)** a plane **(c)** two lines **(d)** a cylindrical surface.

9. The locus of points at a given distance from a plane is **(a)** a plane **(b)** two planes **(c)** a sphere **(d)** a cylindrical surface.

10. The locus of points equidistant from the faces of a dihedral angle is (**a**) a half-plane (**b**) a line (**c**) a plane (**d**) a ray.

11. The locus of points equidistant from the vertices of a tetrahedron is (**a**) the center of the inscribed sphere (**b**) the center of the circumscribed sphere (**c**) a line (**d**) none of these.

12. The locus of points equidistant from the vertices of a regular quadrangular prism is (**a**) a plane (**b**) a line (**c**) a point (**d**) none of these.

13. The locus of points common to the graphs of $3x - 2y = 4$ and $x - 2 = 2y$ on a rectangular coordinate system is (**a**) the null set (**b**) a point (**c**) a line (**d**) none of these.

14. The locus of points common to the lines containing the altitudes of obtuse $\triangle ABC$ (**a**) is on the triangle (**b**) is in the interior of the triangle (**c**) is in the exterior of the triangle (**d**) is the null set.

15. The center of a sphere inscribed in a tetrahedron (**a**) is equidistant from the vertices (**b**) is equidistant from the faces (**c**) lies on the altitude of the tetrahedron (**d**) none of these.

16. The measure of an altitude of an equilateral triangle is 9 inches. The locus of points common to the medians of this equilateral triangle is (**a**) 3 inches from each vertex (**b**) 6 inches from each vertex (**c**) $4\frac{1}{2}$ inches from each vertex (**d**) none of these.

17. The locus of points equidistant from three non-collinear points is (**a**) a point (**b**) two planes (**c**) a plane (**d**) a line.

18. The locus of points equidistant from four non-coplanar points is (**a**) a point (**b**) a line (**c**) a plane (**d**) two planes.

19. The locus of points 5 inches from a given point and equidistant from this point and from a second point 8 inches from it is (**a**) a plane (**b**) a line (**c**) a circle (**d**) a sphere.

20. The measure of the diameter of a given sphere is 12 inches. If the diameter of the sphere is the hypotenuse of the set of right triangles the measure of whose altitude upon the hypotenuse is 3 inches, the locus of the vertex of the right angle of the triangle is (**a**) a circle (**b**) the sphere (**c**) a plane (**d**) none of these.

16

Ratio and Proportion

RATIO

Comparison of *like* quantities is part of our everyday life. We compare such things as salaries, cars, weights, and experiences. If one man weights 200 pounds and another 150 pounds we say the *ratio* of their weights is 4 to 3. If we compare twenty-five cents to one dollar the ratio is 1 to 4.

16.00 A ratio is the indicated quotient of two non-zero real numbers; $\frac{a}{b}$ is called the ratio of a to b.

16.01 Properties of ratios

1. A ratio is meaningful only when the units of measure represented by the numbers of the ratio are the same. For example, the comparison of the value of two dollars to two cents is meaningful only when we convert dollars into cents, or cents into dollars. The ratio of pounds to miles is meaningless, since they cannot be converted to a common unit.

2. A ratio may be represented by a fraction.
3. It is often convenient to write a ratio in its simplest equivalent form. A ratio of 6 to 12 may be written 1 to 2. Any of the following forms are satisfactory; 1 : 2, $\frac{1}{2}$, or 1 ÷ 2.

Oral Exercises

Give the ratio of each of the following.

1. 36 in. to 12 in.
2. $20 to $180
3. 4 ft. to 4 yds.
4. 600 lbs. to 2 tons
5. 80¢ to $2.40
6. $2b$ in. to $3b$ in.
7. $\frac{1}{2}$ to $\frac{3}{4}$
8. $5\frac{1}{2}$ to $7\frac{1}{2}$
9. 6.25 to $\frac{1}{4}$

Exercises

1. What is the ratio of the measures of two segments, one 7 inches and the other 21 inches?
2. What is the ratio of the measures of two segments, one 3 centimeters and one 18 centimeters?
3. Draw a short segment of length x. Construct segments of length $2x$ and $3x$. What is the ratio of their lengths?
4. A segment 25 inches long is divided by a point so that the parts are in the ratio 2 :3. How long is each part?
5. If a segment 30 inches long is divided by a point 6 inches from one end, what is the ratio of the measures of the longer part to the shorter part?
6. Find the lengths of the parts of a 60 inch segment that is so divided by a point that the ratio of the measures of the parts is 1 :3. So that the ratio is 3 :2.
7. Find the lengths of two segments, the sum of whose measures is 28 inches and the ratio of whose measures is 4 :3.
8. If a segment is 33 inches long, how far from one end must a point be placed so that the measures of the parts of the segment may be in the ration 1 :2?
9. Draw a segment and with a straight edge and compasses divide it into two parts whose measures are in the ratio 2 :3.

10. What is the ratio of the measure of a central angle of a circle to the measure of an inscribed angle of the circle intercepting the same arc?

11. The following are the vibrations per second (scientific pitch) for the eight notes of the major musical scale of *C*, beginning at middle *C*:

C	*D*	*E*	*F*	*G*	*A*	*B*	*C*
256	288	320	$341\frac{1}{3}$	384	$426\frac{2}{3}$	480	512

(a) What is the ratio of the vibrations per second of middle *C* to the *C* which is an octave higher in pitch?
(b) What is the ratio of the vibrations per second of middle *C* to *E* of the major scale?
(c) What is the ratio of the vibrations per second of *F* to *G*?

(Concordant harmonies and discordant harmonies depend upon a mathematical law of ratio. This would be an interesting subject for a student report.)

PROPORTION

Figure 16–1

When a photographer enlarges a photograph, the corresponding parts of the two pictures are in *proportion*. A 3″ by 4″ picture can be enlarged to a 6″ by 8″ picture without distortion. Can a

$2\frac{1}{4}''$ by $3\frac{1}{4}''$ picture be enlarged to 8″ by 10″ without distortion or loss of part of the picture?

16.02 The equality of two ratios is called a proportion.

The statements $\frac{1}{2} = \frac{2}{4}$ (or $1:2 = 2:4$) and $\frac{a}{b} = \frac{c}{d}$ (or $a:b = c:d$) are proportions. They are usually read: "1 is to 2 as 2 is to 4," and "a is to b as c is to d."

A proportion is an equation, and all properties of equations apply.

Terms of a Proportion

16.03 In the proportion $\frac{x}{y} = \frac{m}{n}$ (or $x:y = m:n$), x is called the *first term*, y the *second term*, m the *third term*, and n the *fourth term*. The first and third terms, x and m, are the *numerators*; y and n are the *denominators*.

The second and third terms, y and m, are called the means, and the first and fourth terms, x and n, are called the extremes.

Exercises

Which of the following are proportions?

1. $\frac{3}{5} = \frac{12}{20}$
2. $\frac{3}{15} = \frac{15}{75}$
3. $\frac{1}{x} = \frac{6}{6x}$
4. $\frac{12}{11} = \frac{11}{10}$
5. $\frac{8}{9} = \frac{4}{3}$
6. $\frac{xy}{y^2} = \frac{x}{y}$

Find the value of x in each of the following proportions:

7. $\frac{x}{6} = \frac{4}{3}$
8. $\frac{4}{x} = \frac{1}{7}$
9. $\frac{x + 3}{4} = \frac{5}{2}$
10. $\frac{x}{5 - x} = \frac{7}{8}$
11. $\frac{21 - x}{x} = \frac{1}{2}$
12. $\frac{3x + 2}{5} = \frac{x - 4}{4}$
13. Tell what are the first, second, third, and fourth terms of Exercise 7.
14. Tell what are the means and extremes of Exercise 1; of Exercise 8.

INDUCTIVE PROCEDURE WITH PROPORTIONS

$$\frac{5}{10} = \frac{15}{30} \qquad \frac{3}{8} = \frac{6}{16} \qquad \frac{4}{7} = \frac{20}{35}$$

Let us experiment with the previous proportions, and any others you wish to use, to develop tentative or probable relationships concerning proportions.

(a) Does $5 \cdot 30 = 10 \cdot 15$? Does $3 \cdot 16 = 8 \cdot 6$? Does $4 \cdot 35 = 7 \cdot 20$?

Does it seem probable that the product of the means and the product of the extremes of a proportion will always be equal?

(b) Does $\frac{5}{15} = \frac{10}{30}$? Does $\frac{30}{10} = \frac{15}{5}$? Does $\frac{16}{8} = \frac{6}{3}$? Does $\frac{4}{20} = \frac{7}{35}$? Will the interchange (alternation) of the means or the extremes of a proportion produce another proportion?

(c) Will the interchange of the numerators, or the denominators, of the terms of a proportion produce a proportion?

Does $\frac{15}{10} = \frac{5}{30}$? Does $\frac{6}{8} = \frac{3}{16}$? Does $\frac{4}{35} = \frac{20}{7}$?

(d) Does $\frac{10}{5} = \frac{30}{15}$? Does $\frac{8}{3} = \frac{16}{6}$? Does $\frac{7}{4} = \frac{35}{20}$?

Will inverting the terms of a proportion produce a proportion? Will inverting only one term of a proportion produce a proportion? For instance, does $\frac{5}{10} = \frac{30}{15}$?

(e) Since a proportion is the equality of two ratios, if the denominators of a porportion are equal must the numerators be equal? Example: If $\frac{x}{12} = \frac{4}{12}$, what is the value x?

(f) If $\frac{3}{11} = \frac{x}{5}$ and $\frac{3}{11} = \frac{y}{5}$, must $x = y$? If the three terms of one proportion are equal to the three corresponding terms of another, must the fourth terms be equal?

(g) Consider the equal ratios $\frac{3}{6}, \frac{6}{12}, \frac{18}{36}$. Compare the sum of the numerators to the sum of the denominators, $\frac{27}{54}$. Is $\frac{27}{54}$ equivalent to the original ratios? Will this always be true? Try other examples.

(h) If $\frac{5}{6} = \frac{10}{12}$, does $\frac{5}{6} + 1 = \frac{10}{12} + 1$? Does it follow that $\frac{11}{6} = \frac{22}{12}$? Does $\frac{5}{6} - 1 = \frac{10}{12} - 1$ or $\frac{-1}{6} = \frac{-2}{12}$? Which axiom of mathematics justifies these procedures?

Now let us test by deductive procedures the generality of the tentative conclusions developed in (a) through (h). We will represent any proportion by the general expression $\frac{a}{b} = \frac{c}{d}$.

PROPORTION THEOREMS

In the following theorems (16.04–16.11) $a, b, c, d, e, f, x, y,$ denote non-zero real numbers.

Given: $\frac{a}{b} = \frac{c}{d}$ **Conjecture:** $ad = bc$

Proof: If $\frac{a}{b} = \frac{c}{d}$, then $\cancel{b}d \cdot \frac{a}{\cancel{b}} = b\cancel{d} \cdot \frac{c}{\cancel{d}}$, or $ad = bc$.

16.04 THEOREM

In any proportion, the product of the extremes is equal to the product of the means.

Given: $ad = bc$ **Conjecture:** $\frac{a}{c} = \frac{b}{d}$

Proof: If $ad = bc$, then $\frac{a\cancel{d}}{c\cancel{d}} = \frac{b\cancel{c}}{\cancel{c}d}$, or $\frac{a}{c} = \frac{b}{d}$.

16.05 THEOREM

If the product of two numbers is equal to the product of two other numbers, either pair may be made the means of a proportion and the other pair the extremes.

Given: $\frac{a}{x} = \frac{a}{y}$ **Conjecture:** $y = x$

Proof: If $\frac{a}{x} = \frac{a}{y}$, then $ay = ax$ (16.04) and $y = x$ (Division Axiom).
The proof of the converse is left to the student.

16.06 THEOREM

If the numerators of a proportion are equal, the denominators are equal; and conversely.

Given: $\frac{a}{b} = \frac{x}{c}$ and $\frac{a}{b} = \frac{y}{c}$ **Conjecture:** $x = y$

Proof: Solve for x in one proportion and y in the other. The rest of the proof is left to the student.

16.07 THEOREM

If three terms of one proportion are equal respectively to the three corresponding terms of another proportion, the remaining terms are equal.

Given: $\frac{a}{b} = \frac{c}{d}$ **Conjecture:** $\frac{a}{c} = \frac{b}{d}$

Proof: If $\frac{a}{b} = \frac{c}{d}$, then $\frac{a}{b} \cdot \frac{b}{c} = \frac{c}{d} \cdot \frac{b}{c}$ or $\frac{a}{c} = \frac{b}{d}$.

16.08 THEOREM

The terms of a proportion are in proportion by alternation; that is, the first term is to the third term as the second is to the fourth.

Given: $\frac{a}{b} = \frac{c}{d}$ **Conjecture:** $\frac{b}{a} = \frac{d}{c}$

Proof: If $\frac{a}{b} = \frac{c}{d}$, then $ad = bc$ (16.04)

and $\frac{\not{a}d}{\not{a}c} = \frac{b\not{c}}{a\not{c}}$ or $\frac{d}{c} = \frac{b}{a}$.

16.09 THEOREM

The terms of a proportion are in proportion by inversion; that is, the second term is to the first as the fourth is to the third.

Given: $\frac{a}{b} = \frac{c}{d}$

Conjecture: $\frac{a + b}{b} = \frac{c + d}{d}$ and $\frac{a - b}{b} = \frac{c - d}{d}$

Proof: If $\frac{a}{b} = \frac{c}{d}$, then $\frac{a}{b} + 1 = \frac{c}{d} + 1$ or $\frac{a + b}{b} = \frac{c + d}{d}$.

Also $\frac{a}{b} - 1 = \frac{c}{d} - 1$ or $\frac{a - b}{b} = \frac{c - d}{d}$

16.10 THEOREM

The terms of a proportion are in proportion by addition or subtraction; that is, the sum of (or difference between) the first and second terms is to the second term as the sum of (or difference between) the third and fourth terms is to the fourth term.

Given: $\frac{a}{b} = \frac{c}{d} = \frac{e}{f}$, etc. **Conjecture:** $\frac{a + c + e}{b + d + f} = \frac{a}{b}$

Proof:
1. Let $\frac{a}{b} = r$; then $\frac{c}{d} = r$ and $\frac{e}{f} = r$. Why?
2. Then $a = br$, $c = dr$, and $e = fr$. Why?
3. $a + c + e = br + dr + fr = (b + d + f)r$. Why?
4. $\frac{a + c + e}{b + d + f} = r$. Why?
5. $\therefore \frac{a + c + e}{b + d + f} = \frac{a}{b}$. Why?

16.11 THEOREM

In a series of equal ratios, the sum of the numerators is to the sum of the denominators as any numerator is to its denominator.

Exercises

Express the following as the product of the means equaling the product of the extremes. The variables denote non-zero real numbers.

1. $\frac{x}{y} = \frac{m}{n}$ **2.** $m:n = s:t$ **3.** $\frac{a-b}{b} = \frac{a-x}{x}$

Find the ratio of x to y:

4. $3x = 4y$ **6.** $ax = by$ **8.** $ax + bx = cy + dy$

5. $5x = 6y$ **7.** $\frac{x+y}{x-y} = \frac{1}{2}$ **9.** $ax + by = cx + dy$

10. Write each of the following proportions by (**a**) inversion, (**b**) alternation, (**c**) addition, and (**d**) subtraction:

$$\frac{2}{5} = \frac{10}{25};\ m:n = s:t.$$

11. If $\frac{a}{b} = \frac{x}{y}$ and $\frac{a}{b} = \frac{x}{z}$, what is your conclusion?

12. If $\frac{ab}{cd} = \frac{xy}{mn}$ and $\frac{ab}{gh} = \frac{xy}{mn}$, what is your conclusion?

13. Use 16.04 to solve the following proportions for x.

(**a**) $\frac{2}{x} = \frac{3}{7}$ (**b**) $\frac{a}{x} = \frac{b}{c}$ (**c**) $\frac{x-5}{x+3} = \frac{1}{3}$

(**d**) $\frac{a+x}{x} = \frac{b}{c}$ (**e**) $\frac{a+b}{b} = \frac{a+x}{x}$ (**f**) $\frac{3x-2a}{2x} = \frac{2b}{3a}$

14. If $\frac{2}{3} = \frac{m}{n}$, then $\frac{2}{m} = \frac{3}{n}$. Why? Also $\frac{3}{2} = \frac{n}{m}$. Why?

15. If $\frac{5}{3} = \frac{x}{y}$, then $\frac{8}{3} = \frac{x+y}{y}$. Why? Also $\frac{2}{3} = \frac{x-y}{y}$. Why?

SQUARE ROOT

16.12 A square root of a positive number is one of the two equal factors of the number. In symbolic form this means that a number x is the *square root* of the number n if it satisfies $x^2 = n$. It follows that if $n > 0$ then $x < 0$ or $x > 0$. Also if $n = 0$, $x = 0$. To indicate the principal (positive) square root of a positive real number we write $\sqrt{n}$. This symbol is called a *radical*, and n is called the *radicand*. The negative root is indicated by $-\sqrt{n}$. The symbol $\pm\sqrt{n}$ is used to indicate both roots.

> In general, $\sqrt{n} \cdot \sqrt{n} = n$, and
> $-\sqrt{n} \cdot -\sqrt{n} = n$.

Finding square root

16.13 A table of square roots of numbers from 1 to 100 is provided on page 580. The square roots are given to the nearest thousandth.

Find the square roots of the following numbers to the nearest hundredth: 7, 12, 63, 720, 940.*

Clearly the table does not contain the square roots of all numbers. Certain methods for extracting square roots have occasionally been taught. Since the advent of calculating machines, logarithms, and slide rules has simplified the finding of roots, we will not burden the student with a laborious calculating process.

If none of the computational aids are available we can resort to the trial and error method more easily than we can memorize a complicated process. For example, let us find the square root of 313. The root falls between 17 and 18 since $17^2 = 289$ and $18^2 = 324$. Since 313 is closer to 324 than to 289, the root of 313 will probably be closer to 18 than to 17. The root falls between 17.6 and 17.7 since $17.6^2 = 309.76$ and $17.7^2 = 313.29$. Will the root of 313 be closer in value to 17.6 or 17.7? Further trials with such values as 17.69^2 will give the answer to the degree of accuracy desired.

Square root of a fraction

16.14 For example, $\sqrt{\frac{16}{25}} = \frac{\sqrt{16}}{\sqrt{25}} = \frac{4}{5}$. The square root of a fraction is obtained by dividing the square root of the numerator by the square root of the denominator. See the following example.

---

*In rounding off to the nearest hundredth, the digit following the hundredth digit is inspected and, if it is greater than 5, the preceding digit is increased by 1; if it is less than 5, the inspected digit is dropped. If the digit inspected is 5 followed by no other digits or by zero only, the preceding digit is left unchanged if it is an even number but it is increased by 1 if it is an odd number. If the digit inspected is 5 but other non-zero digits follow, the preceding digit is increased by 1.

$\sqrt{\frac{2}{3}} = \frac{1.414}{1.732} = .82$ approximately. This is a rather long process and can be simplified as shown below.

Rationalizing: To rationalize $\sqrt{\frac{2}{3}}$, change $\frac{2}{3}$ to an equivalent fraction whose denominator is the square of an integer, as below.

$$\sqrt{\frac{2}{3}} = \sqrt{\frac{6}{9}} = \frac{\sqrt{6}}{3} = \frac{2.449}{3} = .82\text{, approximately.}$$

> **Rule:** To find the square root of a fraction, change it, *if necessary*, to an equivalent fraction whose denominator is the square of an integer.

Quite often, for quick computation, we will leave the fraction in a form known as the simplest equivalent radical form. That is, $\sqrt{\frac{2}{3}}$ would be written as $\frac{\sqrt{6}}{3}$; $\sqrt{\frac{1}{2}}$ as $\frac{\sqrt{2}}{2}$; $\sqrt{\frac{2}{5}}$ as $\frac{\sqrt{10}}{5}$.

Fourth proportional

16.15 The fourth term of a proportion is called the fourth proportional to the other three terms.

In the proportion $\frac{2}{5} = \frac{6}{15}$, 15 is the fourth proportional to 2, 5, and 6 in the order named. To find the fourth proportional to 2, 3, and 4, solve the proportion $\frac{2}{3} = \frac{4}{x}$ for x. In this case $x = 6$.

Mean proportional

16.16 When the means of a proportion are the same, either of them is called the mean proportional between the other two terms.

In the proportion $\frac{2}{6} = \frac{6}{18}$, 6 is the mean proportional between 2 and 18. If $\frac{a}{b} = \frac{b}{c}$, then b is the mean proportional between a and c. To find the mean proportional between 3 and 12, solve $\frac{3}{x} = \frac{x}{12}$ for x. This gives us $x^2 = 36$ and $x = \pm 6$. In geometry we generally take only the positive root. The third proportional is the fourth term in a proportion having means which are the same number. In $\frac{a}{b} = \frac{b}{x}$, x is the third proportional to a and b.

Exercises

1. Find the square root of the following fractions to the nearest hundredth.

   (a) $\frac{1}{2}$ (b) $\frac{2}{5}$ (c) $\frac{3}{5}$ (d) $\frac{5}{12}$ (e) $\frac{3}{10}$

   (f) $\frac{7}{15}$ (g) 2.6 $\left(\text{change to } \frac{26}{10}\right)$ (h) 2.3

2. Leave the following in simplest radical form for quick computation (integral denominators).

   (a) $\sqrt{\frac{3}{5}}$ (c) $\sqrt{\frac{2}{7}}$ (e) $\sqrt{\frac{3}{10}}$ (g) $\sqrt{\frac{5}{3}}$ (i) $\sqrt{\frac{2}{6}}$

   (b) $\sqrt{\frac{5}{12}}$ (d) $\sqrt{\frac{1}{6}}$ (f) $\sqrt{\frac{7}{15}}$ (h) $\sqrt{\frac{3}{8}}$ (j) $\sqrt{\frac{4}{10}}$

3. If the second and third terms of a proportion are equal, what name denotes this?

4. Write a proportion stating that x is the mean proportional between 4 and 9.

5. Give the mean proportional between each of the following pairs of numbers. Give the answers to the nearest hundredth.

   (a) 4, 16 (c) 11, 7 (e) 12, 6 (g) 37, 10

   (b) 3, 32 (d) 25, 2 (f) 9, 15 (h) 68, 15

6. Find the mean proportional between each of the following. Leave the answers in simplest radical form for quick computation.

   (a) 3, 4 (b) 5, 4 (c) 2, 9 (d) 5, 9 (e) 3, 9 (f) 8, 5

   (g) $\frac{2}{3}, \frac{3}{4}$ (h) $\frac{1}{5}, \frac{5}{3}$ (i) $\frac{3}{2}, \frac{1}{8}$ (j) $\frac{5}{3}, \frac{2}{5}$ (k) $\frac{1}{6}, \frac{5}{3}$ (l) $\frac{7}{12}, \frac{1}{4}$

7. Find the fourth proportional to:

   (a) 4, 6, 10 (b) 3, 4, 7 (c) 2, 3, 4 (d) 2, 5, 10

   (e) 7, 14, 1 (f) 7, 8, 10 (g) 14, 24, 18 (h) $1\frac{1}{2}, 2\frac{1}{3}, 3$

8. Solve for x. Leave the answers in radical form with rational denominators. $x \neq 0$.

(a) $\frac{3}{x} = \frac{x}{5}$ (b) $\frac{2}{x} = \frac{3x}{5}$ (c) $\frac{5}{2x} = \frac{3x}{7}$

(d) $\frac{2x}{11} = \frac{4}{5x}$ (e) $\frac{7x}{3} = \frac{3}{4x}$ (f) $\frac{2x^2 + 5}{2x} = \frac{7x}{3}$

THE GOLDEN SECTION (Optional)

16.17 A segment is divided by the golden section into extreme and mean ratio if the measure of the shorter portion is to the measure of the longer portion as the measure of the longer portion is to the measure of the entire segment.

$$\frac{a}{b} = \frac{b}{a + b}$$

Figure 16–2

It is claimed that a segment is most harmoniously divided when it is either bisected or divided into extreme and mean ratio by the golden section.

If we use these sections to construct a rectangle, artists claim this figure to be more pleasing to the eye than any other rectangle.

It can be shown that the ratio

$$\frac{b}{a} = \frac{1 + \sqrt{5}}{2} = 1.618 \cdots \text{ and that}$$

$$\frac{a}{b} = \frac{1 - \sqrt{5}}{2} = .618 \cdots.$$

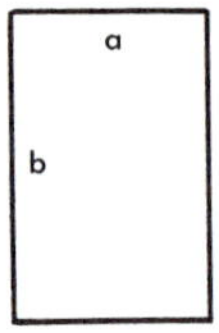

Figure 16–3

The Greeks credit Pythagoras with the discovery of the golden section.

Many famous painters have used the golden section in their work. Leonardo da Vinci used and taught this method. Michelangelo's "Holy Family," Botticelli's "Magnificat," and Salvador Dali's "Corpus Hypercubus" (Hypercubic Body) are famous examples.

The Egyptians used it in the triangular profile of the Great Pyramid of Cheops. It is used by architects today to beautify their designs.

Mother Nature has used the golden number in the design of the sunflower, the starfish, and spider webs, in the placement of buds on a branch, and on many flowers.

PROPORTIONAL LINE SEGMENTS

16.18 Two segments are divided proportionally when the measure of the segments of one have the same ratio as the measure of the *corresponding* segments of the other. In Figure 16-4 $\overline{AC}$ and $\overline{BD}$ are divided proportionally if

$$\frac{m\overline{AE}}{m\overline{EC}} = \frac{m\overline{BF}}{m\overline{FD}}.$$

Figure 16–4

Proportional segments are common in our environment. Examples are the ladderlike structure of a television station's signal tower or a railroad tower. See if you can find other examples.

Measure the segments in the following figures and construct others of your own design to try to determine the conditions which produce proportional segments.

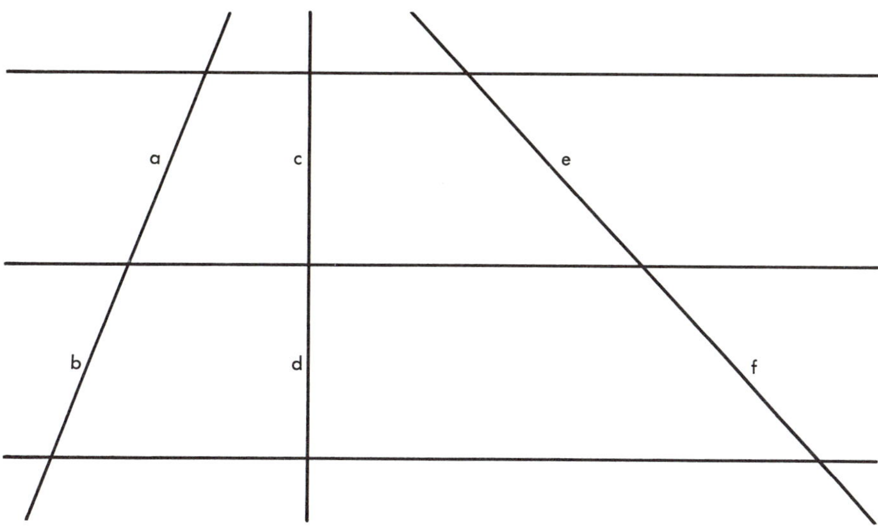

Figure 16–5

In the triangles below, which segments have measures which are in proportion? Can you tell why?

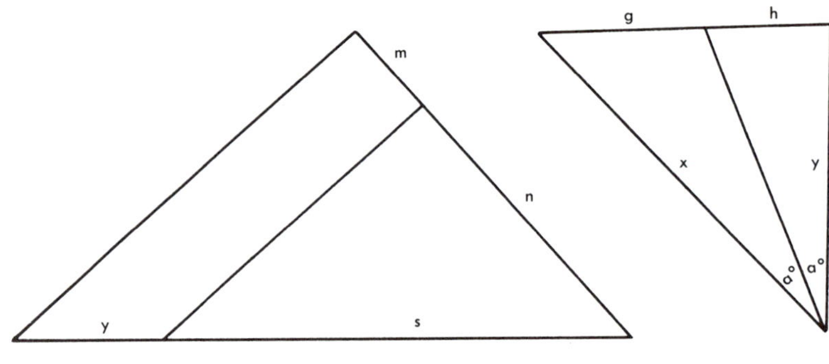

Figure 16–6

We shall now examine deductively several conjectures which might have been reached through your experimentation.

Given: $\triangle ABC$, with $\overleftrightarrow{DE} \parallel \overleftrightarrow{AB}$

Conjecture: $\dfrac{m\overline{CD}}{m\overline{DA}} = \dfrac{m\overline{CE}}{m\overline{EB}}$

Plan: Select a unit $\overline{CX}$ whose measure is an integral factor of $m\overline{CD}$ and $m\overline{DA}$. Let the number of congruent divisions of $\overline{CD}$ be m and of $\overline{AD}$ be n. Then use 9.28.

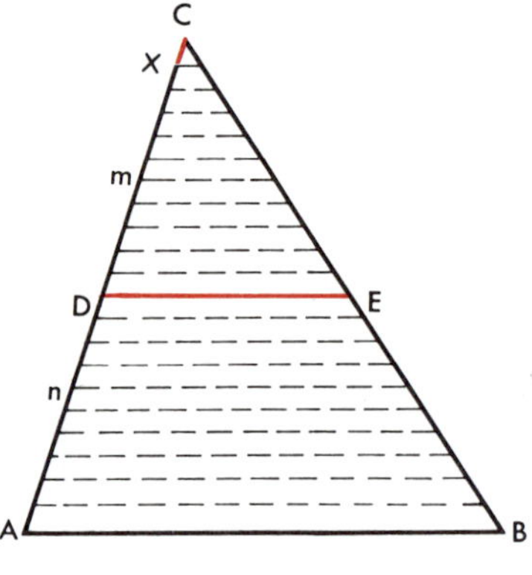

Figure 16–7

Proof:

Statements	*Reasons*
1. $\dfrac{m\overline{CD}}{m\overline{DA}} = \dfrac{m}{n}$	1. Construction, and def. of ratio
2. Through the points of division on $\overline{CA}$, draw lines parallel to $\overline{AB}$, cutting $\overline{CE}$ and $\overline{EB}$ into m and n congruent parts respectively.	2. 9.28

3. $\frac{m\overline{CE}}{m\overline{EB}} = \frac{m}{n}$	3. Def. of ratio
4. $\therefore \frac{m\overline{CD}}{m\overline{DA}} = \frac{m\overline{CE}}{m\overline{EB}}$	4. Why?

16.19 THEOREM

A line parallel to one side of a triangle and intersecting the other two sides divides these sides proportionally.

NOTE: This proof is for the commensurable case only; that is where a unit, $\overline{CX}$ can be found whose measure is an integral factor of both $m\overline{CD}$ and $m\overline{DA}$. Suppose $m\overline{CD} = 1$ and $m\overline{DA} = \sqrt{2}$. Then this method will fail because no unit can be found whose measure is an integral factor of both $m\overline{CD}$ and $m\overline{DA}$.

The proof of the incommensurable case is assumed. It is not given here because it involves ideas not yet considered. (See 18.01 and 18.02)

When two sides of a triangle are divided into segments by a line parallel to a third side, a ratio can also be formed comparing the measure of a segment of a side to the measure of the entire side. Would these ratios be the same for each side if corresponding segments were used?

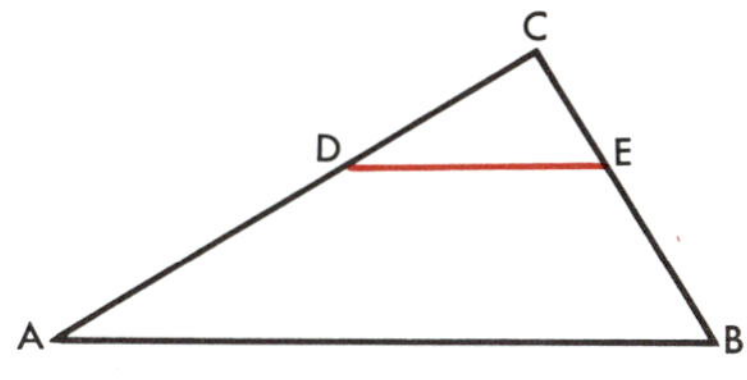

Figure 16–8

Given: $\overline{DE} \parallel \overline{AB}$ in $\triangle ABC$

Conjecture: $m\overline{AC}:m\overline{DC} = m\overline{BC}:m\overline{EC}$

Proof: *Statements*	*Reasons*
1. $m\overline{AD}:m\overline{DC} = m\overline{BE}:m\overline{EC}$	1. Why?
2. $(m\overline{AD} + m\overline{DC}):m\overline{DC} = (m\overline{BE} + m\overline{EC}):m\overline{EC}$	2. 16.10
3. $m\overline{AD} + m\overline{DC} = m\overline{AC}$ and $m\overline{BE} + m\overline{EC} = m\overline{BC}$	3. Why?
4. $\therefore m\overline{AC}:m\overline{DC} = m\overline{BC}:m\overline{EC}$	4. Why?

We have now established the following theorem.

16.20 THEOREM

A line parallel to one side of a triangle divides the other two sides so that the measure of either side is to the measure of one of its segments as the measure of the other side is to the measure of the corresponding segment.

If two parallel lines are cut by a transversal line, how many segments are intercepted on the transversal? How many segments are intercepted on a transversal by three parallel lines? Is there a relationship between the measures of the segments formed on one transversal and the measures of the segments on another transversal of the same three parallel lines?

Given: $\overleftrightarrow{AD} \parallel \overleftrightarrow{BE} \parallel \overleftrightarrow{CF}$. Transversals $\overleftrightarrow{AC}$ and $\overleftrightarrow{DF}$.

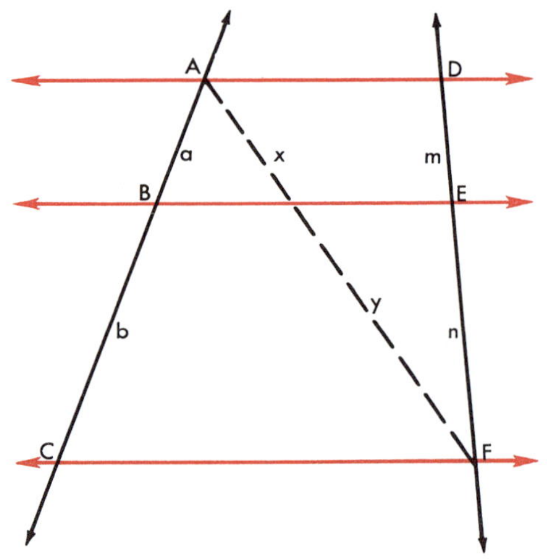

Figure 16–9

Conjecture: $a:b = m:n$

Plan: Prove that $a:b = x:y$ and $m:n = x:y$

Proof: The proof is left to the student.

16.21 THEOREM

Three or more parallel lines intercept proportional segments of transversals.

Let us see if there is an analogy to theorem 16.21 in three dimensional space.

Given: Parallel planes m, n, and r, cutting transversals $\overleftrightarrow{AB}$ and $\overleftrightarrow{CD}$ at A, E, B, and C, F, D, respectively.

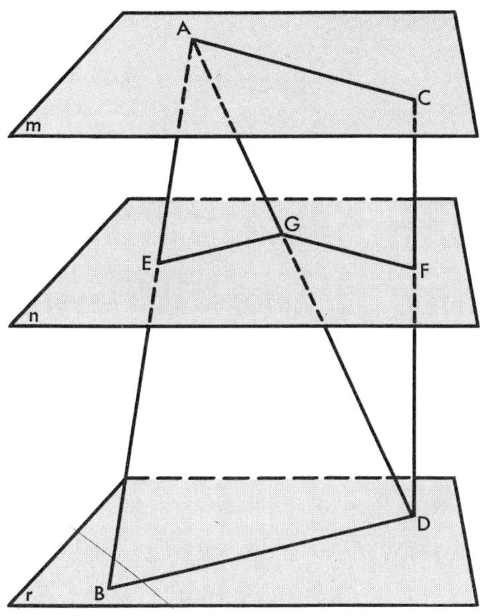

Figure 16–10

Conjecture: $m\overline{AE}:m\overline{EB} = m\overline{CF}:m\overline{FD}$

Plan: Form $\triangle$s ABD and ADC with common side $\overline{AD}$. Show that $\overleftrightarrow{EG} \parallel \overleftrightarrow{BD}$ and $\overleftrightarrow{GF} \parallel \overleftrightarrow{AC}$. Then use 16.19.

Proof:

Statements	*Reasons*
1. Draw $\overline{AD}$ intersecting n at G. Through $\overline{AB}$ and $\overline{AD}$ pass a plane intersecting n in $\overleftrightarrow{EG}$ and r in $\overleftrightarrow{BD}$. Through $\overleftrightarrow{AD}$ and $\overleftrightarrow{CD}$ pass a plane intersecting n in $\overleftrightarrow{GF}$ and m in $\overleftrightarrow{AC}$.	1. Why possible?
2. $\overleftrightarrow{EG} \parallel \overleftrightarrow{BD}$. $\overleftrightarrow{FG} \parallel \overleftrightarrow{AC}$.	2. Why?
3. $m\overline{AE}:m\overline{EB} = m\overline{AG}:m\overline{GD}$ $m\overline{CF}:m\overline{FD} = m\overline{AG}:m\overline{GD}$	3. Why?
4. $\therefore m\overline{AE}:m\overline{EB} = m\overline{CF}:m\overline{FD}$	4. Why?

We have now established the following theorem.

16.22 THEOREM

Three parallel planes intercept proportional segments on two transversals.

Exercises

1. $m\overline{AD} = 3''$, $m\overline{DC} = 2''$, $m\overline{BE} = 6''$. Find $m\overline{EC}$.
2. $m\overline{AD} = 5''$, $m\overline{DC} = 4''$, $m\overline{BE} = 12''$. Find $m\overline{EC}$.
3. $m\overline{CD} = 3''$, $m\overline{AD} = 5''$, $m\overline{CE} = 9''$. Find $m\overline{BE}$.
4. $m\overline{AD} = \frac{4'}{3}$, $m\overline{DC} = \frac{1'}{2}$, $m\overline{BE} = 4'$. Find $m\overline{EC}$.
5. $m\overline{AD} = \frac{3'}{2}$, $m\overline{BE} = 5'$, $m\overline{EC} = \frac{4'}{9}$. Find $m\overline{DC}$.
6. $m\overline{AC} = 18''$, $m\overline{AD} = 5''$, $m\overline{BC} = 24''$. Find $m\overline{BE}$.
7. $m\overline{AC} = 22''$, $m\overline{CD} = 15''$, $m\overline{BC} = 33''$. Find $m\overline{BE}$.
8. $m\overline{AC} = 14$ cm, $m\overline{AD} = 6$ cm, $m\overline{BC} = 10$ cm. Find $m\overline{BE}$.
9. $m\overline{AC} = 8$ cm, $m\overline{BC} = 12$ cm, $m\overline{AD} = m\overline{CE}$. Find $m\overline{BE}$.
10. If $m\overline{BE}:m\overline{CE} = 3:4$ and $m\text{AC} = 9''$, find $m\overline{AD}$ and $m\overline{CD}$. Find $m\overline{CE}$ and $m\overline{BE}$.
11. Is $m\overline{AD}:m\overline{BE} = m\overline{CD}:m\overline{CE}$?

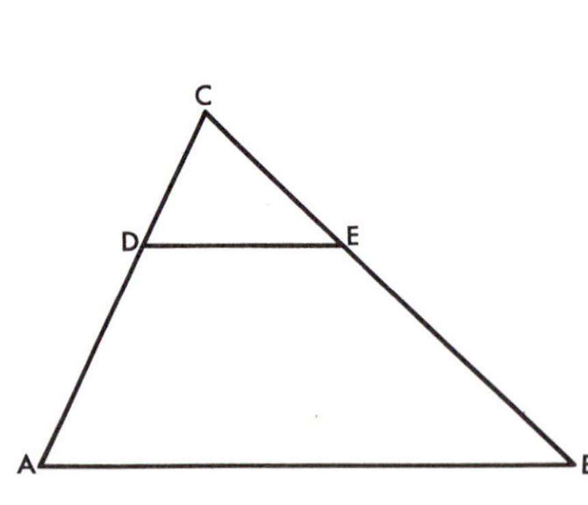

$\overline{DE} \parallel \overline{AB}$
(Exercises 1-11)

Figure 16–11

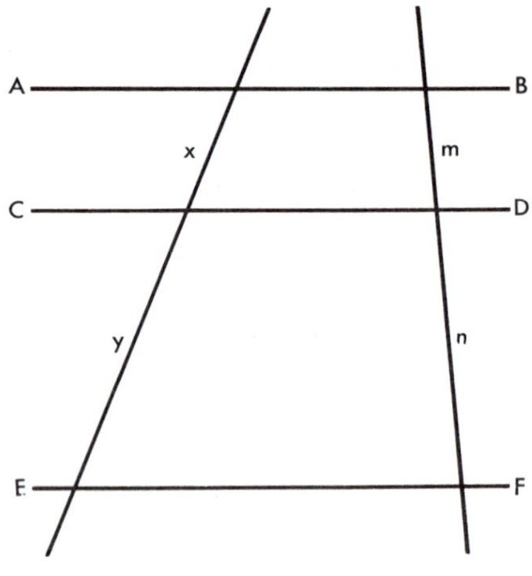

$\overleftrightarrow{AB} \parallel \overleftrightarrow{CD} \parallel \overleftrightarrow{EF}$
(Exercises 12-13)

Figure 16–12

12. $x = 4''$, $m = 6''$, $n = 8''$. Find y.

13. $x = 5''$, $y = 6''$, $m = 4''$. Find $m + n$.

14. Using the figure of 16.22 and the values $m\overline{AE} = 10'$, $m\overline{BE} = 7'$, $m\overline{CF} = 15'$, find $m\overline{FD}$.

Given: $\overleftrightarrow{DE}$ intersecting $\overline{AC}$ and $\overline{BC}$ in $\triangle ABC$ so that $m\overline{AD}:m\overline{CD} = m\overline{BE}:m\overline{CE}$

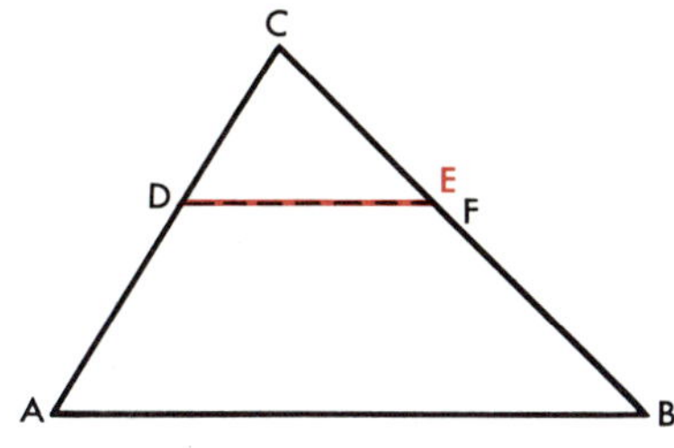

Figure 16–13

Conjecture: $\overleftrightarrow{DE} \parallel \overline{AB}$

Plan: Use indirect proof.

Proof:

Statements	*Reasons*
1. Assume $\overline{DE}$ is not parallel to $\overline{AB}$. Through D construct $\overline{DF} \parallel \overline{AB}$.	1. Why possible?
2. then $m\overline{AC}:m\overline{CD} = m\overline{BC}:m\overline{CF}$	2. Why?
3. but $m\overline{AD}:m\overline{CD} = m\overline{BE}:m\overline{CE}$	3. Why?
4. so $m\overline{AC}:m\overline{CD} = m\overline{BC}:m\overline{CE}$	4. Why?
5. $\therefore m\overline{CE} = m\overline{CF}$	5. Why?
6. So $E = F$	6. 2.38
7. $\therefore \overleftrightarrow{DE} \parallel \overline{AB}$	7. Why?

16.23 THEOREM

A line which divides two sides of a triangle proportionally is parallel to the third side.

Given: $\triangle ABC$ with $\overline{CD}$ bisecting $\angle C$

Prove: $m\overline{AC}:m\overline{BC} = m\overline{AD}:m\overline{DB}$

Plan: Construct a line through point A parallel to $\overline{DC}$, meeting $\overrightarrow{BC}$ at E. Show that $m\overline{BC}:m\overline{CE} = m\overline{BD}:m\overline{DA}$ and that $m\overline{CA} = m\overline{CE}$. Then substitute $m\overline{CA}$ for $m\overline{CE}$.

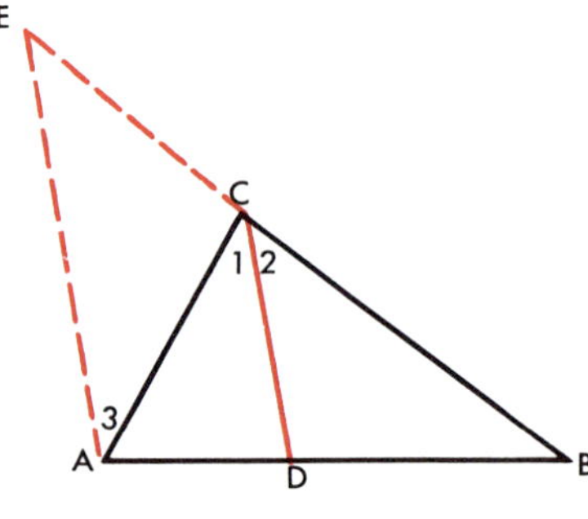

Figure 16–14

16.24 THEOREM

The bisector of an angle of a triangle divides the opposite side into segments which are proportional to the adjacent sides.

Exercises

Use the figure for 16.24.

1. $m\overline{AC} = 3''$, $m\overline{BC} = 4''$, $m\overline{AD} = 2''$. Find $m\overline{DB}$.
2. $m\overline{AC} = 5''$, $m\overline{AD} = 3''$, $m\overline{BC} = 9''$. Find $m\overline{DB}$.
3. $m\overline{AC} = 8''$, $m\overline{DB} = 6''$, $m\overline{CB} = 10''$. Find $m\overline{AD}$.
4. $m\overline{AC} = 16$ cm, $m\overline{BC} = 20$ cm, $m\overline{AB} = 32$ cm. Find $m\overline{AD}$ and $m\overline{DB}$.

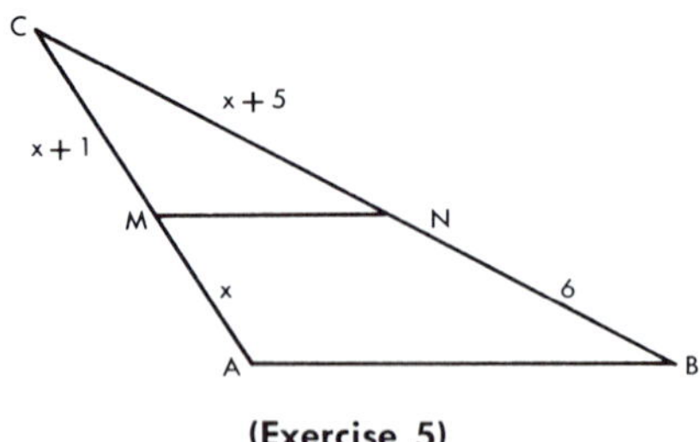

(Exercise 5)

Figure 16–15

5. Place conditions upon x such that $\overline{MN} \parallel \overline{AB}$ if $m\overline{CM} = x + 1$, $m\overline{MA} = x$, $m\overline{CN} = x + 5$, and $m\overline{BN} = 6$.

6. The measures of the sides of a triangle are 8, 12, 15 units. Find the measures of the segments into which the side opposite the largest angle is divided by the bisector of that angle.
7. If a plane contains the midpoints of the lateral edges of any pyramid prove that it also contains the midpoint of the altitude of the pyramid.
8. If a plot of ground is divided into three lots as pictured, what are the measurements of x, y, and z?

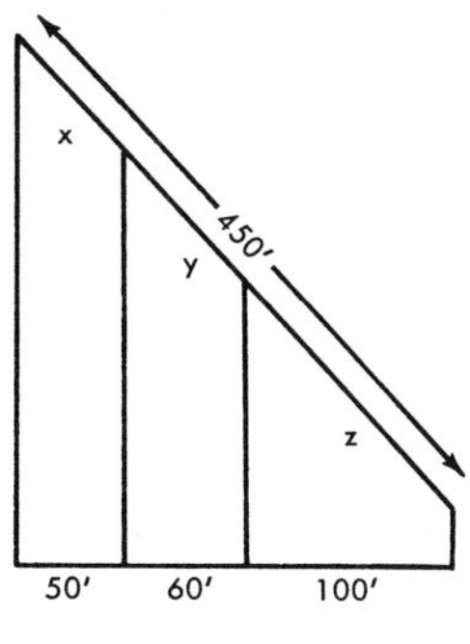

(Exercise 8)

Figure 16–16

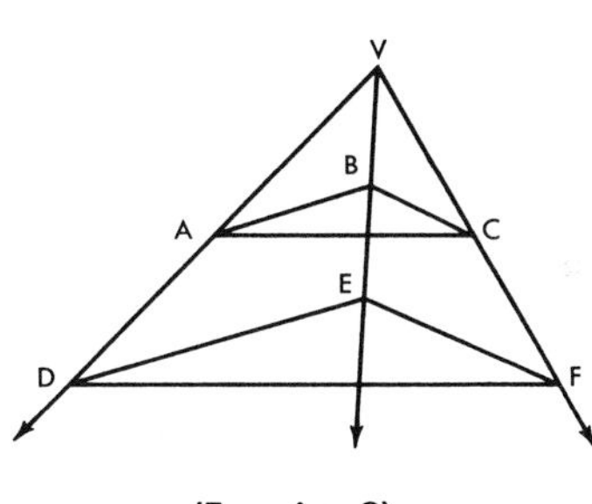

(Exercise 9)

Figure 16–17

9. Triangles ABC and DEF are drawn so that $\overline{AB} \parallel \overline{DE}$, $\overline{BC} \parallel \overline{EF}$, and $\overleftrightarrow{DA}$, $\overleftrightarrow{EB}$, and $\overleftrightarrow{FC}$ are concurrent in V. Prove $\overline{AC} \parallel \overline{DF}$.

10. Given $\triangle ABC$ with $\overline{CD}$ and $\overrightarrow{CD'}$ as the bisectors of the "internal" and "external" angles at C. Prove

$$\frac{m\overline{AD'}}{m\overline{D'B}} = \frac{m\overline{AD}}{m\overline{DB}}.$$ (*Hint*: Construct $\overline{BF} \parallel \overline{D'C}$. F is on $\overline{AC}$.)

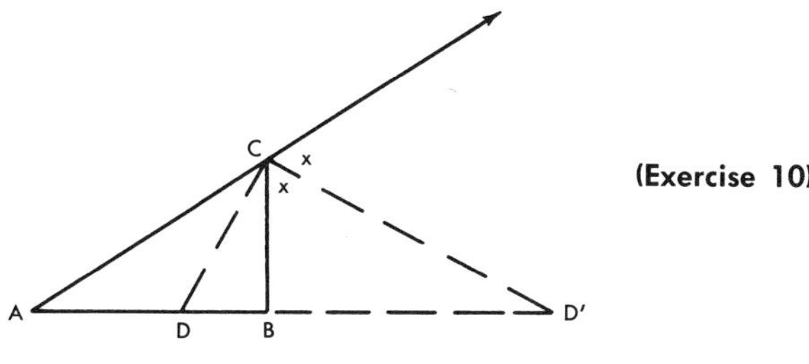

(Exercise 10)

Figure 16–18

Vocabulary List

ratio	radical
proportion	root
means	rationalize
extremes	proportional

Chapter Review

1. Express $7x = 3y$ as a proportion in which 7 and x are the extremes.
2. Name the ratio of x to y if $5x = 8y$.
3. Name the ratio of x to y if $\frac{2x + 3y}{x} = 7$.
4. Find the square root of $\frac{4}{7}$ to the nearest hundredth.
5. Leave the following in simplest radical form for quick computation. (a) $\sqrt{\frac{3}{7}}$ (b) $\sqrt{\frac{6}{30}}$ (c) $\sqrt{\frac{1}{12}}$ (d) $\sqrt{\frac{4}{5}}$
6. Name the mean proportional between 2 and 32.
7. Name the third proportional to 3 and 9.
8. Find the fourth proportional to 6, 15, and 8.

Use Figure 16-19 for Exercises 9–13. $\overline{AB} \parallel \overline{FC} \parallel \overline{ED}$.

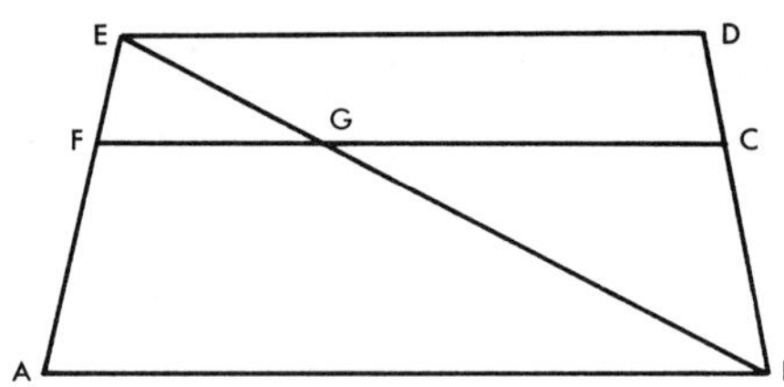

Figure 16–19

9. If $m\overline{AF} = 5$, $m\overline{FE} = 7$, and $m\overline{BC} = 6$, find $m\overline{CD}$.
10. If $m\overline{AE} = 10$, $m\overline{AF} = 8$, and $m\overline{BC} = 6$, find $m\overline{BD}$.
11. If $m\overline{AE} = 12$, $m\overline{FE} = 4$, and $m\overline{BD} = 8$, find $m\overline{BC}$.
12. If $m\overline{AB} = 20$, $m\overline{ED} = 14$ and $AF \cong FE$, find $m\overline{FC}$.

13. If $\overline{AF} \cong \overline{FE}$, and m$\overline{EG}$ = 5, find m$\overline{GB}$.

14. In $\triangle ABC$, m$\overline{AC}$ = 8, m$\overline{BC}$ = 12 and m$\overline{AB}$ = 10. $\overline{CD}$ bisects $\angle C$ and D is between A and B. Find m$\overline{AD}$ and m$\overline{BD}$.

15. If $\frac{m}{n} = \frac{x}{y} = \frac{r}{t}$ where m, n, x, y, r and t are non-zero real numbers, is $\frac{m + x + r}{n + y + t} = \frac{x}{y}$? Justify your response.

Chapter 16 Test

1. Name the ratio of 6 feet to 3 yards in its simplest form.

2. Name the replacements for x that will satisfy the following proportions:

   (a) $\frac{x}{4} = \frac{7}{9}$ (b) $\frac{3}{10} = \frac{7}{x}$

3. Use the proportion $\frac{X}{Y} = \frac{R}{T}$ to name the following:

   (a) the means.
   (b) the extremes.
   (c) an equivalent proportion using the Principle of Inversion.
   (d) an equivalent proportion using the Principle of Alternation.

4. Express $3b = 4a$ as a proportion in which 3 and b are the means.

5. Name the ratio of x to y if:

   (a) $5x = 15y$ (b) $\frac{x + y}{y} = 3$

6. Name the mean proportional between:

   (a) 4 and 36 (b) 18 and 3

7. Name the fourth proportional to 1, 2, and 3.

8. Name the third proportional to 1 and 2.

9. The measures of the nonparallel sides of a trapezoid are 12 and 15 units. A line parallel to a base of the trapezoid divides one of the diagonals of the trapezoid into segments which are in the ratios 2:3. Find the measures of the segments of the nonparallel sides determined by this parallel line.

10. If $\frac{6}{a} = \frac{5}{b} = \frac{3}{c} = \frac{2}{7}$, then $\frac{6 + 5 + 3}{a + b + c} = ?$. (Fig. 16-20)

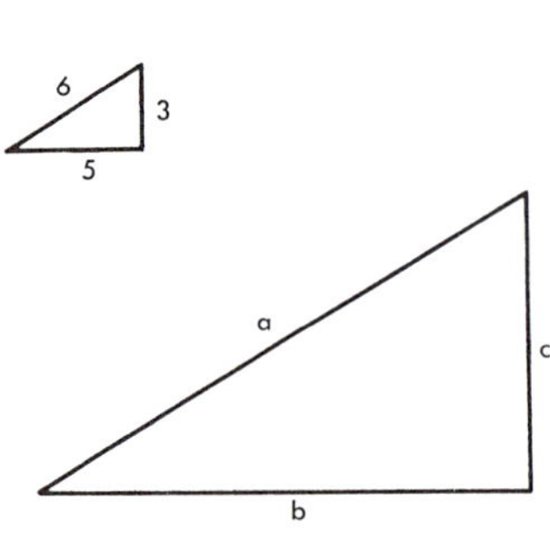

(Exercise 10)

Figure 16–20

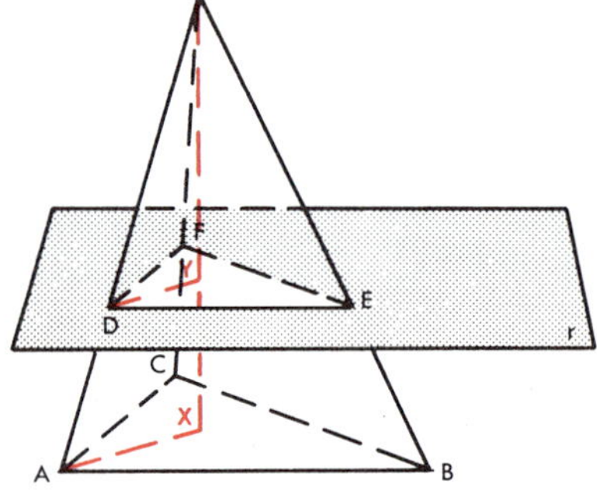

(Exercise 11)

Figure 16–21

11. In Fig. 16-21 plane r is parallel to the plane of $\triangle ABC$ and intersects the lateral edges of the pyramid at D, E, and F. The plane r intersects $\overline{VX}$, the altitude of the pyramid at Y. $m\overline{AV} =$ 10 inches, $m\overline{BV} =$ 12 inches, and $m\overline{BE} =$ 4 inches.

(a) Find $m\overline{DV}$. (b) If $m\overline{VX} =$ 8 inches, find $m\overline{VY}$.

12. The measures of the sides of a triangle are 6, 12, and 15 units. Find the measure of the segments of one side made by the bisector of the angle with the greatest measure.

17

Similar Polygons

17.00 Similar polygons are polygons which have their corresponding angles congruent and the measures of their corresponding sides proportional. The symbol $\sim$ replaces the words *similar to* or *is similar to.*

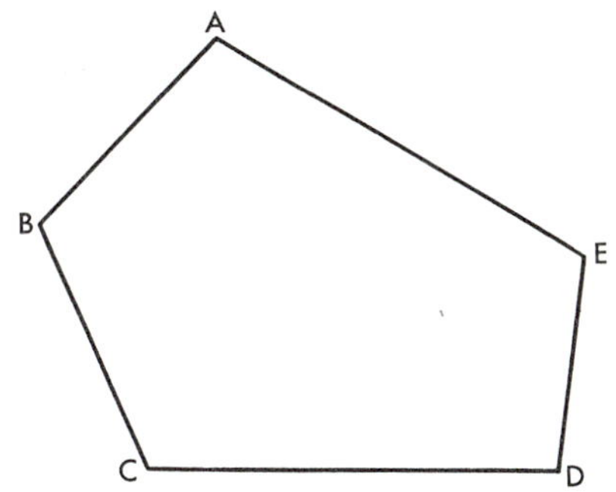

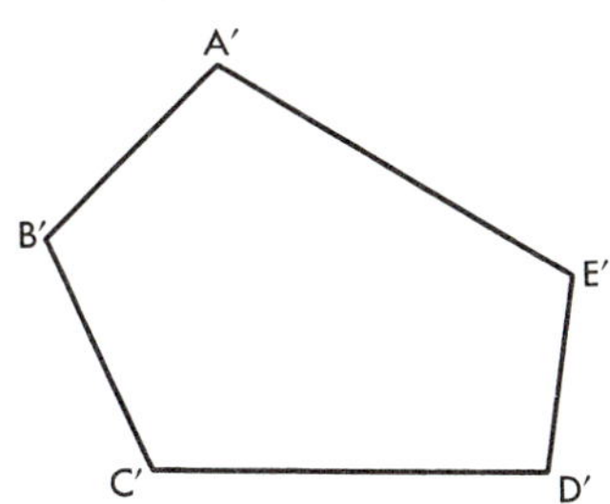

Figure 17–1

Figure $ABCDE \sim A'B'C'D'E'$ if $\angle A \cong \angle A'$, $\angle B \cong \angle B'$, $\angle C \cong \angle C'$, $\angle D \cong \angle D'$, $\angle E \cong \angle E'$, and $\frac{m\overline{AB}}{m\overline{A'B'}} = \frac{m\overline{BC}}{m\overline{B'C'}} = \frac{m\overline{CD}}{m\overline{C'D'}} = \frac{m\overline{DE}}{m\overline{D'E'}} = \frac{m\overline{EA}}{m\overline{E'A'}}$.

Two polygons may have the measures of their corresponding sides proportional and not be similar, as in Figures I and II, or they may have their corresponding angles congruent, as in Figures II and III, and not be similar.

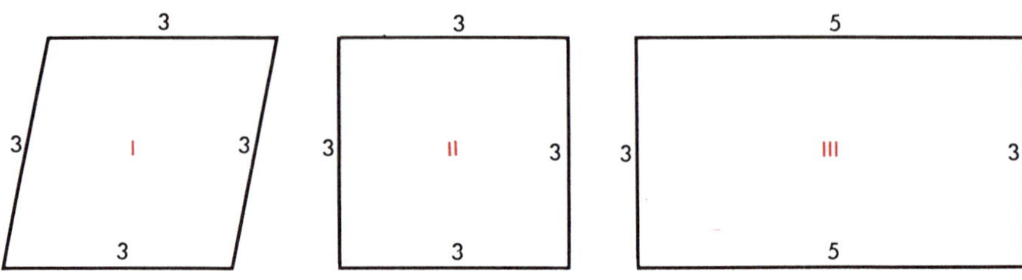

Figure 17–2

17.01 Since a definition is reversible:

(1) Corresponding angles of similar polygons are congruent.
(2) The measures of the corresponding sides of similar polygons are in proportion.

Exercises

1. Must similar polygons have the same number of sides?
2. Are similar triangles congruent? Are congruent triangles similar?
3. Are all right triangles similar? If a figure is similar to a right triangle, is it a right triangle?
4. Are all squares similar? Are all rectangles similar?
5. Are all regular pentagons similar? Are all regular polygons with the same number of sides similar?
6. If two polygons are similar, must they have the same shape?
7. The measures of the sides of a triangle are 3, 4, and 5 inches. Find the measures of the sides of a similar triangle if the measure of the shortest side is 6 inches.
8. The measures of the sides of a triangle are 7, 9, and 10 inches. Find the measures of the sides of a similar triangle if the measure of the longest side is 10 centimeters.

9. The measures of the sides of a pentagon are 4, 5, 6, 8, and 12 inches. If the length of the longest side of a similar polygon is 18, find the lengths of the other sides.
10. State the conditions which must be met before polygons are similar.
11. Prove that all equilateral triangles are similar polygons.
12. Does *similarity of polygons* have reflexive, symmetric, and transitive properties? Explain.

SIMILAR TRIANGLES

Find the lengths of the sides of the triangles below to see if the measures of the corresponding sides form equal ratios. Are the triangles then similar? Examine other sets of triangles whose corresponding angles are congruent, to see if they are similar.

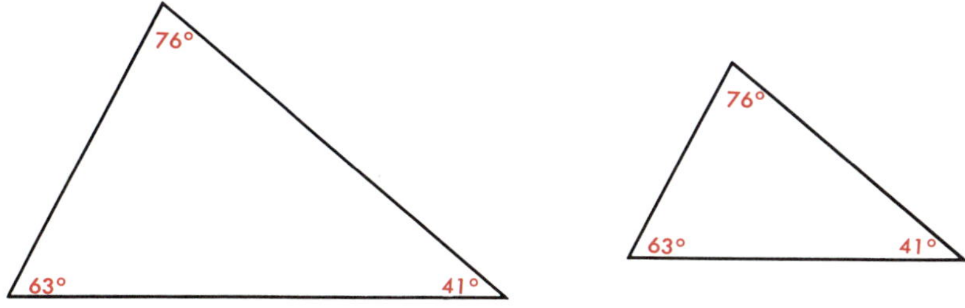

Figure 17–3

Find the measures of the angles of the triangles below, to see if the corresponding angles are congruent. Are the triangles then similar? Repeat the experiment with triangles of your own design.

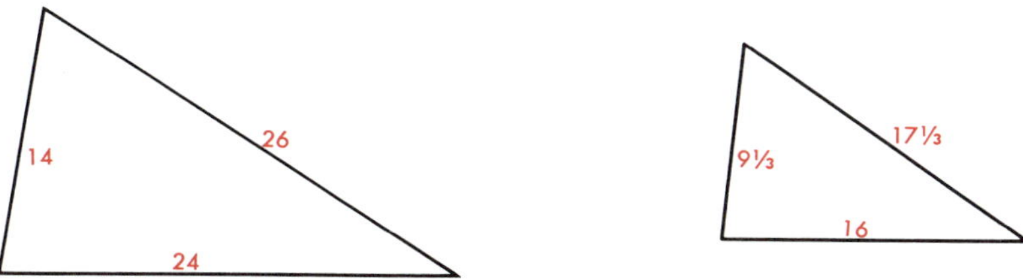

Figure 17–4

Find the measures of the unlabeled sides and angles of the triangles below and determine whether the triangles are similar. Draw other triangles using similar conditions and repeat the experiment.

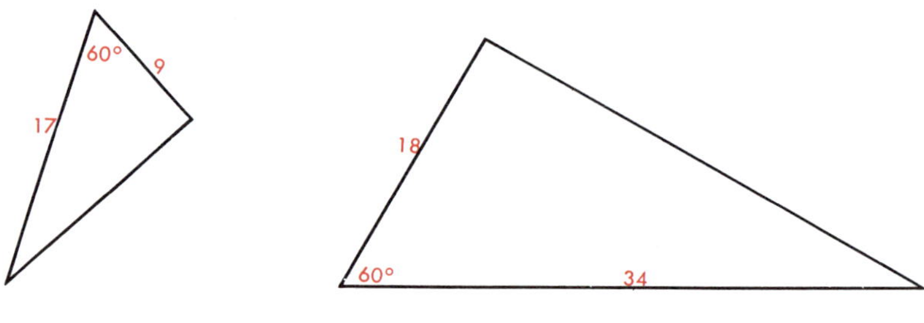

Figure 17–5

Upon the basis of the preceding experimentation complete the following statements listing your tentative conclusion or conjectures.

Two triangles are similar if ____?____ of one triangle are congruent respectively to ____?____ of the other triangle. How many different ways can the blanks be correctly filled?

We will now test our conjectures by means of deductive reasoning.

Given: $\triangle ABC$ and $\triangle DEF$ with $\angle A \cong \angle D$ and $\angle B \cong \angle E$

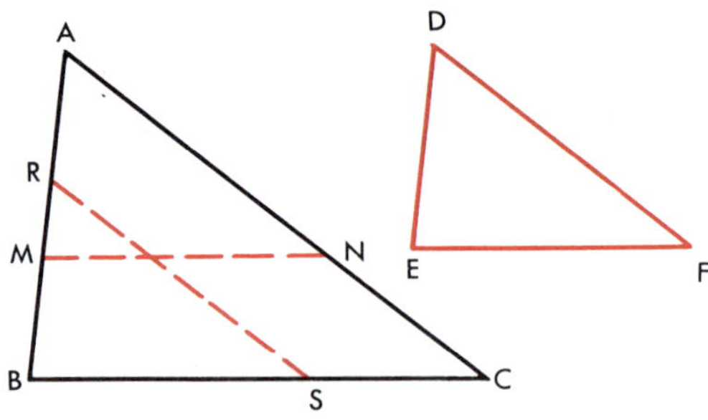

Figure 17–6

Conjecture: $\triangle ABC \sim \triangle DEF$

Plan: To show the conditions of 17.00 are satisfied.

Proof: *Statements*	*Reasons*
1. Construct $\overline{AM} \cong \overline{DE}$, $\overline{AN} \cong \overline{DF}$; draw $\overline{MN}$.	1. 2.38 and 2.10
2. $\angle A \cong \angle D$	2. Given
3. $\triangle AMN \cong \triangle DEF$	3. *s.a.s.*
4. $\angle AMN \cong \angle E$	4. Why?
5. $\angle B \cong \angle E$	5. Given
6. $\angle AMN \cong \angle B$	6. Why?
7. $\overline{MN} \parallel \overline{BC}$	7. Why?
8. $m\overline{AB}:m\overline{AM} = m\overline{AC}:m\overline{AN}$ or $m\overline{AB}:m\overline{DE} = m\overline{AC}:m\overline{DF}$	8. 16.20 and 5.07 using step 1
9. By constructing $\overline{BR} \cong \overline{DE}$ and $\overline{BS} \cong \overline{EF}$ it can be proved that $m\overline{AB}:m\overline{DE} = m\overline{BC}:m\overline{EF}$	9. Steps 1–8
10. $\therefore \triangle ABC \sim \triangle DEF$	10. Steps 2, 5, 8, 9; Sect. 6.20, 17.00

17.02 THEOREM

If two angles of one triangle are congruent respectively to two angles of a second triangle, the two triangles are similar.

17.03 THEOREM

The measures of the corresponding altitudes and the corresponding sides of two similar triangles have the same ratio.

Since the measures of the corresponding sides of similar polygons form equal ratios, use 16.11 to prove the following theorem.

17.04 THEOREM

The perimeters of two similar polygons have the same ratio as the measures of any two corresponding sides.

It can be shown that the ratios of the measures of any two pairs of corresponding linear parts of similar polygons are equal.

Exercises

1. Angle B is a right angle and $\overline{DE} \perp \overline{AC}$. Prove $\triangle ABC \sim \triangle AED$.

2. Prove that the line joining the midpoints of two sides of a triangle cuts off a triangle similar to the given triangle.

3. A tower casts a shadow 125 ft long when a 10 ft vertical pole casts a 4 ft shadow. How high is the tower?

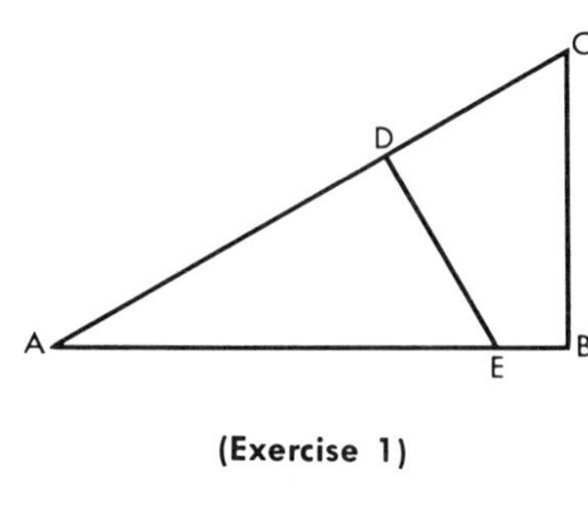

(Exercise 1)

Figure 17–7

Use Figure 17-8 for Exercises 4–6.

4. $\overline{DC} \parallel \overline{AB}$. Prove $\triangle DOC \sim \triangle AOB$.

5. Three holes, E, A, and B, have been made in the bases of a cylinder with a slit between A and B. Explain how you could measure the distance from E to D by measuring $\overline{EB}$, $\overline{AB}$, and $\overline{CD}$. Assume $\overline{AB} \parallel \overline{CD}$.

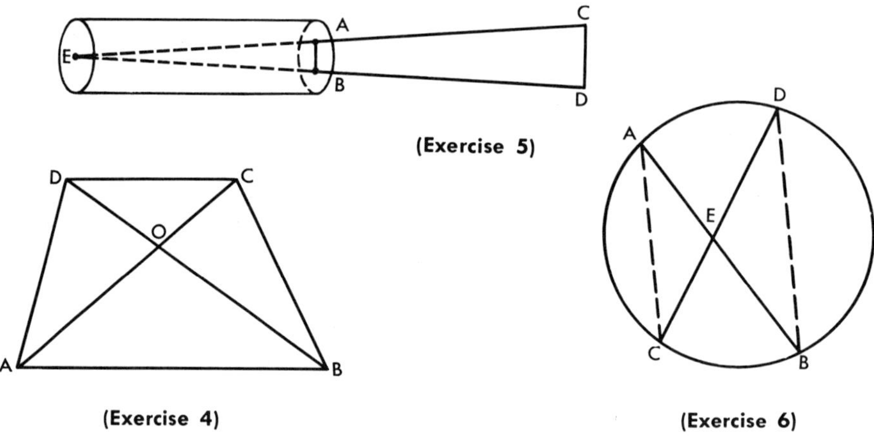

(Exercise 5)

(Exercise 4)

(Exercise 6)

Figure 17–8

6. Two chords of a circle, $\overline{AB}$ and $\overline{CD}$, intersect at E. Prove $\triangle ACE \sim \triangle BDE$. State the three equal ratios formed by the measures of the corresponding sides.

Use Figure 17-9 for Exercises 7–9.

7. Line $\overleftrightarrow{ABC}$ is a secant through the center of the circle. $\overline{DBE}$ is a secant drawn so $\overline{EC} \perp \overline{AC}$. Point B is on the circle. Prove $\triangle ABD \sim \triangle ECB$.

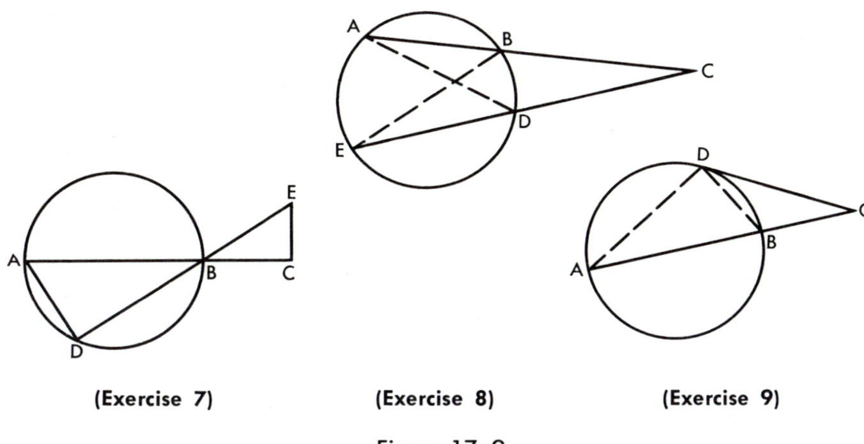

Figure 17–9

8. $\overline{ABC}$ and $\overline{EDC}$ are secants. Prove $\triangle ACD \sim \triangle ECB$. State the three equal ratios formed by the measures of the corresponding sides.

9. $\overline{DC}$ is a tangent, $\overline{ABC}$ a secant. Prove $\triangle ACD \sim \triangle DCB$. State the three equal ratios formed by the measures of the corresponding sides.

10. Prove: two right triangles are similar if a side and the hypotenuse of one are respectively perpendicular to a side and the hypotenuse of the other. (Figure 17-10)

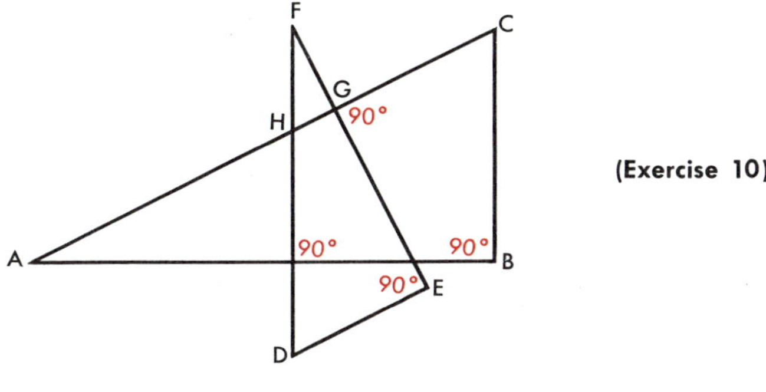

Figure 17–10

11. In $\triangle ABC$, D and E are on $\overline{AC}$ and $\overline{BC}$ respectively, so $\overline{DE} \parallel \overline{AB}$. Prove $m\overline{AC}:m\overline{DC} = m\overline{AB}:m\overline{DE}$.

12. Two circles are internally tangent at T with the radius of the larger congruent to the diameter of the smaller. Prove that any chord of the larger circle drawn from T is bisected by the smaller circle.

17.05 THEOREM

If two chords intersect in the interior of a circle, the product of the measures of the segments of one is equal to the product of the measures of the segments of the other.

Hint: See Figure 17-11. Use Exercise 6, from the preceding set, to show that $\dfrac{m\overline{AE}}{m\overline{ED}} = \dfrac{m\overline{CE}}{m\overline{EB}}$; then $m\overline{AE} \cdot m\overline{EB} = m\overline{CE} \cdot m\overline{ED}$ by 16.04.

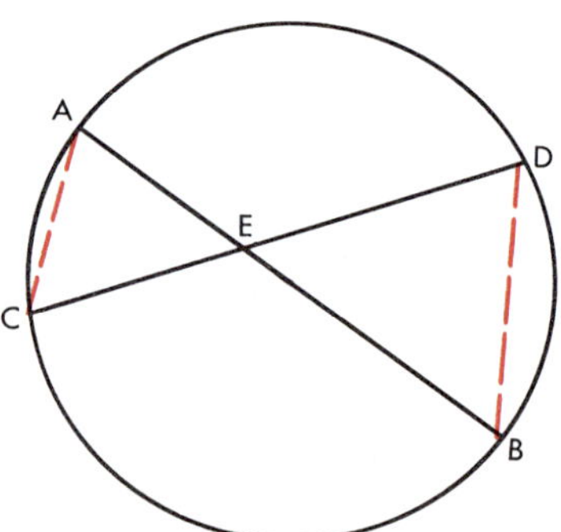

Figure 17–11

Exercises

Use Figure 17-11.

1. If $m\overline{AE} = 2$, $m\overline{CE} = 3$, and $m\overline{ED} = 4$ inches, find $m\overline{EB}$.

2. If $m\overline{CE} = 5$, $m\overline{ED} = 4$, and $m\overline{EB} = 10$ inches, find $m\overline{AE}$.

3. If $m\overline{CE} = 2\frac{1}{2}$, $m\overline{ED} = 3\frac{1}{3}$, and $m\overline{EB} = 6$ inches, find $m\overline{AE}$.

4. If $m\overline{DE} = 4$ inches, $m\overline{CE} = 3$ inches, and $\overline{AE} \cong \overline{EB}$, find $m\overline{AE}$.

5. If $m\overline{BE} = \frac{3}{5}$ inches, $m\overline{AE} = \frac{2}{3}$ inches, and $\overline{CE} \cong \overline{ED}$, find $m\overline{ED}$. (Leave the answer in radical form for quick computation.)

17.06 THEOREM

If from a point outside a circle two secant segments are drawn, the product of the measure of one secant and its external segment is equal to the product of the measures of the other secant and its external segment.

Hint: Use Exercise 8, page 427, to show that $\frac{m\overline{AC}}{m\overline{EC}} = \frac{m\overline{DC}}{m\overline{BC}}$; then $m\overline{AC} \cdot m\overline{BC} = m\overline{EC} \cdot m\overline{DC}$ by 16.04.

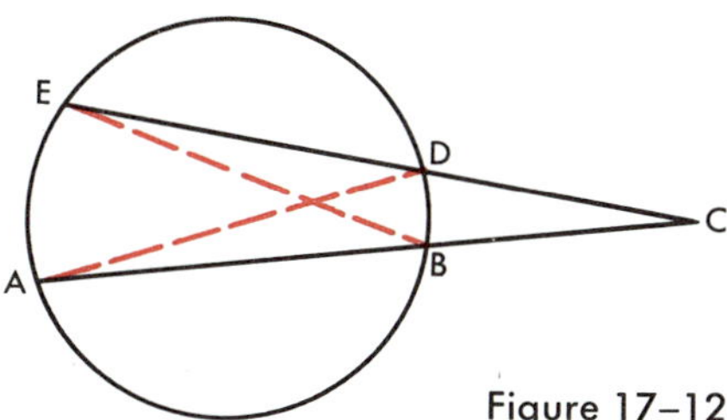

Figure 17–12

Exercises

Use Figure 17-12.

1. If $m\overline{EC} = 9$, $m\overline{DC} = 3$, and $m\overline{AC} = 12$ inches, find $m\overline{BC}$.
2. If $m\overline{EC} = 9$, $m\overline{DC} = 4$, and $m\overline{BC} = 2$ inches, find $m\overline{AC}$.
3. If $m\overline{AB} = 6$, $m\overline{BC} = 4$, and $m\overline{DC} = 3$ inches, find $m\overline{EC}$.
4. If $m\overline{AB} = 7$, $m\overline{BC} = 15$, and $m\overline{DC} = 11$ inches, find $m\overline{ED}$.
5. If $m\overline{ED} = 2\frac{2}{3}$, $m\overline{DC} = 2\frac{1}{2}$, and $m\overline{BC} = 1\frac{3}{4}$ inches, find $m\overline{AB}$.

17.07 THEOREM

If from a point outside a circle a secant segment and a tangent segment are drawn, the product of the measures of the secant and its external segment is equal to the square of the measure of the tangent.

Hint: Use Exercise 9, page 427, to show that $\frac{m\overline{DC}}{m\overline{AC}} = \frac{m\overline{BC}}{m\overline{DC}}$; then $(m\overline{DC})^2 = m\overline{AC} \cdot m\overline{BC}$ by 16.04.

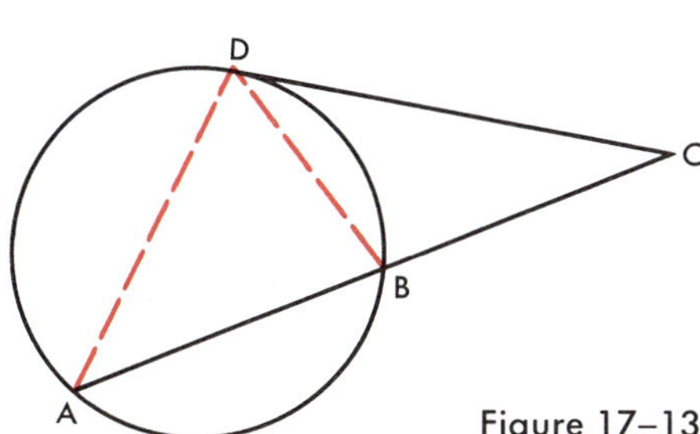

Figure 17–13

Exercises

Use Figure 17-13.

1. If $m\overline{AC} = 9$ inches and $m\overline{BC} = 4$ inches, find $m\overline{DC}$.
2. If $m\overline{AC} = 20$ inches and $m\overline{BC} = 5$ inches, find $m\overline{DC}$.
3. If $m\overline{AB} = 12$ inches and $m\overline{BC} = 4$ inches, find $m\overline{DC}$.
4. If $m\overline{AB} = 15$ inches and $m\overline{BC} = 5$ inches, find $m\overline{DC}$.
5. If $m\overline{AB} = 10$ feet and $m\overline{BC} = 5$ feet, find $m\overline{DC}$.
6. If $m\overline{AC} = 9$ feet and $m\overline{AB} = 4$ feet, find $m\overline{DC}$.
7. If $m\overline{AC} = \frac{5}{4}$ feet and $m\overline{BC} = \frac{2}{3}$ feet, find $m\overline{DC}$. (Leave answer in simplest radical form.)

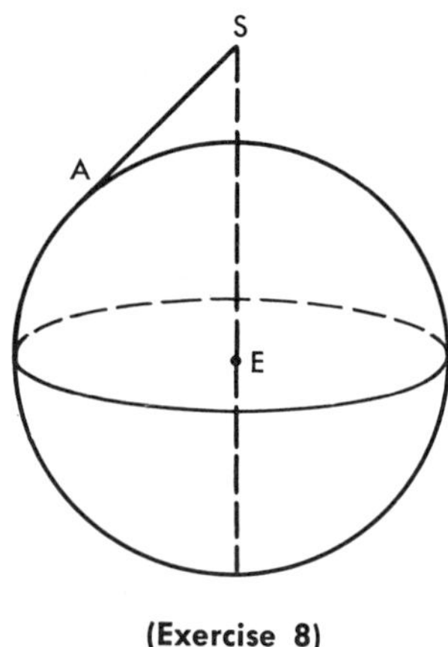

(Exercise 8)

Figure 17–14

8. If a space ship (S) is 50 miles above the earth (E), what is the length of its longest line of sight ($\overline{SA}$) to the earth? (Consider the diameter of the earth to be 8000 mi.)
9. A television tower on the plains of the midwest is 300 ft high. An observer on the ground with a powerful telescope can see only the top of the tower above the curvature of the earth. How far is he from the top of the tower?
10. Assuming that a television signal can be transmitted only in a straight line, from what maximum distance can a television antenna 100 feet high receive signals from the transmitter described in Exercise 9?

Given: $\triangle ABC$ and $\triangle DEF$, with $\angle A \cong \angle D$ and $\dfrac{m\overline{AB}}{m\overline{DE}} = \dfrac{m\overline{AC}}{m\overline{DF}}$

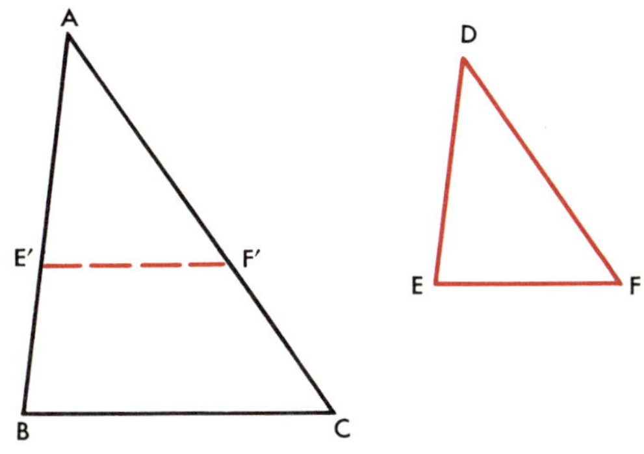

Figure 17–15

Conjecture: $\triangle ABC \sim \triangle DEF$

Plan: Use 16.23 and 17.02.

Proof:

Statements	*Reasons*
1. Construct $\overline{AE'} \cong \overline{DE}$ $\overline{AF'} \cong \overline{DF}$; draw $\overline{E'F'}$	1. 2.38 and 2.10
2. $m\overline{AB}:m\overline{DE} = m\overline{AC}:m\overline{DF}$	2. Given
3. $m\overline{AB}:m\overline{AE'} = m\overline{AC}:m\overline{AF'}$	3. Substitution Axiom
4. $\overline{E'F'} \parallel \overline{BC}$	4. 16.23
5. $\angle B \cong \angle AE'F'$	5. 5.24
6. $\angle A \cong \angle D$	6. Given
7. $\triangle AE'F' \cong \triangle DEF$	7. *s.a.s.*
8. $\angle AE'F' \cong \angle E$	8. *c.p.c.t.c.*
9. $\angle B \cong \angle E$	9. Substitution Axiom
10. $\triangle ABC \sim \triangle DEF$	10. 17.02

17.08 THEOREM

Two triangles are similar if an angle of one is congruent to an angle of the other, and the measures of the sides including these angles are proportional.

Given: $\triangle ABC$ and $\triangle DEF$ with $\dfrac{m\overline{AB}}{m\overline{DE}} = \dfrac{m\overline{BC}}{m\overline{EF}} = \dfrac{m\overline{CA}}{m\overline{FD}}$

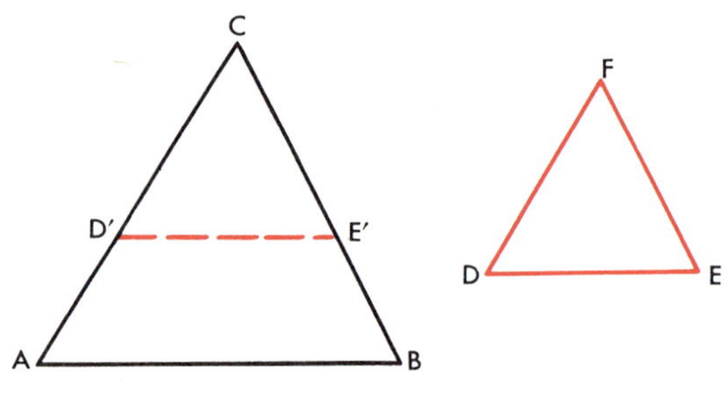

Figure 17–16

Conjecture: $\triangle ABC \sim \triangle DEF$

Plan: Make $\overline{CD'} \cong \overline{FD}$ and $\overline{CE'} \cong \overline{FE}$ and prove that $\triangle D'E'C \sim \triangle ABC$; then prove that $\triangle DEF \cong \triangle D'E'C$.

Proof: *Statements*	*Reasons*
1. $\overline{CD'} \cong \overline{FD}$ and $\overline{CE'} \cong \overline{FE}$. Draw $\overline{D'E'}$.	1. 2.38 and 2.10
2. $\dfrac{m\overline{CA}}{m\overline{FD}} = \dfrac{m\overline{BC}}{m\overline{EF}}$ or $\dfrac{m\overline{CA}}{m\overline{CD'}} = \dfrac{m\overline{BC}}{m\overline{E'C}}$	2. Given and Substitution Axiom
3. In $\triangle ABC$ and $\triangle D'E'C$, $\angle C \cong \angle C$	3. Reflexive Axiom
4. $\triangle D'E'C \sim \triangle ABC$	4. 17.08
5. $\dfrac{m\overline{CA}}{m\overline{CD'}} = \dfrac{m\overline{AB}}{m\overline{D'E'}}$	5. 17.01
6. $\dfrac{m\overline{CA}}{m\overline{FD}} = \dfrac{m\overline{AB}}{m\overline{DE}}$	6. Given
7. $\overline{D'E'} \cong \overline{DE}$	7. 16.07, Steps 1, 5, 6
8. $\triangle D'E'C \cong \triangle DEF$	8. Why?
9. $\triangle ABC \sim \triangle DEF$	9. Transitive Axiom

17.09 THEOREM

Two triangles are similar if the measures of their corresponding sides are proportional.

Exercises

1. $\overline{AB} \parallel \overline{DE}$. Prove $m\overline{AB}:m\overline{DE} = m\overline{AC}:m\overline{CD}$. (Figure 17-17)

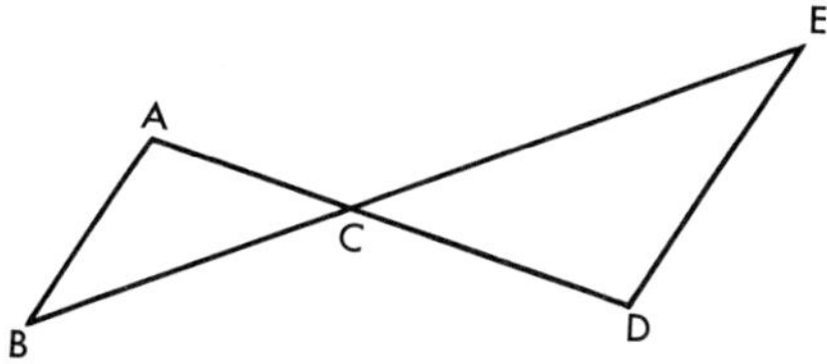

Figure 17–17

2. In the acute triangle CDE, the altitudes $\overline{CG}$ and $\overline{DH}$ meet at F. Prove that the following proportions are true:
   (a) $m\overline{DH}:m\overline{CG} = m\overline{EH}:m\overline{EG}$ (d) $m\overline{ED}:m\overline{DH} = m\overline{DF}:m\overline{DG}$
   (b) $m\overline{ED}:m\overline{DH} = m\overline{CE}:m\overline{CG}$ (e) $m\overline{CF}:m\overline{CH} = m\overline{DE}:m\overline{DH}$
   (c) $m\overline{CF}:m\overline{DF} = m\overline{HF}:m\overline{GF}$ (f) $m\overline{EH}:m\overline{HF} = m\overline{DH}:m\overline{CH}$

3. Prove that two right triangles are similar if the measures of the sides adjacent to the right angles are proportional.

4. Given: $\triangle ABC$ with $m\angle C = 90°$. $\overline{CD}$ is an altitude. Prove: (a) $m\overline{AD}:m\overline{CD} = m\overline{CD}:m\overline{DB}$. (b) $m\overline{AB}:m\overline{AC} = m\overline{AC}:m\overline{AD}$. (c) $m\overline{AB}:m\overline{CB} = m\overline{CB}:m\overline{DB}$. Does $m\overline{CD}^2 = m\overline{AD} \cdot m\overline{DB}$? Why? (Figure 17-18 left)

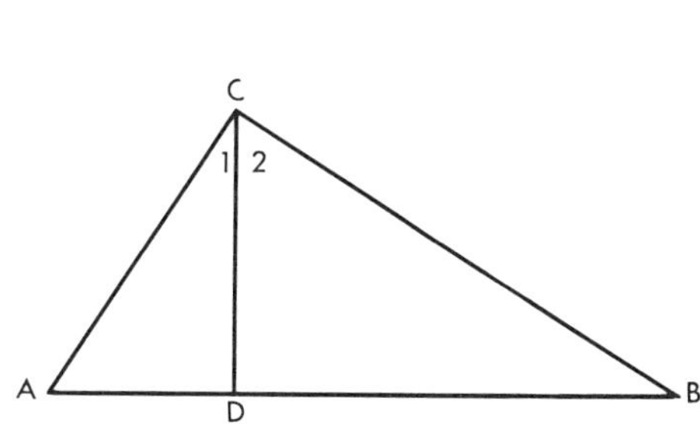

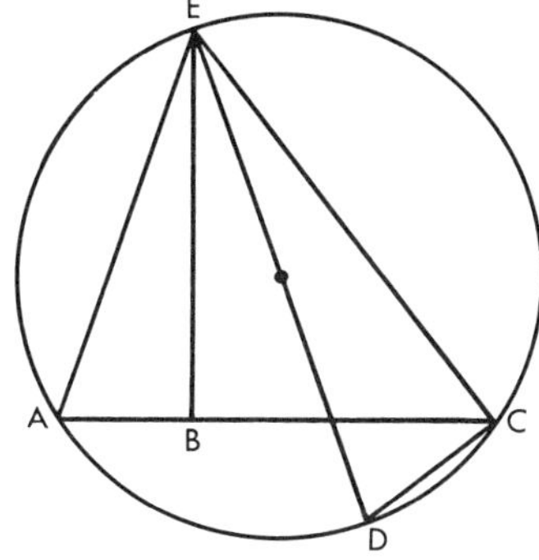

Figure 17–18

5. Triangle ACE is inscribed in a circle, $\overline{DE}$ is a diameter, and $\overline{EB} \perp \overline{AC}$. Prove $m\overline{AE} \cdot m\overline{EC} = m\overline{EB} \cdot m\overline{ED}$. (*Suggestion*: First prove that $m\overline{AE}:m\overline{ED} = m\overline{EB}:m\overline{EC}$.) (Figure 17-18 right)

6. Prove that the measures of the corresponding angle bisectors of two similar triangles have the same ratio as the measures of any corresponding sides.
7. Prove that the measures of the corresponding medians of two similar triangles have the same ratio as the measures of any two corresponding sides.

QUADRATIC EQUATIONS

17.10 A quadratic equation in x is an equation of the form

$$ax^2 + bx + c = 0,$$

where a, b, and c are real numbers and $a \neq 0$.

This section discusses quadratic equations where $b = 0$. In this case, the quadratic may be written in the equivalent form $x^2 = \frac{-c}{a}$, where $\frac{-c}{a} \geq 0$. There are exactly two real numbers whose squares are $\frac{-c}{a}$, so we can write $x = \sqrt{\frac{-c}{a}}$ and $x = -\sqrt{\frac{-c}{a}}$.

For example, if $x^2 = 9$, then the solution set is $\{3, -3\}$. If $2x^2 = 50$, $x^2 = 25$ and the solution set is $\{5, -5\}$. If the right hand member is not the square of an integer we will leave the answer in *simplest radical form.*

EXAMPLE: $7x^2 - 9 = 2x^2 + 21$

Step 1	$5x^2 = 30$	*Step 1*	Subtract $2x^2$ from and add 9 to each member.
Step 2	$x^2 = 6$	*Step 2*	Divide by 5.
Step 3	$x = \pm\sqrt{6}$	*Step 3*	Find the square roots of 6.

The solution set is $\{\sqrt{6}, -\sqrt{6}\}$.

Simplest Radical Form

A radicand may be simplified if any of its factors are squares of integers. (Review 16.14.)

EXAMPLES:

1. $\sqrt{12} = \sqrt{4 \cdot 3} = \sqrt{4}\sqrt{3} = 2\sqrt{3}$
2. $\sqrt{50} = \sqrt{25}\sqrt{2} = 5\sqrt{2}$
3. $\sqrt{27} = \sqrt{9}\sqrt{3} = 3\sqrt{3}$

Exercises

Solve for x where x is a real number. When the square roots cannot be found as integers or common fractions, leave the answers in simplest radical form.

1. $x^2 = 144$
2. $x^2 = 156$
3. $3x^2 = 75$
4. $x^2 + 7 = 34$
5. $5x^2 + 2 = 3x^2 + 6$
6. $2x^2 - 9 = x^2 + 45$
7. $100 = 2x^2 + 3x^2$
8. $4x^2 = x^2 + 1$
9. $3x^2 - 5 = 10$
10. $5x^2 = 3x^2 + 96$
11. $4x^2 = x^2 + 25$
12. $3x^2 = 54$

17.11 Solving Quadratic Equations of the form $ax^2 + bx + c = 0$

I. *Solve by factoring when possible.*

1. $x^2 + 4x = 12$
$x^2 + 4x - 12 = 0$
$(x - 2)(x + 6) = 0$
If $x - 2 = 0$ and $x + 6 = 0$
then $x = 2$ and -6

2. $2x^2 - 5x = 3$
$2x^2 - 5x - 3 = 0$
$(2x + 1)(x - 3) = 0$
If $2x + 1 = 0$ and $x - 3 = 0$
then $x = -\frac{1}{2}$ and 3

To solve a quadratic equation by factoring, write an equivalent equation in which one member is zero. Factor the other member, set each factor equal to zero, and solve. This is based upon the principle that if $ab = 0$, then $a = 0$ or $b = 0$ or both.

II. *Completing the square.*

$$3x^2 + 5x + 1 = 0$$

An equation such as the one above is not factorable over the set of integers, so we must develop another method of solution. In this method we write an equivalent equation such that one member is the square of a binomial, and then solve by means of previously established principles of algebra.

Solution:

1. Write an equivalent equation in which all terms involving the variable are in one member and all other terms in the other member.

$$3x^2 + 5x = -1$$

2. Divide both members of this equation by the coefficient of x^2 (if other than 1).

$$x^2 + \frac{5x}{3} = -\frac{1}{3}$$

3. Now add some value to both members that will make $x^2 + \frac{5x}{3} + \underline{?}$ the square of a binomial. (Do you see that this value will always be the square of $\frac{1}{2}$ the coefficient of x in step 2?)

$$x^2 + \frac{5x}{3} + \frac{25}{36} = \frac{25}{36} - \frac{1}{3} \qquad \text{or} \qquad \left(x + \frac{5}{6}\right)^2 = \frac{13}{36}$$

4. Find the square root of both members of the equation and solve.

$$\sqrt{\left(x + \frac{5}{6}\right)^2} = \pm\sqrt{\frac{13}{36}}$$

$$x + \frac{5}{6} = \pm\frac{\sqrt{13}}{6}$$

$$x = \frac{-5 \pm \sqrt{13}}{6}$$

$$x = -1.434 \text{ and } -0.232$$

EXAMPLES:

(a) Solve $2x^2 - 5x - 3 = 0$

Step 1 $2x^2 - 5x = 3$

Step 2 $x^2 - \frac{5x}{2} = \frac{3}{2}$

Step 3 $x^2 - \frac{5x}{2} + \frac{25}{16} = \frac{3}{2} + \frac{25}{16}$

$$\left(x - \frac{5}{4}\right)^2 = \frac{49}{16}$$

Step 4 $x - \frac{5}{4} = \pm\frac{7}{4}$

$$x = 3 \text{ and } -\frac{1}{2}$$

(b) Solve $x^2 + 3x - 5 = 0$

Step 1 $x^2 + 3x = 5$

Step 3 $x^2 + 3x + \frac{9}{4} = 5 + \frac{9}{4}$

$$\left(x + \frac{3}{2}\right)^2 = \frac{29}{4}$$

Step 4 $x + \frac{3}{2} = \pm\frac{\sqrt{29}}{2}$

$$x = \frac{-3 \pm \sqrt{29}}{2}$$

$$x = -4.192 \text{ and } +1.192$$

Exercises

Solve by factoring or by completing the square.

1. $x^2 + 5x = 14$
2. $x^2 + 3x = 10$
3. $x^2 + 6x = -8$
4. $2x^2 - 3x = -1$
5. $x(8 - x) = -33$
6. $x^2 = 2x + 63$
7. $6x^2 + 5x - 6 = 0$
8. $x^2 + 2x = 5$
9. $2x^2 + x = 1$
10. $x^2 - 3x - 9 = 0$
11. $2x^2 - x - 5 = 0$
12. $x(x - 2) - 4 = 0$
13. $x^2 - 2x - 15 = 0$
14. $y^2 - 4y + 4 = 0$
15. $6y^2 - 8y = 30$
16. $6x^2 + 13x + 6 = 0$
17. $8y^2 = 15 - 2y$
18. $y + 3 = 2y^2$
19. $10x^2 - 11x = 6$
20. $a^2 + 12a + 27 = 0$
21. $z^2 + 4z + 2 = 0$
22. $2t^2 = 9t - 3$
23. $x^2 - 4x - 12 = 0$
24. $3m^2 + 6m + 1 = 0$
25. $3a^2 - 13a + 4 = 0$
26. $y^2 - 4y - 12 = 0$
27. $x^2 + 6x + 8 = 0$
28. $y^2 - 6y + 2 = 0$
29. $a^2 = 5a + 14$
30. $2b(b - 2) = 1$

17.12 THEOREM

In any right triangle the measure of the altitude to the hypotenuse is the mean proportional between the measures of the segments of the hypotenuse.

Plan: Use Exercise 4a, page 433, to show $m\overline{CD}^2 = m\overline{AD} \cdot m\overline{DB}$ by 16.04.

17.13 THEOREM

If the altitude upon the hypotenuse of a right triangle is drawn, the measure of either leg is the mean proportional between the measures of the whole hypotenuse and the segment of the hypotenuse adjacent to the leg.

Plan: Use Exercises 4b and 4c, page 433, to show $m\overline{AC}^2 = m\overline{AD} \cdot m\overline{AB}$ and $m\overline{BC}^2 = m\overline{DB} \cdot m\overline{AB}$.

Exercises

1. In $\triangle DEF$, $\angle F$ is a right angle and $\overline{FG} \perp \overline{DE}$.

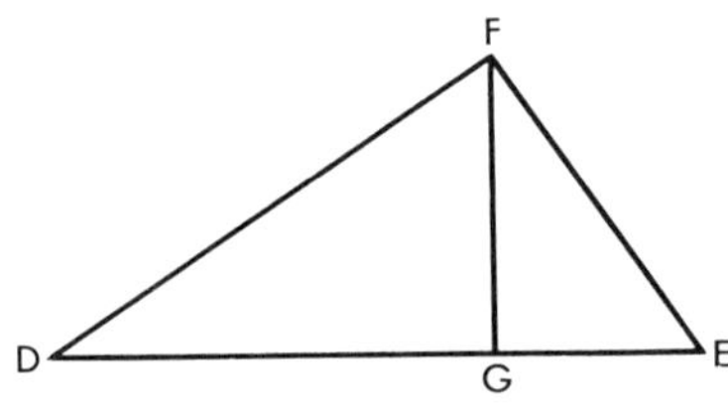

Figure 17–19

(a) If $m\overline{DG} = 9$, $m\overline{GE} = 4$, find $m\overline{FG}$.

(b) If $m\overline{FG} = 25$, $m\overline{GE} = 18$, find $m\overline{DG}$.

(c) If $m\overline{DE} = 12$, $m\overline{DG} = 3$, find $m\overline{DF}$.

(d) If $m\overline{DE} = 16$, $m\overline{GE} = 4$, find $m\overline{FE}$.

(e) If $m\overline{GE} = 20$, $m\overline{FG} = 10$, find $m\overline{DG}$.

(f) If $m\overline{GE} = 9$, $m\overline{FE} = 15$, find $m\overline{DG}$.

(g) If $m\overline{DE} = 20$, $m\overline{GE} = 5$, find $m\overline{FG}$.

(h) If $m\overline{GE} = 9$, $m\overline{DG} = 4$, find $m\overline{DF}$ and $m\overline{FE}$.

(i) If $m\overline{GE} = 18$, $m\overline{DG} = 2$, find $m\overline{FG}$, $m\overline{DF}$, and $m\overline{FE}$.

(j) If $m\overline{DE} = 13$, $m\overline{FG} = 6$, find $m\overline{DG}$ and $m\overline{GE}$.

2. $\overline{AB}$ is a diameter of a circle, C is a point on the circle, and $\overline{CD} \perp \overline{AB}$. If $m\overline{AD} = 3$ inches and $m\overline{DB} = 27$ inches, find $m\overline{CD}$.

3. $\overline{RS}$ with a length of 8 ft represents the diameter of a circle. Point T is on the circle and $\overline{TP}$ is an altitude drawn to $\overline{RS}$ intersecting $\overline{RS}$ three-fourths of the way from R to S. Find $m\overline{TP}$.

Review Exercises

Rationalize and simplify the following radicals.

1. $\sqrt{8}$	2. $\sqrt{18}$	3. $\sqrt{32}$	4. $\sqrt{40}$	5. $\sqrt{48}$
6. $\sqrt{60}$	7. $\sqrt{2x^2}$	8. $\sqrt{xy^2}$	9. $\frac{1}{\sqrt{2}}$	10. $\frac{1}{\sqrt{3}}$
11. $\frac{2}{\sqrt{3}}$	12. $\frac{7}{\sqrt{2}}$	13. $\frac{8}{\sqrt{2}}$	14. $\frac{6}{\sqrt{3}}$	15. $\sqrt{\frac{1}{3}}$
16. $\sqrt{\frac{1}{2}}$	17. $\sqrt{\frac{3}{2}}$	18. $\sqrt{\frac{5}{3}}$	19. $\sqrt{\frac{7}{2}}$	20. $\sqrt{\frac{8}{3}}$

RIGHT TRIANGLES

When art and mathematics were flourishing during the Egyptian dynasties, much was learned about the right triangle. It was learned that certain combinations of lengths of the sides of a triangle always formed a right triangle. This may have followed from some ancient scholar's placing the sides of various-sized squares together to form a right triangle as shown below. *Is there a relationship between the areas of the square regions?*

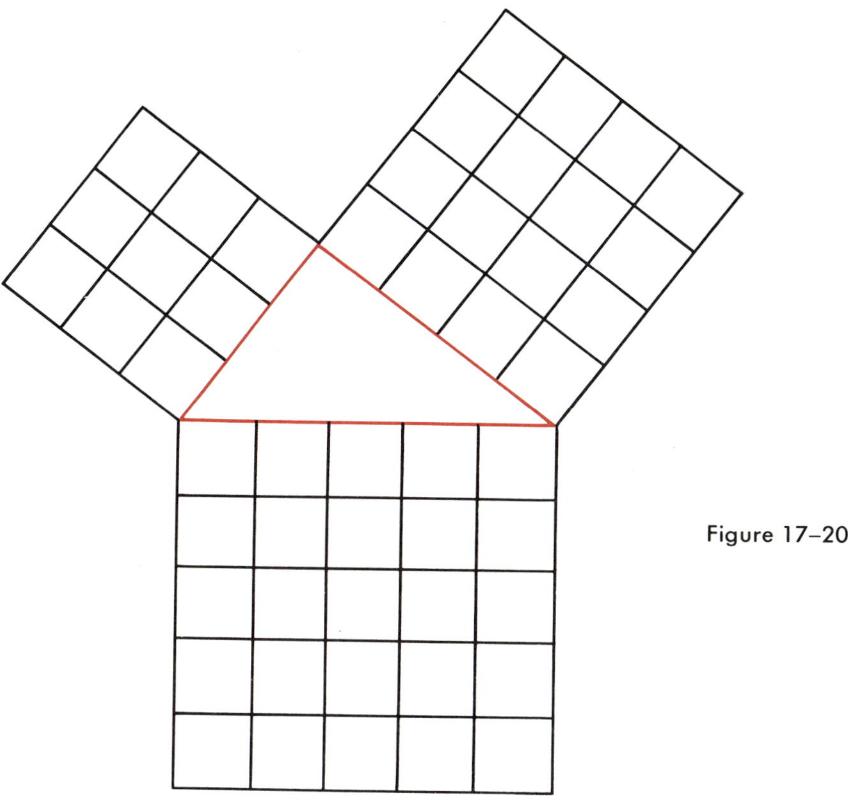

Figure 17–20

When the art of deductive proof was established, it was natural for mathematicians to attempt to generalize and prove the conjecture that the area of the square upon the hypotenuse of a right triangle is equal to the sum of the areas of the squares upon the legs of the right triangle. The original proof is believed to have been made by the Greek philosopher and mathematician, Pythagoras, about 525 B.C.

Many other proofs of this interesting relationship have been developed. Most of them are algebraic in character and not purely geometric as was the original proof by Pythagoras. See page 576 of the appendix for the geometric proof. Attempt to find examples of other proofs. There is one credited to President Garfield.

Given: $\triangle ABC$, $m\angle C = 90°$, and a, b, c, the measures of the sides.

Figure: 17–21

Conjecture: $c^2 = a^2 + b^2$

Plan: Draw altitude $\overline{CD}$ and use 17.13.

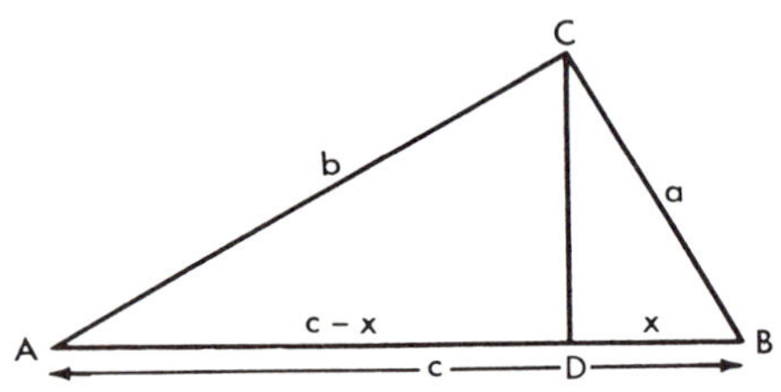

Figure 17–21

Proof: *Statements*	*Reasons*
1. $\overline{CD} \perp \overline{AB}$ in $\triangle ABC$, $m\angle C = 90°$	1. Given
2. $c:a = a:x$	2. 17.13
3. $a^2 = cx$	3. Why?
4. $c:b = b:(c - x)$	4. Why?
5. $b^2 = c^2 - cx$	5. Why?
6. $a^2 + b^2 = c^2$	6. Why?

17.14 PYTHAGOREAN THEOREM

The square of the measure of the hypotenuse of a right triangle is equal to the sum of the squares of the measures of the legs.

EXAMPLE: In Figure 17-21, if $a = 9$ inches and $b = 12$ inches, find c.

$$\begin{aligned} \text{Since } a^2 + b^2 &= c^2 \\ 9^2 + 12^2 &= c^2 \\ 225 &= c^2 \\ 15 &= c \end{aligned}$$

THE CONVERSE OF THE PYTHAGOREAN THEOREM

Knowing that a triangle is a right triangle if the sum of the squares of the measures of two sides of the triangle is equal to the square of the measure of the third side is almost as important as knowing the Pythagorean Theorem.

Given: $\triangle ABC$ in which $\text{m}\overline{AC} = \text{b}$, $\text{m}\overline{BC} = \text{a}$, $\text{m}\overline{AB} = c$ and $a^2 + b^2 = c^2$.

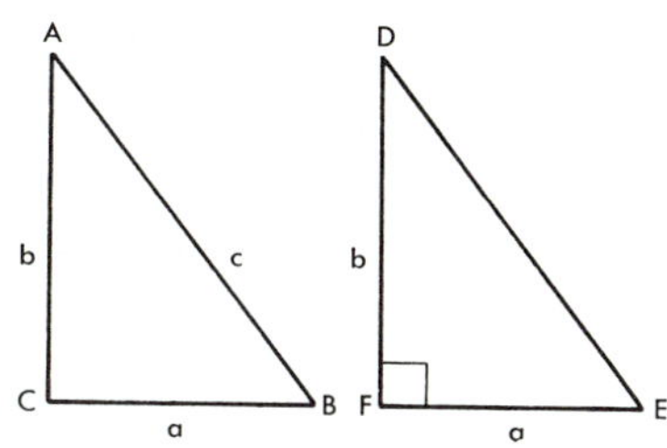

Figure 17–22

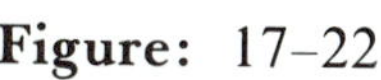

Figure: 17–22

Conjecture: $\triangle ABC$ is a right triangle.

Proof:

Statements	*Reasons*
1. Let $\angle F$ be a right angle.	1. Right angles exist. (3.01 and 3.09)
2. Let D and E be points on the sides of $\angle F$ such that $\text{m}\overline{FE} = a$ and $\text{m}\overline{FD} = b$.	2. 2.38
3. $(\text{m}\overline{DE})^2 = a^2 + b^2$	3. Why?
4. $a^2 + b^2 = c^2$	4. Why?
5. $(\text{m}\overline{DE})^2 = c^2$	5. Why?
6. $\text{m}\overline{DE} = c$	6. Why?
7. $\overline{DE} \cong \overline{AB}$	7. Why?
8. $\overline{FE} \cong \overline{CB}$, $\overline{FD} \cong \overline{CA}$	8. Why?
9. $\triangle ABC \cong \triangle DEF$	9. Why?
10. $\angle C \cong \angle F$	10. Why?
11. $\triangle ABC$ is a right triangle	11. Why?

17.15 THEOREM

If the sum of the squares of the measures of two sides of a triangle is equal to the square of the measure of the third side, the triangle is a right triangle.

Exercises

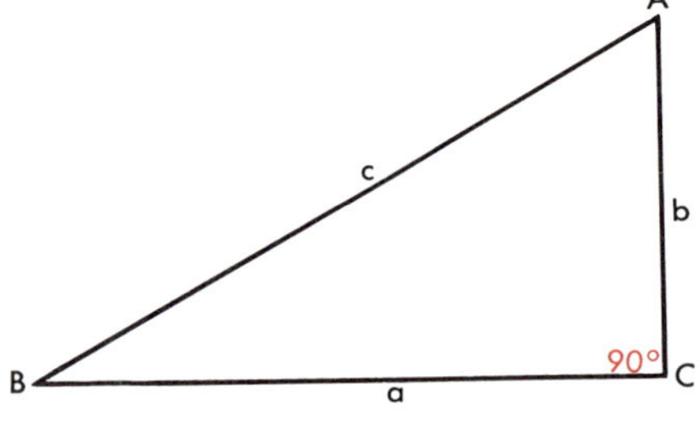

Figure 17–23

(Exercise 1)

1. Find the missing elements. (a, b, c are the measures of the sides.)
   (a) $a = 6, b = 8, c = ?$
   (b) $a = 24, b = 10, c = ?$
   (c) $a = ?, b = 2, c = 6$
   (d) $a = 2, b = ?, c = 4$
   (e) $a = ?, b = 4, c = 8$
   (f) $a = 1, b = 1, c = ?$
   (g) $a = ?, b = 2, c = 3$
   (h) $a = 8, b = ?, c = 17$

2. Show that a triangle the measures of whose sides are 3, 4, and 5 inches is a right triangle. Do the same for a triangle the measures of whose sides are 5, 12, and 13 inches. Can you find other combinations of integers that could be the lengths of the sides of a right triangle? (A Pythagorean triple.)

3. Find the height of an equilateral triangle the length of whose side is 12 inches; whose side is 6 inches; whose side is x inches.

4. Find the length of the chord of a circle of 25-inch radius if the chord is 15 inches from the center of the circle.

5. Find the length of the diagonal of a square the measure of whose side is 2 inches; 3 inches; x inches.

6. An isosceles triangle the length of whose base is 4 inches has an altitude upon the base whose measure is 4 inches. What is the measure of the congruent legs?

7. The measures of the congruent legs of an isosceles triangle are 6 inches and the measure of the base is 4 inches. Find the length of the altitude to the base.

8. The measure of a chord is 24 inches and is five inches from the center of the circle. Find the measure of the radius of the circle.

9. A point is 39 inches from the center of a circle, and the length of the tangent from this point to the point of tangency is 36 inches. What is the measure of the radius of the circle?

10. The lengths of the diagonal and a side of a rectangle are 17 feet and 15 feet respectively. Find the width of the rectangle.

11. The measure of the diameter of a small circle is 6 inches and the measure of the radius of its sphere is 5 inches. What is the distance of the small circle of the sphere from the center of the sphere?

How many different letters would be needed to represent the measures of the sides of an isosceles right triangle? By using one letter to represent the measure of the two equal sides, several theorems become evident.

Given: Isosceles right triangle with a the measure of the legs.

Find: Measure of the hypotenuse c.

Proof:
$$a^2 + a^2 = c^2$$
$$2a^2 = c^2$$
$$a\sqrt{2} = c$$

17.16 THEOREM

In an isosceles right triangle the measure of the hypotenuse is equal to the product of the measure of one of the equal legs and $\sqrt{2}$.

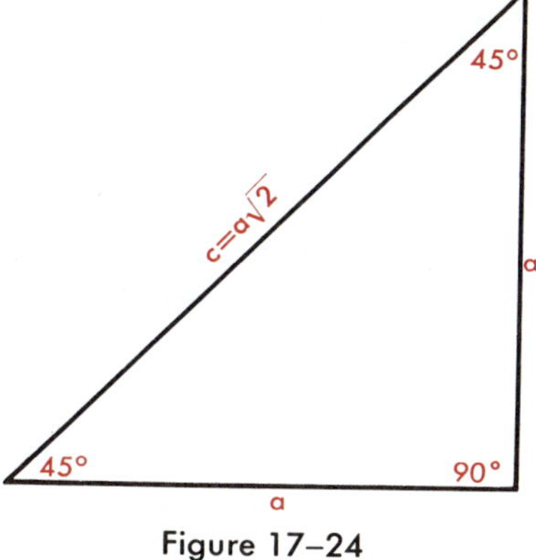

Figure 17–24

Given: Isosceles right triangle with c the measure of the hypotenuse.

Find: Measure of one of the legs a.

Proof: Since $c = a\sqrt{2}$ (by Theorem 17.16); then $a = \dfrac{c}{\sqrt{2}}$

17.17 THEOREM

In an isosceles right triangle the measure of either of the congruent legs is equal to the measure of the hypotenuse divided by $\sqrt{2}$.

Oral Exercises

1. In Figure 17-24, find c when a is: 1, 3, 5, 7, 9, x.
2. In Figure 17-24, find a when c is: 1, 2, 3, 4, 6, 8, x.

Given: Right $\triangle ABC$, with $m\angle B = 90°$ and $m\angle C = 30°$

Conjecture: $m\overline{AC} = 2 \cdot m\overline{AB}$

Plan: Make $\angle BCD \cong \angle BCA$. $\overrightarrow{AB}$ intersects $\overrightarrow{CD}$ at D. Show $\triangle ADC$ is equilateral. Prove $m\overline{AD} = 2 \cdot m\overline{AB}$ and substitute $m\overline{AC}$ for $m\overline{AD}$.

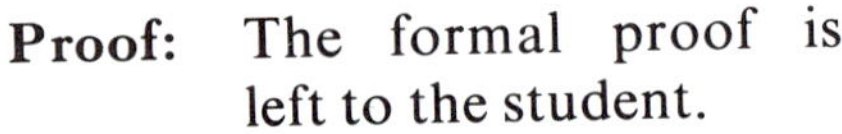

Proof: The formal proof is left to the student.

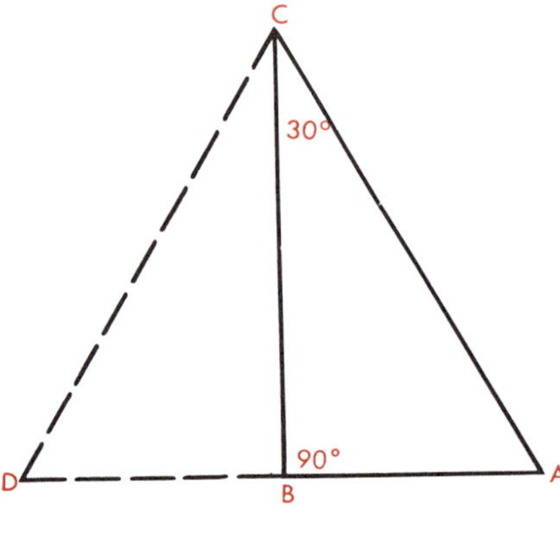

Figure 17–25

17.18 THEOREM

In a right triangle, if the measure of one angle is 30°, the measure of the hypotenuse is twice the measure of the side opposite the 30° angle.

17.19 CONVERSE THEOREM

In a right triangle, if the measure of the hypotenuse is twice the measure of one of the legs, the measure of the angle opposite that leg is 30°.

By using algebra along with some previous theorems, other relationships between the sides of this special triangle can be proved.

Let $m\overline{AC} = x$. Then $m\overline{AB} = 2x$ (17.18). Using 17.14, $m\overline{BC} = \sqrt{(2x)^2 - x^2} = \sqrt{3x^2} = x\sqrt{3}$.

EXAMPLE: If $m\overline{AC} = 5$, $m\overline{BC} = 5\sqrt{3}$.

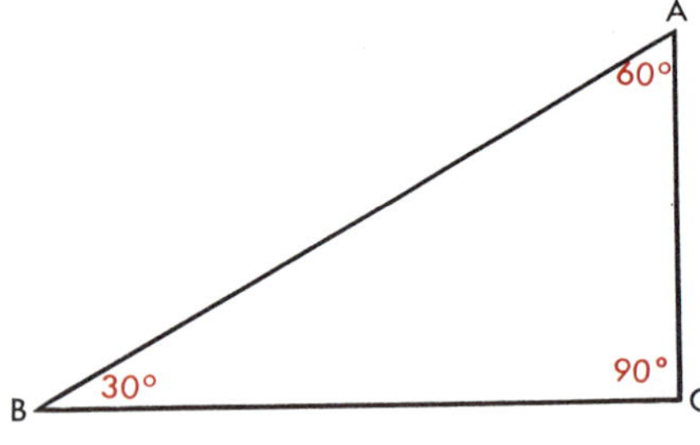

Figure 17–26

17.20 THEOREM

If the measure of one angle of a right triangle is 30°, the measure of the side opposite the 60° angle is equal to the product of $\sqrt{3}$ and the measure of the side opposite the 30° angle.

Using 17.20, $m\overline{BC} = m\overline{AC}\sqrt{3}$. Divide by $\sqrt{3}$. Then $\dfrac{m\overline{BC}}{\sqrt{3}} = m\overline{AC}$.

EXAMPLE: If $m\overline{BC} = 6$, $m\overline{AC} = \dfrac{6}{\sqrt{3}} = 2\sqrt{3}$.

17.21 THEOREM

If the measure of one angle of a right triangle is 30°, the measure of the side opposite the 30° angle is equal to the measure of the side opposite the 60° angle divided by $\sqrt{3}$.

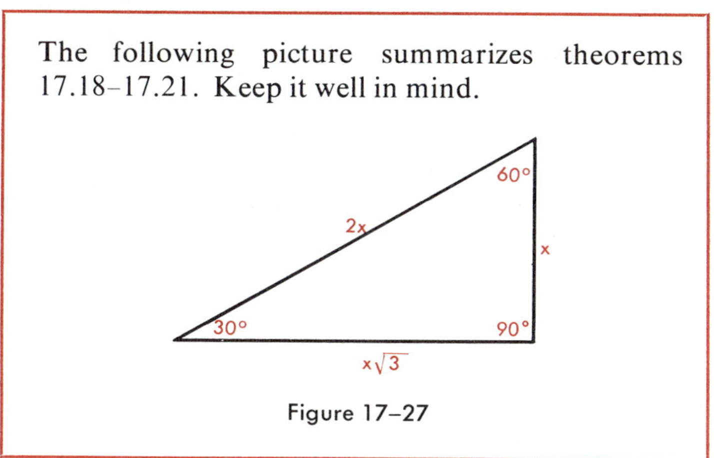
The following picture summarizes theorems 17.18–17.21. Keep it well in mind.

Figure 17–27

Oral Exercises

1. Find c and a if b is 1, 2, 3, 4, 6, 8.
2. Find b and a if c is 2, 6, 8, 10, 5, 3, 1.
3. Find b and c if a is 3, 1, 4, 6, $\sqrt{3}$, $2\sqrt{3}$.

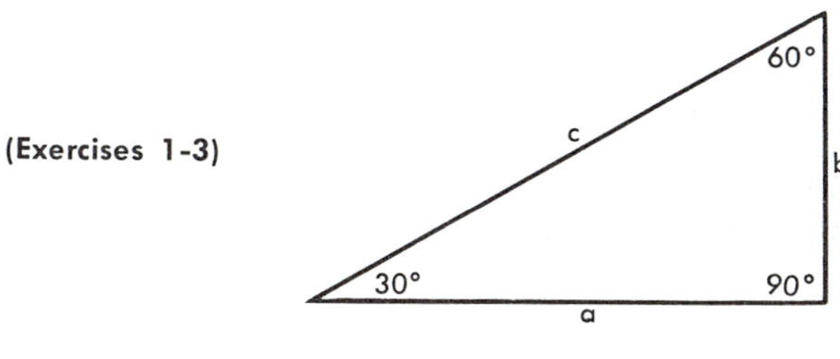

Figure 17–28

4. Give answers to the following at sight:

(a) If $c = 3$, $b = ?$ and $a = ?$
(b) If $c = 5$, $b = ?$ and $a = ?$
(c) If $a = 2$, $b = ?$ and $c = ?$
(d) If $a = 6$, $b = ?$ and $c = ?$
(e) If $b = 9$, $a = ?$ and $c = ?$
(f) If $b = 7$, $a = ?$ and $c = ?$

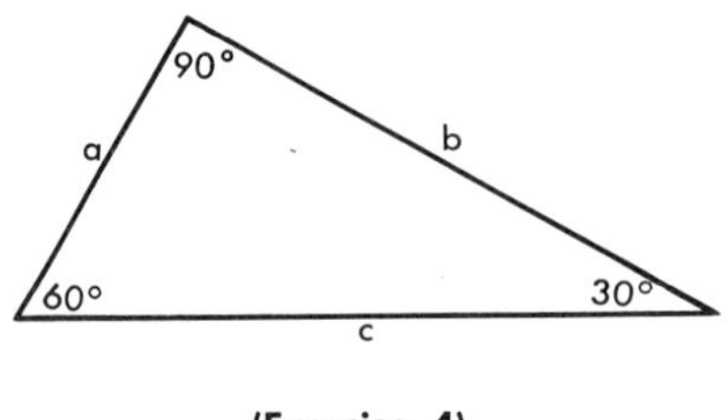

(Exercise 4)

Figure 17–29

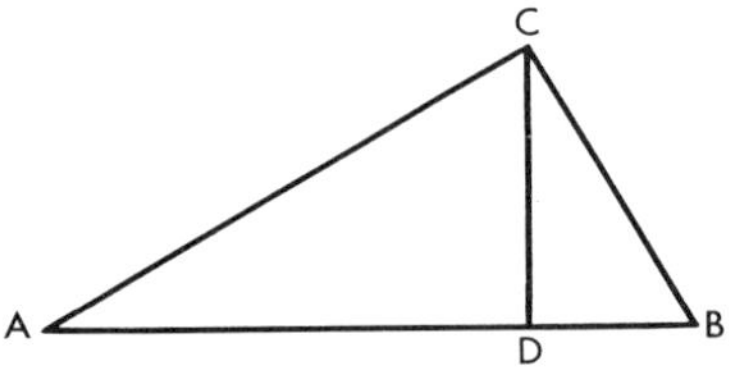

(Exercise 5)

Figure 17–30

5. $m\angle A = 30°$, $\angle ACB$ is a right angle, and $\overline{CD} \perp \overline{AB}$. Give the lengths of $\overline{AB}$, $\overline{BC}$, $\overline{AC}$, $\overline{DB}$, $\overline{AD}$, and $\overline{CD}$ at sight if:

(a) $m\overline{BC} = 4$ (b) $m\overline{BC} = 7$ (c) $m\overline{AC} = 6$ (d) $m\overline{AC} = 8$
(e) $m\overline{AB} = 10$ (f) $m\overline{AB} = 5$ (g) $m\overline{AD} = 12$ (h) $m\overline{AD} = 10$
(i) $m\overline{DB} = 4$ (j) $m\overline{DB} = 3$ (k) $m\overline{CD} = 8$ (l) $m\overline{CD} = 3$

Exercises

1. Find the length of a side of a square the length of whose diagonal is $3\sqrt{2}$ inches.
2. A ladder makes an angle with the side of a house whose measure is 30° and rests 10 feet from the base of the house. What is the length of the ladder? How high on the house is it placed?
3. Find the length of the diagonal of a square the measure of whose side is 5 feet.
4. Find the distance from home plate to second base on a baseball diamond. (A baseball diamond is square, 90 feet on each side.)
5. Find the measure of the altitude of an equilateral triangle the measure of whose side is 9 feet.
6. The length of the median of an equilateral triangle is $6\sqrt{3}$ feet. Find the length of a side of the triangle.
7. The measure of the altitude of an equilateral triangle is 3 feet. Find the length of a side of the triangle.

8. The measure of the diagonal of a rectangle is 104 feet and the width of the rectangle is 40 feet. Find the length of the rectangle.
9. Find the measure of the radius of the circle circumscribed about a equilateral triangle the length of whose side is 12 inches. Find the measure of the radius of the inscribed circle.
10. The centers of two circles the measure of whose radii are 6 inches and 8 inches are 21 inches apart. Find the length of the segment of the common internal tangent.
11. Find the measure of the altitude of the frustum of a right circular cone the measures of whose base diameters are 8 inches and 18 inches and whose slant height is 13 inches.
12. Find the slant height of a regular quadrangular pyramid the length of whose base edge is 6 inches and the measure of whose altitude is 4 inches.
13. Find the measure of the altitude of a regular quadrangular pyramid the measures of whose lateral edge is 4 inches and the measure of whose base edge is 4 inches.
14. Find the measure of the altitude of a regular triangular pyramid the measures of whose lateral edge is 6 inches and of whose base edge is 2 inches.
15. Find the slant height of a regular hexagonal pyramid the measure of whose altitude is 12 inches and the measure of whose base perimeter is 36 inches.
16. Find the measure of the altitude of a regular tetrahedron the length of whose edge is 8 inches.
17. The point of concurrency of the altitudes of a regular tetrahedron is what portion of the length of the altitude from the vertex? Explain your answer.

TRIGONOMETRY (Optional)

Trigon in Greek means three angles; *metry* means measurement. Trigonometry was originally devised to measure angles and distances by using the concept of similar triangles. This is the presentation used in this text. Modern trigonometry deals with circular functions and functions of numbers, but this will be reserved for a more advanced course.

Trigonometric Ratios

We know that if two angles of one triangle are congruent to two angles of another, the triangles are similar and the measures of their sides are proportional. (17.02 and 17.01).

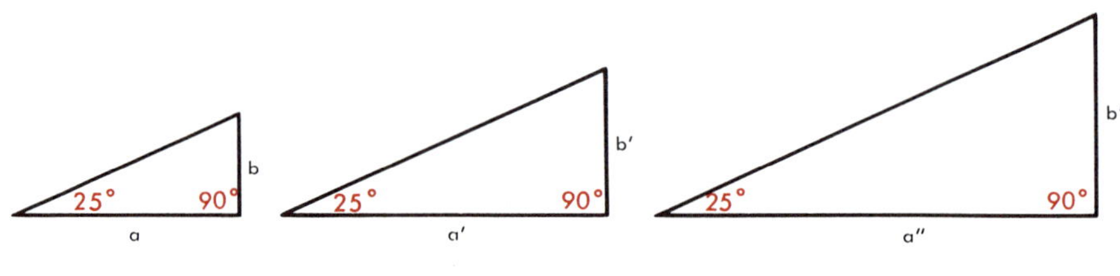

Figure 17–31

17.22 In Figure 17-31, then, $a:b = a':b' = a'':b''$. We can see that the ratios are constant for a given set of angle values in similar triangles. The size of the triangle does not change the ratio. For example, in right triangles, we know the ratio of the measure of the side opposite the 30° angle to the measure of the side adjacent to the 30° angle (but not the hypotenuse) is always $\frac{1}{\sqrt{3}}$ (by 17.20). If we know the measure of the side opposite the 30° angle is 10 feet, then the measure of the side adjacent to the angle is $10\sqrt{3}$ feet.

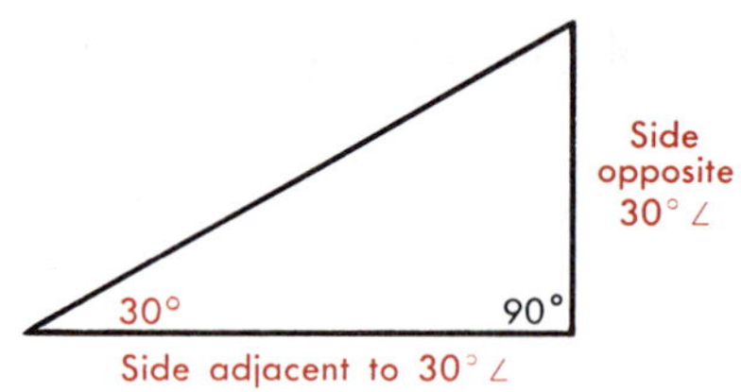

Figure 17–32

By developing the ratios of the measures of any two sides of a *right triangle* for any given angle from 0° to 90°, we may "solve" any right triangle if we know the measures of one side and an acute angle. Later work in trigonometry will deal with all shapes of triangles, but our work here will be confined to right triangles.

THE TANGENT RATIO

17.23 The following ratio of lengths has been defined as the tangent ratio of an acute angle of a right triangle.

$$\frac{\text{the measure of the side opposite } \angle B}{\text{the measure of the side adjacent to } \angle B} = \text{tangent } \angle B$$

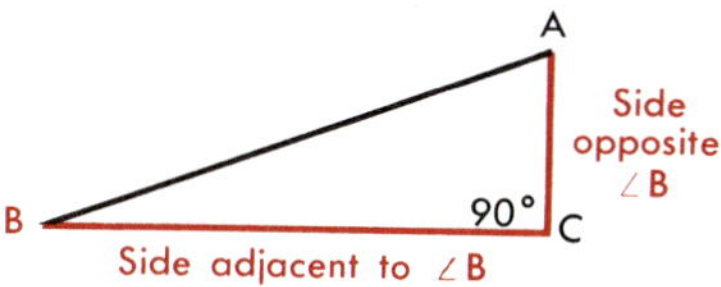

Figure 17–33

The values of the tangent ratio may be estimated by using Figure 17-34, where $\overleftrightarrow{AC}$ is a tangent and $\overline{BC}$ a radius. If $m\overline{BC} = 1$ unit, then $m\overline{DC}$, $m\overline{EC}$, etc., represent the values of the tangent ratio.

tangent 30° = .577 tangent 50° = 1.19
tangent 45° = 1 tangent 70° = 2.75

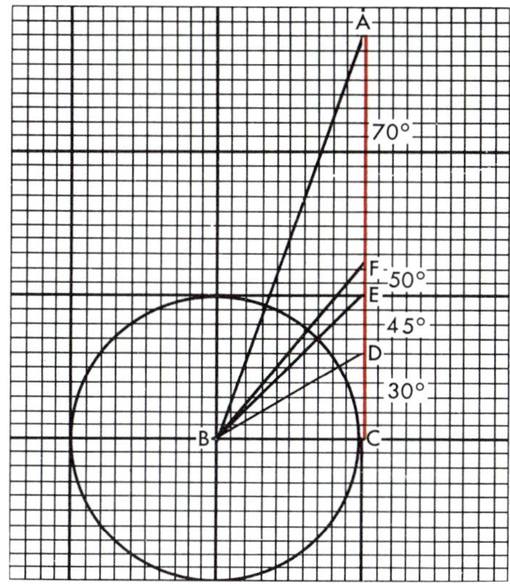

Figure 17–34

A table of trigonometric ratios appears on page 581. The tangents of angles from 0° to 90° are given in the column headed "tan." Check the values obtained in Figure 17-34 against the values in the tables. The tables are correct to the fourth decimal place.

Exercises

Using the table, complete the following.

1. (a) $\tan 28° = ?$ (b) $\tan 42° = ?$ (c) $\tan 85° = ?$
2. (a) If $\tan \angle A = .2493$, $m\angle A = ?$
   (b) If $\tan \angle A = 2.3559$, $m\angle A = ?$
3. Find the nearest angle value.
   (a) If $\tan \angle B = .7008$, $m\angle B = ?$
   (b) If $\tan \angle X = 1.1301$, $m\angle X = ?$

Using the Tangent Ratio

EXAMPLE 1. At a point 120 feet from the base of a flagpole, the measure of the *angle of elevation* of the top of the pole is 75°. Find the height of the flagpole. (Figure 17-35)

Solution: $\frac{h}{120} = \tan 75°$. Using the table, $\frac{h}{120} = 3.7321$.

then $h = 120 \cdot 3.7321$
$\therefore h = 448$ feet (to the nearest foot).

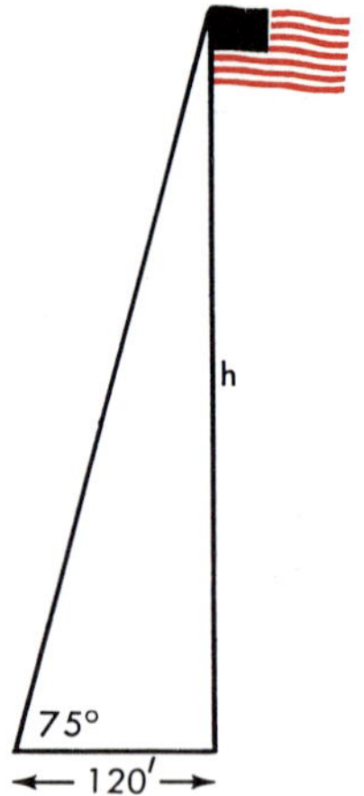

Figure 17–35

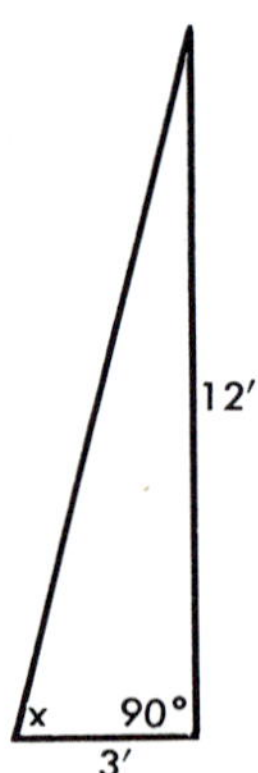

Figure 17–36

EXAMPLE 2. What is the degree measure of the slope of a river bank, if it rises 12 feet vertically for every 3 feet horizontally? (Figure 17-36)

Solution: $\tan x = \frac{12}{3} = 4.0000$

Using the table, $x = 76°$ (to the nearest degree).

Exercises

1. Find d to the nearest whole number when:
 (a) $e = 124$ ft, $m\angle E = 35°$
 (b) $e = 321$ ft, $m\angle E = 29°$
 (c) $e = 227$ ft, $m\angle D = 52°$
 (d) $e = 432$ ft, $m\angle D = 74°$

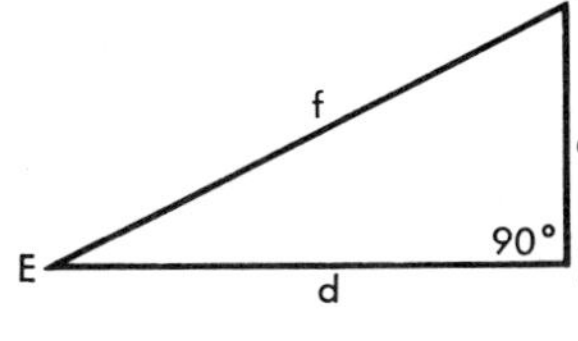

Figure 17–37

2. Is $\tan 75° = \tan 45° + \tan 30°$?
3. From a point 60 feet from the foot of a tree, the measure of the angle of elevation of the top of the tree is 38°. What is the height of the tree? ($\angle E$ is the *angle of elevation* of T from E. $\angle ETA$ is the *angle of depression* of E from T.)

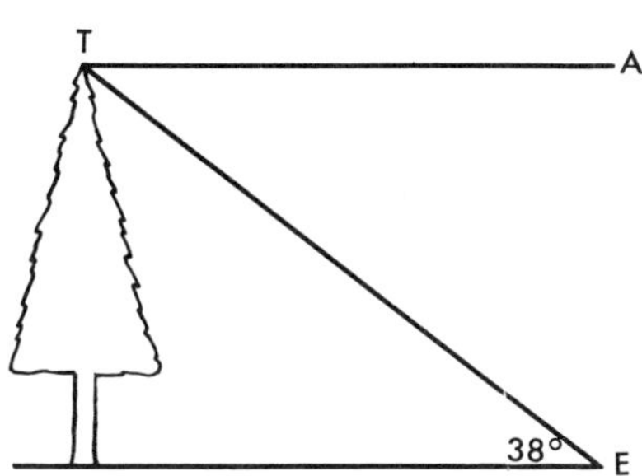

Figure 17–38

4. If an airplane rises 1 mile for every 40 miles of horizontal motion, what is the measure of its angle of rise, to the nearest degree?
5. A building 100 feet high casts a shadow 112 feet long. What is the measure of the angle of elevation of the sun, to the nearest degree?

THE SINE RATIO

17.24 The sine ratio (sin) of an acute angle of a right triangle is the ratio of the measure of the side opposite the acute angle to the measure of the hypotenuse.

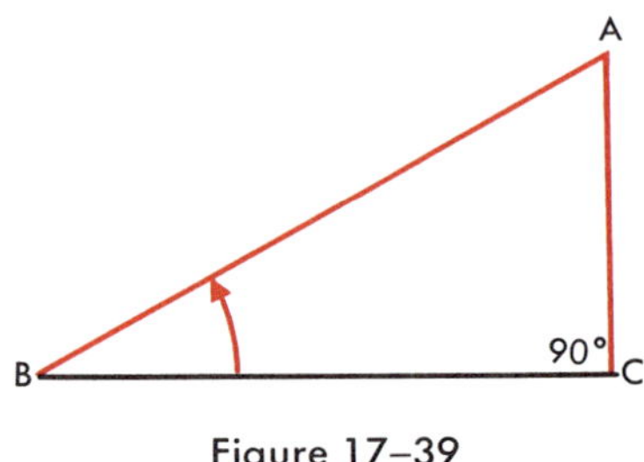

Figure 17–39

$$\frac{\text{the measure of the side opposite } \angle B}{\text{the measure of the hypotenuse}} = \sin \angle B$$

The values of the sine ratio may be estimated by using Figure 17-40, where $\overline{AB}$ is the radius of the circle. If m$\overline{AB}$ = 1 unit, m$\overline{AC}$, m$\overline{A'C'}$, and m$\overline{A''C''}$ represent the values of the sine ratio. Sin 30° = .5000, sin 60° = .8660, sin 75° = .9659, etc. Check these values in the table on page 581, using the column headed "sin."

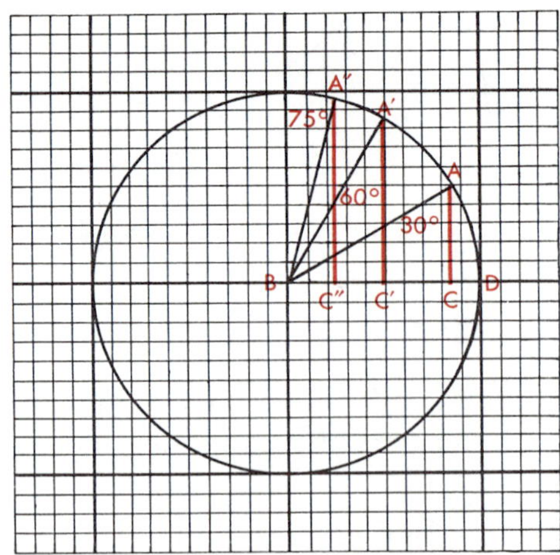

Figure 17–40

Using the Sine Ratio

EXAMPLE 1. A kite flying with 100 feet of string makes an angle with the ground whose measure is 41°. Assuming the string is straight, how high is the kite above the ground?

Solution: $\frac{h}{100} = \sin 41°$

$$\frac{h}{100} = 0.6561$$

$$h = 100 \cdot 0.6561$$

$$h = 65.6 \text{ ft}$$

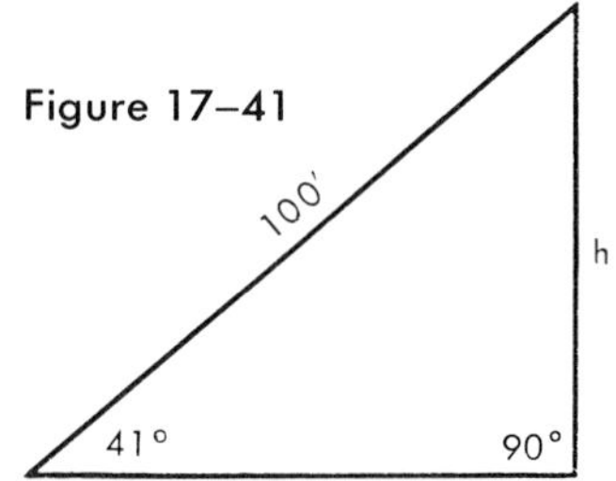

Figure 17–41

Exercises

1. Find e when:

(a) $f = 112$ ft, $m\angle E = 16°$
(b) $f = 68$ ft, $m\angle E = 23°$
(c) $f = 208$ ft, $m\angle E = 7°$
(d) $f = 315$ ft, $m\angle E = 33°$

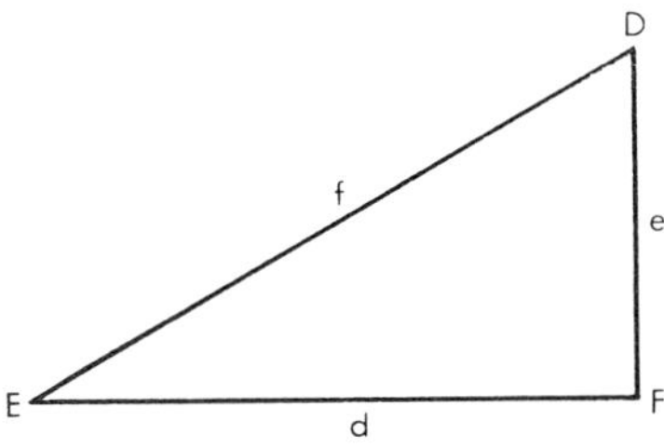

Figure 17–42

2. Is sin 60° twice as great as sin 30°?
3. A plane takes off and flies 10,500 feet along a straight line inclined at an angle to the horizon whose measure is 22°. How high is the plane above its starting point?
4. A guy wire runs from a tower 100 feet high forming an angle with the ground whose measure is 52°. What is the length of the wire?
5. What is the measure of the angle of elevation of a ski run that is 300 feet long and 100 feet high?

THE COSINE RATIO

17.25 In a right triangle, the cosine ratio (cos) of an acute angle is the ratio of the measure of the side adjacent to the angle (but not the hypotenuse) to the measure of the hypotenuse.

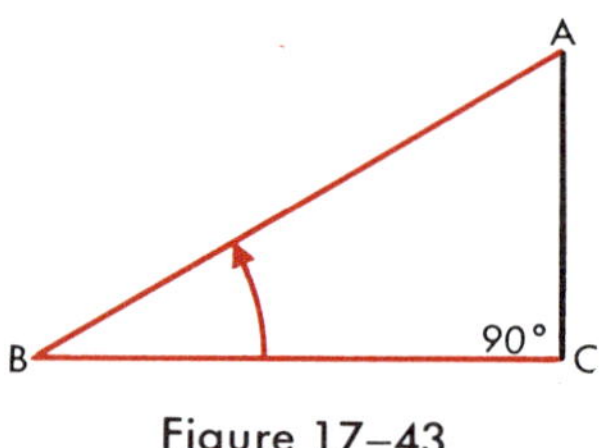

Figure 17–43

$$\frac{\text{the measure of the side adjacent to } \angle B}{\text{the measure of the hypotenuse}} = \cos \angle B$$

The values of the cosine ratio may be estimated by using Figure 17-44, where $\overline{AB}$ is the radius of the circle. If m$\overline{AB}$ = 1 unit, m$\overline{BC}$, m$\overline{BC'}$, and m$\overline{BC''}$ represent the values of the cosine ratio. Cos 60° = .5000, cos 10° = .98, cos 85° = .08, etc.

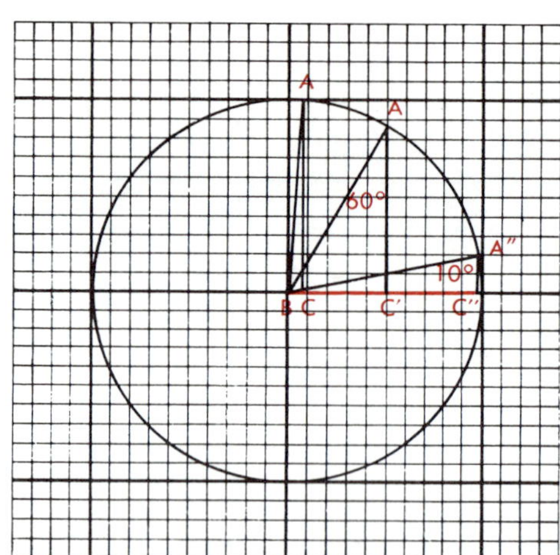

Figure 17–44

Check the above values in the table on page 581, using the column headed "cos."

Exercises

1. Find e to the nearest whole number when:

(a) $f = 128$ feet, $m\angle D = 32^\circ$
(b) $f = 322$ feet, $m\angle D = 24^\circ$
(c) $f = 257$ feet, $m\angle E = 56^\circ$
(d) $f = 472$ feet, $m\angle E = 71^\circ$
(e) $f = 235$ feet, $m\angle E = 37^\circ$
(f) $f = 440$ feet, $m\angle D = 55^\circ$

Figure 17–45

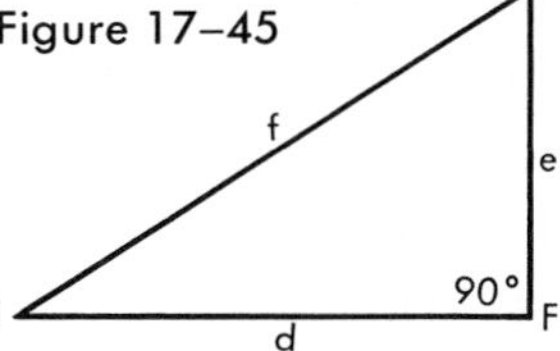

2. In right triangle ABC with C as the right angle, does $\cos \angle A = \sin \angle B$?

3. The measure of the base of an isosceles triangle is 16 feet. The measure of the vertex angle is 44°. Find the length of the congruent sides.

4. A tangent to a circle is connected to the center of the circle by a segment 18 inches long, making an angle with the tangent whose measure is 41°. What is the length of the tangent?

5. If the measure of the radius of a circle is 15 feet, what is the distance of the chord from the center if it intercepts an arc whose measure is 56°?

Miscellaneous Exercises

1. From the accompanying figures, give the sine, cosine, and tangent of each acute angle in terms of the measures of the sides of the triangle. That is, $\sin \angle B = \dfrac{b}{c}$, etc.

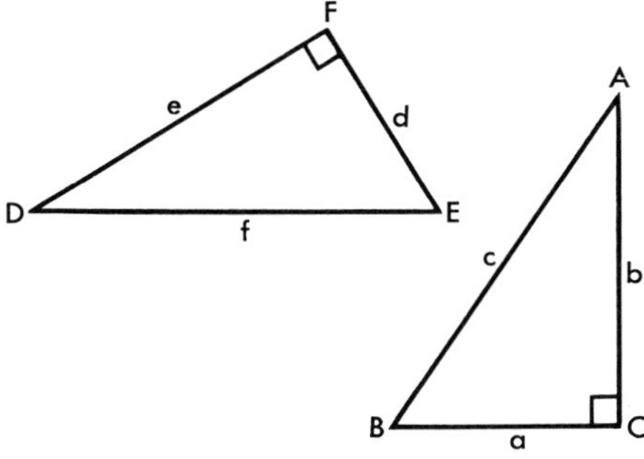

Figure 17–46

2. The Washington Monument is 555 feet high. What is the measure of the angle of elevation of the top when viewed from a distance of one mile? (Assume that you and the base of the tower are in the same horizontal plane.)

3. The measure of the angle of depression of a ship at sea from a cliff 520 feet high is 7°. How far is the ship from the base of the cliff?

4. The measure of the average slope of a mountain road is 22°. Find the measure of its fall in a distance of half a mile.

5. If the walls of a valley have a slope whose angle measure is 10°, how far up the wall from the floor of the valley will a dam 50 feet high force the water?

6. The lengths of the sides of a right triangle are 3, 4, and 5 feet. Find the measure of each acute angle.

7. Find the length of a side of an isosceles triangle if the measure of the vertex angle is 50° and of the altitude is 4 feet.

8. Does $\sin 40° = 2 \cdot \sin 20°$?

9. Does $\tan 50° = \tan 30° + \tan 20°$?

10. To find the inaccessible distance $\overline{AB}$, some scouts found that $m\angle A = 58°$ and $m\angle B = 17°$, $m\overline{AC} = 253$ feet and $m\overline{BC} = 410$ feet. What was the length $\overline{AB}$? (*Hint*: Construct the per-perpendicular from C to $\overline{AB}$ in Figure 17-47.

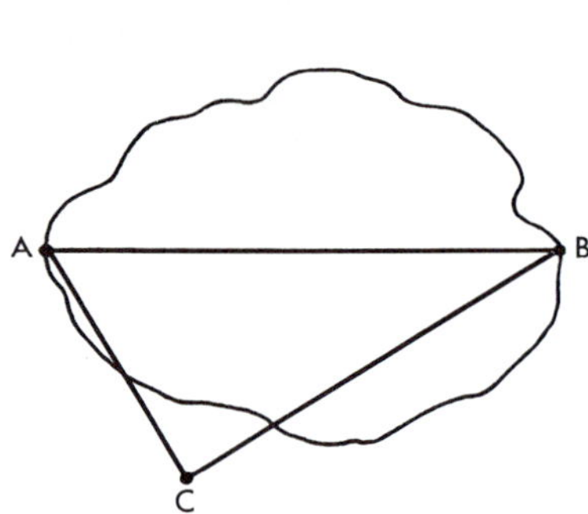

Figure 17–47

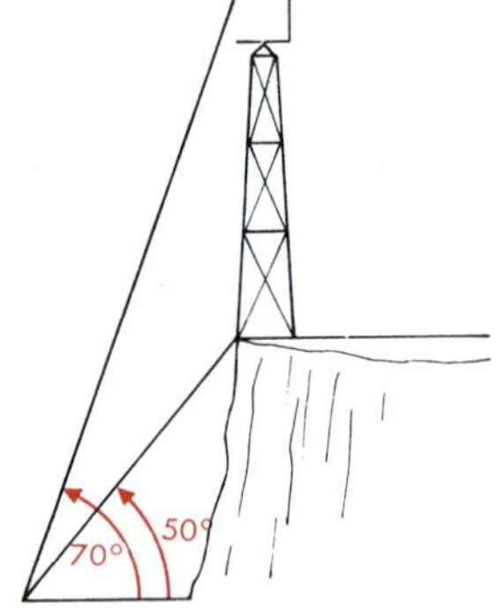

Figure 17–48

11. A tower on a cliff is 50 feet high. An observer notes that the measures of the angles of elevation (from ground level) of the top and bottom of the tower are 70° and 50° respectively. How high is the cliff? (See Figure 17-48.)

12. Two men 5 miles apart note the measures of the angles of elevation of an airplane directly between them to be 35° and 80°. What is the height of the airplane above the straight line connecting the men?

13. The measure of a chord is 1.8 feet and makes an angle of 18° with the radius drawn to one end of the chord. Find the measure of the radius of the circle, and the distance of the chord from the center.

14. Find the length of the shorter altitude of a parallelogram if the measures of the sides are 10 feet and 12 feet and the measure of the angle determined by the sides is 24°.

15. Find the measure of a dihedral angle of a regular tetrahedron.

16. Two chords are drawn in a circle such that one is half the length of the other. Is the shorter chord twice as far from the center as the longer chord? Prove your response.

INTERESTING APPLICATIONS OF SIMILAR POLYGONS

The pantograph is an instrument used to draw a plane figure similar to a known plane figure. It is used to reduce or enlarge maps or drawings. The figure is basically a parallelogram with two sides extended. Point E is fixed, and as F traces the given figure, it is enlarged by point A.

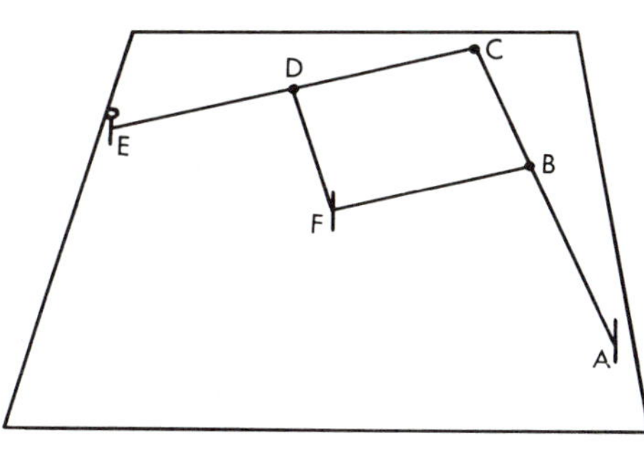

Figure 17–49

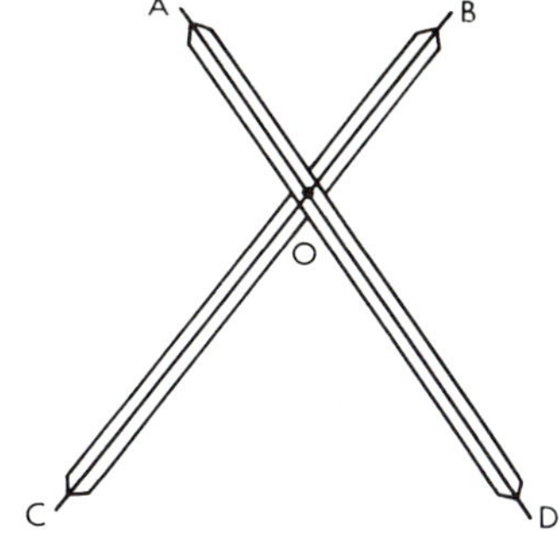

Figure 17–50

The bars of the proportional compasses are adjustable at O. Keep $\overline{AO} \cong \overline{BO}$ and $\overline{CO} \cong \overline{DO}$. If the compasses are set so that $m\overline{AO} = \frac{1}{3}m\overline{DO}$, then $m\overline{AB} = \frac{1}{3}m\overline{CD}$. Why?

A diagonal scale on a ruler provides a way to measure with great accuracy. The measure of the base of $\triangle DEF$ = .1 inch. The measures of the segments parallel to $\overline{FE}$ and ending in sides $\overline{DE}$ and $\overline{DF}$ are .01, .02, .03 inches, etc., reading up from D.

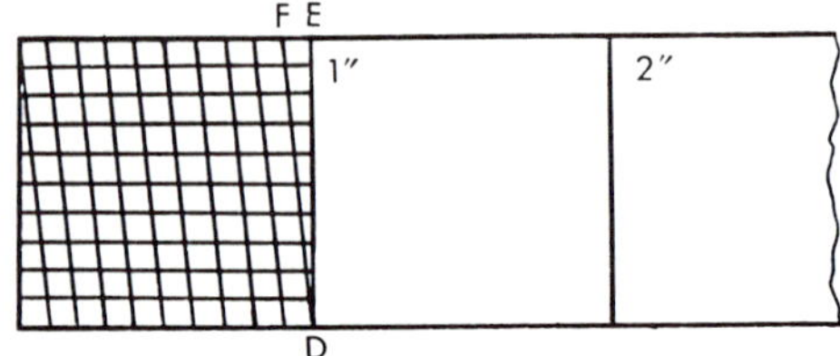

Figure 17–51

Instruments for Indirect Measurement

The hypsometer is used, as diagrammed, to find a ratio for similar triangles.

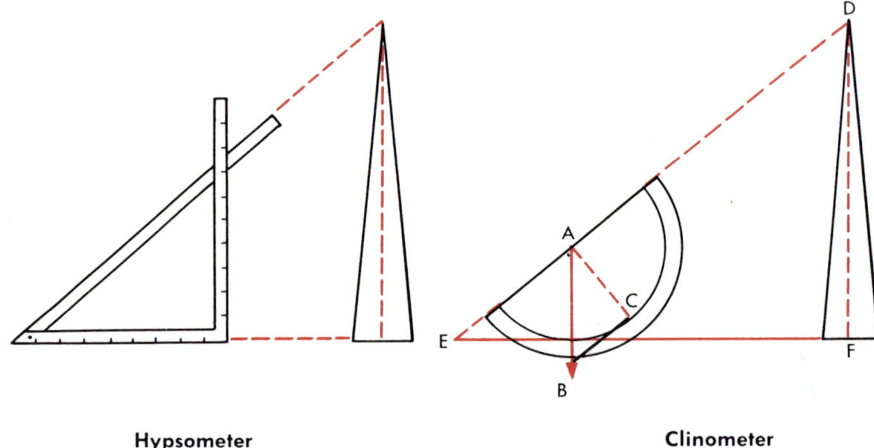

Figure 17–52

The clinometer is used on ships to find the angle of elevation of the sun. To understand how to measure the height of an unknown object using the clinometer, study the following procedure. From C, the 90° mark on the protractor, draw a line parallel to the line-of-sight. It intersects the plumb line in B. The m$\overline{BC}$ and m$\overline{AC}$ can be used in a proportion to find the height. $\triangle ABC \sim \triangle EDF$, so $\frac{\overline{\overline{BC}}}{\overline{\overline{DF}}} = \frac{\overline{\overline{AC}}}{\overline{\overline{EF}}}$. (See Exercise 10, page 427.)

Vocabulary List

similar	trigonometry	pantograph
quadratic	sine	hypsometer
Pythagorean	cosine	clinometer

Chapter Review

1. The measures of the sides of a triangle are 5, 9, and 12 inches. Find the measures of the sides of a similar triangle whose perimeter is 78 inches.
2. Are all regular polygons similar? Explain.
3. To know that two pentagons are similar we must know (1) _?_, and (2) _?_.
4. Name three methods, other than the definition, to prove two triangles are similar.
5. In circle O, chords $\overline{AB}$ and $\overline{CD}$ intersect at E. If $m\overline{AE} \cdot m\overline{EB} = 12$ and $m\overline{CE} = 4$, find $m\overline{ED}$.
6. Secants to circle O are drawn from P to intersect the circle at A, B, D and E so that B is between A and P and D is between E and P. If $m\overline{PE} = 20$, $m\overline{PA} = 12$, $m\overline{DE} = 18$, find $m\overline{PB}$.
7. $\overline{AC}$ is tangent to circle O at C. $\overline{AB}$ intersects circle O at D and B with D between A and B. If $m\overline{AB} = 18$ and $m\overline{AD} = 6$, find $m\overline{AC}$.
8. Solve for x: (a) $3x^2 - 17x + 10 = 0$ (b) $2x^2 - 4x + 1 = 0$.
9. If $m\overline{DE} = 33$, $m\overline{DF} = 65$ and $m\overline{EF} = 56$, is $\triangle DEF$ a right triangle? Explain.
10. In $\triangle ABC$, $m\overline{AC} = 10$, $m\angle B = 90°$ and $m\angle C = 60°$. Find $m\overline{AB}$ and $m\overline{BC}$.
11. In $\triangle RST$, $m\overline{RT} = 8$, $m\angle S = 90°$ and $\overline{RS} \cong \overline{ST}$. Find $m\overline{RS}$.
12. Draw a triangle and a segment in the exterior of the triangle on your paper. Using the segment as a side, construct a triangle similar to the given triangle.
13. Draw two collinear non-congruent segments on your paper then use Theorem 17.12 to construct the segment whose measure is the mean proportional between the measures of the given segments.

14. If two isosceles triangles have congruent vertex angles, are the triangles similar? Prove your response.

15. If two triangles have their sides respectively parallel, are the triangles similar? Prove your response.

Optional (Use Figure 17-53 for Exercises 16–20, where $m\angle Y = 90°$ and x, y and z represent the measures of the sides of $\triangle XYZ$.)

16. Name sine $\angle X$.

17. Name cosine $\angle Z$.

18. Name tangent $\angle Z$.

19. If $m\angle X = 28°\ 10'$, find tangent $\angle Z$.

20. If $m\angle Z = 70°$ and $x = 10$, find y.

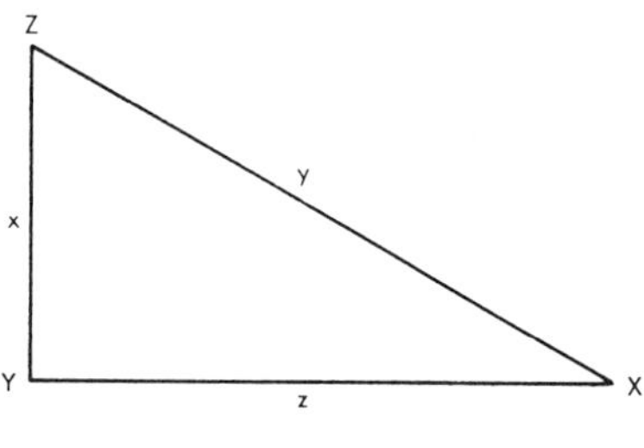

Figure 17–53

Chapter 17 Test

1. State which of the following statements is always true.

a. All squares are similar polygons.
b. All rectangles are similar polygons.
c. All regular hexagons are similar polygons.
d. All regular polygons are similar polygons.
e. The measure of an angle of rhombus A is 55°. The measure of an angle of rhombus B is 125°. $A \sim B$.
f. Two right triangles are similar if an acute angle of one is congruent to an acute angle of the other.
g. Two right triangles are similar if the hypotenuse of one is congruent to the hypotenuse of the other.
h. The diagonals and the bases of a trapezoid form a pair of similar triangles.
i. Two isosceles triangles are similar if an angle of one is congruent to an angle of the other.
j. The measures of the bases of a trapezoid are 12 in. and 30 in., and the altitude is 6 in. The lines containing the nonparallel sides of the trapezoid meet at a point 3 inches above the shorter base.

2. State the conditions which must be satisfied for two polygons to be similar.

3. A tower casts a shadow 64 feet long while a 6 ft high pole near the tower casts a shadow 8 feet long. How high is the tower?

4. If two triangles are similar and the measures of the sides of one triangle are 6, 7, and 8, what is the measure of the shortest side of the second triangle if the measure of the longest side is 10?

5. A man drives 5 miles due north and 12 miles due east. How far is he from his starting point?

6. Find the measure of the altitude of an equilateral triangle if the measure of a side is 6 inches.

7. The measure of a side of a square is 5 inches. What is the measure of the diagonal?

8. Find the measure of a diagonal of a cube if the measure of an edge is 3 cm.

9. Find the ratio of the measures of the diagonals of a rhombus if the measure of an angle of the rhombus is 60°.

10. Find the measure of the altitude of a regular tetrahedron if the measure of an edge is 12 inches.

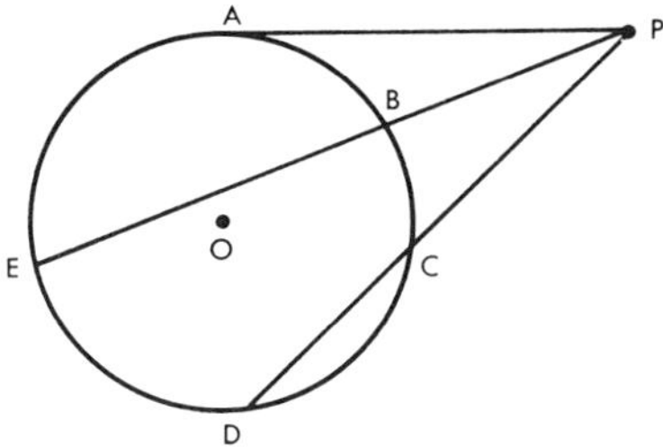

Figure 17–54

11. Given: Circle O with $\overline{PA}$ tangent to the circle at point A. Secants $\overline{PE}$ and $\overline{PD}$ are drawn intersecting the circle at B, E, and C, D, respectively.

a. If $m\overline{PA} = 8$ in., $m\overline{PB} = 2$ in., find $m\overline{PE}$.

b. If $m\overline{PB} = 3$ in., $m\overline{BE} = 7$ in., $m\overline{PC} = 5$ in., find $m\overline{CD}$.

12. A triangle is inscribed in a circle. The measure of a chord of the circle containing the median of the triangle is 10 inches. The measure of the side which the chord intersects is 6 inches. Find the measure of the median of the triangle.

(Use table 2, page 581.)

13. In $\triangle ABC$, $m\angle C = 90°$, $m\angle A = 15°$, $m\overline{AC} = 8$ in. Find $m\overline{BC}$.

14. In $\triangle DEF$, $m\angle D = 35°$, $m\angle F = 90°$, $m\overline{DF} = 10$ in. Find $m\overline{DE}$.

15. In $\triangle MNP$, $m\angle N = 90°$, $m\overline{MN} = 12$ in., $m\overline{MP} = 13$ in. Find $m\angle P$.

18

Areas of Polygonal Regions

In previous mathematics courses, you have found the areas of the common polygonal regions (Section 6.04). You were given a formula, told what the symbols meant, and shown how to substitute and solve. Here we will examine how these formulas were derived. In other words, we are now developing an understanding of mathematics rather than just skill in manipulating the symbols of mathematics. We will check both skill and understanding by applying the area relationships to practical problems.

UNITS OF MEASUREMENT

18.00 The choice of a unit of measure can be made to fit our convenience; for an angle it is usually the degree; for a segment it is sometimes the inch. Essentially it is the smallest unit used in making a measurement.

Commensurable Quantities

18.01 If a common unit of measurement exists for two segments, they are said to be commensurable; that is, $\overline{AB}$ and $\overline{CD}$ must each be an integral (whole) number of times as long as some segment $\overline{RT}$.

Thus a segment whose measure is 2 feet 5 inches and a segment whose measure is 3 feet are commensurable, for each is an integral number of inches in length. In like fashion, segments whose measures are $\frac{5}{16}$ inches and $\frac{7}{8}$ inches are commensurable, since each is an integral number of sixteenths of an inch in length.

Incommensurable Quantities

18.02 If no common unit of measurement exists for two segments, they are said to be incommensurable. Thus a segment whose measure is 3 feet and a segment whose measure is $\sqrt{2}$ feet are incommensurable, since there is no segment whose measure is an integral factor of 3 and $\sqrt{2}$.

However, for all practical purposes, any two like quantities are commensurable, for, by using sufficiently small units, they may be measured to any desired degree of precision.

(In this text we prove theorems for the commensurable case only and assume they are true for the incommensurable case. The proof of the latter involves the theory of limits and will not be considered here.)

Unit of area

18.03 The area of a polygonal region is determined by counting the number of distinct unit regions that would fill the region. The polygonal region formed by a square whose side is a unit of length is used as the *unit of area*. If the unit of length is an inch, the unit of area is a square inch, and so forth.

18.04 *Assumption: To every polygonal region there corresponds a unique positive number called the area of the region.*

Area of a Rectangular Region

Find the area of the figures in 18-1 by counting the squares.

While we must count fractional squares for the parallelogram, triangular, and trapezoid regions, the area of the rectangular region poses no difficulty since the length and width are both integral terms. As you can readily see, the number of unit square regions filling the rectangular region is the product of its length and width. We will abbreviate "the area of a rectangular region" to "the area of a rectangle," etc.

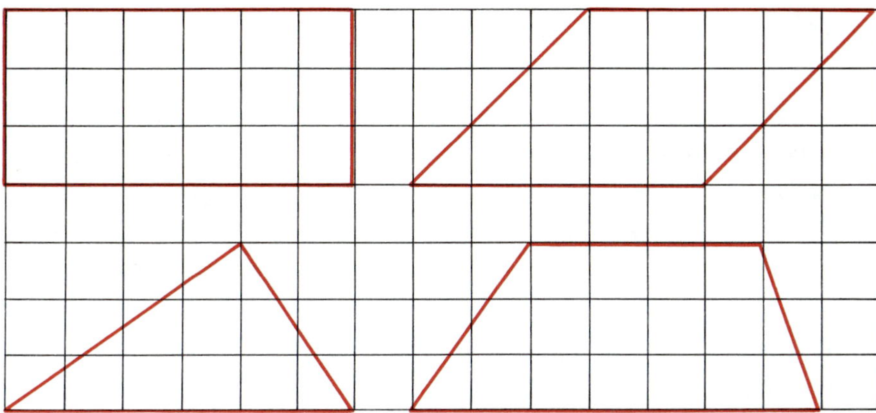

Figure 18–1

18.05 *Assumption: The area of a rectangle is equal to the product of the lengths of the base "b" and of the altitude "h" to that base.**

$$A = bh$$

18.06 THEOREM

The area of a square is equal to the square of the length of one of its sides.

Exercises

1. Find the areas of rectangles the measures of whose bases and altitudes are as follows:

(a) 10 cm, 5 cm (b) 3 in., $1\frac{1}{2}$ in. (c) 5.53 ft, 3.28 ft

(d) 5 ft, 5 in. (e) $a, a + 3$ (f) $a - 3, a + 3$

---

* We will assume a common unit of measure is represented by the variables of any formula in this text.

2. Find the height of the rectangles having the following areas and bases:

(a) 30 sq in., 3 in.
(b) $18\frac{3}{4}$ sq ft, $3\frac{1}{2}$ ft
(c) 19.82 sq ft, 8.3 ft
(d) 25 sq ft, 15 in.
(e) 16.8 sq in., 3.2 ft
(f) $a^2 - 2a$ sq units, a units

3. If you double the length of the altitude of a rectangle and leave the base unchanged, how is the area changed?

4. If the lengths of the altitude and base of a rectangle are both doubled, how is the area changed?

5. How many 3-inch square tiles does it take to cover a floor 12 feet 3 inches long by 10 feet 9 inches wide?

6. Find the area of a square whose perimeter is 18 feet.

7. The perimeters of a square and of a rectangle are 24 rods. The rectangle is 7 by 5 rods. Which has the greater area, the square or the rectangle?

18.07 *Assumption: Congruent simple closed curved regions (2.49, 6.04) have equal areas.*

18.08 THEOREM

Two polygonal regions are equal in area if they can be separated into the same number of respectively congruent parts.

Exercises

1. $ABCD$ is a parallelogram. $\overline{BH} \perp \overline{FC}$ and $\overline{AF} \perp \overline{FC}$. Prove that the area of $ABHF$ is equal to the area of $ABCD$.

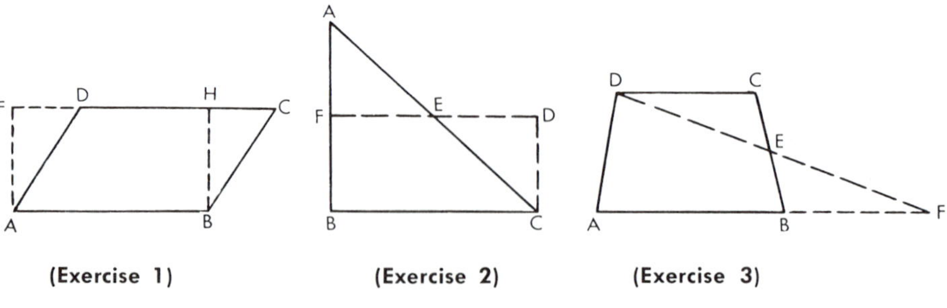

(Exercise 1) (Exercise 2) (Exercise 3)

Figure 18–2

2. $\overline{AB} \perp \overline{BC}$. F and E are the midpoints of $\overline{AB}$ and $\overline{AC}$ respectively. $\overline{CD} \perp \overline{FD}$. Prove that the areas of ABC and $BCDF$ are equal.

3. Given: $ABCD$ is a trapezoid, E is the midpoint of $\overline{BC}$, $\overline{DEF}$ and $\overline{ABF}$ are segments of straight lines. Prove that the areas of $ABCD$ and AFD are equal.

4. Construct a rectangle with an area equal to the area of a given scalene triangle. (*Hint*: The diagram for Exercise 2 may suggest a solution.)

Re-examine Exercise 1. Is $ABHF$ a rectangle? How can you find the area of $ABHF$? Is the area of $ABHF$ equal to the area of parallelogram $ABCD$?

Has this exercise shown that the area of any parallelogram can be found by finding the area of a rectangle whose base and altitude upon that base are congruent to those of the parallelogram? This proposition can be restated as shown in the following theorem.

Given: Parallelogram $ABCD$, with base b, altitude h, and area A

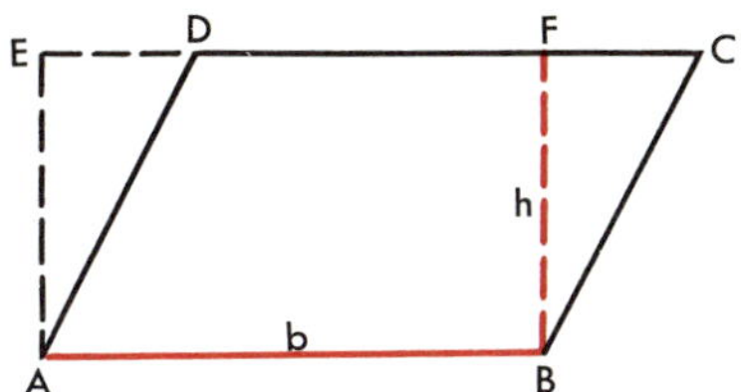

Figure 18–3

Conjecture: $A = bh$

Plan: Show that the area of $ABCD$ = the area of rectangle $ABFE$, whose area is bh, by 18.05.

Proof: The proof is left to the student.

18.09 THEOREM

The area of a parallelogram is equal to the product of the length of the base and the height upon that base.

Exercises

1. $ABDE$ is a parallelogram, with altitude $\overline{AC}$. Given the following unit lengths, find the area when:

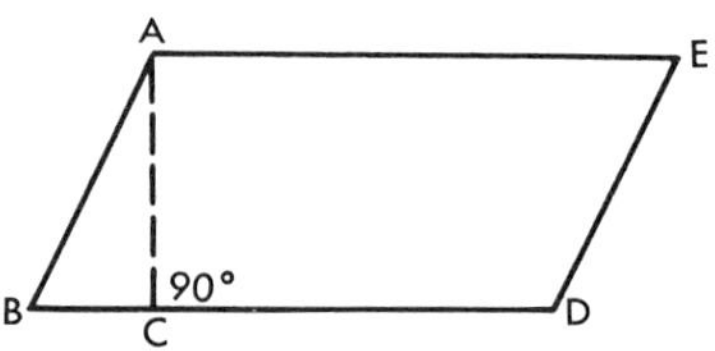

Figure 18–4

(a) $m\overline{BD} = 12, m\overline{AC} = 6$ (b) $m\overline{AE} = 5\frac{1}{3}, m\overline{AC} = 3\frac{3}{8}$

(c) $m\overline{BD} = 10, m\overline{AB} = 8, m\angle B = 30°$

(d) $m\overline{BD} = 14, m\overline{AB} = 10, m\angle B = 60°$ (Leave in simplest radical form.)

(e) $m\overline{BD} = 8, m\overline{AB} = 7, m\angle B = 45°$

(f) $m\overline{BD} = 12, m\overline{DE} = 30, m\angle B = 30°$

(g) $m\overline{BD} = 9, m\overline{AB} = 2, m\angle B = 60°$

(h) $m\overline{AE} = 16, m\overline{AB} = 6, m\angle B = 45°$

(i) $m\overline{BD} = 12, m\overline{AB} = 6, m\angle D = 120°$

(j) $m\overline{BD} = 6, m\overline{DE} = 5, m\angle D = 150°$

Study Exercise 2, page 467. Is the area of ABC equal to the area of $BCDF$? Is $BCDF$ a rectangle? Is the area of $BCDF$ equal to the $m\overline{FB} \times m\overline{BC}$? Is $m\overline{FB}$ equal to $\frac{1}{2}$ $m\overline{AB}$? Does it follow that the area of triangle ABC is equal to $\frac{1}{2}$ $(m\overline{AB})(m\overline{BC})$?

Suppose the triangle is a general triangle (see Figure 18-5) and not a right triangle, as in Exercise 2. How can we find the area? If we construct $\overline{CE}$ and $\overline{BE}$ so that $ABEC$ is a parallelogram, what will be the relationship of the area of $\triangle ABC$ to the area of parallelogram $ABEC$? Would this show that the area of any triangle is one-half the area of a parallelogram with congruent base and altitude upon that base?

Given: $\triangle ABC$, with base length b, altitude h, and area A

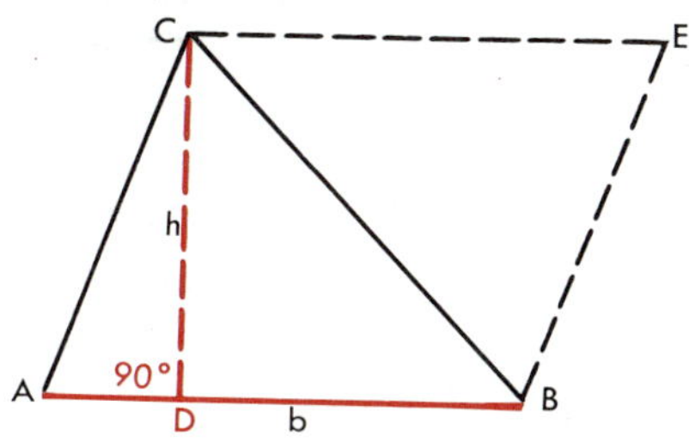

Figure 18–5

Conjecture: $A = \frac{1}{2}bh$

Plan: Draw $\overrightarrow{CE} \parallel \overline{AB}$ and $\overrightarrow{BE} \parallel \overline{AC}$, meeting at E. Show that $ABEC$ is a parallelogram and use 9.02.

Proof: The proof is left to the student.

18.10 THEOREM

The area of a triangle is equal to one-half the product of the lengths of the base and of the altitude upon that base.

Exercises

1. Find the area of $\triangle ABC$ with altitude $\overline{CD}$ if:

(a) $m\overline{AB} = 8', m\overline{CD} = 5'$

(b) $m\overline{AB} = 18'\,6'', m\overline{CD} = 12'\,3''$

(c) $m\overline{AB} = 3', m\overline{CD} = 7''$

(d) $m\overline{AB} = 12'', m\overline{AC} = 6''$, $m\angle A = 30°$

(e) $m\overline{AB} = 10'', m\overline{AC} = 4''$, $m\angle A = 45°$

(f) $m\overline{AB} = 10'', m\overline{AC} = 8''$, $m\angle A = 60°$

(g) $m\overline{AB} = 20'', m\overline{AC} = 14''$, $m\angle A = 90°$

(h) $m\overline{AB} = 10', m\overline{AC} = 10'$, $m\angle A = 120°$

(i) $m\overline{AB} = 8', m\overline{AC} = 14'$, $m\angle A = 135°$

(j) $m\overline{AB} = 14', m\overline{AC} = 20'$, $m\angle A = 150°$

2. Find the area of $\triangle ABC$, with right angle C, if $m\overline{AC} = 10$ feet and $m\angle B = 15°$. (Use trigonometric ratios.)

3. Find the area of $\triangle ABC$, in Exercise 2, if $m\overline{AC} = 12$ feet and $m\angle A = 73°$.

4. If the length of the side of an equilateral triangle is s, show that the area is $\frac{s^2}{4}\sqrt{3}$.

Examine Exercise 3, page 467. Is it apparent from this exercise that the area of any trapezoid can be found by constructing a triangle whose area is equivalent to that of the trapezoid and finding the area of this constructed triangle?

In this exercise the area of $\triangle AFD$ is clearly $\frac{1}{2}h(m\overline{AF})$, by 18.10. Is $\overline{BF} \cong \overline{DC}$? If so, the area of the $\triangle AFD$—or of the trapezoid $ABCD$—could be found by $\frac{1}{2}h(m\overline{AB} + m\overline{DC})$, by substitution. See whether you can prove the relationship stated in Theorem 18.11 by using the following general figure and conditions.

Given: Trapezoid $ABCD$, with altitude h, and base lengths b and b'

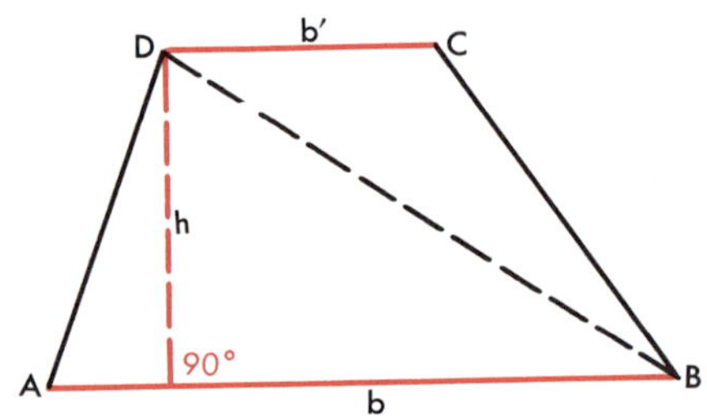

Figure 18–6

Conjecture: $A = \frac{1}{2}h(b + b')$

Plan: Show that the area of $ABCD$ equals the sum of the areas of two triangles with base lengths b and b' and altitude h.

Proof: The formal proof is left to the student.

18.11 THEOREM

The area of a trapezoid is equal to half the product of the length of the altitude and the sum of the lengths of the bases, or $A = \frac{1}{2}h(b + b')$.

Exercises

1. Using the following values for the preceding figure, find the value of the missing element:

	b	b'	h	A
(a)	18″	10″	6″	?
(b)	3″	2″	5″	?
(c)	18″	12″	?	135 sq in.
(d)	?	10″	7″	84 sq in.
(e)	40″	34″	?	148 sq in.

2. Show that the formula for the area of a trapezoid can be used to find the area of a parallelogram and a triangle.

3. Show that the formula for the area of a trapezoid can be used to find the sum of the series of numbers formed by adding a given number to each succeeding member of the series (*e.g.*, 5, 7, 9, 11, 13, . . .). Let A represent the sum of the numbers, h represent the number of terms in the series, and b and b' represent respectively the first and last terms of the series.

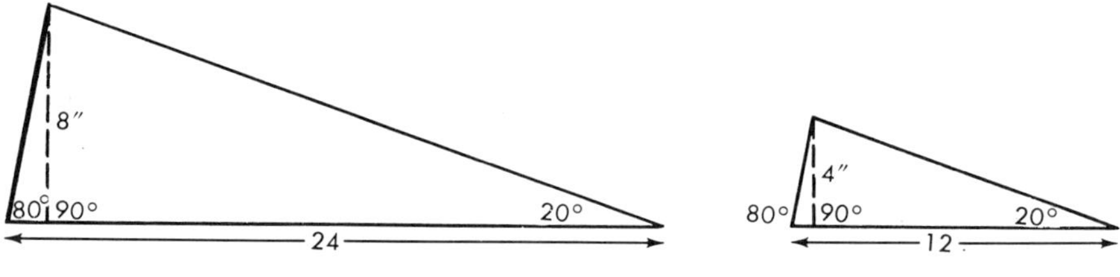

Figure 18–7

4. Are the two triangles similar? What is the ratio of their areas? What is the ratio of the lengths of their sides? Their altitudes? (Figure 18-7)

5. Using Figure 18-8, answer the questions of Exercise 4.

6. After completing Exercises 4 and 5, can you state a relationship concerning the areas of similar triangles?

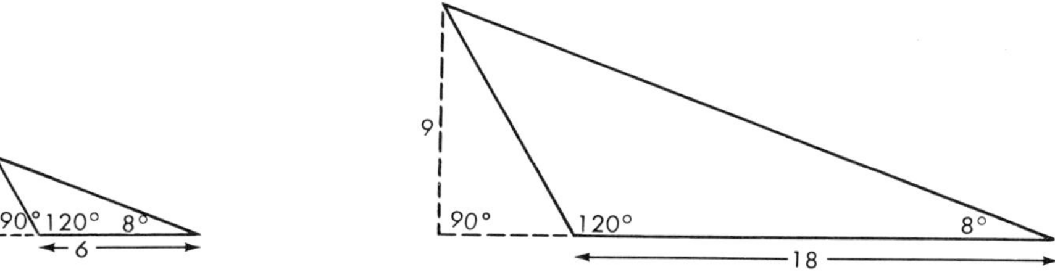

Figure 18–8

Given: $\triangle ABC \sim \triangle A'B'C'$, with a and a' the lengths of corresponding sides

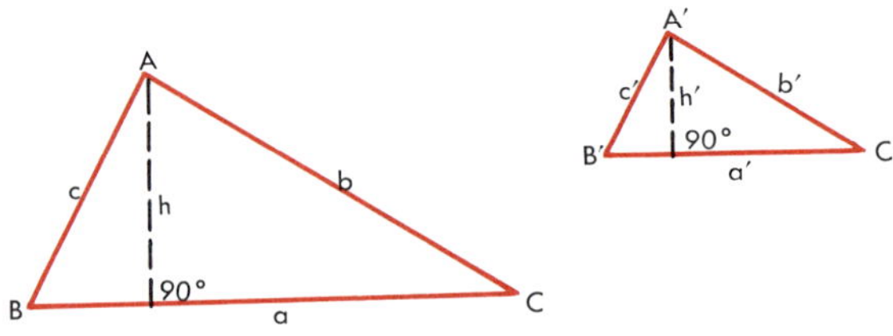

Figure 18–9

Conjecture: $\dfrac{\text{area } \triangle ABC}{\text{area } \triangle A'B'C'} = \dfrac{a^2}{(a')^2}$

Construction: Draw the corresponding altitudes h and h'.

Proof: *Statements*	*Reasons*
1. area $\triangle ABC = \dfrac{1}{2}ah$ area $\triangle A'B'C' = \dfrac{1}{2}a'h'$	1. Why?
2. $\dfrac{\text{area } \triangle ABC}{\text{area } \triangle A'B'C'} = \dfrac{ah}{a'h'} = \dfrac{a}{a'} \cdot \dfrac{h}{h'}$	2. Division Axiom
3. $\dfrac{h}{h'} = \dfrac{a}{a'}$	3. 17.03
4. $\dfrac{\text{area } \triangle ABC}{\text{area } \triangle A'B'C'} = \dfrac{a}{a'} \cdot \dfrac{a}{a'} = \dfrac{a^2}{(a')^2}$	4. Substitution Axiom

18.12 THEOREM

The areas of two similar triangles have the same ratio as the squares of the lengths of any two corresponding sides.

18.13 THEOREM

The areas of two similar triangles have the same ratio as the squares of the lengths of any two corresponding altitudes (or the squares of the lengths of any two corresponding linear dimensions). See 17.03 and 17.04.

18.14 THEOREM

The areas of two similar polygons have the same ratio as the squares of the lengths of any two corresponding sides (or the squares of the lengths of any two corresponding linear dimensions).

Exercises

1. Two similar triangles have corresponding sides whose lengths are 8 inches and 4 inches. If the area of the first is 52 square inches, find the area of the second.
2. Two similar triangles have corresponding sides whose lengths are 5 inches and 3 inches. If the area of the first is 100 square inches, find the area of the second.
   Suggestion: Let x represent the unknown area; then
   $$\frac{5^2}{3^2} = \frac{100}{x}$$
3. Two similar triangles have areas of 98 and 50 square inches respectively. The length of a side of the first is 6 inches. Find the length of the corresponding side of the second.
4. The lengths of the corresponding sides of two similar polygons are 6 and 18 inches respectively. What is the ratio of their areas? Of their perimeters?
5. Two similar pentagons have areas of 200 square inches and 1152 square inches respectively. If the length of a side of the first is 20 inches, find the length of the corresponding side of the second.
6. If the ratio of the areas of two similar triangles is 4:1, what is the ratio of the measures of the corresponding sides?
7. If the areas of two similar polygons are in the ratio 2:1, what is the ratio of the lengths of the corresponding sides?
8. If the lengths of the sides of an equilateral triangle are increased by a factor of 3, what is the increase in area?
9. If the ratio of the areas of two squares is 1:2, what is the ratio of the measures of their sides?
10. The side of one equilateral triangle is congruent to the altitude of another equilateral triangle. What is the ratio of their areas?

Since pyramids are composed of triangular faces (14.40) and certain sections (14.03) of pyramids are triangles, we are able to establish several ratios and proportions that will prove useful in our study of solids. The proof of the following theorem is presented for your study and understanding.

Given: Pyramid $V\text{-}ABCDE$, with section $A'C'$ parallel to AC and altitude $\overline{VP}$ intersecting plane $A'C'$ at P'

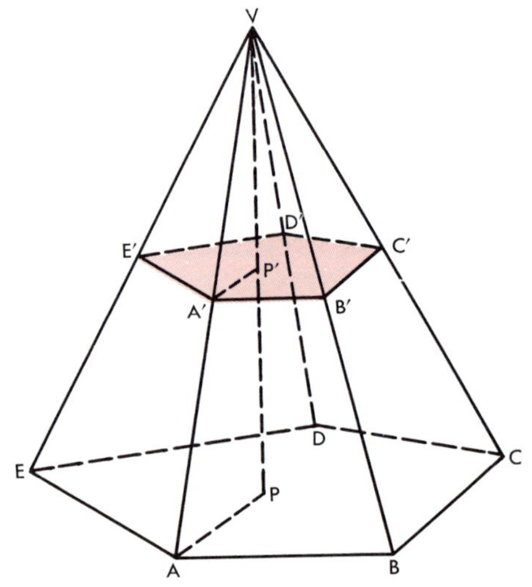

Figure 18–10

Conjecture: (a) $\dfrac{m\overline{VA}}{m\overline{VA'}} = \dfrac{m\overline{VB}}{m\overline{VB'}} = \dfrac{m\overline{VC}}{m\overline{VC'}}$, etc.

(b) Polygon $ABCDE \sim$ polygon $A'B'C'D'E'$

(c) $\dfrac{\text{Area } AC}{\text{Area } A'C'} = \dfrac{m\overline{VP}^2}{m\overline{VP'}^2}$

Proof:

Statements	*Reasons*
1. $\overline{AP} \parallel \overline{A'P'}$, $\overline{AB} \parallel \overline{A'B'}$, $\overline{BC} \parallel \overline{B'C'}$, $\overline{CD} \parallel \overline{C'D'}$, etc.	1. 11.16
2. $\dfrac{m\overline{VP}}{m\overline{VP'}} = \dfrac{m\overline{VA}}{m\overline{VA'}} = \dfrac{m\overline{VB}}{m\overline{VB'}} = \dfrac{m\overline{VC}}{m\overline{VC'}}$, etc.	2. 16.20

3. $\triangle VAP \sim \triangle VA'P'$ $\triangle VAB \sim \triangle VA'B'$, $\triangle VBC \sim \triangle VB'C'$, etc.	3. 17.08
4. $\frac{m\overline{VP}}{m\overline{VP'}} = \frac{m\overline{VA}}{m\overline{VA'}} = \frac{m\overline{AB}}{m\overline{A'B'}} = \frac{m\overline{VB}}{m\overline{VB'}} = \frac{m\overline{BC}}{m\overline{B'C'}} = \frac{m\overline{VC}}{m\overline{VC'}} = \frac{m\overline{CD}}{m\overline{C'D'}}$, etc.	4. 17.01 & Transitive Axiom
5. $\angle ABC \cong \angle A'B'C'$, $\angle BCD \cong \angle B'C'D'$, etc.	5. 11.33
6. Polygon $ABCDE \sim$ polygon $A'B'C'D'E'$	6. 17.00
7. $\frac{\text{Area }(AC)}{\text{Area }(A'C')} = \frac{m\overline{AB}^2}{m\overline{A'B'}^2} = \frac{m\overline{VP}^2}{m\overline{VP'}^2}$	7. 18.14, Transitive Axiom, Step 4

18.15 THEOREM

If a pyramid is intersected by a plane parallel to the base:

(a) The lateral edges and altitude are divided proportionally.

(b) The section is a polygon similar to the base.

(c) The area of the section is to the area of the base as the square of the distance of the plane from the vertex is to the square of the measure of the altitude of the pyramid.

Vocabulary List

area	parallelogram	altitude
commensurable	trapezoid	height
incommensurable	pyramid	ratio
	base	

Chapter Review

1. Draw two non-congruent parallelograms that have congruent bases and congruent altitudes.
2. Find the area of the polygon $ABCDE$. (A common way to measure irregular polygons is to enclose them in a rectangle as shown.)

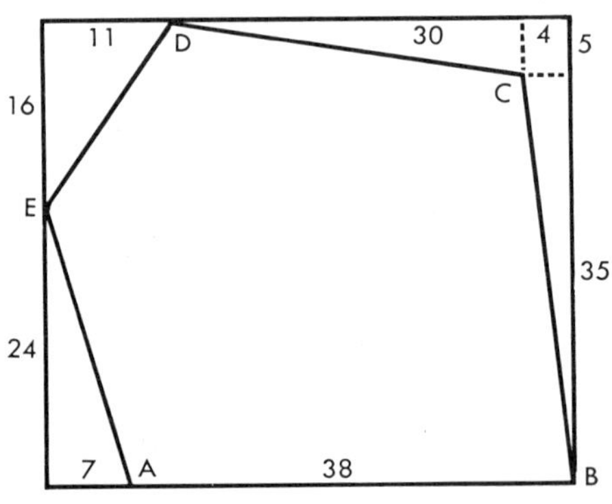

Figure 18–11

3. Find the area of a square the measure of whose side is 5 inches.
4. Find the area of a square the measure of whose diagonal is $2\sqrt{2}$ inches.
5. Find the area of an equilateral triangle the length of whose side is 8 inches.
6. Find the area of an equilateral triangle whose height is $5\sqrt{3}$ inches.
7. Find the area of an equilateral triangle the length of whose median is 6 inches.
8. The length of the side of an equilateral triangle is 12 inches. What is the length of the side of an equilateral triangle with twice the area? With three times the area?
9. The length of the congruent sides of an isosceles triangle is 10 inches and the base is 16 inches. Find the area of the triangle.
10. The length of the congruent sides of an isosceles triangle is 17 inches and the base is 16 inches. Find the area of the triangle.

11. The lengths of the diagonal and of one side of a rectangle are 13 inches and 5 inches respectively. Find the area of the rectangle.

12. The lengths of the diagonal and of one side of a rectangle are $\sqrt{56}$ inches and 5 inches respectively. Find the area of the rectangle.

13. Find the area of $\triangle ABC$, in which $\overline{CD} \perp \overline{AB}$, when:
(a) $m\overline{AC} = 16$ inches, $m\overline{AB} = 11$ inches, and $m\angle A = 30°$.
(b) $m\overline{AC} = 10$ inches, $m\overline{AB} = 8$ inches, and $m\angle A = 60°$.
(c) $m\overline{AC} = 4$ inches, $m\overline{AB} = 3$ inches, and $m\angle A = 45°$.
(d) $m\overline{AC} = 12$ inches, $m\overline{AB} = 8$ inches, and $m\angle A = 150°$.
(e) $m\overline{AC} = 10$ inches, $m\overline{AB} = 7$ inches, and $m\angle A = 120°$.
(f) $m\overline{AC} = 5$ inches, $m\overline{AB} = 8$ inches, and $m\angle A = 135°$.

14. Find the area of parallelogram $ABCD$, in which $\overline{DE} \perp \overline{AB}$, when:
(a) $m\overline{AD} = 10$ inches, $m\overline{AB} = 15$ inches, and $m\angle A = 30°$.
(b) $m\overline{AD} = 15$ inches, $m\overline{DC} = 10$ inches, and $m\angle A = 60°$.
(c) $m\overline{BC} = 8$ inches, $m\overline{AB} = 12$ inches, and $m\angle A = 45°$.
(d) $m\overline{AD} = 6$ inches, $m\overline{AB} = 18$ inches, and $m\angle A = 120°$.
(e) $m\overline{AD} = 9$ inches, $m\overline{AB} = 15$ inches, and $m\angle A = 150°$.
(f) $m\overline{AD} = 12$ inches, $m\overline{AB} = 16$ inches, and $m\angle A = 135°$.

15. What is the area of a rhombus the lengths of whose diagonals are 10 and 14 inches?

16. A rectangle has an area of 90 square feet and a perimeter of 42 feet. What are the dimensions?

17. The ratio of the height to the length of the base in a rectangle is 2:3. The area is 96 square feet. Find the dimensions.

18. The perimeter of a square is 12 inches. Find its area.

19. The length of a diagonal of a square is 6 inches. Find its area.

20. Find the area of a square in terms of the length of the diagonal.

21. The area of the base of a pyramid is 16 square inches and its height is 6 inches. What is the area of a section made by a plane that is parallel to the base and 3 inches from the vertex?

22. The area of the base of a pyramid is 64 square inches and its height is 8 inches. Find the distance from the vertex to the plane parallel to the base if the area of its section is 9 square inches.

23. Find the area of $\triangle ABC$ in the accompanying cube if the measure of its edge is 16 inches and A is the midpoint of an edge.

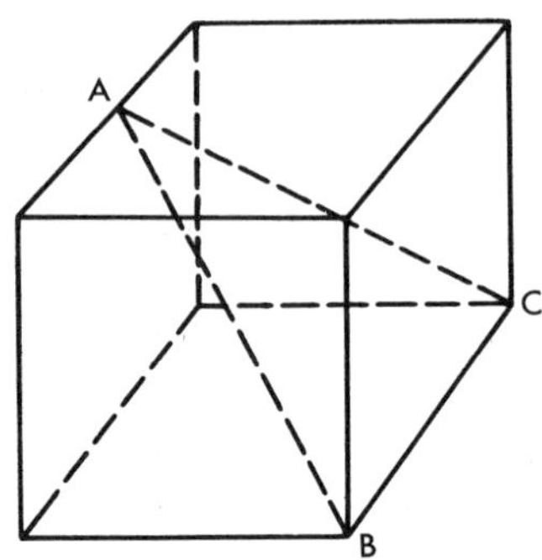

Figure 18–12

Determine whether each proposition below is true or false. Prove those that you believe to be true.

24. A median of a triangle divides it into two equivalent (equal in area) triangles.
25. The area of an isosceles right triangle is equal to one-fourth the square of the length of the hypotenuse.
26. The area of a trapezoid is equal to the product of the lengths of the altitude and of the median.
27. The diagonals of a parallelogram divide it into four equivalent triangles.
28. The area of a rhombus is equal to one-half the product of the lengths of its diagonals.

Chapter 18 Test

1. Find the area of

   (a) a triangle, the measures of whose base is 15 inches and the altitude upon the base is 8 inches.

   (b) a parallelogram, the measures of whose base is 15 inches and the altitude upon the base is 8 inches.

   (c) a square, the measure of whose side is 9 inches.

   (d) a trapezoid, the measures of whose bases are 5 inches and 10 inches and the altitude is 6 inches.

2. Find the ratio of the areas of two similar triangles if the ratio of the measures of two corresponding sides is 4 : 5.

3. The measures of the diagonals of a rhombus are 10 and 24 inches. Find the area of the rhombus.

4. The measure of the median of a trapezoid is 12 inches and the altitude of the trapezoid is 10 inches. Find the area of the trapezoid.

5. The measures of the sides of a triangle are 3,4,5 inches.
   (a) Find the area of the triangle.
   (b) The median is drawn to the 4 inch side of this triangle. Find the area of each of the two triangles thus formed.
   (c) Show that $\sqrt{s(s - a)(s - b)(s - c)}$, where a, b, c are the measures of the sides of a triangle, and $s = \frac{1}{2}(a + b + c)$, will also give the area of the triangle.

19

Regular Polygons and the Circle

This chapter examines the relationship of a circle to the regular polygons which may be inscribed in, and circumscribed about, the circle (12.33–12.34). This work is directed toward developing an expression for the area of a circle.

First, let us review our knowledge of polygons and circles by answering the following questions.

1. State the conditions necessary for a polygon to be regular.
2. Are all equiangular polygons regular? All equilateral polygons?
3. State the conditions necessary for polygons to be similar.
4. Are all equiangular polygons similar? All equilateral polygons?
5. When is a polygon said to be inscribed in a circle? Circumscribed about a circle?

6. When is a circle said to be circumscribed about a polygon? Inscribed in a polygon?
7. What is the relationship of the side of a circumscribed polygon to the radius of the circle drawn to the point of tangency?
8. What is a regular hexagon?
9. Is a rhombus a regular polygon?
10. What is the sum of the measures of the exterior angles of a polygon?
11. State the relationship for the sum of the measures of the angles of a polygon.
12. State two theorems concerning congruent chords in a circle.
13. State the theorem concerning the perimeters of two similar polygons.
14. State the theorem concerning the areas of two similar polygons.

In our attempt to develop expressions for the area and circumference of a circle, we will compare a circle to a regular polygon with a very large number of sides. Toward this end, let us study the possibility of inscribing and circumscribing regular polygons and circles.

We have defined (12.33–12.34) inscribed and circumscribed polygons and shown (15.09 and 15.12) that any triangle may be circumscribed about, or inscribed within, a circle. It is quite obvious that this is not true of all polygons, as shown below.

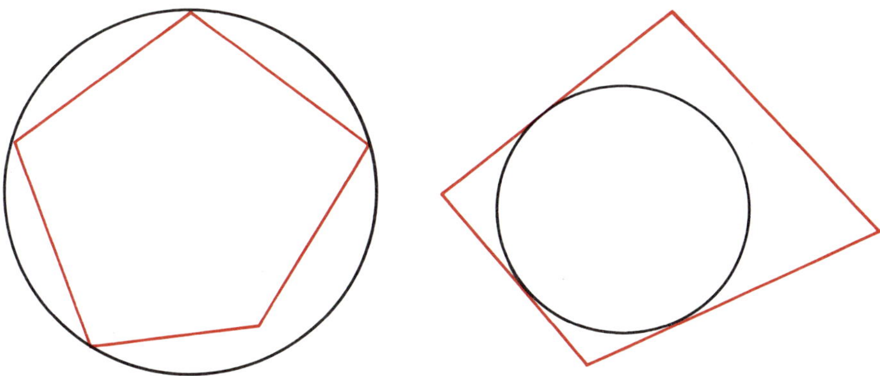

Figure 19–1

The question is, can we inscribe any *regular* polygon in, or circumscribe it about, a circle? We shall examine this proposition deductively.

Given: Any regular polygon $ABCD\ldots$

Conjecture: A circle can be circumscribed about $ABCD\ldots$

Plan: A circle can be drawn through any three vertices. Show that the center of this circle is the same distance from a fourth vertex as it is from any of the three.

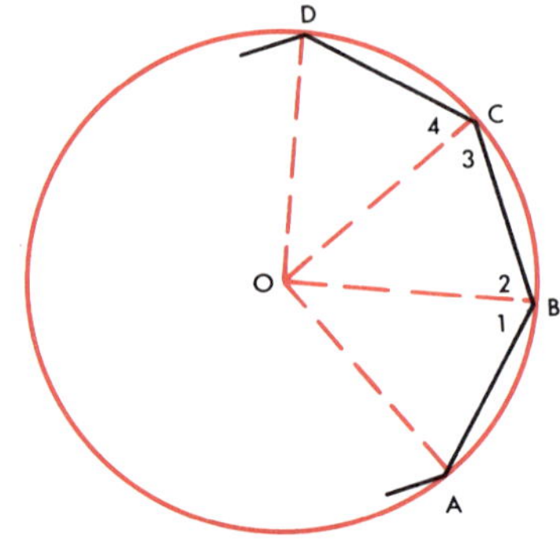

Figure 19–2

Proof:

Statements	*Reasons*
1. Construct a circle through points A, B, and C. Draw radii $\overline{OA}$, $\overline{OB}$, $\overline{OC}$. Draw $\overline{OD}$.	1. Why possible?
2. $\overline{AB} \cong \overline{CD}$	2. Why?
3. $\overline{OB} \cong \overline{OC}$	3. Why?
4. $m\angle 1 + m\angle 2 = m\angle 3 + m\angle 4$	4. Why?
5. $m\angle 2 = m\angle 3$	5. Why?
6. $m\angle 1 = m\angle 4$ and $\angle 1 \cong \angle 4$	6. Why?
7. $\triangle AOB \cong \triangle COD$	7. Why?
8. $\overline{OD} \cong \overline{OA}$	8. Why?

Therefore circle O passes through D, and it can be shown to pass through any other vertex of the regular polygon.

19.00 THEOREM

A circle can be circumscribed about any regular polygon.

Given: Regular polygon $ABCDE$

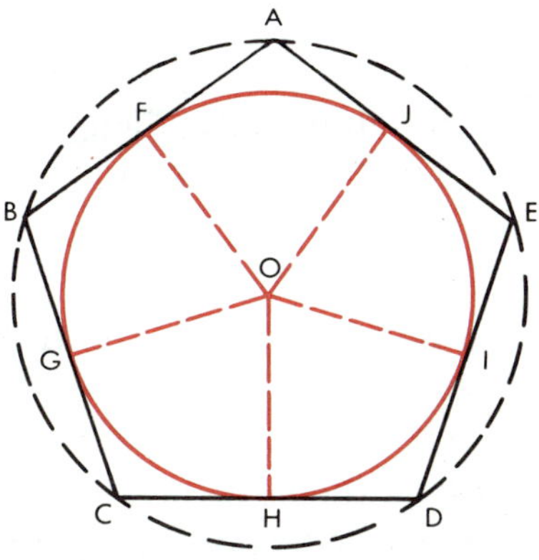

Figure 19–3

Conjecture: A circle can be inscribed in $ABCDE$.

Plan: Show that the sides of the polygon are tangent to the circle.

Proof: *Statements*	*Reasons*
1. Circumscribe circle O about $ABCDE$	1. 19.00
2. $\overline{AB} \cong \overline{BC} \cong \overline{CD} \cong \overline{DE} \cong \overline{EA}$	2. Why?
3. Construct $\perp$ s from O to each side of $ABCDE$.	3. Why possible?
4. $\overline{OF} \cong \overline{OG} \cong \overline{OH} \cong \overline{OI} \cong \overline{OJ}$	4. 12.38
5. With O as center and $\overline{OF}$ as radius, construct a circle.	5. Why possible?
6. The sides of $ABCDE$ are tangent to the circle in step 5.	6. 2.47, 12.55
7. $\therefore$ this circle is inscribed in $ABCDE$.	7. 12.34

19.01 THEOREM

A circle can be inscribed in any regular polygon.

19.02 The center of a regular polygon is the common center of its inscribed and circumscribed circles.

19.03 The radius of a regular polygon is the radius of its circumscribed circle. We will abbreviate "length of the radius" to "radius." The meaning will be obvious from the context.

19.04 THEOREM

The radius of a polygon bisects the angle at the vertex to which it is drawn.

19.05 The apothem (ăp′ ō thĕm) of a regular polygon is the radius of its inscribed circle drawn to the point of contact (the perpendicular bisector of its side). We will also use "apothem" to mean "length of the apothem."

19.06 The central angle of a regular polygon is an angle formed by two radii drawn to the extremities of a side.

19.07 THEOREM

The measure of a central angle of a regular polygon of n sides is equal to $360° \div n$.

Exercises

1. Find the measure of the central angle and an angle of each of the following regular polygons:

   (a) Triangle (b) Quadrilateral (c) Pentagon
   (d) Hexagon (e) Octagon (f) Dodecagon

   Do you note a relationship between the central angle and an angle of a regular polygon? Is this inductive, or deductive reasoning?

2. Find the apothem of a square the measure of whose side is 4 inches. What is the length of the radius of the square?

3. An equilateral triangle is inscribed in a circle whose radius is 6 inches. Find the lengths of the apothem and of a side of the triangle.

4. What is the ratio of the apothem of an equilateral triangle to the radius of its circumscribed circle? (Let the apothem be x units.)

5. What is the length of the apothem of a regular hexagon inscribed in a circle whose radius is 8 inches?
6. Find the area of a square circumscribed about a circle whose radius is 4 inches. What is the length of the diagonal of this square?
7. Find the area of a regular hexagon the measure of each of whose sides is 4 inches.

Draw a circle and select three points on the circle which divide it into three congruent arcs. Join these points in order by segments, each of which is a chord of the circle. What is the figure formed by the chords? If five points on the circle are selected so that they divide the circle into five congruent arcs, and the five chords are drawn so as to join the points in order, what figure is formed? Do the chords have to be congruent? Why? Does the figure lie within the circle?

Given: Circle O, with $\widehat{AB} \cong \widehat{BC} \cong \widehat{CD} \cong \widehat{DE} \cong \widehat{EA}$, and chords $\overline{AB}$, $\overline{CD}$, $\overline{BC}$, $\overline{DE}$, and $\overline{EA}$ (Figure 19-4)

Conjecture: $ABCDE$ is a regular inscribed polygon.

Plan: Show that the chords are congruent. Then show that $\widehat{BCDE} \cong \widehat{CDEA} \cong \widehat{DEAB}$, etc., so $\angle A = \angle B \cong \angle C$, etc.

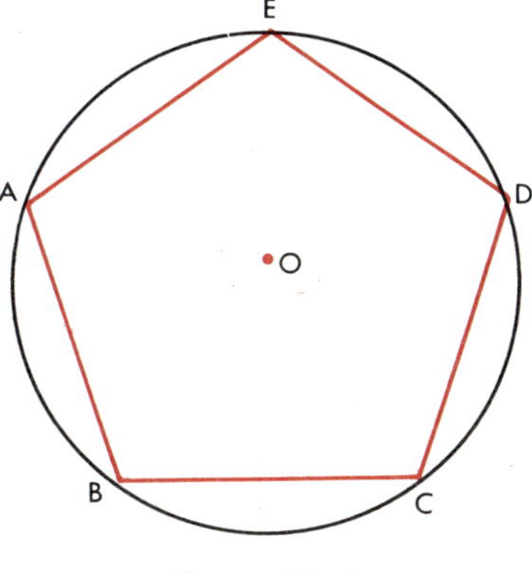

Figure 19–4

Proof: The formal proof is left to the student.

19.08 THEOREM

If a circle is divided into three or more congruent arcs, the chords of these arcs form a regular inscribed polygon.

19.09 THEOREM

An equilateral polygon inscribed in a circle is a regular polygon.

Draw a circle and select three points on the circle which divide it into three congruent arcs. To these points draw the corresponding three radii. Construct lines perpendicular to each radius at the endpoint on the circle; what relation do these lines have to the circle? Do they intersect so as to form a geometric figure? What is the figure? If five points are selected dividing a circle into five congruent arcs, will the tangents to the circle at these points form a special figure? What is the name of the figure? Does the figure lie entirely in the exterior of the circle?

Given: Circle O, with congruent arcs $\widehat{AB}$, $\widehat{BC}$, $\widehat{CD}$, etc., and tangents at points A, B, C, etc., intersecting at F, G, H, etc.

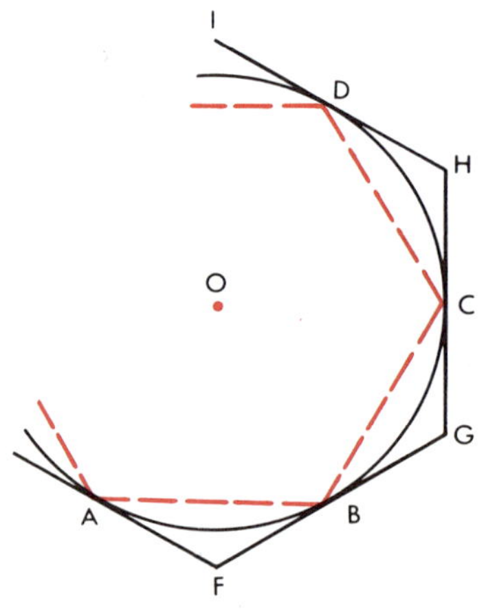

Figure 19–5

Conjecture: $FGHI \cdots$ is a regular circumscribed polygon.

Plan: Draw chords $\overline{AB}$, $\overline{BC}$, $\overline{CD}$, etc., and then prove that ⓢAFB, BGC, CHD, etc., are congruent and isosceles. Therefore $\overline{FG} \cong \overline{GH} \cong \overline{HI} \cong \cdots$ and $\angle F \cong \angle G \cong \angle H \cong \cdots$.

Proof: The formal proof is left to the student.

19.10 THEOREM

If a circle is divided into three or more congruent arcs, the tangents at the points of division form a regular circumscribed polygon.

What are the requirements for two polygons to be similar? Do two regular polygons with the same number of sides satisfy these requirements?

Given: Regular polygons $ABCD \cdots$ and $A'B'C'D' \cdots$

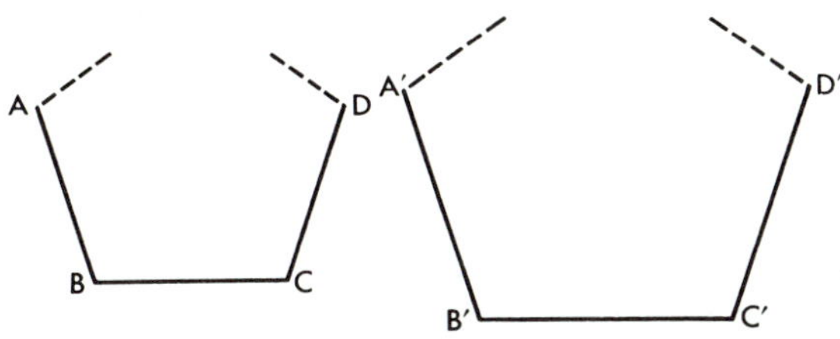

Figure 19–6

Conjecture: $ABCD \cdots \sim A'B'C'D' \cdots$

Plan: Use 17.00.

Proof: *Statements*	*Reasons*
1. $\overline{AB} \cong \overline{BC} \cong \overline{CD}$, etc. $\overline{A'B'} \cong \overline{B'C'} \cong \overline{C'D'}$, etc.	1. 6.07
2. $\dfrac{m\overline{AB}}{m\overline{A'B'}} = \dfrac{m\overline{BC}}{m\overline{B'C'}} = \dfrac{m\overline{CD}}{m\overline{C'D'}}$, etc.	2. Division Axiom
3. The measure of each angle of both polygons equals $\dfrac{(n-2)180°}{n}$.	3. 6.22
4. $ABCD \cdots \sim A'B'C'D' \cdots$	4. 17.00

19.11 THEOREM

Regular polygons with the same number of sides are similar.

The measures of the corresponding sides of two regular polygons with the same number of sides are proportional. Are the measures of the perimeters, apothems, or radii of such polygons proportional? Let us test this question deductively.

Given: Regular polygons $ABCD\cdots$ and $A'B'C'D'\cdots$, each having n sides, with centers O and O', radii r and r', apothems a and a', and perimeters p and p'

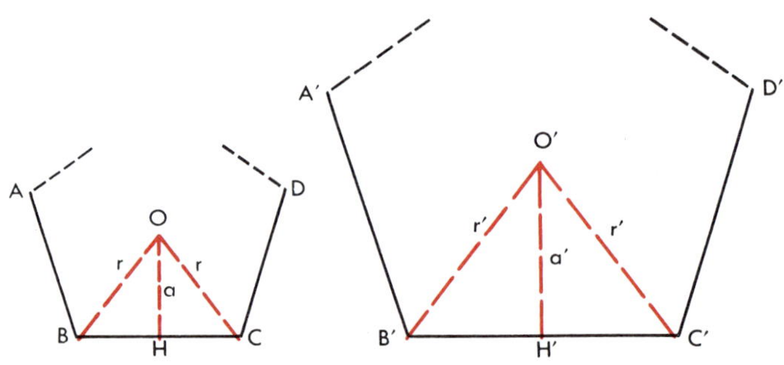

Figure 19–7

Conjecture: $\dfrac{p}{p'} = \dfrac{r}{r'} = \dfrac{a}{a'}$

Plan: Show that $\triangle BOH \sim \triangle B'O'H'$. Then use 17.04.

Proof:

Statements	*Reasons*
1. $\angle OBH \cong \angle O'B'H'$	1. 6.07, 19.04
2. $\overline{OH} \perp \overline{BC}$, and $\overline{O'H'} \perp \overline{B'C'}$	2. 19.05
3. $\triangle BHO \sim \triangle B'H'O'$	3. 17.02
4. $\dfrac{r}{r'} = \dfrac{a}{a'} = \dfrac{m\overline{BH}}{m\overline{B'H'}}$	4. 17.01
5. $m\overline{BC} = 2 \cdot m\overline{BH}$, $m\overline{B'C'} = 2 \cdot m\overline{B'H'}$	5. Why?
6. $\dfrac{m\overline{BH}}{m\overline{B'H'}} = \dfrac{m\overline{BC}}{m\overline{B'C'}}$	6. Why?
7. $ABCD\cdots \sim A'B'C'D'\cdots$	7. Why?
8. $\dfrac{p}{p'} = \dfrac{m\overline{BC}}{m\overline{B'C'}}$	8. 17.04
9. $\dfrac{p}{p'} = \dfrac{r}{r'} = \dfrac{a}{a'}$	9. Transitive axiom

This proof has established the following theorem.

19.12 THEOREM

The ratio of the perimeters of two regular polygons with the same number of sides equals the ratio of the radii or of the apothems.

In what ways can a regular polygonal region be divided into smaller regions? How could the center of the polygon be used to form regions whose area can be found?

Given: Regular polygon $ABCD\cdots$, with n sides, center O, perimeter p, and apothem a

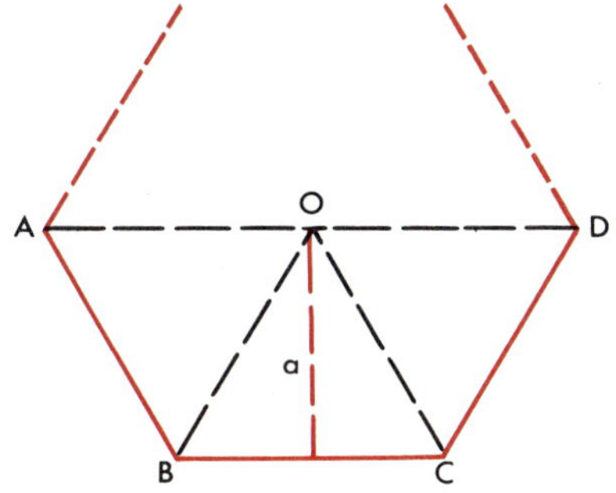

Figure 19–8

Conjecture: The area of $ABCD\cdots = \frac{1}{2}ap$

Plan: Draw radii $\overline{AO}$, $\overline{BO}$, $\overline{CO}$, $\overline{DO}$, etc., forming n triangles. Prove the triangles congruent. Find the area of one triangle and multiply by n. Now we see that $m\overline{AB} + m\overline{BC} + m\overline{CD} + \cdots = n(m\overline{BC}) =$ perimeter (p), so $A = \frac{1}{2}ap$.

19.13 THEOREM

The area of a regular polygon is half the product of its apothem and its perimeter.

Spanish National Tourist Office and Iberia Air Lines

The geometric forms and interlocking lines in this design from the Alhambra in Granada, Spain, are typical of Moorish decoration. Can you find any regular polygons within tne design?

Since for two similar regular polygons the ratio of the apothems and the ratio of the perimeters is the same, it seems likely that the areas of the polygons will also be related. Let us see whether this relationship can be proved.

Outline of Proof:

Let the areas of the polygons be A and A', the radii r and r', the apothems a and a', and the perimeters p and p'.

Then $A = \frac{1}{2}ap$ and $A' = \frac{1}{2}a'p'$.

Now $\frac{A}{A'} = \frac{\frac{1}{2}ap}{\frac{1}{2}a'p'} = \frac{ap}{a'p'} = \frac{a}{a'} \cdot \frac{p}{p'}$. Why?

Since $\frac{p}{p'} = \frac{a}{a'} = \frac{r}{r'}$, therefore $\frac{A}{A'} = \frac{a^2}{a'^2} = \frac{r^2}{r'^2}$. Why?

19.14 THEOREM

The areas of two regular polygons of the same number of sides are to each other as the squares of their radii or as the squares of their apothems.

Exercises

1. In circles whose radii are 2, 8, 12, and x centimeters, find the length of the sides of the inscribed equilateral triangles. What is the area of the triangle in each case?
2. In circles with radii of 2, 4, 6, and x inches, what would be the length of the apothem of a regular hexagon inscribed in each? What would be the area in each case?
3. Find the areas of the regular hexagons circumscribed about circles whose radii are 2, 6, 10, and x inches.
4. Find the area of the equilateral triangles circumscribed about circles whose radii are 3, 6, 12, and x centimeters.
5. Compare the measures of the sides and areas of two equilateral triangles that are respectively inscribed in, and circumscribed about, a given circle.
6. If the area of one regular polygon is 36 square inches, what is the area of another regular polygon of the same number of sides and half the perimeter?
7. If the area of one regular polygon is 10 square centimeters what is the area of another regular polygon having the same number of sides and three times the perimeter?
8. If the radius of a circle is tripled, what is the effect upon the area of a regular polygon inscribed within it?
9. What is the ratio of the perimeters of two regular polygons inscribed in circles whose radii are 3 inches and 5 inches? What is the ratio of the areas of the polygons?
10. The area of one regular pentagon is 212 square inches, and the area of another regular pentagon is 53 square inches. Find the ratio of their radii. Find the ratio of the lengths of their sides.
11. Find the ratio of the apothems of two regular octagons whose areas are 25 and 50 square feet respectively.
12. In order to double the area of an inscribed regular pentagon, in what ratio must the radius of the circle be increased?
13. The area of a right section of a metal beam in the form of a regular pentagonal prism is 28 square inches. A second regular pentagonal prismatic beam has a base edge half as long as

the base edge of the first. What is its cross-sectional (right-section) area?

14. The area of one equilateral triangle is 16 times that of another. What is the ratio of their perimeters?

15. What is the area of the cross section of the largest possible square piece of timber that can be sawed from a log in the form of a circular cylinder 18 inches in diameter?

16. The radius of a circle inscribed in a regular octagon is $7\frac{1}{4}$ inches, and the length of a side of the octagon is 6 inches. What is the area of the octagon?

17. The radius of a circle inscribed in a regular quadrilateral is $7\frac{1}{4}$ inches, and the length of a side of the quadrilateral is $14\frac{1}{2}$ inches. What is the area of the quadrilateral?

MEASUREMENT OF THE CIRCLE

19.15 One of the most interesting mathematical relationships is that of the circumference of a circle to the diameter. This ratio $\left(\frac{c}{d}\right)$ is called π (pi). It is amazing that Euclid and his contemporaries evidently never calculated π. At various times crude estimates of the value had been made. The Old Testament gives an estimate of 3 for the value. This may have been satisfactory for the roughly circular wheels of the carts of that day but would hardly be suitable for today's wheels.

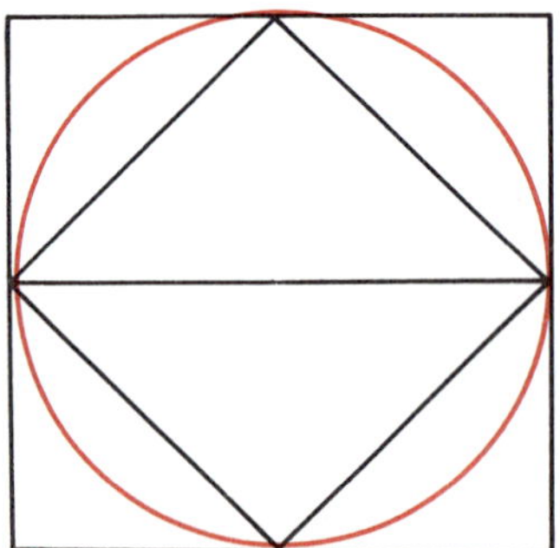
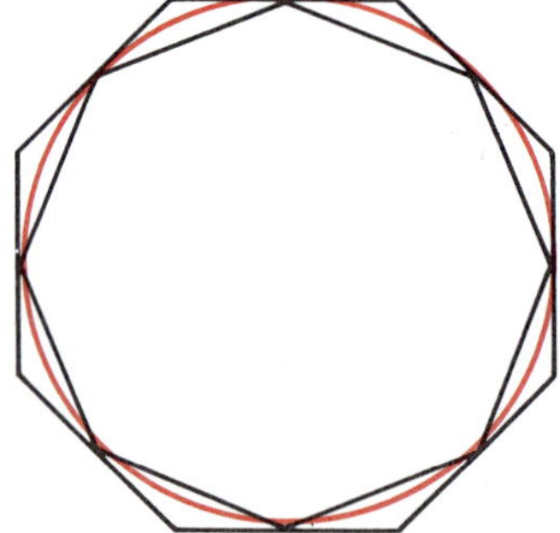

Figure 19–9

Archimedes, in the century following Euclid, developed the following method of finding π: Take a circle whose diameter is one. Inscribe a square in the circle and circumscribe a square about it. It is quite evident (Figure 18-9) that the circumference of the inscribed square is $2\sqrt{2}$, or approximately 2.828, and the circumference of the circumscribed square is 4. Thus $2.828 < \pi < 4$. Now we can double the number of sides. It is easy to calculate that the circumference of the inscribed regular octagon is 3.0615 and the circumference of the circumscribed regular octagon is 3.3137. Thus $3.0615 < \pi < 3.3137$. If we continue to double the number of sides of the regular polygons, we obtain the results in the following table.

Number of sides	*Perimeter of inscribed polygon divided by the diameter of the circle*	*Perimeter of circumscribed polygon divided by the diameter of the circle*
4	2.828427	4.000000
16	3.121445	3.182597
64	3.140331	3.144118
256	3.141513	3.141750
1024	3.141587	3.141602
4096	3.141592	3.141593

As we increase the number of sides, the circumference of the inscribed (or circumscribed) regular polygon approaches the circumference of the circle as a limit. We may define the circumference (or area) of the circle as the limit of the circumference (or area) of the inscribed regular n-gon as n increases indefinitely. By making n large enough we can approximate π as closely as we wish.

Archimedes' method is called the method of exhaustion because the expanding inscribed regular polygon gradually "exhausts the area" of the circle. This method contains the beginning of a type of mathematics called integral calculus.

19.16 The number 3.14159+ is an approximation for

$$\frac{\text{circumference}}{\text{diameter}} = \frac{c}{d} = \pi$$

$$\text{or } c = \pi d = 2\pi r.$$

Since the ratio of circumference to diameter is constant for all circles, (it equals *pi*), the circumferences of any two circles are in the same ratio as their diameters (or radii).

Proof: Let c_1 and c_2 represent the circumferences, d_1 and d_2 represent the diameters, and r_1 and r_2 represent the radii of two circles. (In the expression c_1 and c_2, 1 and 2 are subscripts. The expression is read "c sub-one and c sub-two.")

Since $c_1 = \pi d_1 = 2\pi r_1$,
and $c_2 = \pi d_2 = 2\pi r_2$ by 19.14,

therefore $\frac{c_1}{c_2} = \frac{r_1}{r_2}$.

19.17 THEOREM

The circumferences of two circles are to each other as the radius of one is to the radius of the other.

Exercises

1. The radii of two circles are 1 and 3 inches. What is the ratio of their circumferences?
2. Find the circumferences of circles whose radii, in inches, are as follows: (Use 3.14 as the value for π.)

(a) 14 (b) 7 (c) 2.45 (d) 42.7

(e) $5\frac{1}{2}$ (f) $3\frac{1}{7}$ (g) $6\frac{1}{4}$ (h) 3.33

3. Find the circumference of a circle whose diameter is 12.4 inches.
4. Find the length of the arc of a circle with a radius of 6 inches, intercepted by a central angle whose measure is 60°.
5. Find the length of the arc of a circle with a 15-inch radius, intercepted by a central angle whose measure is 120°.
6. Find the length of the arc of a circle with a 12-inch radius, intercepted by a central angle whose measure is 75°.
7. What is the length of the equator to the nearest thousand miles if we use 8000 miles as the earth's diameter?
8. What is the ratio of the circumferences of a small circle and a great circle of a sphere if the diameter of the sphere is 10 inches

and the plane of the small circle is 3 inches from the center of the sphere?

9. Find the circumference of a circle inscribed in an isosceles trapezoid the lengths of whose bases are 6 inches and 18 inches.

THE AREA OF A CIRCLE

19.18 In 19.15 we see that the area of an inscribed regular polygon increases as the number of its sides increases, and approaches the area of the circle. In 19.13 the area of a regular polygon was proved to be $\frac{1}{2}ap$, where a is the apothem and p is the perimeter. As the number of sides of the polygon increases, its apothem approaches the length of the radius of the circle and its perimeter approaches the length of the circumference of the circle.

Hence $\frac{1}{2}ap$ becomes $\frac{1}{2}rc$ for a circle. If $A = \frac{1}{2}rc$, and $c = 2\pi r$ (See 19.16), then $A = \frac{1}{2}r \cdot 2\pi r$.

$$\therefore A = \pi r^2$$

Thus we have established the following theorem.

19.19 THEOREM

The area of a circle is πr^2.

Since *pi* is constant, the area of a circle ($A = \pi r^2$) depends on the square of the radius. Thus the areas of any two circles can be compared by comparing the squares of their radii.

Proof: Let A_1 and A_2 be the areas of the circles, and r_1 and r_2 be the radii of the circles. Then $A_1 = \pi r_1^2$ and $A_2 = \pi r_2^2$, therefore $\frac{A_1}{A_2} = \frac{r_1^2}{r_2^2}$

19.20 THEOREM

The areas of two circles are to each other as the square of the radius of the first circle is to the square of the radius of the second.

Exercises

1. Find the areas of the circles whose radii, in inches, are as follows: (Use $\pi = \frac{22}{7}$.)

   (a) 7 (b) 14 (c) 42.7 (d) 2.45
   (e) 12.8 (f) 2 (g) 5 (h) 9

2. If the radii of two circles are 2 and 3 inches respectively, what is the ratio of their areas?
3. If the radii of two circles are 1 and 4 inches respectively, what is the ratio of their circumferences and their areas?
4. How much must the radius of a circle be increased to make its area four times as large?
5. The areas of two circles are 9 and 16 square inches respectively. What is the ratio of their radii?
6. What is the ratio of the radii of two circles if the area of one is twice the area of the other?
7. What is the ratio of the areas of two circles if the radius of one is twice the radius of the other?

SECTOR OF A CIRCLE

19.21 A sector of a circle is the region determined by two radii and their intercepted arc. Region OAB is a sector of circle O.

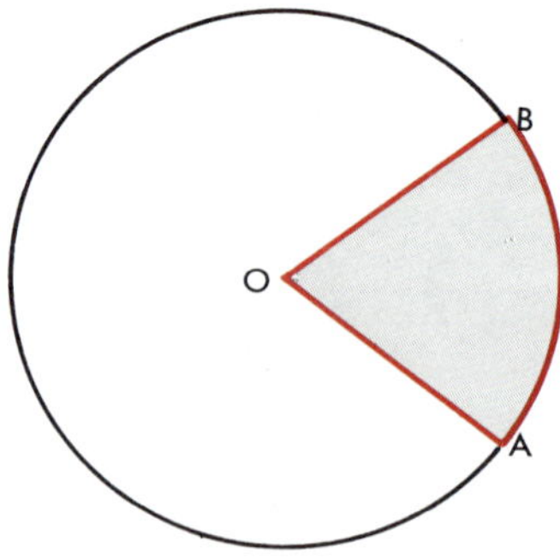

Figure 19–10

19.22 THEOREM

The area of a sector of a circle is to the area of the circle as the angle of the sector is to 360°.

EXAMPLE: If $m\angle AOB$ is 40°, then the area of sector AOB is $\frac{40}{360}$, or $\frac{1}{9}$, of circle O.

SEGMENT OF A CIRCLE

19.23 A segment of a circle is the region determined by a chord and its arc. A *minor segment* is formed by a chord and its *minor arc*. A *major segment* is formed by a chord and its *major arc*.

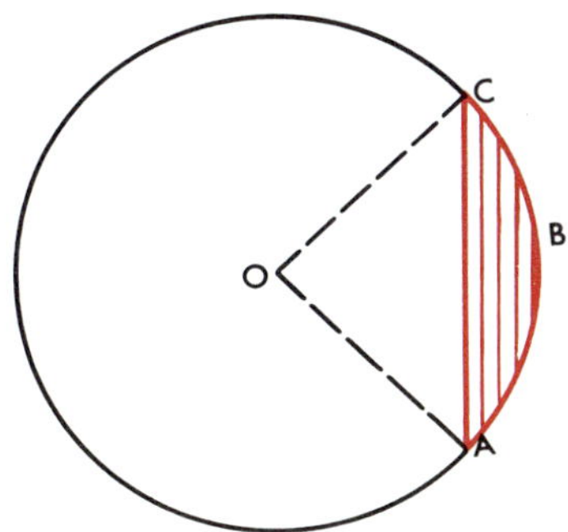

Figure 19–11

The area of segment ABC in circle O may be found by subtracting the area of $\triangle OAC$ from the area of sector $OABC$.

Exercises

1. In a circle with a 6-inch radius find the area of a segment whose arc has a central angle whose measure is 60°.
2. In a circle with a 2-inch radius find the area of the minor segment cut off by the side of an inscribed equilateral triangle.
3. In a circle with a 3-inch radius find the area of the minor segment cut off by the side of an inscribed regular octagon. (Use trigonometry.)

Vocabulary List

apothem	sector
Archimedes	segment of a circle
inscribe	central angle
circumscribe	radius
regular polygon	pi

Chapter Review

1. Find the diameter of a circle whose circumference is 10 π feet.
2. Find the radius of a circle whose area is 225 π square inches.
3. Find the area of the region of a ring made by two concentric circles of radii 6 and 8 inches respectively.
4. The length of a side of an equilateral triangle is 9 inches. Find the area of the circumscribed circle.
5. If each diameter of the three mutually tangent circles is 4 inches, what is the area of x, the region between the three circles?

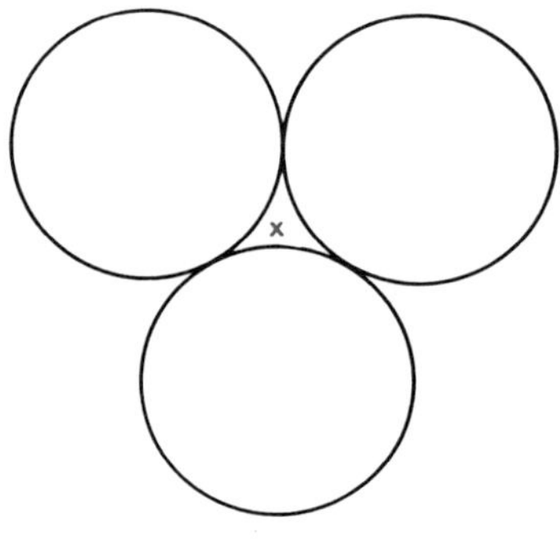

(Exercise 5)

Figure 19–12

6. The halves of a 12-inch diameter of a circle are bisected. From these points as centers, circles are drawn whose diameters are congruent to the radii of the original circle. Find the difference between the areas of the small circles and the original circle.

7. What is the area and circumference of a circle of a sphere having a 5-inch radius, if the center of the circle is 3 inches from the center of the sphere?

8. A small circle of a sphere has a radius of 5 inches, and its center is 12 inches from the center of the sphere. How far from the center of the sphere is the center of a circle whose radius is 12 inches?

9. Two spheres with centers 10 inches apart have radii of 6 and 8 inches respectively. Find the circumference of their intersection.

10. A cone is made from a sector of a circle whose central angle measures 240°. If the radius of the sector is 10 inches, find the altitude, slant height, and radius of the base of the cone.

11. Two circular rods (right circular cylinders) having radii of 4 inches and 12 inches lie side by side. What is the length of the shortest piece of wire which will just go around both of them?

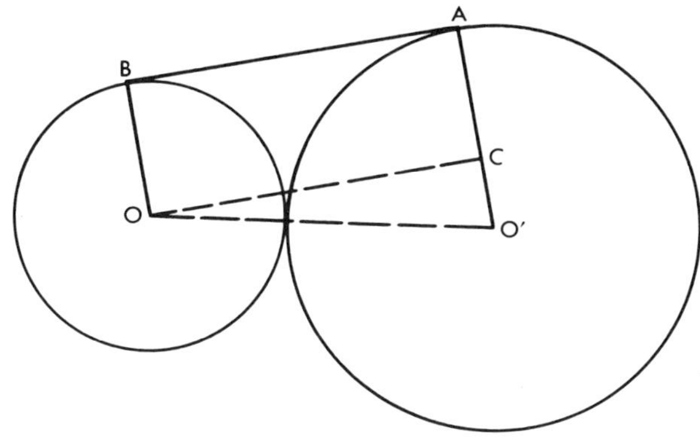

Figure 19–13

12. Find the area of a regular hexagon with a radius 7 inches in length. (Leave your answer in simplest radical form.)

13. Find the area of a regular pentagon (a) with an apothem $3\frac{7}{16}$ inches in length and a side 5 inches in length. (b) with a side 6 inches in length. *Hint:* Use 19.12, 19.14, and your answer for (a).

14. Find the area of a square which is inscribed in a circle whose area is 18π square inches.

Chapter 19 Test

In questions 1–10 place the letter of a statement in Column II after the number of the corresponding statement in Column I.

Column I

1. The central angle of a regular polygon.
2. The apothem of a regular polygon.
3. The radius of a regular polygon.
4. The center of a regular polygon.
5. Area of a regular polygon.
6. Area of a circle.
7. A segment of a circle.
8. A sector of a circle.
9. Circumference of a circle.
10. Regular polygon.

Column II

A. πr

B. πd

C. Equilateral and equiangular.

D. The union of a chord, the arc determined by the chord, and the interior of the circle bounded by the chord and its arc.

E. Radius of its inscribed circle.

F. Determined by two chords.

G. Radius of its circumscribed circle.

H. Determined by two radii drawn to the endpoints of a side.

I. πr^2

J. $\frac{1}{2}ap$

K. Center of its inscribed circle.

L. The union of two radii, the arc determined by the radii, and the interior of the circle bounded by the radii and the arc.

11. Find the apothem **(a)** of a square, the measure of whose side is 8 inches. **(b)** of an equilateral triangle, the measure of whose side is 6 inches. (Leave answer in simplest radical form.)

12. Find the radius

    (a) of a regular hexagon, the measure of whose side is 9 inches.

    (b) of a square, the measure of whose side is 8 inches.

13. The ratio of the measures of the sides of two regular pentagons is 2:3.

(**a**) what is the ratio of their perimeters?

(**b**) what is the ratio of their areas?

14. The measure of the circumference of a circle is $10\,\pi$.

(**a**) Find the radius of the circle.

(**b**) Find the area of the circle (in terms of π).

15. Given a circle whose radius is 6 inches.

(**a**) Find the area of a sector of the circle if the measure of the arc of the sector is 60°.

(**b**) Find the area of the segment of the circle if the measure of the arc of the segment is 60°.

20

Measurement of Surfaces and Solids

This chapter is devoted to determining the area and volume of the common geometric surfaces and solids we studied in Chapter 14: prisms, cylinders, pyramids, cones, and spheres.

AREA OF PRISMS

20.00 The lateral area of a prism is the sum of the areas of the lateral faces.

20.01 The total area of a prism is the sum of its lateral area and the area of its bases.

20.02 The lengths of the three edges meeting at any vertex of a rectangular solid are called its dimensions—or *length, width,* and *height*.

Facts to remember about prisms (14.07–14.26):

1. The lateral edges are congruent and parallel.
2. The bases are congruent, parallel polygonal regions.
3. The lateral faces are parallelogram regions.
4. The edges of a right prism are perpendicular to the bases.
5. The lateral faces of a right prism are rectangular regions.
6. A regular prism is a right prism whose bases are regular polygonal regions.
7. A parallelepiped is a prism whose bases are parallelogram regions.
8. A rectangular solid is a right parallelepiped solid whose bases are rectangular regions.
9. A cube is a rectangular parallelepiped whose edges are congruent.
10. The altitude of a prism is the segment perpendicular to and terminated by the planes of the bases. The measure of the altitude is called the height.
11. The height of a right prism is the length of a lateral edge.
12. A right section of a prism is the intersection of the prism and a plane perpendicular to one of the lateral edges.

Rather than find the area of each lateral face of a prism and then add the results to find the lateral area, it is sometimes possible to simplify the procedure. Let us examine the following figures with this in mind.

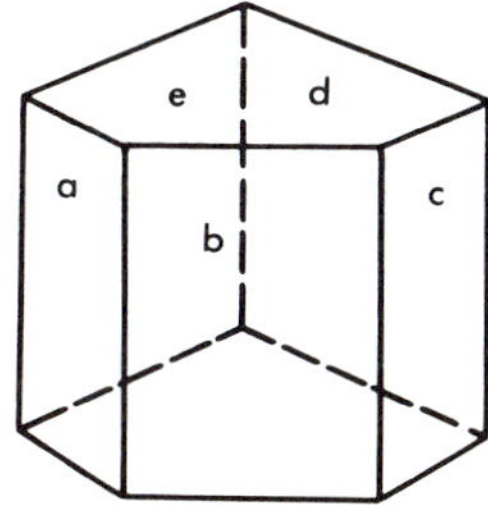

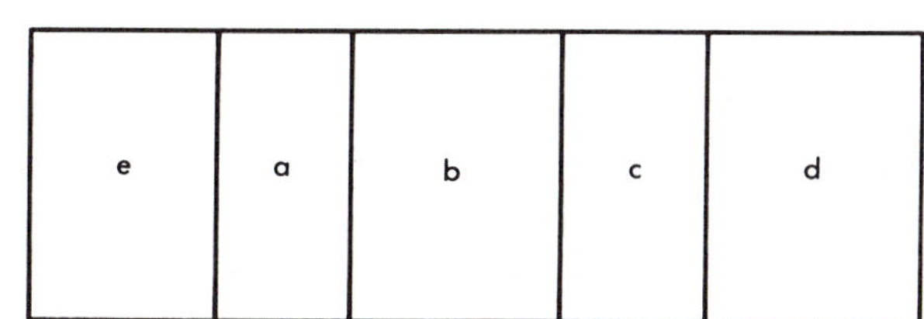

Figure 20–1

Construct a right prismatic surface similar to that shown in Figure 20-1. Cut the surface along one of the edges and observe inductively that this surface can be flattened to a plane surface, a rectangle. The lateral area then is found by determining the product of the height (an edge) and the perimeter of the base of the right prismatic surface ($L = ep$).

How could lateral area be determined if the prismatic surface were oblique?

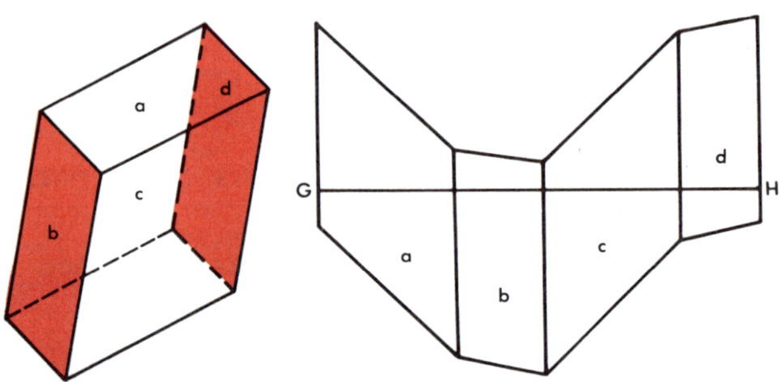

Figure 20–2

Make an oblique prismatic surface like that shown in Figure 20-2. Cut along one edge and flatten to form a plane surface. Draw a straight line on the surface, perpendicular to one edge. (See $\overline{GH}$ in Figure 20-2.)

Would the lateral area equal the length of $\overline{GH}$ times the length of a lateral edge? Reform the figure as a prismatic surface. Does $\overline{GH}$, now a broken line, represent a right section of the prism?

We are now ready to examine deductively the tentative conclusions of the preceding experiment.

Given: Prism $ABCDE\text{-}A'B'C'D'E'$ with right section $LMNOP$. The perimeter of the right section is p, e is the length of a lateral edge, and L is the lateral area of the prism.

Conjecture: $L = ep$

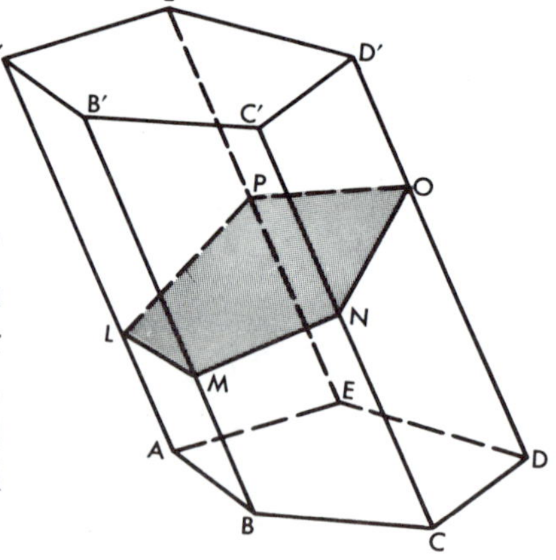

Figure 20–3

Proof: *Statements*	*Reasons*
1. $\overline{AA'} \cong \overline{BB'} \cong \overline{CC'} \cong \overline{DD'} \cong \overline{EE'}$	1. 14.10
2. $\overline{MN} \perp \overline{BB'}$, $\overline{NO} \perp \overline{CC'}$, $\overline{OP} \perp \overline{DD'}$, etc.	2. 14.14, 3.23, 11.31
3. area $(BCC'B') = e \times \mathrm{m}\overline{MN}$, area $(CDD'C') = e \times \mathrm{m}\overline{NO}$, area $(DEE'D') = e \times \mathrm{m}\overline{OP}$, etc.	3. Why?
4. area $(BCC'B')$ + area $(CDD'C')$ + area $(DEE'D') + \ldots = e(\mathrm{m}\overline{MN} + \mathrm{m}\overline{NO} + \mathrm{m}\overline{OP} + \ldots)$	4. Why?
5. $L = ep$	5. Why?

20.03 THEOREM

The lateral area of a prism is equal to the product of the length of a lateral edge and the perimeter of its right section.

If the prism is a right prism, a similar argument establishes the following:

20.04 THEOREM

The lateral area of a right prism is the product of the length of a lateral edge and the perimeter of a base.

$$L = ep$$

Exercises

1. Find the area of a rectangular solid whose dimensions are 9 inches by 10 inches by 11 inches.
2. Find the lateral area of a regular quadrangular prism if its height is 10 inches and the length of one side of the base is 4 inches. Find the total area.

3. Find the lateral area of the following right prisms:
   (a) perimeter = 8 inches; height = 14 inches.
   (b) perimeter = 3 feet 8 inches; height = 2 feet 4 inches.
   (c) perimeter = 4 feet; height = 18 inches.
   (d) perimeter = 6 feet; height = 3 feet 3 inches.
   (e) perimeter = 2 feet 2 inches; height = 2 feet 10 inches.
4. If the dimensions of a rectangular solid are 6, 8, and 4 inches, what is the total area?
5. Find the area of a cube the length of each of whose edges is 1 foot, 3 inches.
6. If the lateral area of a right prism is 429.04 square feet and the height is 12.4 feet, find the perimeter of one base.
7. If the area of a cube is a square yard, find the length of an edge of the cube in inches.
8. The length of a lateral edge of an oblique prism is 8 inches. The distances between the consecutive lateral edges are 3 inches, 5 inches, and 7 inches. Find the lateral area of the prism.
9. The distances between the lateral edges of an oblique triangular prism are 2 inches, 3 inches, and 4 inches. The lateral area is 45 square inches. What is the length of the lateral edge?
10. The base of a right prism is a right triangular region with legs whose lengths are 6 inches and 8 inches. The height of the prism is 12 inches. What is the lateral area and the total area?
11. The base of a right prism is a regular hexagonal region the measure of whose side is 7 inches. The height of the prism is 16 inches. Find the lateral area and the total area.
12. The length of an edge of the base of a regular triangular prism is 6 inches. The height of the prism is 14 inches. Find the lateral area and the total area.
13. Find the lateral area of a right triangular prism whose height is 12 inches and whose base is a right triangular region. The length of the hypotenuse of the base is 13 inches and of one leg is 12 inches.
14. A prism has a rectangular base whose dimensions are 8 and 6 inches. The lateral edges, 10 inches long, are perpendicular to the base. Find the total area of the prism. Find the length of a diagonal of the prism.

15. A prism is 18 inches high, and has a right section which is a rectangle having a diagonal of 15 inches and a side of 9 inches. Find the lateral area.

16. The length of each side of the base of a regular hexagonal prism is 10 inches. The height of the prism is 8 inches. What is the total area?

17. An oblique prism has as a right section a pentagon the measure of whose sides are respectively 3, 4, 5, 6, and 7 inches. The measure of a lateral edge of the prism is 7 inches. What is the lateral area?

18. A certain triangular prism has the lateral edges perpendicular to the base, the lengths of the sides of the triangular base are 6, 6, and 8 inches respectively, and the length of the lateral edge is 10 inches. Find the area of each of the lateral faces and the total area of the prism.

19. Find the lateral area of a right triangular prism whose height is 14 inches and whose base is a right triangle the lengths of whose hypotenuse is 17 inches and of one leg is 15 inches.

20. Find the lateral area of a right prism whose height is 10 inches and whose base is a rhombus region with diagonals whose measures are 6 inches and 8 inches. Find the total area.

21. Two prisms have congruent bases and congruent altitudes. Do the prisms have equal lateral areas?

22. The ratio of the lengths of the edges of two cubes is 3 : 1. What is the ratio of their areas?

23. If two of the dimensions of a rectangular solid are 8 and 12 inches respectively, and the length of the diagonal is 16 inches, find the length of the third dimension.

24. If the length of an edge of a cube is 8 inches, what is the length of the diagonal? What is the total area of the cube?

25. If the length of the diagonal of a cube is 6 feet, find the length of an edge.

26. What is the sum of the measures of all the face angles of a parallelepiped?

27. In a parallelepiped the three edges meeting at C are $\overline{BC}$, $\overline{DC}$, and $\overline{EC}$: $m\angle BCE = 60°$, $m\angle ECD = 120°$, $m\angle DCB = 90°$, $m\overline{BC} = 6$ inches, $m\overline{CD} = 8$ inches, and $m\overline{CE} = 4$ inches. Find the total area of the parallelepiped.

American Airlines

Almost all building architecture uses prisms. New York Hospital, located in Manhattan, is a good example of a building which includes many prisms in its structure.

28. The length of the diagonal of a rectangular solid is 6 inches. The base is a square region and the height is one-half the length of a side of the base. Find the dimensions and the area of the solid.

29. What is the area of the diagonal section of a cube the length of whose edge is 5 inches?

30. What is the area of a cube if the area of the diagonal section is $100\sqrt{2}$ square inches?

31. What is the length of an edge of a cube if its total surface area is 12 square feet?

32. The area of the largest diagonal section containing two lateral edges of a regular right prism with a hexagonal base region is 1 square foot. What is the lateral area of the prism?

VOLUME OF RECTANGULAR SOLIDS

20.05 The volume of a geometric solid is the number of distinct unit volumes that would fill the region. In discussing volume, we will use the names of geometric surfaces, such as cube, prism, or sphere, to refer to the solid determined by that surface.

The polyhedral solid (or region) determined by a cube, the measure of whose edge is one unit, is used as the unit of volume. If the unit length is an inch, then the volume will be in cubic inches; if the unit is a centimeter, the volume is in cubic centimeters.

20.06 Assumption: *To every three-dimensional region there corresponds a unique positive real number called the volume of the region.*

In Figure 20-4, the bottom layer contains 5×3 or 15 cubic units. Since there are 3 such layers, the solid contains 15×3 or 45 cubic units. ∴ Volume $= 5 \times 3 \times 3$ cubic units.

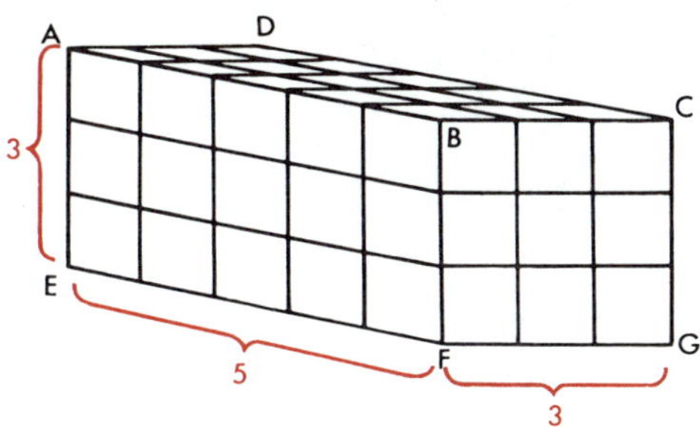

A rectangular parallelepided with dimensions as shown

Figure 20–4

20.07 Assumption: *The volume of a rectangular solid equals the product of the dimensions.*

$$V = lwh$$

20.08 THEOREM

The volume (V) of a rectangular solid is equal to the product of the area of its base (B) and its height (h).

$$V = Bh$$

In the preceding formulas it is assumed that the dimensions are commensurable. The formulas are equally true, however, if the dimensions are incommensurable. (See 18.01 and 18.02.)

20.09 THEOREM

The volume (V) of a cube is equal to the cube of the length of an edge (e).

$$V = e^3$$

Working with volumes will involve the cubes and cube roots of numbers. The following exercises are designed to review and extend your knowledge of roots and powers. As you remember, the *square root* of a positive real number is defined as *one of the two equal factors* of the number. The *cube root* of a real number is defined as *one of the three equal factors* of the number. In general, where $x \geq 0$ and n is a positive even number, $\sqrt[n]{x^n} = |x|$. Where x is a real number and n is odd, $\sqrt[n]{x^n} = x$.

EXAMPLES:

$\sqrt[3]{8} = 2$. *Check:* $2 \cdot 2 \cdot 2 = 8$. Does $(-2)(-2)(-2) = 8$?

$\sqrt[3]{343} = 7$. *Check:* $7 \times 7 \times 7 = 343$.

Review Exercises

Expand:

1. 9^2
2. 15^2
3. 4^3
4. $(x - 1)^2$
5. 7^3
6. $(x^2 + 1)^2$

Find the root:

7. $\sqrt{625}$
8. $\sqrt[3]{27}$
9. $\sqrt[3]{125}$
10. $\sqrt{121x^2y^6}$
11. $\sqrt[3]{216}$
12. $\sqrt[3]{64y^6}$

Find the solution set over the set of positive real numbers:

13. $x^3 = 8$
14. $6e^2 = 54$
15. $2e^3 = 128$
16. $\frac{1}{2}e^3 = 108$
17. $\sqrt[3]{V} = 5$
18. $8e^3 = -64$

Exercises

Volumes of Rectangular Solids.

1. A cube, the length of whose edge is 4 inches, is to be cut into small cubes whose edge length is 1 inch. How many 1-inch cubes will there be?
2. The dimensions of a classroom are 24 feet by 17 feet by 12 feet. Find the number of cubic feet of air space it contains.

3. Find the volumes of rectangular solids whose three dimensions are as follows:
   (a) 10, 8, and 5 inches.
   (b) $2\frac{1}{3}$, $1\frac{3}{4}$, and $5\frac{1}{2}$ inches.
   (c) $10\frac{5}{8}$, $6\frac{2}{3}$, and $4\frac{1}{2}$ inches.
   (d) 3.24, 2.62, and 1.34 feet.
4. Solve the formula $V = lwh$ for h.
5. Find the heights of rectangular solids having the following volumes and areas of bases:
   (a) 50 cubic inches, 10 square inches.
   (b) 64 cubic inches, 15 square inches.
   (c) $30\frac{2}{3}$ cubic feet, $6\frac{1}{4}$ square feet.
   (d) 32.7 cubic feet, 9.28 square feet.
   (e) $25\frac{3}{4}$ cubic feet, $5\frac{1}{2}$ square inches.
6. What is the ratio of the volumes of two rectangular solids whose dimensions are 4, 3, and 2 inches and 5, 6, and 2 inches respectively?
7. What is the effect upon the volume of a rectangular solid when one dimension is doubled and the other two remain the same? When two dimensions are doubled? When all three are doubled? If one dimension is doubled, another tripled, and the third quadrupled?
8. What is the effect upon the volume of a cube if the length of each edge is increased in the ratio 1 : 3?
9. Find the volume of a cube whose surface area is 54 square inches.
10. The volume of one rectangular solid is twice that of another. If they have congruent bases, what is the ratio of their heights?
11. One acre of land is to be irrigated by an amount of water equivalent to a 2-inch depth. This irrigation is to be done in 7 days, with the water running 8 hours each day. The rate of flow should be how many gallons per minute? (1 cu ft of water $= 7\frac{1}{2}$ gal; 1 acre = 43,560 sq ft) What is the total weight of this water? (1 cu ft of water weighs approximately 62.4 lbs.)

VOLUME OF PRISMS

Let us suppose that we could form a right prism (Figure 20-5 left) by placing thin sheets of paper together. Let us then tilt the axis of this prism to the shape of Figure 20-5 center. We have merely changed the positions of each sheet except the base. The bases are still congruent and parallel. The prisms are still the same height. Do they contain the same amount of space (or sheets of paper)?

Let us form a rectangular solid (Figure 20-5 right) whose base area and height is equal to the base areas and heights of Figure 20-5 left and Figure 20-5 center. Does this figure have the same volume as the other two?

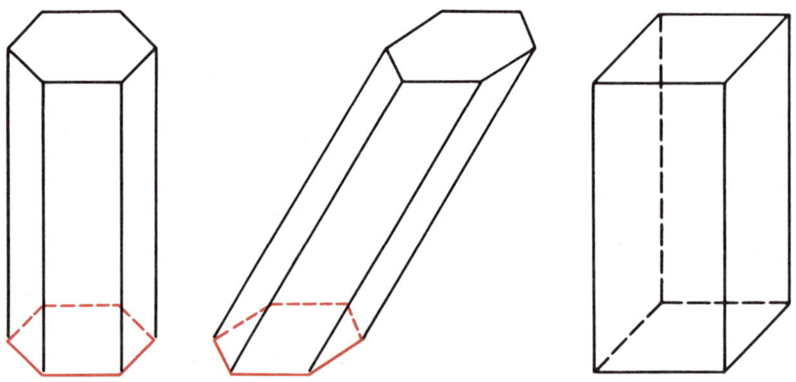

Figure 20–5

Bonaventura Cavalieri (1598–1647), a famous mathematician, developed a proof of this proposition. It is now called Cavalieri's Theorem, which we shall accept without proof. It, like Archimedes' work with the circle and parabola, was a forerunner of integral calculus.

20.10 CAVALIERI'S THEOREM

If two solids are included between the same two parallel planes and if any plane parallel to these planes forms sections of the solids having equal area, the solids have equal volumes.

The two solids lie between the parallel planes I and III. The theorem states that if all planes such as II, parallel to I and III, form sections with equal areas such as *ABCD* and *EFG*, the solids are equal in volume.

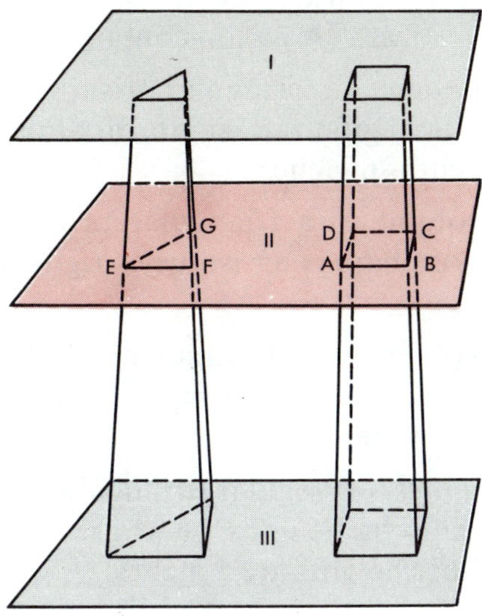

Figure 20–6

Any plane passed parallel to the bases of the figures in 20-5 (page 512) will form a section congruent to the base (14.12 and 14.09). Using 20.08 and Cavalieri's Theorem to compare any prism to a rectangular solid with equal base area and height, we may state the following theorem:

20.11 THEOREM

The volume (V) of a prism equals the product of the area of the base (B) and the height (h).

$$V_{\text{prism}} = Bh$$

Exercises

1. What is the volume of a prism whose base area is 24 square inches and whose height is 8 inches?
2. Find the volume of a cube if the measure of an edge is 15 inches.

3. An irregular-shaped piece of metal is dropped into a prismatic container whose base dimensions are 4 inches by 4 inches and whose height is 6 inches. The water level in the container rose 2 inches. What is the volume of the metal?

4. How many gallons of water will a tank hold whose dimensions are $12 \times 10 \times 8$ feet? (A gallon contains 231 cubic inches.)

5. What is the volume of a triangular prism whose base is an equilateral triangular region the length of whose side is 10 inches and whose height is 6 inches?

6. What is the volume of a triangular prism whose height is 6 inches and whose base is an isosceles right triangle with legs 8 inches long?

7. What is the volume of a triangular prism whose height is 6 inches and whose base is an isosceles triangle with legs 5 inches long and base 4 inches long?

8. Find the volumes of regular triangular, quadrangular, and hexagonal prisms the lengths of whose base edges are each 6 inches and whose heights are 5 inches.

9. A swimming pool is 75 feet long and 25 feet wide. It is 2 feet deep at one end and 8 feet deep at the other end. The base is a plane surface. How many gallons of water are required to fill it? (One cubic foot contains seven and one-half gallons of water.)

10. Find the volume of a prism whose height is 16 inches and whose base is an equilateral triangular region the measure of whose sides are 5 inches.

11. Find the volume of a prism whose height is 10 inches and whose bases are regular hexagonal regions the measure of whose sides are 9 inches.

12. Find the volume of a prism with trapezoid base regions and height of 10 feet. The lengths of the bases of the trapezoid are 1 foot and 3 feet, and its height is 6 inches.

13. A certain gasoline tank is a right prism whose base has an area of 77 square inches. If it holds 20 gallons of gasoline, how long is the tank? (one gallon = 231 cubic inches)

14. A house attic is in the shape of a right prism 50 feet long. The base is an isosceles triangle which has legs 25 feet long and a base 40 feet long. Find the volume of the attic.

15. Find the volume of a right prism 30 in. high, whose base is a rhombus 17 in. on a side, and whose long diagonal is 30 in.

16. A prism has a square base and a volume of 605 cubic inches. If it is 5 inches high, how long is an edge of the base?

17. A certain lead box has inside dimensions of 4, 6, and 9 in., and outside dimensions of 5, 7, and 10 in. If lead weighs .41 lbs per cubic inch, what is the weight of the box?

18. What does the lead box of Exercise 17 weigh when filled with material which weighs .06 lbs per cubic inch?

19. A rectangular solid is 13 inches high and has a volume of 858 cubic inches. If the length of the base is 5 inches more than the width, what are the dimensions of the base?

20. Find the volume of a right prism 15 inches high, whose base is a square with a diagonal 4 inches long.

21. A board is $\frac{3}{8}$ inches thick and the edges of one face form a right triangle with legs 3 inches and 4 inches long. Find its volume.

22. In a triangular prism the measures of the sides of the base are 6, 8, and 10 inches. The height of the prism is 1 foot. Find its volume.

23. A prism has a base which is a regular hexagonal region. The prism's height is 9 inches and the length of a side of the base is 10 inches. Find the volume.

24. A metal bar is formed as a prism 3 feet long with isosceles trapezoid base regions. The lengths of the bases of the trapezoid are 8 inches and 14 inches and the lengths of the congruent arms are 5 inches. Find the volume of the bar.

25. Find the volume of a right triangular prism whose height is 14 inches and whose base is a right triangular region with a hypotenuse 17 inches long and one leg 15 inches long.

26. Three metal cubes the lengths of whose edges are 3, 4, and 5 inches are melted into one cube. What is the length of the edge of the new cube?

27. The ratio of the lengths of the edges of two cubes is 3 : 1. What is the ratio of their volumes?

28. The ratio of the volumes of two cubes is 64 : 1. What is the ratio of the lengths of their edges?

AREA AND VOLUME OF CYLINDERS

Facts to remember about cylinders: (14.27–14.35)

1. The elements of a cylinder are congruent and parallel.
2. The bases are congruent and parallel.
3. The altitude is a segment perpendicular to and terminated by the planes of the bases. The measure of the altitude is called the height.
4. The elements of a right cylinder are perpendicular to the bases.
5. A cylinder of revolution is a right circular cylinder.

INSCRIBED AND CIRCUMSCRIBED PRISMS

20.12 A plane tangent to a cylinder (Figure 20-7 left) contains one and only one element of the cylinder.

A prism circumscribed about a cylinder (Figure 20-7 center) is a prism whose lateral faces are tangent to the cylinder and whose polygonal bases are circumscribed about the bases of the cylinder.

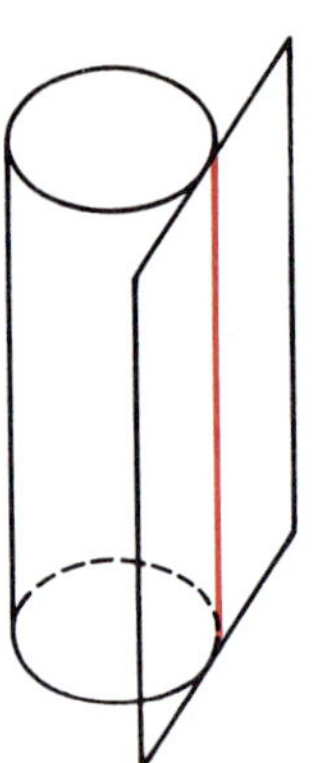

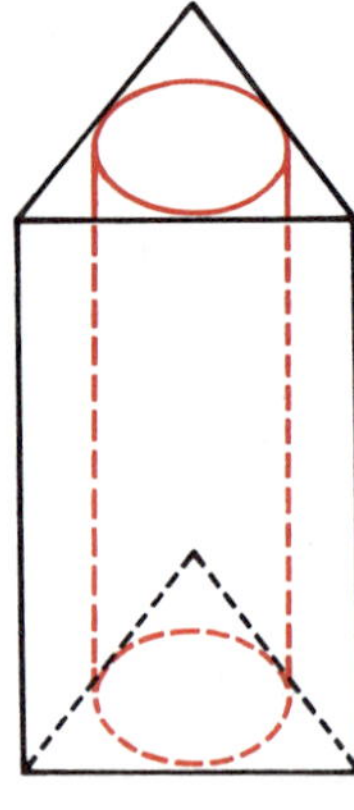

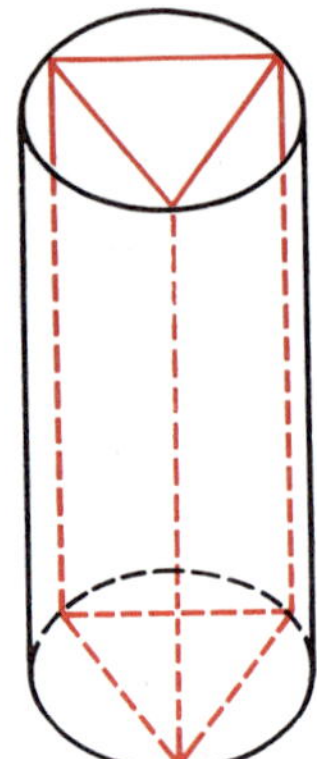

Figure 20–7

A prism inscribed in a cylinder (Figure 20-7 right) is a prism whose lateral edges are elements of the cylinder and whose bases are inscribed in the bases of the cylinder.

Exercises

1. Draw the figure and find the lateral area of a regular triangular prism circumscribed about a circular cylinder whose height is 6 inches and whose base radius is 3 inches.
2. Find the lateral area of a regular triangular prism inscribed in a circular cylinder whose height is 5 inches and the radius of whose base is 2 inches.

A circular cylinder may be thought of as a prism with a regular polygon as its base, a given perimeter, and an infinite number of lateral faces. The area of each face approaches the area of an element of the cylinder.

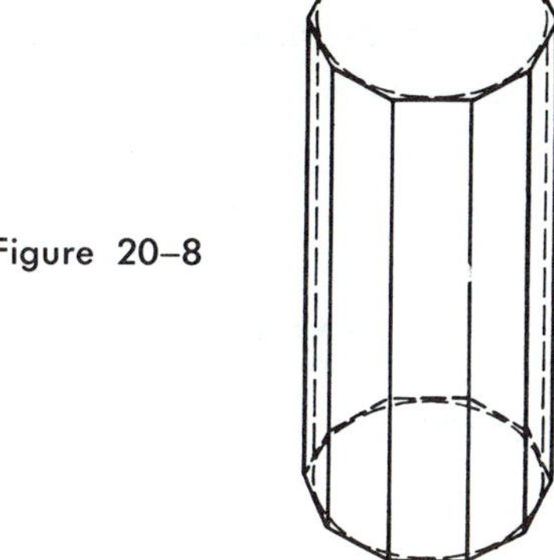

Figure 20–8

20.13 *Assumption: Properties of a circular cylinder are the same as those for a prism with regular bases—except where the properties of a prism depend upon the number of lateral faces.*

20.14 THEOREM

The lateral area (L) of a circular cylinder is equal to the product of the measure of an element (e) and the perimeter (p) of a right section.

$$L = ep$$

20.15 THEOREM

The lateral area (L) of a right circular cylinder is equal to the product of its height (h) and the circumference (C) of its base.

$$L = hC = 2\pi rh$$

20.16 The total area of a cylinder equals the sum of its lateral area and the area of its bases.

Using Cavalieri's Theorem, 20.11 and 20.13 we can state the following relationship:

20.17 THEOREM

The volume (V) of a circular cylinder is equal to the product of the area of its base (B) and its height (h).

$$V = Bh$$

20.18 THEOREM

The volume (V) of a cylinder of revolution whose height is h and whose base has a radius of r is given by the formula:

$$V = \pi r^2 h$$

Exercises

Leave the answer in terms of π unless otherwise indicated. (Use $\frac{22}{7}$ for the value of π.)

1. A cylindrical can is 6 inches deep and the radius of its base is 3 inches. Find the lateral area and total area. What is the volume of the can?
2. Find the total area and volume of a circular cylindrical gas tank which is 30 feet long and the measure of whose base diameter is 10 feet.
3. If the volume of a cylinder is 72.072 cubic inches and the radius of the base is 2.1 inches, find the height to the nearest tenth of an inch.
4. A square, the length of whose side is 5 inches, is revolved about one of the sides. What is the total area and volume of the cylinder of revolution?
5. Find the total area and volume of a cylinder of revolution whose height is 16 feet and whose base has a radius of 8 feet.
6. A rectangle 4×12 inches is rotated about a 12-inch side as axis. What is the area of the surface generated by the other 12-inch side?

7. A cylindrical tank to hold 1,540 cubic feet is to be built on a circular base the length of whose diameter is 7 feet. What must be its height to the nearest foot?

8. What is the ratio of the volume of a regular triangular prism to the volume of its circumscribed circular cylinder? Write the answer in simplest radical form, then write an approximation to three decimal places.

9. What is the ratio of the volume of a regular quadrangular prism to the volume of its circumscribed circular cylinder?

10. A circular cylindrical lawn roller has a length of 4 feet and a circumference of $6\frac{2}{7}$ feet. Find how many rotations it must make to roll 1,056 square feet of lawn.

11. A circular cylindrical silo has a 12 foot diameter and is 24 feet high. How many cubic feet of silage could it hold?

12. How many gallons of water may be contained by a circular cylindrical well which is $5\frac{1}{4}$ feet in diameter and 8 feet deep? (1 cubic foot contains $7\frac{1}{2}$ gallons.)

13. If a cubic foot of brass is drawn out into a circular cylindrical wire of $\frac{1}{30}$ inch diameter, how long will the wire be, to the nearest foot?

14. A tunnel is 450 feet long and has a semicircular cross-section with a 15-foot radius. How many cubic yards of earth were removed from the tunnel?

15. What is the weight to the nearest pound of a 10-foot length of steel pipe that has an inside diameter of 5 inches and outside diameter of 6 inches? (1 cubic inch of steel weighs 0.29 pounds.)

16. Find the approximate length of garden hose of outside diameter 1 inch which can be wound flush with the outer rim on the reel shown in Figure 20-9.

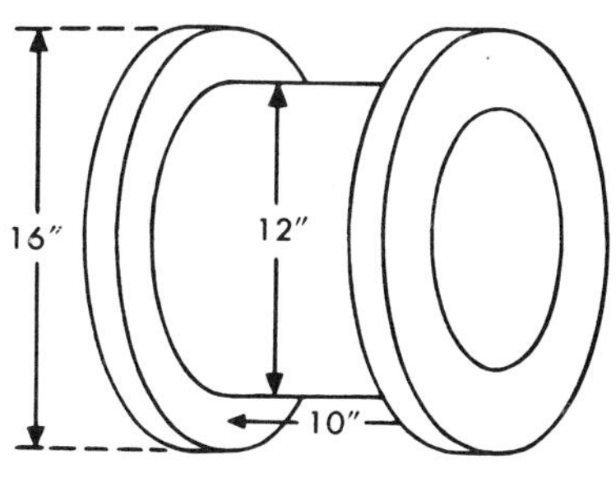

Figure 20–9

AREA OF PYRAMIDS

Facts to remember about pyramids: (14.36–14.48)

1. The altitude of a pyramid is the segment from the vertex perpendicular to and terminated by the plane of the base. The measure of the altitude is called the *height*.
2. A regular pyramid is a pyramid whose base is a regular polygonal region and whose altitude contains the center of the base.
3. The lateral edges of a regular pyramid are congruent.
4. The lateral faces of a regular pyramid are congruent isosceles triangular regions.
5. The slant height of a regular pyramid is the height of any of the lateral faces.

20.19 The lateral area of a pyramid is the sum of the areas of its lateral faces.

20.20 The total area of a pyramid is the sum of the lateral area and the area of the base.

We will now examine deductively the formula for the lateral area of a pyramid.

Given: A regular pyramid, with s the slant height, b the length of one side of the base, p the perimeter of the base, and the lateral area L

Conjecture: $L = \frac{1}{2} sp$

Outline of Proof: The area of a lateral face equals $\frac{1}{2} sb$. There are n congruent triangles, so $L = \frac{1}{2} sbn$. Since the bases are congruent, bn is the perimeter p of the base of the pyramid; $\therefore L = \frac{1}{2} sp$.

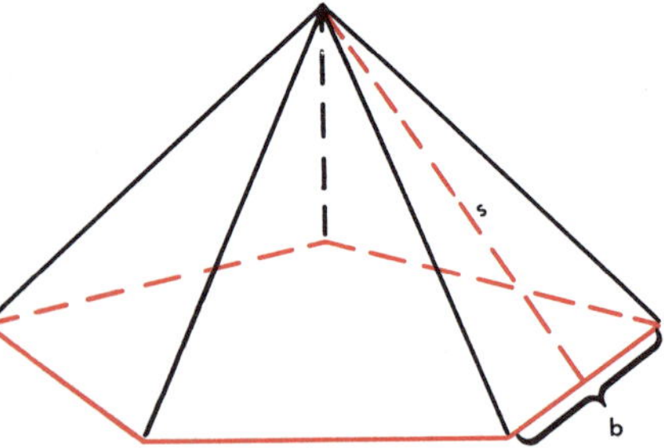

Figure 20–10

20.21 THEOREM

The lateral area of a regular pyramid is equal to half the product of its slant height and the perimeter of its base.

$$L = \frac{1}{2} sp$$

Exercises

1. Find the lateral area of a regular pyramid with the following dimensions:
   (a) $p = 18$ inches, $s = 5\frac{1}{2}$ inches
   (b) $p = 2$ feet 3 inches, $s = 17$ inches
   (c) $s = 2\frac{1}{2}$ feet, $p = 2\frac{1}{2}$ yards
2. Find the lateral area of a regular triangular pyramid if the length of a side of the base is 8 inches and the slant height is 5 inches.
3. Find the lateral area of a regular quadrangular pyramid if the measure of each side of its base is 6 inches and its height is 4 inches.
4. Find the slant height of a regular pyramid with a lateral area of 80 square inches and a base perimeter of 20 inches.
5. Find the lateral area and height of a regular quadrangular pyramid if the length of an edge of the base is 10 inches and the slant height is 13 inches.
6. In a regular pyramid with a square base region, find the length of an edge of the base if the height is 15 feet and the measure of a lateral edge is 17 feet.
7. Find the height of a regular quadrangular pyramid the measure of whose lateral edge is 25 inches and whose base edge is 14 inches.
8. Find the area of a section of a pyramid parallel to the base and 24 inches from the base, if the height of the pyramid is 36 inches and the area of the base is 576 square inches.
9. The measure of an edge of a regular octahedron is 6 inches. What is its total surface area? Leave the answer in simplest radical form.

AREA OF FRUSTUMS OF PYRAMIDS

Facts to remember about frustums of pyramids: (14.49–14.53)

1. The height of the frustum is the length of a segment perpendicular to and terminated by the planes of the bases.
2. The bases of a frustum of a pyramid are similar and parallel polygonal regions.
3. The lateral faces of a frustum of a pyramid are trapezoid regions.
4. The lateral faces of a frustum of a regular pyramid are congruent isosceles trapezoid regions.
5. The slant height of the frustum of a regular pyramid is the height of any lateral face.

Given: A frustum of a regular pyramid, s the slant height, p the perimeter of the lower base, p' the perimeter of the upper base, and L the lateral area

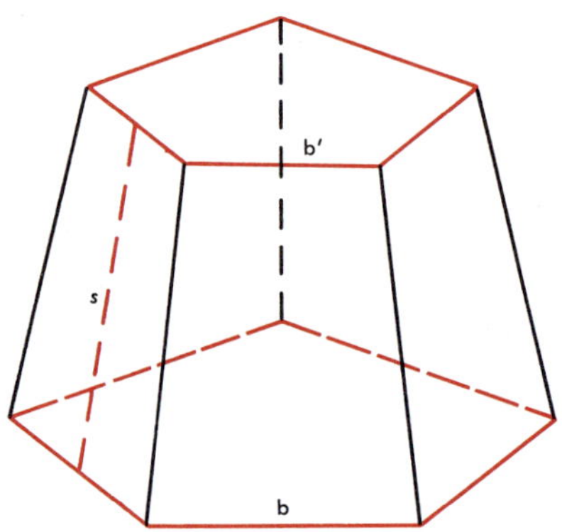

Figure 20–11

Conjecture: $L = \frac{1}{2}s(p + p')$

Outline of Proof: The area of a lateral face (trapezoid region) = $\frac{1}{2}s(b + b')$. The sum of the areas of n congruent lateral faces equals $\frac{1}{2}s(b + b')n$. The sum of the bases of n congruent trapezoids equals the perimeters of the bases of the frustum, or $p + p' = nb + nb' = (b + b')n$. Therefore, $L = \frac{1}{2}s(p + p')$.

20.22 THEOREM

The lateral area of a frustum of a regular pyramid is equal to half the product of the slant height and the sum of the perimeters of the bases.

Exercises

1. Find the lateral area of the frustum of a pyramid whose perimeters of the bases and slant height are as follows:
   (a) $p = 12$ inches, $p' = 15$ inches, $s = 8$ inches.
   (b) $p = 12\frac{1}{4}$ feet, $p' = 13\frac{7}{8}$ feet, $s = 2$ feet.
   (c) $p = 2.34$ feet, $p' = 5.18$ feet, $s = 0.83$ feet.
   (d) $p = 17$ inches, $p' = 2\frac{3}{8}$ feet, $s = 4$ feet.
2. The bases of a frustum of a regular pyramid are square regions. The measures of the base edges are 8 inches and 6 inches. The slant height of the frustum is 4 inches. Find the lateral area and the total area.
3. The lateral area of a flower pot in the form of a frustum of a regular hexagonal pyramid is 108 square inches. The lengths of the sides of the bases are 2 and 4 inches. Find the slant height.
4. The bases of a frustum of a regular pyramid are square regions. The measures of the edges of the bases are 11 inches and 5 inches. The height of the frustum is 4 inches. What is the lateral area and the total area?

VOLUME OF PYRAMIDS

Following the reasoning of Cavalieri's Theorem (20.10), we may accept the following assumption:

20.23 *Assumption: Two pyramids having equal heights and equal base areas are equal in volume.*

A relationship between the volume of a prism and the volume of a pyramid can be found by constructing a prism having the same height and the same base area as the pyramid. A proof of this relationship is given on the next page.

Given: Triangular pyramid $E\text{-}ABC$

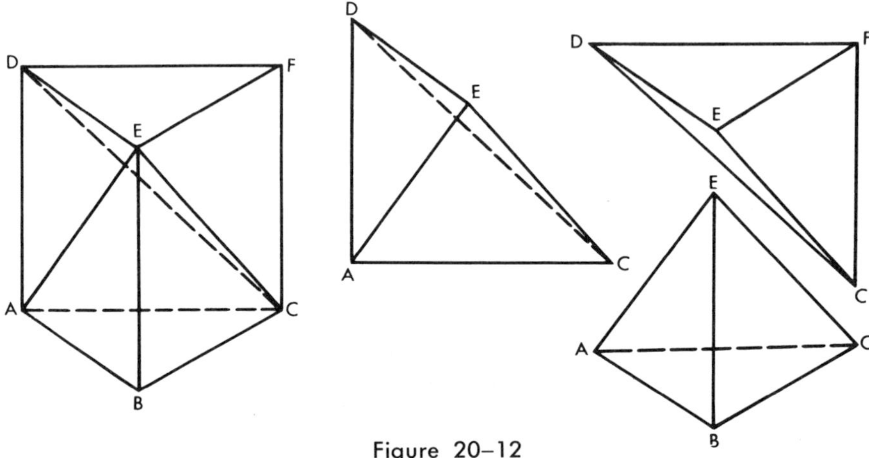

Figure 20–12

Conjecture: $V = \frac{1}{3} Bh$

Plan: Construct the triangular prism $DEF\text{-}ABC$ and show that the volume of $E\text{-}ABC$ is equal to one-third the volume of the prism. (Shown in Figure 20-12 is the prism separated into the three parts being considered.)

Proof:

Statements	*Reasons*
1. $DEF\text{-}ABC$ is a triangular prism.	1. Construction
2. $ABED$ is a parallelogram.	2. 14.10 and 9.09
3. Base(ADE) of pyramid $C\text{-}ADE \cong$ Base(EBA) of pyramid $C\text{-}ABE$	3. 9.02
4. Pyramids $C\text{-}ADE$ and $C\text{-}ABE$ have the same height.	4. 14.42
5. $\therefore$ Volume ($C\text{-}ADE$) = Volume ($C\text{-}ABE$)	5. 20.23
6. Volume ($C\text{-}ABE$) = Volume ($E\text{-}ABC$)	6. Reflexive Axiom

Statement	Reason
7. Base(DEF) of pyramid C-$DEF \cong$ Base(ABC) of pyramid E-ABC	7. 14.12
8. Pyramids C-DEF and E-ABC have the same height.	8. 14.42
9. $\therefore$ Volume (C-DEF) = Volume (E-ABC)	9. 20.23
10. Then Volume (C-ADE) = Volume (E-ABC) = Volume (C-DEF)	10. Transitive Axiom, steps 5, 6, 9.
11. Volume (C-ADE) + Volume (E-ABC) + Volume (C-DEF) = Volume (DEF-ABC)	11. The three regions fill the region of DEF-ABC.
12. So Volume (E-ABC) = $\frac{1}{3}$ Volume (DEF-ABC)	12. It is one of the three regions of equal measure filling the prism.

20.24 THEOREM

The volume (V) of a triangular pyramid is equal to one-third of the product of the area of its base (B) and height (h).

Using Theorem 20.24, we can develop a formula for the volume of any pyramid.

Reasoning: Any pyramid can be divided into a whole number of triangular pyramids. The sum of the volumes is $\frac{1}{3}B_1h + \frac{1}{3}B_2h + \frac{1}{3}B_3h + \cdots$. Since this equals $\frac{1}{3}h(B_1 + B_2 + B_3 + \cdots)$, and $B_1 + B_2 + B_3 + \ldots =$ the base (B) of any pyramid, then $V = \frac{1}{3}Bh$

20.25 THEOREM

The volume (V) of any pyramid is equal to one-third of the product of the height (h) and the area of the base (B).

$$V = \frac{1}{3}Bh$$

Exercises

1. Find the volumes of the pyramids whose base areas and heights are as follows:

   (a) 10 square inches; 3 inches. **(b)** 25 square inches; 6 inches.

   (c) $10\frac{1}{2}$ square inches; $6\frac{3}{4}$ inches. **(d)** 3 square feet; 8 inches.

2. Find the volume of a regular quadrangular pyramid the measure of whose base edge is 10 inches and whose height is 15 inches.

3. Find the volume of a triangular pyramid whose height is 18 inches and whose base is an equilateral triangular region, the measure of whose sides is 12 inches.

4. A regular hexagonal pyramid has a height of 18 inches and a base edge whose measure is 6 inches. Find its volume.

5. A pyramidal stone monument has a base area of 100 square feet. The height of the monument is 12 feet, and the stone weighs about 165 pounds per cubic foot. How many tons does the monument weigh?

6. Find the volume of a regular tetrahedron the measure of whose edge is 18 inches.

7. The length of the edge of a regular octahedron is 10 inches. What is the area and volume of the octahedron?

8. Find the lateral area and volume of a regular quadrangular pyramid whose slant height is 13 inches and whose height is 5 inches.

9. The area of the base of a pyramid is 81 square inches. The height of the pyramid is 6 inches. What is the area of a section of the pyramid parallel to the base and 2 inches from the base?

10. Find the height of a regular quadrangular pyramid whose volume is 160 cubic inches and the length of whose base edge is 10 inches.

11. *EFGH-ABCD* is a cube. A plane is passed through F and the midpoints L and M of $\overline{AB}$ and $\overline{BC}$. What is the ratio of the volume of *F-LBM* to the volume of the cube? (Figure 20-13)

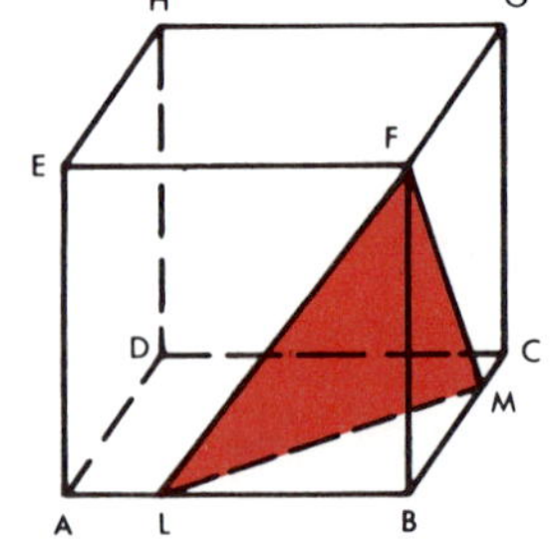

Figure 20–13

VOLUME OF A FRUSTUM OF A PYRAMID

By using the formula for the volume of a pyramid, we can also develop a formula for the volume of a frustum of a pyramid.

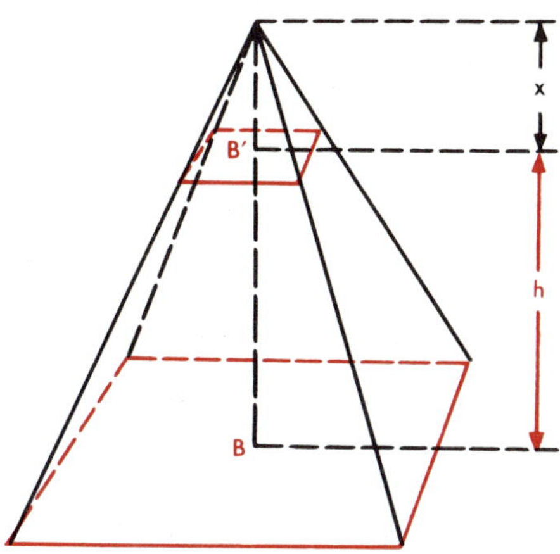

Figure 20–14

Outline of Proof (optional): Consider the complete pyramid from which the frustum was formed. Let V_1 represent the volume of the whole pyramid and V_2 the volume of the smaller pyramid.

$V_1 = \frac{1}{3}B(h + x)$ and $V_2 = \frac{1}{3}B'x$; then $V = V_1 - V_2 =$

$\frac{1}{3}\left[Bh + x(B - B')\right]$.

$\frac{B'}{B} = \frac{x^2}{(x + h)^2}$, by 18.15, or $\frac{\sqrt{B'}}{\sqrt{B}} = \frac{x}{x + h}$.

Solving for x, $x\sqrt{B} = x\sqrt{B'} + h\sqrt{B'}$.

Collecting terms and factoring, $x(\sqrt{B} - \sqrt{B'}) = h\sqrt{B'}$.

Therefore, $x = \frac{h\sqrt{B'}}{\sqrt{B} - \sqrt{B'}}$.

Rationalizing, $x = \frac{h\sqrt{B'}}{\sqrt{B} - \sqrt{B'}} \cdot \frac{\sqrt{B} + \sqrt{B'}}{\sqrt{B} + \sqrt{B'}} = \frac{hB' + h\sqrt{BB'}}{B - B'}$,

so $x(B - B') = hB' + h\sqrt{BB'}$.

Therefore, $V = \frac{1}{3}(Bh + hB' + h\sqrt{BB'}) = \frac{1}{3}h(B + B' + \sqrt{BB'})$.

We have now established the following theorem.

20.26 THEOREM

The volume of a frustum of a pyramid is given by the formula $V = \frac{1}{3}h(B + B' + \sqrt{BB'})$ where B and B' are areas of the bases and h is the height.

Exercises

1. Find the volume of the frustums of the pyramids whose dimensions are as follows:
   (a) $h = 3$ inches; $B = 36$ square inches; $B' = 25$ square inches.
   (b) $h = 6$ inches; $B = 16$ square inches; $B' = 9$ square inches.
2. The area of the upper base of a frustum of a pyramid is 3 square inches and the area of the lower base is 12 square inches. If the volume is 42 cubic inches, what is the height?
3. A farmer made a feed "hopper" in the form of a frustum of a pyramid. The closed lower base is a rectangular region 5 feet by 13 feet, and the open base is a rectangle 15 feet by 23 feet. The height is 15 feet. Find its capacity in cubic feet.
4. A hollow frustum of a regular quadrangular pyramid, open at both bases, has walls 3 feet thick. It is 24 feet high and the outside dimensions of its bases are 20 feet and 16 feet. How many cubic feet of mortar and stone were used in the construction?
5. A pyramid 8 feet high has a square base region whose area is 64 square feet. Four feet from the base the pyramid is cut by a plane parallel to the base. Find the volume of the frustum.
6. A pyramid 4 inches high, which has a square base region whose area is 48 square inches, is cut by a plane parallel to the base and 1 inch from the vertex. Find the volume of the frustum.
7. The lower base of a frustum of a pyramid is a square, 2 inches on a side. The area of the upper base is half that of the lower base, and the height of the frustum is 2 inches. What is the volume of the frustum?
8. A frustum of a regular quadrilateral pyramid has the length of an edge of its upper base equal to half that of its lower base. What is the volume of such a frustum if the edges of the original pyramid were each 4 inches?

AREA OF CONES

Facts to remember about cones: (14.54–14.59)

1. The axis of a circular cone is the segment from the vertex to the center of the base.
2. The altitude of a cone is the segment from the vertex perpendicular to and terminated by the plane of the base. The measure of the altitude is called the height.
3. In a right circular cone the axis is perpendicular to the base and coincides with the altitude.
4. The elements of a right circular cone are congruent.
5. The slant height of a right circular cone is the length of one of the elements.
6. A cone of revolution is a right circular cone.

A circular cone may be thought of as a pyramid with a regular base of finite perimeter and in infinite number of lateral faces.

20.27 *Assumption: Everything that is true of a pyramid with a regular polygonal region as its base is true of a circular cone—where the statement does not depend upon the number of lateral faces.*

20.28 THEOREM

The lateral area (L) of a right circular cone equals half the product of the slant height (s) and the circumference (C) of the base.

$$L = \frac{1}{2}sC = \pi rs$$

20.29 The total area of a cone is equal to the lateral area plus the area of the base.

$$T = \pi r^2 + \pi rs = \pi r(r + s)$$

Exercises

1. Find the lateral area of the following cones of revolution: $\left(\text{Use } \pi = \frac{22}{7}\right)$

   (a) $r = 4.9$ feet; $s = 10$ feet. (b) $h = 24$ inches; $s = 40$ inches.

2. Find the total area of the following cones of revolution: (a) $C = 48\pi$ inches, $h = 18$ inches. (b) $d = 7$ inches; $h = 3.5$ inches.
3. How many square yards of canvas will it take to make a circular conical tent (without a floor) 14 feet in diameter and 7 feet high?
4. A model of a cone is made from a sector of a circle the measure of whose central angle is 300° and whose radius is 10 inches. Find the lateral area of the cone.
5. Show how to cut from a piece of paper a cone having a slant height of 12 inches and a base diameter of 8 inches.
6. The measures of the sides of a triangle are 3, 4, and 5 inches. The triangle is rotated about the 4-inch side. Find the area of the surface generated by the 5-inch side.
7. The area of the base of a right circular cone is 144π square inches and the height is 5 inches. Find the lateral area.

AREA OF FRUSTUMS OF CONES

Facts to remember about frustums of cones: (14.60–14.62)

1. The bases of a frustum are in parallel planes.
2. The height of a frustum is the length of a segment perpendicular to and terminated by the planes of the bases.
3. The slant height of a frustum of a right circular cone is the length of one of its elements.

Applying the reasoning of 20.27 and 20.22 to compare a frustum of a cone to a frustum of a pyramid, we develop the following relationship.

20.30 THEOREM

The lateral area (L) of a frustum of a right circular cone equals half the product of the slant height (s) and the sum of the circumferences (C and C') of the bases.

$$L = \frac{1}{2}s(C + C') = \pi s(r + r')$$

Exercises

1. Find the lateral area of the frustum of a circular cone whose base dimensions and slant height are as follows:
(a) $C_1 = 25$ inches, $C_2 = 30$ inches, $s = 6$ inches. (b) $C_1 = 18.7$ feet, $C_2 = 2\frac{1}{2}$ feet, $s = 3$ feet. (c) $r_1 = 3$ inches, $r_2 = 5$ inches, $s = 7$ inches. (d) $r_1 = 2.7$ inches, $r_2 = 3.6$ inches, $s = 5$ inches.

2. How many square inches of tin are required to make a circular conical funnel in which the diameters of the ends are 2 inches and 10 inches respectively and whose slant height is 8 inches?

3. Find the lateral area of a frustum of a right circular cone whose slant height is 6 inches and whose bases' radii are 5 and 15 inches.

4. A model of a frustum of a circular cone made from a sector of a circle is shown. CD is an arc of a circle whose radius is 14 inches. AB is an arc of a circle whose radius is 7 inches; $m\angle V = 120°$. What is the lateral area of the frustum?

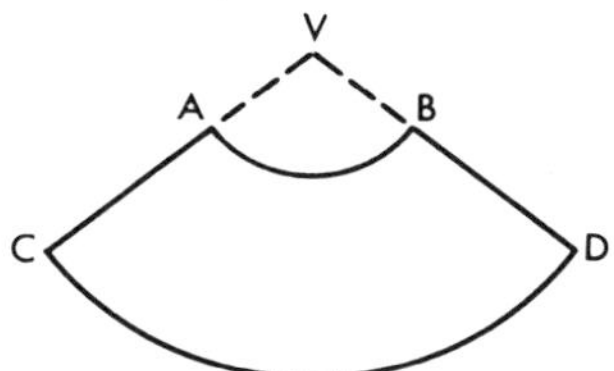

Figure 20–15

5. A lamp shade in the form of a frustum of a right circular cone is made with a lower diameter of 19 inches and an upper diameter of $9\frac{1}{2}$ inches. The height of the shade is $9\frac{1}{2}$ inches. Find, to the nearest square inch, the amount of material in the shade. Do not allow for overlap.

6. An aluminum company is to deliver an order of milk pails with a lower diameter of 12 inches and an upper diameter of 15 inches. The slant height is 14 inches. Allow 5% waste for overlap. How many square feet of aluminum will be required to fill an order for 1,000 pails?

VOLUME OF CONES

Applying the reasoning of 20.27 and 20.25, we develop the following relationship.

20.31 THEOREM

The volume (V) of a circular cone is equal to one-third the product of the area of its base (B) and its height (h).

$$V = \frac{1}{3}Bh = \frac{\pi r^2 h}{3}$$

Exercises

Use $\frac{22}{7}$ for the value of π.

1. Find the volume of a cone whose radius, or area of the base, and height are as follows:
   (a) $A = 36$ square inches, $h = 8$ inches. (b) $A = 57.74$ square inches, $h = 15.3$ inches. (c) $r = 7$ inches, $h = 10$ inches. (d) $r = \frac{3}{4}$ feet, $h = 21$ inches.
2. A pile of sand is roughly conical in shape. The diameter of the pile is about 5 feet, and the height of the pile is about $4\frac{1}{2}$ feet. About how many cubic yards of sand are in the pile?
3. A cone-shaped cup has an inside base diameter of $2\frac{3}{4}$ inches and a height of $3\frac{3}{4}$ inches. What is the capacity of the cup, to the nearest cubic inch?
4. A quantity of dry loose soil is piled against the side of a building. The highest point of the soil is 5 feet. The base of the pile is semicircular in shape and 12 feet in width along the building. About how many cubic feet of soil are in the pile?
5. A piece of metal in the form of a sector of a circle with a radius of 12 inches is rolled into a circular cone. If the measure of the central angle of the sector is 240°, what is the volume of the cone? What is the slant height of the cone?

VOLUME OF FRUSTUMS OF CONES

Applying the reasoning of 20.27 and 20.26, we develop the following relationship.

20.32 THEOREM

The volume (V) of a frustum of a circular cone is given by the formula $V = \frac{1}{3}h(B + B' + \sqrt{BB'})$ where B and B' are the areas of the bases and h is the height.

$$\text{or } V = \frac{\pi h}{3}(r^2 + r'^2 + rr')$$

Exercises

1. Find the volume of the frustum of a circular cone with the following base dimensions and height:
   (a) $c = 10\pi$ inches $c' = 14\pi$ inches $h = 10$ inches
   (b) $c = 6.4\pi$ inches $c' = 9.4\pi$ inches $h = 7.5$ inches
   (c) $d = 3\frac{1}{2}$ feet $d' = 6\frac{2}{3}$ feet $h = 4\frac{1}{2}$ feet
   (d) $r = 2.3$ inches $r' = 3.2$ inches $h = 5.7$ inches
2. A well is dug in the form of an inverted frustum of a circular cone. Its depth is 20 feet, and the diameters of its bases are 3 feet and 2 feet. How many gallons of water could the well contain? (1 cubic foot $= 7\frac{1}{2}$ gallons.)
3. How many cubic yards of earth will be required to fill 1500 flower pots the inside diameters of whose upper and lower bases are 10 inches and 6 inches respectively, and whose depth is 10 inches?
4. Find the volume, in gallons, of a pail which has a 12-inch top diameter and 10-inch bottom diameter with a 12-inch slant height. (231 cubic inches contains 1 gallon.)
5. The bottom of a drinking glass has a radius of one inch, and the glass is filled to a height of 2.5 inches where the radius of the surface is 1.1 inches. What volume of water does the glass contain?

AREA OF SPHERES

The proof of the following proposition is beyond the scope of this course. We shall accept the statement without proof.

20.33 *Assumption: The area (S) of a sphere is equal to four times the area of a great circle region of the sphere.*

$$S = 4\pi r^2$$

20.34 THEOREM

The ratio of the areas of two spheres equals the ratio of the squares of the radii or diameters of the sphere.

$$\frac{S_1}{S_2} = \frac{r_1^2}{r_2^2} = \frac{d_1^2}{d_2^2}$$

Reasoning: $\frac{S_1}{S_2} = \frac{4\pi r_1^2}{4\pi r_2^2} = \frac{r_1^2}{r_2^2} = \frac{(\frac{1}{2}d_1)^2}{(\frac{1}{2}d_2)^2} = \frac{d_1^2}{d_2^2}$

Exercises

1. Find the surface of the sphere whose radius or diameter is as follows:
   (a) $r = 7$ inches (b) $r = 14$ inches
   (c) $d = 13.34$ inches (d) $d = \frac{14}{15}$ inches
2. Find the radii of spheres whose areas are as follows:
   (a) 2464 square inches (b) 12.57 square feet
3. If the hemispherical dome of the Iowa State Capitol is 80 feet in diameter, find its area in square yards.
4. Consider the earth as a sphere, with radius 4000 miles, and find its area.
5. What is the ratio of the radii of two spheres if the area of one is 4 times the area of the other?

VOLUME OF A SPHERE

Let us suppose that we could divide the region of a sphere into a great number of congruent solids resembling pyramids, the center of the sphere being the vertex of each pyramid. The bases of the pyramids, of course, are not plane regions. If the sphere is divided into an infinite number of such pyramids, the base of each will be infinitely small (a point?) and approach a plane region as a limit.

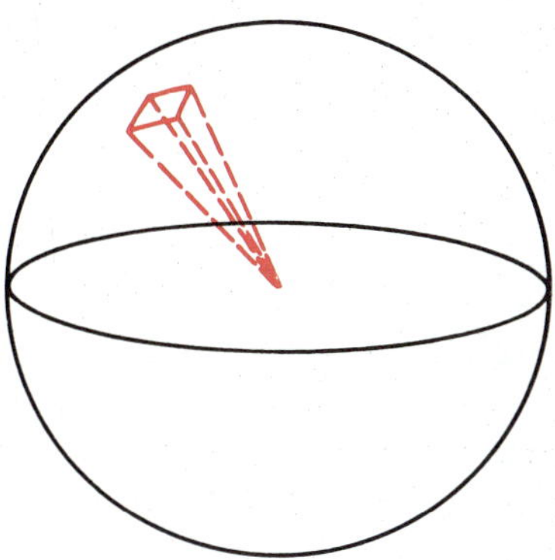

Figure 20–16

The volume of each spherical pyramid approaches $\frac{1}{3}Bh$, where h approaches in length the radius of the sphere. The sum of the small pyramids $= \frac{1}{3}B_1r + \frac{1}{3}B_2r + \frac{1}{3}B_3r + \ldots =$

$$\frac{1}{3}r(B_1 + B_2 + B_3 + \ldots) = \frac{1}{3}rS.$$

Since $S = 4\pi r^2$, the sum of the volume of all the spherical pyramids approaches the value of $\left(\frac{1}{3}r\right)(4\pi r^2)$.

20.35 *Assumption: The volume (V) of a sphere is equal to its area ($4\pi r^2$) multiplied by one-third of its radius $\left(\frac{r}{3}\right)$.*

$$V = \frac{4}{3}\pi r^3$$

20.36 THEOREM

The ratio of the volumes of two spheres is equal to the cube of the ratio of their respective radii.

$$\frac{V_1}{V_2} = \frac{(r_1)^3}{(r_2)^3}$$

Exercises

In the following exercises use $\frac{22}{7}$ for the value of π.

1. Find the volume of the spheres whose dimensions are as follows:
   (a) $r = 6$ inches (b) $r = 6\frac{7}{8}$ inches
   (c) $d = 18$ inches (d) $d = 9.24$ feet
2. Solve $V = \frac{4}{3}\pi r^3$ for r.
3. If the volume of a sphere is $\frac{32}{3}\pi$ cubic inches, what is its radius?
4. If the circumference of a great circle of a sphere is 18.9 inches, what is the volume of the sphere?
5. If the radius of a sphere is 3 times that of another what is the ratio of their volumes?
6. Find the diameter of a hemispherical container if it is to hold 2 gallons of water.

Vocabulary List

volume	right section	cylinder
lateral area	frustum	right cylinder
surface area	Cavalieri	prism

Chapter Review

Prisms

1. Find the area of a rectangular solid whose dimensions are 7, 8, and 10 inches.
2. Find the volume of the above solid.
3. Find the area of the polygonal region formed by a diagonal section of a right rectangular prism whose dimensions are 4, 6, and 8 inches, if the section contains the 4 inch edge.
4. The ratio of the areas of two cubes is 9:16.
   (a) What is the ratio of the measures of their edges?
   (b) What is the ratio of their volumes?
5. Find the area and volume of a regular hexagonal prism if the height is 6 inches and the measure of an edge of the base is 4 inches.

Cylinders

6. The height of a right circular cylinder is 1 foot. The measure of the diameter of the base is 10 inches. Find the area and volume of the cylinder. $\left(\text{Use } \pi = \frac{22}{7}.\right)$

Pyramids

7. The height of a regular rectangular pyramid is 50 feet. The measure of an edge of the base is 12 feet. Find the lateral area and volume of the pyramid.

8. Find the volume of a regular tetrahedron, the measure of whose edge is 18 inches.

Cones

9. The height of a right circular cone is 15 inches. The radius of its base is 8 inches. Find the lateral area and volume of the cone (in terms of π).

10. Find the height of the right circular cone whose volume is 924 cubic inches and whose base radius is 7 inches.

Spheres

11. The circumference of a sphere is 12π inches. Find its area and volume.

Chapter 20 Test

Prisms

1. Given the rectangular solid whose dimensions are 4, 5, and 6 inches.
   (a) Find the total area of the prism.
   (b) Find the volume of the prism.
   (c) Find the area of the region formed by a diagonal section of the prism containing the 4 inch edge.

2. If the ratio of the measures of the edges of two cubes is 2:1
   (a) What is the ratio of their areas?
   (b) What is the ratio of their volumes?

3. Find the volume of a right triangular prism whose height is 5 inches and whose base is an equilateral triangle the measure of whose edge is 8 inches.

Cylinders

4. The height of a right circular cylinder is 8 inches. The circumference of the base is 44 inches.
   (a) Find the lateral area of the cylinder.
   (b) Find the volume of the cylinder. $\left(\text{Use } \pi = \frac{22}{7}\right)$

Pyramids

5. The height of a regular hexagonal pyramid is 10 inches. The radius of the base is 4 inches.
   (a) Find the lateral area of the pyramid.
   (b) Find the volume of the pyramid.
6. Find the volume of a regular tetrahedron the measure of whose edge is 12 inches.
7. The height of a regular rectangular pyramid is 12 inches and the measure of an edge of the base is 12 inches.
   (a) Find the volume of the frustum formed by a plane parallel to the base of the pyramid which bisects the altitude.
   (b) Find the lateral area of the frustum.

Cones

8. The radius of the base of a right circular cone is 5 inches. The slant height of the cone is 13 inches.
   (a) Find the lateral area of the cone (in terms of π).
   (b) Find the volume of the cone (in terms of π).
9. Two concentric semicircles are drawn on a piece of paper. Their radii are 6 and 12 inches. The region formed between the semicircles and the segments of the diameter between their endpoints is cut from the paper to form a model of a frustum of a cone.
   (a) Find the lateral area of the frustum (in terms of π).
   (b) Find the volume of the frustum (in terms of π).

Spheres

10. The radius of a sphere is 3 inches.
   (a) Find the area of the sphere (in terms of π).
   (b) Find the volume of the sphere (in terms of π).

21

Coordinate Geometry

The original presentation of geometry by Euclid (300 B.C.) was designed as a preparation for philosophical studies. He was not interested in practical applications or the mathematical approach.

In our more recent history, mathematicians have approached geometry from a practical as well as a philosophical standpoint and have added immeasurably to its usefulness and applications. Descartes (1596–1650) was the most influential of these mathematicians. He was primarily responsible for joining the two streams of mathematical knowledge, algebra and geometry, into what is now called coordinate or Cartesian geometry. A frequently repeated story relates that Descartes watched the path of a fly across the ceiling and walls of his room and developed coordinate geometry in an attempt to describe the path of the fly.

Actually the essentials of Descartes' thinking (coordinate geometry) had been used for many years before his time. Coordinates were used to locate points on the earth for mapmaking and navigational purposes. They were used to lay out towns and farm lands and to locate positions of stars in space.

Let us mentally place one of the descendants of Descartes' fly on the front wall of our classroom. How could you describe its location? Is there one and only one point which fits this description?

If we were to draw a vertical line down the wall and a horizontal line across the wall, how would you describe the position of the fly in relation to these lines?

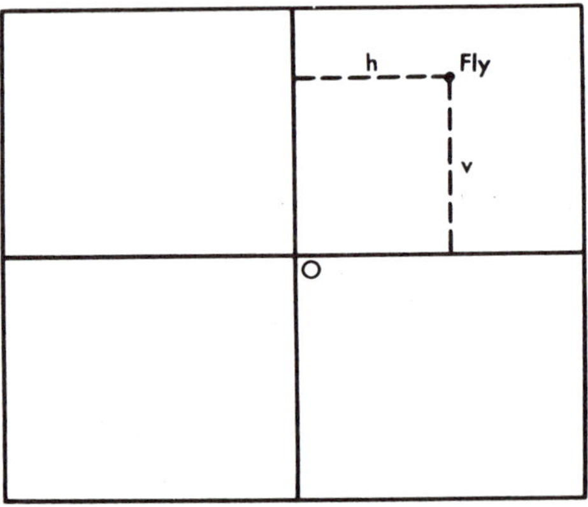

Figure 21–1

Notice that your description of position requires that two distances (coordinates) be stated. For example: (1) The fly is _?_ feet from the ceiling and _?_ feet from the left wall of the room, or (2) The fly is _?_ units to the right of the vertical line and _?_ units above the horizontal line.

Would it help to mark off the two intersecting lines with a scale and number the divisions of this scale?

You realize by now that our front wall with the two reference lines drawn is essentially a piece of graph paper and that the fly represents a point on the plane. Let us review the terminology we used in conjunction with graphing so that we can delve further into coordinate geometry.

COORDINATE SYSTEM

21.00 Our rectangular coordinate system is composed of two axes at right angles to each other. The horizontal axis is commonly called the x-axis and the vertical axis the y-axis. The intersection of the axes is called the origin. Values to the right of the origin on the x-axis are *positive* and to the left of the origin are *negative*;

values above the origin on the y-axis are *positive* and below the origin are *negative*.

The position of a point on the system is determined by an ordered pair of numbers; a number representing the distance of the point from the y-axis (abscissa or x-coordinate) and a number representing its distance from the x-axis (ordinate or y-coordinate). These ordered numbers are called the coordinates of the point. The point is called the graph of the coordinates.

The fact that the coordinates of a point are written in a particular order has led us to call the written designation of such points an "ordered pair." The point (2,3) is a different point than (3,2). By common agreement the x-coordinate is written before the y-coordinate. The general form for labeling a point is $P(x, y)$.

It is most important to realize that there is a *one-to-one correspondence between the set of points of a plane and the set of ordered pairs of real numbers*, called their coordinates.

Figure 21-2 shows the location of points $A(4,5)$; $B(-7,3)$; $C(-4,-3)$; and $D(8,-2)$.

The quadrants are labeled as shown in Figure 21-3.

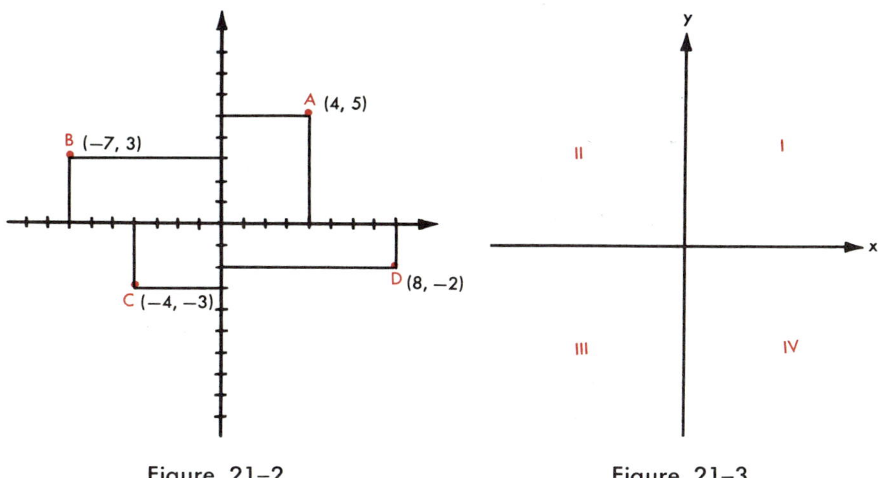

Figure 21–2

Figure 21–3

DIRECTED LINE SEGMENTS

21.01 Euclid failed to consider the idea of the order of points on a line or in space. Descartes introduced the idea of order and of directed lines. The great generality of the analytic methods and formulas is due primarily to the use of directed lines. These are lines on which one direction is regarded as positive and the opposite direction as negative. To pursue the study of coordinate geometry, we need to understand the meaning of directed line segments.

Extending the definitions of 21.00, we designate motion along a line parallel to the x-axis as positive for increasing values of x (from left to right), and as negative for decreasing values of x (from right to left).

EXAMPLE:
$$\begin{aligned} m\overline{AB} &= +10 \text{ units} \\ m\overline{BA} &= -10 \text{ units} \\ \therefore m\overline{AB} &= -m\overline{BA} \\ \text{or } m\overline{BA} &= -m\overline{AB} \end{aligned}$$

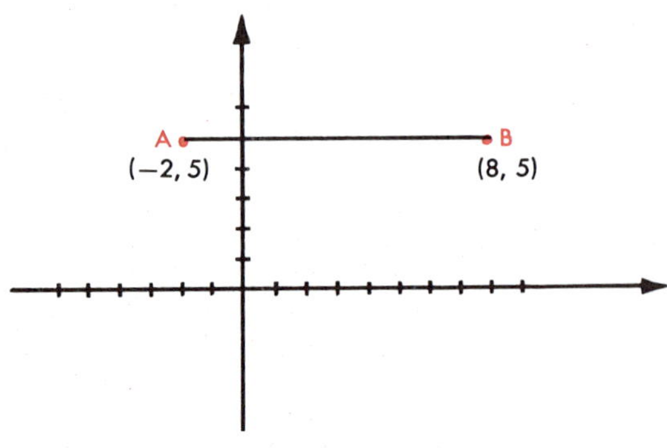

Figure 21–4

Motion along a line parallel to the y-axis is designated as positive for increasing values of y and negative for decreasing values of y.

EXAMPLE:
$$\begin{aligned} \therefore m\overline{CD} &= -m\overline{DC} \\ \text{or } m\overline{DC} &= -m\overline{CD} \end{aligned}$$

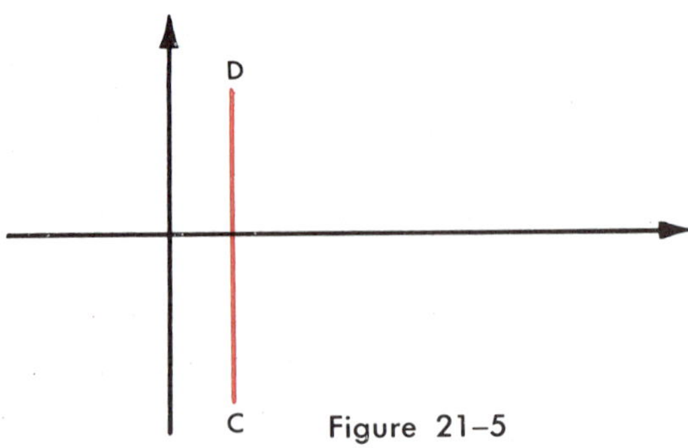

Figure 21–5

The length of a line segment not parallel to either axis is always considered to be positive regardless of the order in which it is read.

In adding segments of directed lines it is understood that the addition is to be performed algebraically. For example, suppose C is any point on the directed line passing through points A and B. (The arrow commonly indicates positive direction.) Then $m\overline{AB} + m\overline{BC} = m\overline{AC}$, since $m\overline{BC} = -m\overline{CB}$ and $m\overline{AB} - m\overline{CB} = m\overline{AC}$.

(1) A C B

Figure 21-6

There are two other arrangements of the points A, B, and C on a directed line which we should consider.

(2) A B C

Figure 21-7

Does $m\overline{AB} + m\overline{BC} = m\overline{AC}$ for Figure 21-7?

(3) C A B

Figure 21-8

Does $m\overline{AB} + m\overline{BC} = m\overline{AC}$ for Figure 21-8? (*Note:* $m\overline{CA} + m\overline{AB} = m\overline{CB}$, $m\overline{CA} = -m\overline{AC}$ and $m\overline{CB} = -m\overline{BC}$, or $-m\overline{AC} + m\overline{AB} = -m\overline{BC}$ so $m\overline{AB} + m\overline{BC} = m\overline{AC}$.)

Exercises

1. Show that $m\overline{AC} + m\overline{CB} = m\overline{AB}$ for any arrangement of the points A, B, and C on a directed line.
2. Repeat Exercise 1 but with the negative of $m\overline{AB}$.

The exercise above shows that the net result of the two motions, from A to C and from C to B, is always the same and that the equation covers all possible cases.

THE DISTANCE FORMULA

Remember the fly on the wall in Figure 21-1? If you know the values of the coordinates, h and v, could you find the distance from point O to the fly? Explain thoroughly. (Refer to the Pythagorean Theorem, page 440.)

If there were two flies on the wall (Figure 21-9) and the horizontal and vertical distances from the lines are known, could you find the distance between the flies? Explain thoroughly.

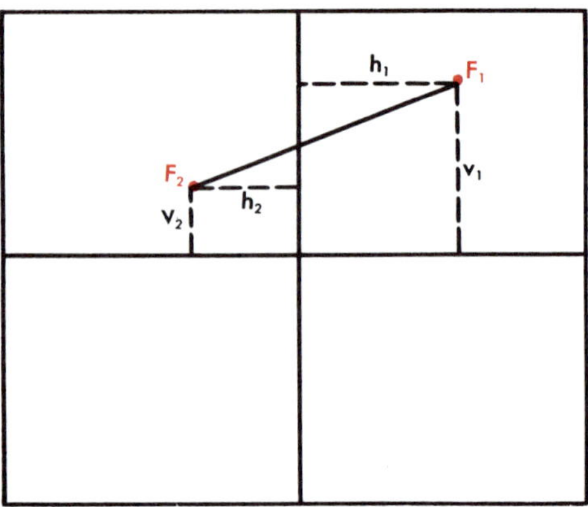

Figure 21–9

Can you express the distance between any two points on a rectangular coordinate system in terms of the coordinates of the points? Let us take the points $P_1(x_1, y_1)$ and $P_2(x_2, y_2)$, as shown below, and determine the distance between them. Draw $\overrightarrow{P_2A}$ parallel to the y-axis and $\overrightarrow{P_1A}$ parallel to the x-axis. $\angle A$ is a right angle. (Why?) $m\overline{P_1A} = x_2 - x_1$ and $m\overline{P_2A} = y_2 - y_1$. (Why?) Consider the previous statement when the points are placed in other quadrants. Now $(m\overline{P_1P_2})^2 = (x_2 - x_1)^2 + (y_2 - y_1)^2$. (Why?)

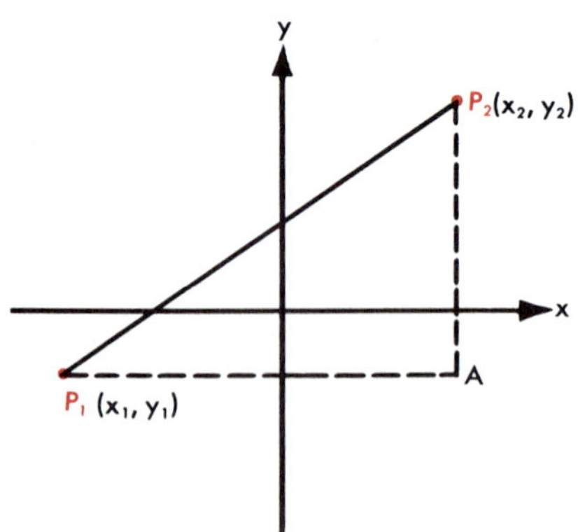

Figure 21–10

21.02 THEOREM

The measure of the distance, d, between two points is given by the formula

$$d = \sqrt{(x_2 - x_1)^2 + (y_2 - y_1)^2}$$

Note that it makes no difference which of the two points is taken as (x_1, y_1) and which as (x_2, y_2), because the differences are squared and you will get the same result whichever choice you make.

The distance formula is very important both in practical mathematics and in more advanced mathematics courses. Its uses are numerous. The following problems have been designed to improve your understanding of this relationship.

EXAMPLE: Show that $(-5,0)$, $(7,-6)$, and $(4,3)$ are the vertices of an isosceles triangle.

Solution: $d_1 = \sqrt{(-5-4)^2 + (0-3)^2} = \sqrt{90}$

$d_2 = \sqrt{(7-4)^2 + (-6-3)^2} = \sqrt{90}$

Since $d_1 = d_2$ the triangle is isosceles.

Exercises

1. Find the lengths of the sides of the triangle whose vertices are $(6,3)$, $(-5,3)$, $(-4,-4)$.
2. Show that the triangle whose vertices are $(-2,2)$, $(10,-3)$, and $(5,9)$ is an isosceles triangle.
3. Show that the diagonals of the rectangle whose vertices are $(-2,10)$, $(-2,-14)$, $(5,-14)$, $(5,10)$ are congruent.
4. Find the perimeter of the quadrilateral whose vertices are $(2,8)$, $(5,7)$, $(3,-1)$, $(-3,3)$.
5. Show that ABC is a right triangle if its vertices are $A(5,0)$, $B(-3,-1)$, $C(3,3)$. (*Hint:* Show that $a^2 + b^2 = c^2$.)
6. If the points $(0,1)$, $(7,0)$, $(12,5)$, and $(5,6)$ are the vertices of a parallelogram compare the lengths of the diagonals of the parallelogram.
7. Show that the points $(5,0)$, $(8,2)$, $(6,5)$, and $(3,3)$ are the vertices of a square.

SLOPE OF A LINE

21.03 Some of you may be familiar with the terms that carpenters use to describe the slope of the rafters of a house. They compare the *rise* of the rafter to the *run* to determine the slope. The accompanying picture illustrates the use of the terms rise and run.

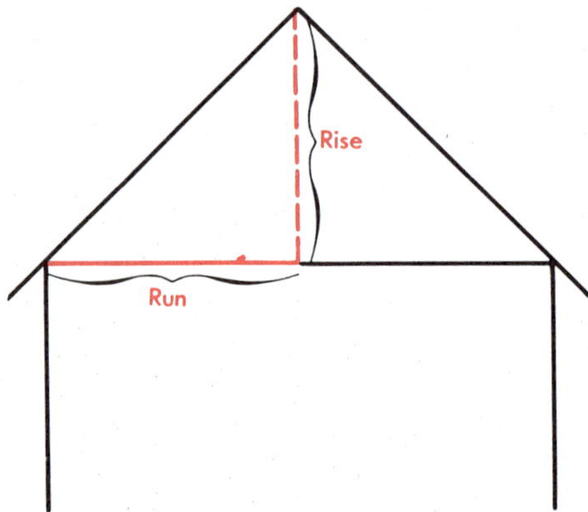

Figure 21–11

Let us apply this idea to a line on a rectangular coordinate system. Let P_1 and P_2 be two points *not on the same vertical line.* The rise of $\overleftrightarrow{P_1P_2}$ (or $\overline{P_1P_2}$) is the length of the directed segment $y_2 - y_1$. The run of $\overleftrightarrow{P_1P_2}$ is the length of the directed segment $x_2 - x_1$.

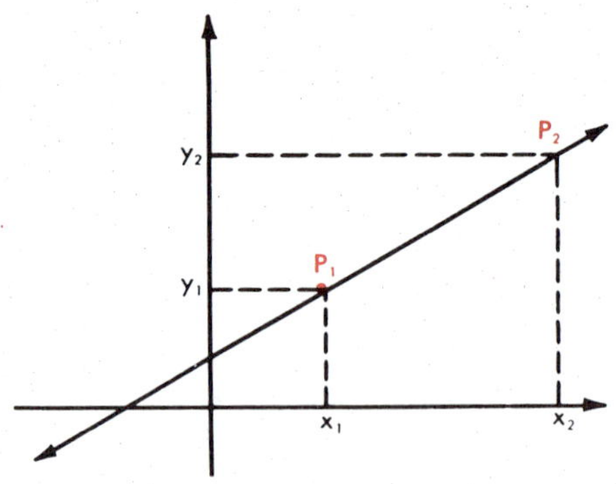

Figure 21–12

Place P_1 and P_2 in different quadrants on this line. Does the preceding statement still hold? Reverse the order of the points to show that the result is the same.

In Figure 21-13, does $y_2 - y_1$ still indicate the rise from P_1 to P_2? Does $x_2 - x_1$ still give the run from P_1 to P_2? Since $x_2 - x_1$ is negative should we say that the line has negative slope?

We *define* the slope m of a line $\overleftrightarrow{P_1P_2}$ to be the rise divided by the run (run $\neq 0$). That is

$$m = \frac{y_2 - y_1}{x_2 - x_1}, \quad x_2 - x_1 \neq 0$$

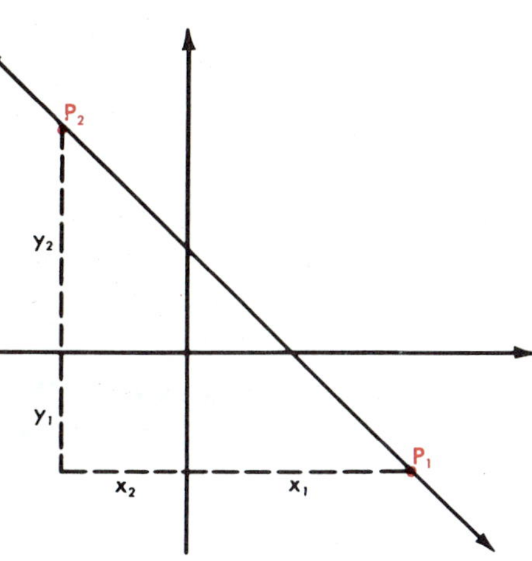

Figure 21–13

The slope may be positive, as in Figure 21-12, or negative, as in Figure 21-13. To summarize, a line with *positive* slope slants *upward* from left to right; a line with *negative* slope slants *downward* from left to right. What is the slope of a line parallel to the x-axis?

(If P_1 and P_2 are on the same vertical line then $x_2 - x_1 = 0$. Since division by zero is undefined, slope is not defined for a vertical line.)

EXAMPLES:

The slope of $\overleftrightarrow{MN}$ is

$$m = \frac{6 - (-4)}{5 - (-3)} = \frac{5}{4}$$

The slope of $\overleftrightarrow{ST}$ is

$$m = \frac{3 - (-4)}{-8 - 0} = \frac{+7}{-8} = -\frac{7}{8}$$

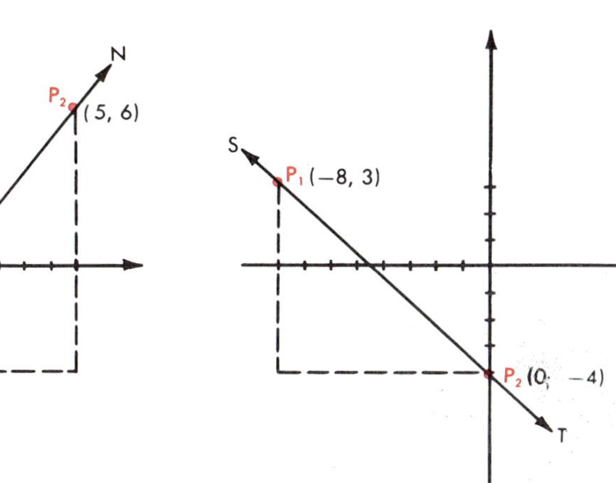

Figure 21–14

Exercises

1. Find the slopes of the lines determined by the following points:
   (a) (2,3) and (−3,5). (c) (0,0) and (8,8).
   (b) (−1,−2) and (5,6). (d) (−8,8) and (0,0).
   What is the probable relationship of lines (c) and (d)?
2. Find the slopes of the nonparallel sides of the following isosceles trapezoid: (−4,8), (2,8), (−7,3), and (5,3).
3. Find the lengths and slopes of the sides of the quadrilateral (0,0), (5,3), (8,8), (3,5). Find the slopes of the diagonals of the quadrilateral. What type of quadrilateral is this?

POINT DIVIDING A LINE SEGMENT IN A GIVEN RATIO

It is probably evident that the ability to determine the coordinates of a point for any desired fraction of the distance between two points is important. For example, it might be applied in the field of physics to determine the position of a point on a spring at a given instant in its motion.

The definition of slope makes it easy to derive expressions for the coordinates (x, y) of a point P that divides $\overline{P_1P_2}$ in a given ratio r $\left(\text{i.e., } r = \dfrac{m\overline{P_1P}}{m\overline{PP_2}}\right)$ if $r \geq 0$.

Using 16.21, $\dfrac{m\overline{A_1A}}{m\overline{AA_2}} = \dfrac{m\overline{P_1P}}{m\overline{PP_2}}$.

By substitution, $\dfrac{x - x_1}{x_2 - x} = r$.

(1) Solving for x, $x = \dfrac{x_1 + rx_2}{1 + r}$.

By drawing parallels to the x-axis from points P, P_1, and P_2 and proceeding as above,

(2) $y = \dfrac{y_1 + ry_2}{1 + r}$.

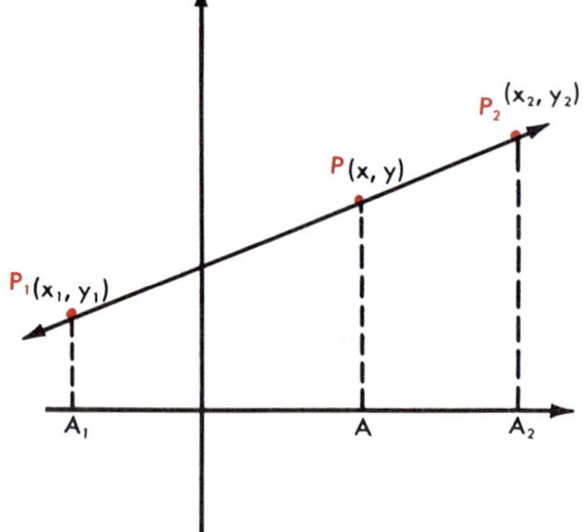

Figure 21–15

These algebraic results may be summarized in the following theorem.

21.04 THEOREM

The coordinates of the point dividing a line segment in a given ratio (r) are

$$\left(\frac{x_1 + rx_2}{1 + r}, \frac{y_1 + ry_2}{1 + r}\right), \quad r \geqq 0$$

For a particular case, if P is the midpoint of $\overline{P_1P_2}$, then $r = 1$ and the following theorem results.

21.05 THEOREM

The coordinates of the midpoint, m, of a segment $\overline{P_1P_2}$ are the ordered pair

$$\left(\frac{x_1 + x_2}{2}, \frac{y_1 + y_2}{2}\right)$$

EXAMPLE 1

Find the midpoint of the segment determined by $(-9,2)$ and $(3, -5)$.

Solution:

$$x = \frac{3 - 9}{2} = -3,$$

$$y = \frac{-5 + 2}{2} = \frac{-3}{2}$$

The coordinates of the midpoint are $\left(-3, -\frac{3}{2}\right)$.

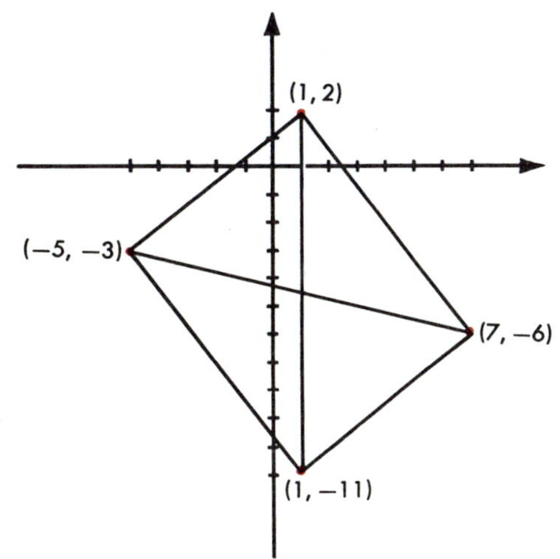

Figure 21–16

EXAMPLE 2

Prove that the diagonals of the parallelogram whose vertices are $(-5, -3)$, $(1, -11)$, $(7, -6)$, and $(1,2)$ bisect each other. (Figure 21-16)

Solution: The midpoint of $d_1 = \left(\frac{1 + 1}{2}, \frac{2 - 11}{2}\right) = \left(1, \frac{-9}{2}\right)$

The midpoint of $d_2 = \left(\frac{-5 + 7}{2}, \frac{-3 - 6}{2}\right) = \left(1, \frac{-9}{2}\right)$

Since d_1 and d_2 have the same midpoint they bisect each other.

Exercises

1. Find the midpoints of the sides of the triangle whose vertices are (3, −5), (−2, −7), and (1,5).
2. The points (−5,5), (9,5), and (1,−5) are the vertices D, E, and F respectively of triangle DEF. Show that m$\overline{AB}$, where $\overline{AB}$ joins the midpoints of $\overline{DE}$ and $\overline{DF}$, is equal to one-half m$\overline{EF}$.
3. The points (−5,6), (−7,−2), (10,−2), and (4,6) are the vertices A, B, C, D respectively of a trapezoid $ABCD$. Show that m$\overline{EF}$, where $\overline{EF}$ joins the midpoints of $\overline{AB}$ and $\overline{CD}$, equals one-half (m$\overline{BC}$ + m$\overline{AD}$).
4. If (1,−4), (7,4), and (4,5) are respectively the vertices of right triangle ABC, with $\angle C$ the right angle, show that the measure of the median (m$\overline{CD}$) is $\frac{1}{2}$ m$\overline{AB}$.
5. Find the slopes of the medians of the triangle $A(0,4)$, $B(2,-3)$, and $C(1,-5)$.
6. Find the coordinates (x,y) of a point P which divides P_1P_2 into the ratio 1:2.
7. Find the coordinates of the points which trisect the segment joining $A(-2,-4)$ and $B(4,8)$.

PARALLEL LINES

In our previous discussion of parallel lines (2.41) the word "slope" was not used. Is it reasonable to assume that parallel lines would have the same slope and that lines in a plane having the same slope are parallel? Let us examine two parallel lines ($\overleftrightarrow{A_1B_1}$ and $\overleftrightarrow{A_2B_2}$) in a rectangular coordinate system. (Figure 21-17.)

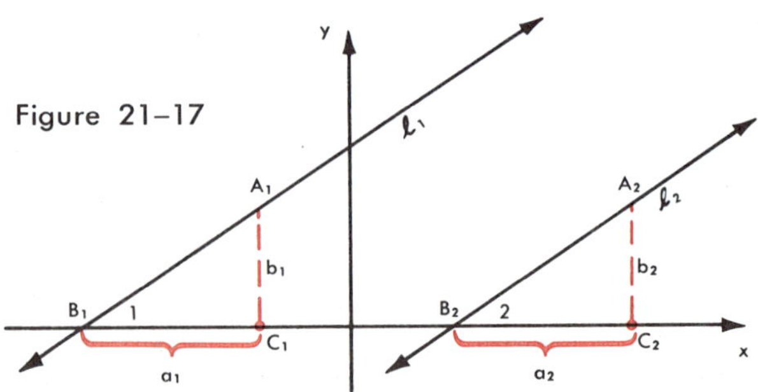

Figure 21–17

If $\overleftrightarrow{A_1B_1}$ is parallel to $\overleftrightarrow{A_2B_2}$, what is true of $\angle 1$ and $\angle 2$? Why? Since $\overline{A_1C_1}$ and $\overline{A_2C_2}$ are perpendicular to the x-axis, are the triangles similar? Why? Does $\frac{b_1}{a_1} = \frac{b_2}{a_2}$? Explain.

Does this show that parallel lines always have the same slope? Can you prove that the converse is true? (*Hint:* Use the definition of slope to prove that the triangles are similar, by 17.08; then use Assumption 5.23) We have now proved the following theorem and its converse.

21.06 THEOREM

Lines in a given plane that have the same slope are parallel. Converse: Parallel lines have the same slope ($m_1 = m_2$).

EXAMPLE 1

Show that $\overleftrightarrow{AB}$, passing through points $(-1,2)$ and $(13,-5)$, is parallel to $\overleftrightarrow{CD}$, passing through points $(-12,5)$ and $(6,-4)$.

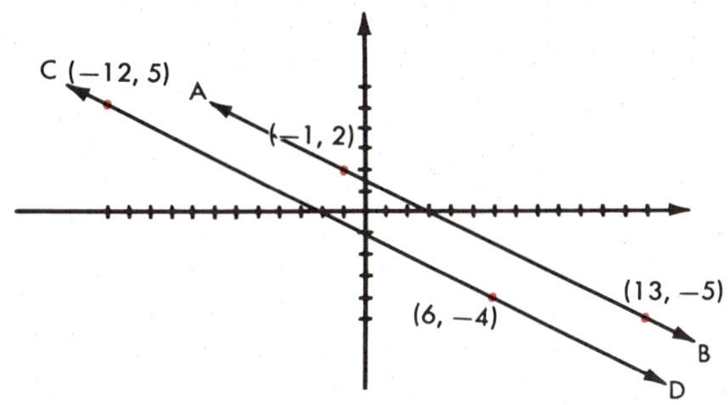

Figure 21–18

Solution: The slope of $\overleftrightarrow{AB} = \dfrac{-5 - 2}{13 - (-1)} = -\dfrac{1}{2}$.

The slope of $\overleftrightarrow{CD} = \dfrac{5 - (-4)}{-12 - 6} = -\dfrac{1}{2}$.

Since the slopes are equal, the lines are parallel.

EXAMPLE 2

Show that $(-4,8)$, $(2,8)$, $(5,3)$, and $(-7,3)$ are the vertices of an isosceles trapezoid.

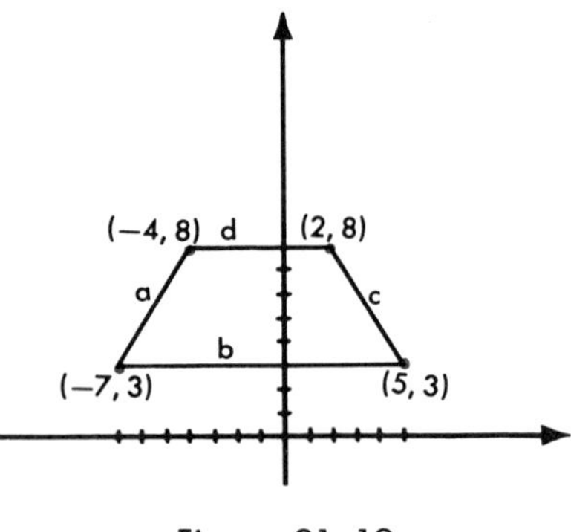

Figure 21–19

Solution:

The slope of $d = \dfrac{8 - 8}{-4 - 2} = 0$

The slope of $b = \dfrac{3 - 3}{-7 - 5} = 0$

The length of $a = \sqrt{(-7 + 4)^2 + (3 - 8)^2} = \sqrt{34}$.
The length of $c = \sqrt{(2 - 5)^2 + (8 - 3)^2} = \sqrt{34}$.

We have shown that $a = c$ and $b \parallel d$; therefore the figure is an isosceles trapezoid (9.19 and 9.22).

PERPENDICULAR LINES

Review Exercises 1 and 3, 21.03. Were the lines perpendicular in both instances? What is the relationship of the slopes of these lines? Let us attempt to prove this relationship.

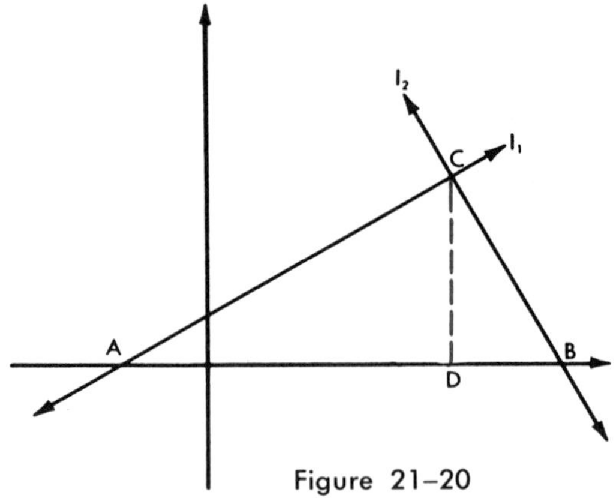

Figure 21–20

Proof:
If two lines, l_1 and l_2, having slopes m_1 and m_2 respectively, are perpendicular to each other, they form with the x-axis a right tri-

angle ABC, with angle C the right angle. If the altitude $\overline{CD}$ is drawn to $\overleftrightarrow{AB}$ from C, we have $\dfrac{m\overline{AD}}{m\overline{DC}} = \dfrac{m\overline{DC}}{m\overline{DB}}$ (Section 17.12), or $\dfrac{m\overline{DC}}{m\overline{DB}} \cdot \dfrac{m\overline{DC}}{m\overline{AD}} = 1$. Since the slope $m_1 = \dfrac{m\overline{DC}}{m\overline{AD}}$ and $m_2 = \dfrac{m\overline{CD}}{m\overline{DB}}$ (Section 21.03), and since $m\overline{CD} = -m\overline{DC}$ (Section 21.01), then, by substitution, $m_1 \cdot m_2 = -1$.

21.07 THEOREM

If two lines are perpendicular, then the slope of one is the negative reciprocal of the slope of the other (The product of the slopes $m_1 \cdot m_2 = -1$), and conversely.

The proof of the converse must wait until certain trigonometric relationships have been established.

EXAMPLE 1

The slope of $\overleftrightarrow{AB}$ is $\dfrac{3}{5}$. The slope of $\overleftrightarrow{CD}$ is $-\dfrac{5}{3}$. Show that $\overleftrightarrow{CD} \perp \overleftrightarrow{AB}$.

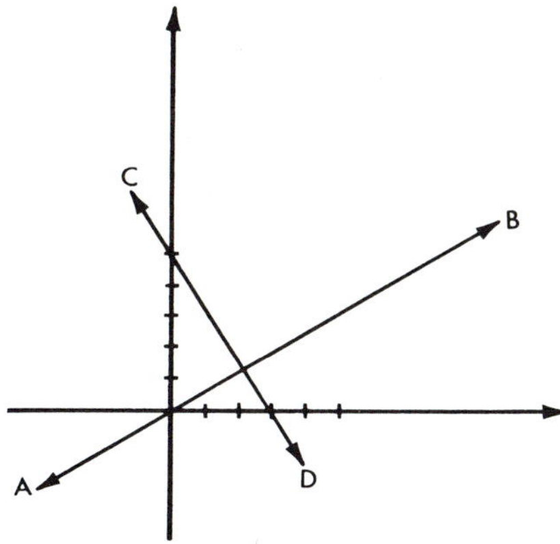

Figure 21–21

Solution:

$\overleftrightarrow{CD}$ and $\overleftrightarrow{AB}$ are perpendicular because $\dfrac{3}{5} \cdot -\dfrac{5}{3} = 1$.

(Figure 21–21)

EXAMPLE 2

Show that line l_1 passing through point (3,5) and intersecting line l_2 at (6,−3) is perpendicular to l_2 if l_2 passes through point (−2,−6). (Figure 21-22)

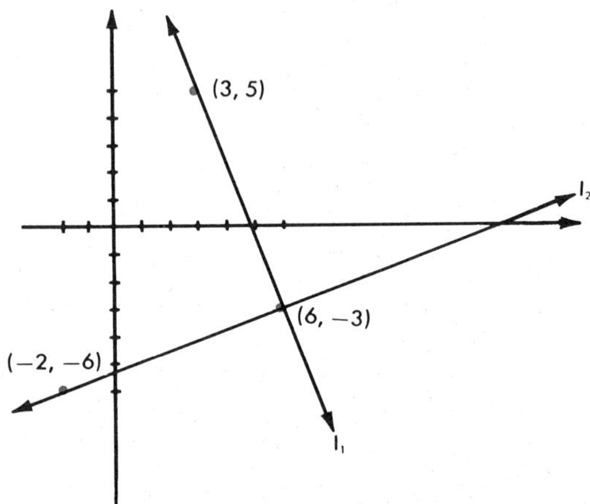

Figure 21–22

Solution:

The slope of $l_1 = \dfrac{5-(-3)}{3-6} = -\dfrac{8}{3}$.

The slope of $l_2 = \dfrac{-6-(-3)}{-2-6} = \dfrac{3}{8}$.

Since the product of the slopes $= -1$, $l_1 \perp l_2$.

Exercises

1. Find the slopes of the lines passing through the following sets of points:
   (a) (5,−2) and (−8,5) (b) (−4,−7) and (6,5)
2. Which of the lines determined by the following points intersect, are parallel, or are perpendicular? (Could any of them be skew?)
   (a) $A(2,3), B\left(3, -\dfrac{1}{2}\right)$ (b) $C(9,6), D(2,4)$
   (c) $E(9,-1),\ F\left(8, \dfrac{5}{2}\right)$ (d) $G(-2,-1), H(5,2)$

3. Do the three points $(-2,-10)$, $(4,8)$, $(2,2)$ lie on the same straight line? (Use slopes to determine your answer.)

4. Show that the points $(5,0)$, $(8,2)$, $(6,5)$, and $(3,3)$ are the vertices of a square.

5. Prove or disprove that $(-9,0)$, $(-10,-6)$, $(4,8)$, and $(5,14)$ are the vertices of a parallelogram.

6. The points $(-6,6)$, $(2,4)$, and $(-2,-8)$ are the vertices of a triangle. Show that the line joining the midpoints of any two sides of this triangle is parallel to the third side.

7. Prove or disprove by means of slopes that $(-7,3)$, $(7,10)$, and $(-3,-5)$ are the vertices of a right triangle.

8. Prove or disprove that the following quadrilateral is a rectangle. The vertices are $(3,3)$, $(6,0)$, $(2,-6)$, and $-1,4)$.

9. The points $(0,0)$, $(5,3)$, $(8,8)$, and $(3,5)$ are the vertices of a rhombus. Show that the diagonals of the rhombus are perpendicular to each other.

10. Show that the median of the trapezoid whose vertices are $(-8,4)$, $(16,-4)$, $(14,8)$, and $(2,12)$ is parallel to the bases and equals half the sum of their lengths.

11. How many lines with a given slope can be drawn through a given point?

12. Find the area of the triangle whose vertices are $(-3,-4)$, $(6,-2)$, and $(-5,5)$.

13. Find the area of the quadrilateral whose vertices are $(0,2)$, $(5,2)$, $(0,-4)$, and $(5,-4)$.

14. Find the area of the quadrilateral whose vertices are $(5,5)$, $(5,-5)$, $(-5,5)$, and $(-5,-5)$.

15. Find the area of the quadrilateral whose vertices are $(-3,3)$, $(-8,-2)$, $(-3,-7)$, and $(2,-2)$.

16. Find the area of the quadrilateral whose vertices are $(0,4)$, $(-3,-4)$, $(5,-1)$, and $(8,7)$.

17. Find the area of the quadrilateral whose vertices are $(1,1)$, $(6,1)$, $(8,4)$ and $(3,4)$.

18. Find the area of the quadrilateral whose vertices are $(-3,-1)$, $(-4,2)$, $(9,3)$, and $(2,4)$.

SUMMARY

We have now established that it is possible to plot geometric figures on a rectangular coordinate system and make use of our knowledge of algebra to prove or disprove certain conjectures. The following list shows the correspondence between certain geometric concepts and their algebraic counterparts.

Geometry	*Algebra*
point P	(x,y)
m $\overline{P_1P_2}$	$\sqrt{(x_2 - x_1)^2 + (y_2 - y_1)^2}$
midpoint of $\overline{P_1P_2}$	$\left(\frac{x_1 + x_2}{2}, \frac{y_1 + y_2}{2}\right)$
slope of $\overleftrightarrow{P_1P_2}$	$m = \frac{y_2 - y_1}{x_2 - x_1}, x_2 - x_1 \neq 0$
$l_1 \parallel l_2$	$m_1 = m_2$
$l_1 \perp l_2$	$m_1 m_2 = -1$

ANALYTIC GEOMETRY PROOFS

21.08 Some of the theorems of geometry can be proved more easily by analytic geometry than by Euclidean methods. The previous proofs of this chapter have been for specific figures. The following proofs show how to apply the methods of analytic geometry to general figures.

Remember that on page 109 we stated that the figure we use to illustrate a proof is a *symbol of the set of all figures* fitting the given conditions. It should not have properties that any other member of the set would not have. For example, an equilateral triangle does not truly represent the set of all triangles. It symbolizes only the set of all equilateral triangles.

In each case we will superimpose the general figure upon a rectangular coordinate system. It is best to have a vertex of the figure coincide with the origin of the coordinate system and one side of the polygon coincide with an axis. The coordinates of the vertices are letters, standing in place of any and all pairs of ordered numbers that would produce the desired figure.

Given: Triangle ABC, with vertices $A(0,0)$, $B(2a,0)$, and $C(2b, 2c)$. D and E the midpoints of $\overline{AC}$ and $\overline{BC}$ respectively. (NOTE: We use $2a$, $2b$, and $2c$ to make our algebra easier.)

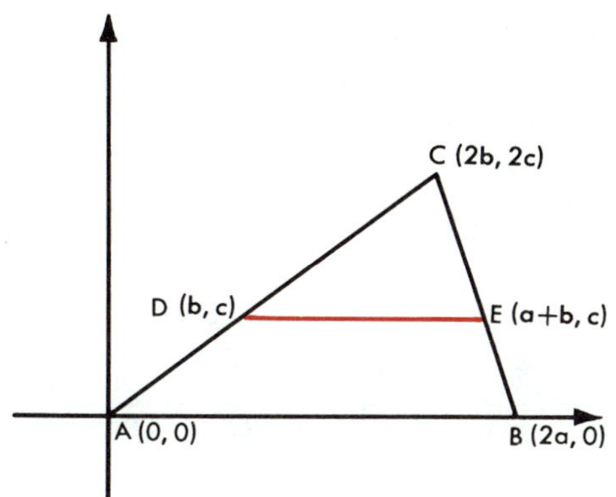

Figure 21–23

Conjecture: $m\overline{DE} = \frac{1}{2} m\overline{AB}$. $\overline{DE} \parallel \overline{AB}$

Proof: The coordinates of D are $x = \frac{0 + 2b}{2} = b$, and $y = \frac{2c + 0}{2} = c$.

The coordinates of E are $x_1 = \frac{2a + 2b}{2} = a + b$, and $y_1 = \frac{2c + 0}{2} = c$.

$m\overline{AB} = \sqrt{(2a - 0)^2 + (0 - 0)^2} = 2a$.

$m\overline{DE} = \sqrt{(a + b - b)^2 + (c - c)^2} = a$. $\therefore m\overline{DE} = \frac{1}{2} m\overline{AB}$. The slope of $\overline{DE} = 0$ as does the slope of $\overline{AB}$.

$\therefore \overline{AB} \parallel \overline{DE}$. Use 21.02 to show that $m\overline{DE} = \frac{1}{2} m\overline{AB}$.

21.09 THEOREM

The segment connecting the midpoints of two sides of a triangle is parallel to the third side and its length is half the length of the third side. (See 9.26)

Given: Parallelogram $ABCD$

Conjecture: $\overline{AB} \cong \overline{CD}$ and $\overline{AD} \cong \overline{BC}$

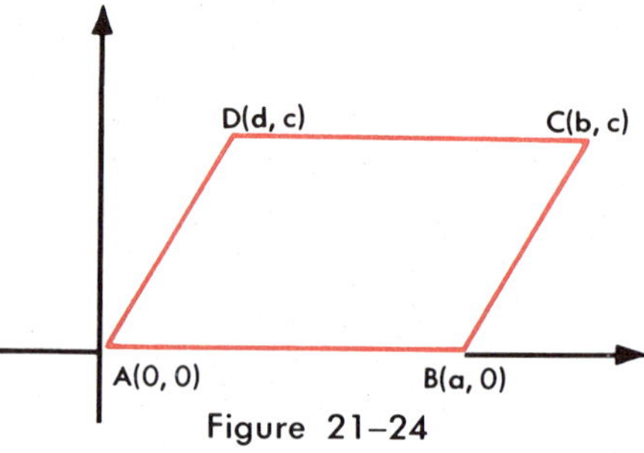

Figure 21–24

Proof: Since $\overline{AB} \parallel \overline{CD}$ the y-coordinates of C and D are equal. Since $\overline{AD} \parallel \overline{BC}$, then $\frac{c}{d} = \frac{c}{b - a}$ and $d = b - a$.

$$m\overline{AD} = \sqrt{(d - 0)^2 + (c - 0)^2} = \sqrt{d^2 + c^2}$$
$$m\overline{CB} = \sqrt{(b - a)^2 + (c - 0)^2} = \sqrt{(d)^2 + (c - 0)^2} = \sqrt{d^2 + c^2}$$
$\therefore \overline{AD} \cong \overline{BC}$.
$$m\overline{AB} = \sqrt{(0 - a)^2 + (0 - 0)^2} = a$$
$$m\overline{CD} = \sqrt{(d - b)^2 + (c - c)^2} = \sqrt{(-a)^2} = a$$
$\therefore \overline{AB} \cong \overline{CD}$.

21.10 THEOREM

The opposite sides of a parallelogram are congruent. (See 9.03.)

Given: Rectangle $ABCD$

Conjecture: $\overline{AC} \cong \overline{BD}$

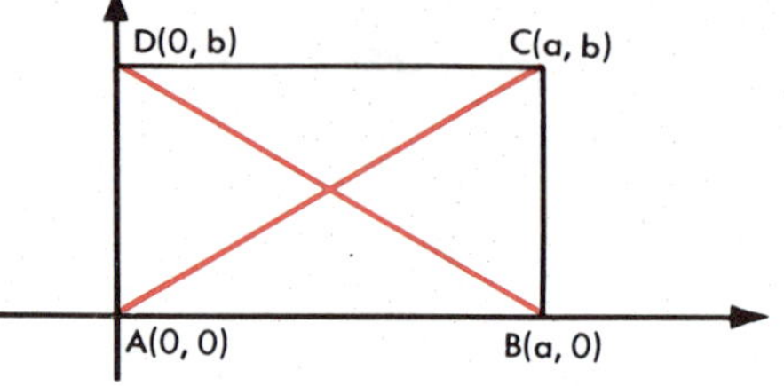

Figure 21–25

Proof: Place vertex A at the origin and side $\overline{AB}$ along the x-axis. Side $\overline{AD}$ then lies along the y-axis. Why? $\overline{AD} \cong \overline{BC}$ and $\overline{CD} \cong \overline{AB}$.

$\therefore$ the vertices may be labeled $A(0,0)$, $B(a,0)$, $C(a,b)$, $D(0,b)$

$$m\overline{AC} = \sqrt{(0 - a)^2 + (0 - b)^2} = \sqrt{a^2 + b^2}.$$
$$m\overline{BD} = \sqrt{(0 - a)^2 + (b - 0)^2} = \sqrt{a^2 + b^2} \quad \therefore \overline{AC} \cong \overline{BD}$$

21.11 THEOREM

The diagonals of a rectangle are congruent. (See 9.13.)

Exercises

Establish the following propositions by using the methods of analytic geometry.

1. The diagonals of a parallelogram bisect each other.
2. If the diagonals of a parallelogram are congruent, the parallelogram is a rectangle.
3. If the diagonals of a quadrilateral bisect each other, the quadrilateral is a parallelogram.
4. If two sides of a quadrilateral are congruent and parallel, the quadrilateral is a parallelogram.
5. The diagonals of an isosceles trapezoid are congruent.
6. The segments joining the midpoints of the four sides of a quadrilateral, taken in order, form a parallelogram.
7. The diagonals of a rhombus are perpendicular to each other.
8. Two lines in the same plane perpendicular to the same line are parallel.
9. The midpoint of the hypotenuse of a right triangle is equidistant from all three vertices of the triangle.
10. Segments joining the midpoints of the sides of an isosceles trapezoid, taken in order, form a rhombus or a square.
11. Segments joining the midpoints of opposite sides of any quadrilateral bisect each other.
12. The medians of a triangle are concurrent in a trisection point of each.

This does not represent a complete list of propositions that can be proved analytically. Select others from your text for analytic proof.

EQUATION OF A STRAIGHT LINE

21.12 We know that two points determine a line (2.10). If we know the slope of a line and one point of this line, have we determined the line? Does one and only one line fit the conditions?

We have established that the slope of a line can be expressed by the formula $m = \frac{y_2 - y_1}{x_2 - x_1}$, which was developed in 21.03. The formula will be easier to use, and will remain equally correct, if we change the subscripts to read as follows, $m = \frac{y - y_1}{x - x_1}$. Let us examine this relationship when we know only the slope and one point of the line.

EXAMPLE 1

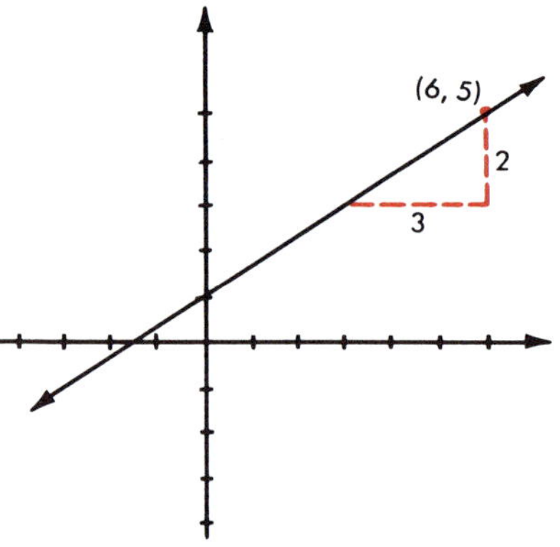

Figure 21–26

If $m = \frac{2}{3}$ and a point of the line with this slope is (6,5), then

$$\frac{2}{3} = \frac{y - 5}{x - 6}$$

or, transformed, $3y = 2x + 3$.

Since (x,y) *represents any point of the line,* we have derived the *equation of the locus of this point*—the equation of the line.

This leads us to use the slope formula as a means for finding the equation of a line, if we know the slope and a point of the line. This is known as the point-slope equation of a line, or

$$y - y_1 = m(x - x_1) \qquad (1)$$

EXAMPLE 2

If the coordinates of a point are (−4,2) and the slope of a line through this point is $-\frac{3}{4}$, find the equation of the line. Using (1),

$y - 2 = -\frac{3}{4}[x - (-4)]$, or $4y = -3x - 4$.

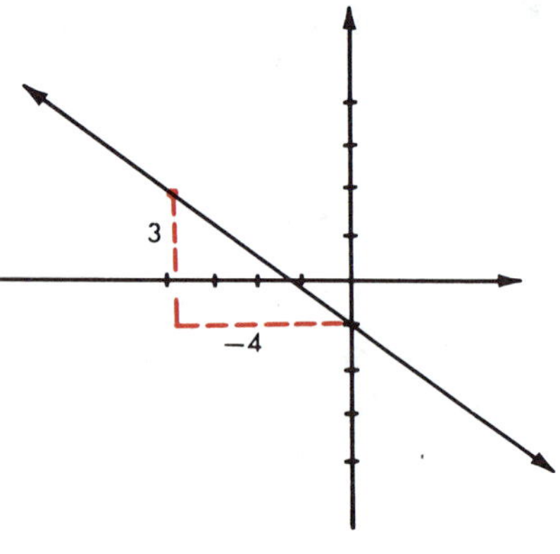

Figure 21–27

Transforming the equations of the lines in Examples 1 and 2 we have $y = \frac{2}{3}x + 1$ and $y = -\frac{3}{4}x - 1$.

Examine the graphs accompanying Examples 1 and 2. At what points do the lines cross the y-axis? What is the x-coordinate of each of these points? Do the rearranged equations show these points clearly? Do these equations also show the slopes of the lines?

In coordinate geometry the point at which a line intersects the y-axis is called the y-intercept and is designated by the letter b. Therefore $(0,b)$ represents the coordinates of the y-intercept of any straight line. If we substitute $(0, b)$ in the point-slope formula

(1) $$y - y_1 = m(x - x_1)$$

then $$y - b = m(x - 0)$$

(2) or $$\boxed{y = mx + b}$$

This called the slope-y-intercept form of the equation of a line.

It should be evident that if a line is parallel to the x-axis the equation of such a line would be $y = b$. If a line is parallel to the y-axis the equation of such a line would be $x = k$ (any real number). To make this clear, the graphs of $y = 5$ and $x = 3$ are shown.

EXAMPLE

Find the equations of the lines drawn on the coordinate system of Figure 21–29.

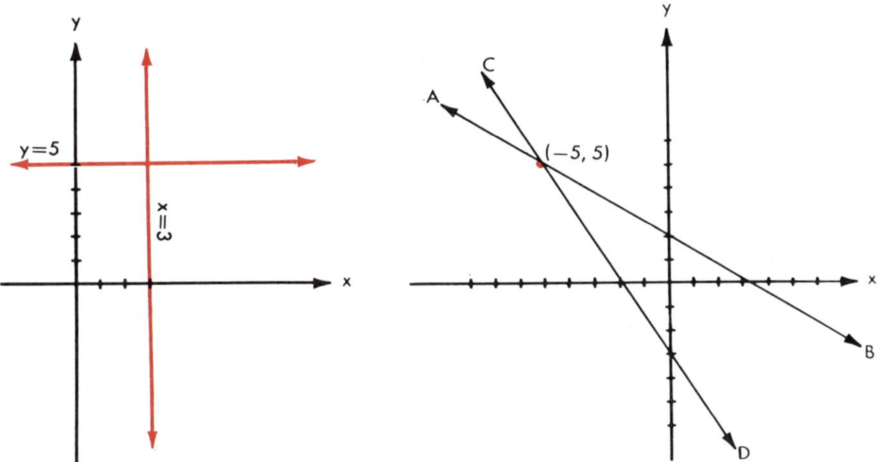

Figure 21–28

Figure 21–29

(a) The slope of $\overleftrightarrow{AB}$ is $-\frac{3}{5}$. Its y-intercept is $+2$.
Using (2), the equation is $y = -\frac{3}{5}x + 2$.

(b) The slope of $\overleftrightarrow{CD}$ is $-\frac{8}{5}$. Its y-intercept is -3.
Using (2), the equation is $y = -\frac{8}{5}x - 3$.

If we know two points, can we determine the equation of the line through these points?

We know the slope of the line through these points is $\frac{y_2 - y_1}{x_2 - x_1}$. Knowing the slope and a point of the line, we can derive the equation of a line by using formula (1). This process is sometimes expressed by the following formula, which is called the two-point form of the equation of a line.

(3)
$$\frac{y - y_1}{x - x_1} = \frac{y_2 - y_1}{x_2 - x_1}$$

Exercises

1. Find the equations of the lines determined by the following pairs of points:
   (a) $(2,-7), (-6,2)$
   (b) $(3,5), (-4,-3)$
   (c) $(0,1), (-8,2)$
   (d) $(-3,0), (-1,-1)$
2. Find the equations of the lines passing through point $(6,8)$ whose slopes are as follows:
   (a) $-\frac{2}{3}$
   (b) $\frac{1}{2}$
   (c) $-\frac{1}{2}$
   (d) $-\frac{3}{4}$

   How many lines could pass through point $(6,8)$?
3. The vertices of a triangle are $(3,5)$, $(-2,-7)$, and $(1,5)$. Find the equations of the lines containing the medians of the triangle.
4. Find the equation of the line containing the altitude of the triangle whose vertices are $(-5,8)$, $(-10,-2)$, and $(4,2)$, from the vertex $(-5,8)$.
5. Find the distance from the point $(-1,6)$ to the line passing through the points $(1,2)$, $(6,3)$.
6. Given the points $(-6,4)$, $(12,-2)$, and $(2,8)$, find a fourth point that, with the previous three, would form the vertices of a parallelogram. How many such points are there?
7. Prove that the lines containing the altitudes of a triangle are concurrent.
8. Prove that the perpendicular bisectors of the sides of a triangle are concurrent.

THE EQUATION OF A CIRCLE

21.13 Let us place a circle on a coordinate system. Represent its radius by r, and its center by the coordinates (h,k). Take any point (x,y) on the circle. Then, using the distance formula,

$$\sqrt{(x-h)^2+(y-k)^2}=r \quad \text{or} \quad (x-h)^2+(y-k)^2=r^2.$$

This is the equation of a circle.

If the center of the circle were at the origin, then $(x-0)^2 + (y-0)^2 = r^2$, or $x^2+y^2=r^2$.

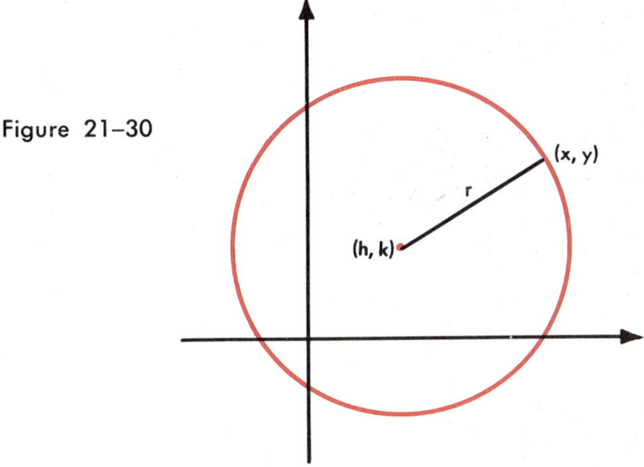

Figure 21–30

EXAMPLE 1

Find the equation of the circle whose center is (0,0) and whose radius is 5.

Since the center is at the origin, the equation is $x^2+y^2=25$.

EXAMPLE 2

Find the equation of a circle whose center is $(-1,3)$ and whose radius is 2.

Since the general formula is $(x-h)^2+(y-k)^2=r^2$,
then $(x+1)^2+(y-3)^2=4$,
or $x^2+2x+1+y^2-6y+9=4$.
Simplified, $x^2+y^2+2x-6y=-6$ is the equation of the circle.

EXAMPLE 3

Given the equation of a circle as $x^2+y^2-4x-6y=12$, find its center and radius.

Arranging in a form similar to the general form $(x-h)^2+(y-k)^2=r^2$, we have $(x^2-4x+?)+(y^2-6y+?)=12$. To form trinomials which are squares of binomials we must add 4 and 9 respectively to the terms. Then, $(x^2 - 4x + 4) + (y^2 - 6y + 9) = 25$, since we must add the same number to both members of the equation. Finally, $(x - 2)^2 + (y - 3)^2 = 5^2$, so the radius is 5 and the center (2,3).

Exercises

1. Write the equations of the following circles and construct their graphs:
   (a) Center (0,0) and radius 3
   (b) Center (0,3) and radius 5
   (c) Center $(-1,3)$ and radius 2
   (d) Center $(2,-4)$ and radius 6
2. Find the center and radius of each of the following circles and construct the graph of the circle:
   (a) $x^2 + y^2 = 81$
   (b) $x^2 + y^2 - 6x + 8y = 0$
   (c) $x^2 + y^2 + 10x - 24y - 27 = 0$
   (d) $x^2 + y^2 - 6y - 7 = 0$
3. Write the equation of a circle whose diameter is the segment joining points $(-3,6)$ and $(3,-2)$.
4. Write the equation of a circle circumscribed about the triangle whose vertices are $(-6,10)$, $(6,4)$, and $(-6,-8)$.

Vocabulary List

coordinate
Descartes
origin
abscissa
ordinate
directed lines
slope
y-intercept

Chapter Review

1. Name the abscissa of the point denoted by the ordered pair $(-4, 7)$.
2. Write the equation of the line whose slope is $-\frac{4}{5}$ and which contains point $(0, -2)$.
3. Write the equation of the line containing point $(-1,4)$ and which is parallel to the line of Exercise 2.

4. Find the length of the segment whose endpoints are $(5,-2)$ and $(-3,-4)$.

5. Find the slope of the line containing points $(1,2)$ and $(-4,8)$.

6. What is the slope of the line which is perpendicular to the line of Exercise 5?

7. Find the midpoint of the segment whose endpoints are $(-6,2)$ and $(4,8)$.

8. Write the equation of the line containing the points of Exercise 7.

9. Write the equation of the line which is the perpendicular bisector of the segment of Exercise 7.

10. Write the equation of the circle whose center is $(2,3)$ and whose radius is 4.

11. Prove that the points $(1,3)$, $(-2,5)$, and $(7,-1)$ lie in a line.

12. Prove that the length of the median to the hypotenuse of a right triangle is half the length of the hypotenuse.

Chapter 21 Test

1. On a rectangular coordinate system
   (a) plot the point whose abscissa is -5 and whose ordinate is 5.
   (b) Plot the point $(5,-2)$

2. Given the equation of a line, $y = \frac{2}{3}x + 4$.
   (a) Name the slope of this line.
   (b) Name the y-intercept of the line.
   (c) Name the slope of a line parallel to the given line.
   (d) Write the equation of the line through $(1,-2)$ parallel to the given line.

3. Given points $(-1,-2)$ and $(3,4)$.
   (a) Find the distance between the points.
   (b) Find the slope of the line containing the points.
   (c) Name the slope of a line perpendicular to the line containing the two given points.
   (d) Find the midpoint of the segment determined by the points.
   (e) Write the equation of the line containing the two given points.
   (f) Write the equation of the line which is the perpendicular bisector of the segment determined by the given points.

4. Write the equation of the circle whose center is (0,0) and whose radius is 5.

5. The coordinates of one endpoint of a segment are (2,4) and the coordinates of the midpoint are (5,1). Name the coordinates of the other endpoint.

6. Is the triangle with vertices A (3,6), B (–1,2) and C (–7,8) a right triangle? Prove your response.

7. Given: Triangle $A(0,0)$, $B(10,0)$, $C(4,8)$. Points D and E on $\overline{AB}$. $m\overline{AD} = \frac{1}{4} m\overline{AB}$. $m\overline{EB} = \frac{1}{4} m\overline{AB}$.

   Prove: The segments are parallel which connect the midpoints of $\overline{AC}$ and $\overline{BC}$ to D and E.

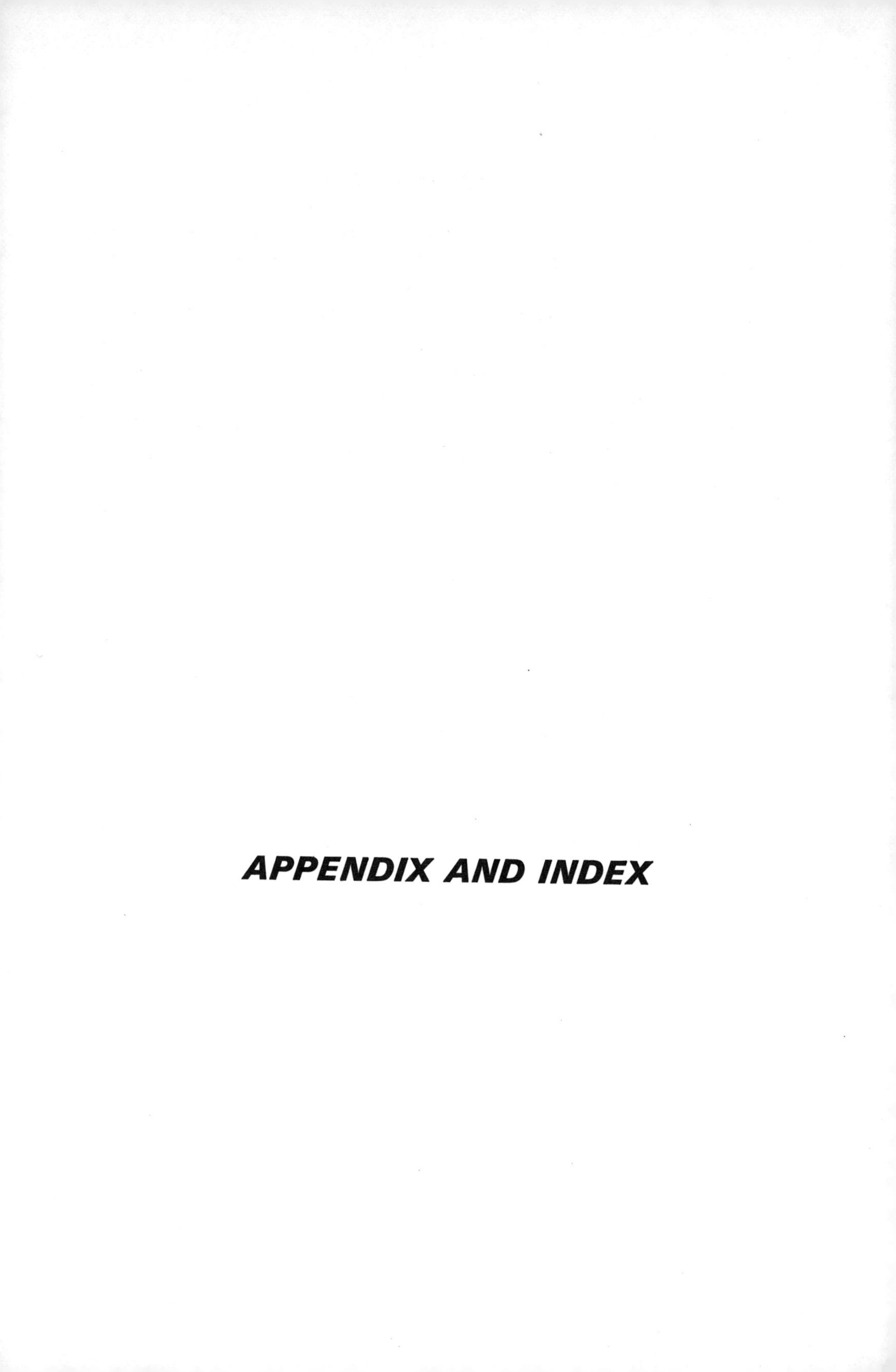

APPENDIX AND INDEX

TRUTH TABLES

We agreed (2.13) that statements are declarative sentences that are either true or false. The truth or falsity of a simple sentence is a judgement depending upon the person, time and place but in most cases is easy to determine. Truth of a simple sentence in this course means consistent with reality or with the structure of the course. For example, "Water freezes" is a statement. Within our experience we believe it to be true. "It is false that water freezes" or "Water does not freeze" is the negation of the first statement. Our experience shows this is false. This leads us to a simple "truth table" for statements and their negations. The letter p will represent a statement while $\sim p$ (read "not p") represents the negation of p.

Negation

p	$\sim p$
T	F
F	T

Determining the truth of compound declarative sentences is not as simple. For example, if we accept as true the simple statement, "The grass is green," and if we also accept as true the statement, "The sky is blue," what would be the truth value of the compound statement, "The sky is blue and the grass is green"? What is the truth value of "The sky is blue and the grass is purple"? What is the truth value of "The grass is purple and trees are blue"?

A compound sentence formed by joining two simple sentences with the connective *and* is called a conjunction. Let "p and q" represent a conjunction, where p represents a simple sentence and q represents a simple sentence. There are only four possible combinations of truth values for p and q:

p is true and q is true;
p is true and q is false;
p is false and q is true;
p is false and q is false.

The word "and" suggests that the conjunction is true if and only if both of the simple sentences are true. The four possible combinations of truth and falsity and the conclusions concerning the truth of these conjunctions are listed in the following truth table.

Conjunction

p	q	p and q
T	T	T
T	F	F
F	T	F
F	F	F

Compound sentences using the connective "if..., then..." present some interesting problems when we consider the truth values of the sentence. We will examine these possibilities in the following exercises. We use the expression "$p \to q$" to represent the conditional, "If p, then q," where p and q are simple sentences.

Oral Exercises

Let A be "Horses can fly." (*False*)
B be "The sun is hot." (*True*)
C be "Trees are purple." (*False*)
D be "Water is wet." (*True*)

Decide upon the truth or falsity of the following implications. (To read the implications, replace the letters with the sentences which they represent.)

1. $B \to D$
2. $B \to A$
3. $A \to B$
4. $A \to C$
5. $D \to B$
6. $D \to C$
7. $C \to A$
8. $C \to B$
9. $C \to D$
10. $B \to C$
11. $A \to D$
12. $D \to A$
13. From your responses to Exercises 1–12, complete the following truth table.

Conditional

p	q	$p \to q$
T	T	?
T	F	?
F	T	?
F	F	?

Most of you will agree concerning the truth values to be assigned to lines 1 and 2 of the above truth table for conditional statements but you may not agree readily with the truth values assigned to lines 3 and 4. If you think carefully and not emotionally about these truth values, you will understand why mathematicians and logicians have assigned the truth values indicated by your teacher.

Exercises

1. Form a truth table for a compound sentence composed of two distinct simple sentences joined by the disjunction "or" in its inclusive form, that is, "p or q, or both."

2. Complete the following truth table.

Hypothetical Syllogism

p	q	r	$p \to q$	$q \to r$	$(p \to q)$ and $(q \to r)$	$p \to r$	$[(p \to q)$ and $(q \to r)] \to (p \to r)$
T	T	T					
T	T	F	*Hint: Use the*		*Hint: Use*	*Hint: Use the conditional*	
T	F	T	*conditional*		*the con-*	*truth table.*	
T	F	F	*truth table*		*junction*		
F	T	T			*truth*		
F	T	F			*table*		
F	F	T					
F	F	F		DO NOT WRITE ON THIS BOOK!			

3. Complete the following truth table. (It is used with the connective "if and only if.")

Biconditional

p	q	$p \to q$	$q \to p$	$(p \to q)$ and $(q \to p)$ Sometimes written $(p \leftrightarrow q)$
T	T			
T	F	DO NOT WRITE ON THIS BOOK!		
F	T			
F	F			

The symbol $\leftrightarrow$ is called an equivalence and is read as "if and only if." The equivalence asserts that if p is true, then q is true, and if p is false, then q is false. Note again that $p \leftrightarrow q$ is an abbreviation for $[(p \to q)$ and $(q \to p)]$. It is interesting and important to note that a biconditional sentence $p \leftrightarrow q$ is true if and only if p and q are both true or both false; otherwise the biconditional sentence is false.

Definitions of the terms of mathematics are expressed in several ways. Sometimes they are written as conditional sentences, such as

"If two angles have the same measure, they are congruent."

Sometimes they are written without connective, such as

"Two angles having the same measure are congruent." Definitions should be expressed as biconditional sentences, such as

"Two angles are congruent *if and only if* they have the same measure,"

since we require (page 21) that the statement be reversible.

Tautology

A compound sentence in logic, whose truth value is *always* true, is called a tautology. Such tautologies are represented by the Law of the Syllogism in Exercise 2, page 570, and in the Law of Contraposition as shown below.

Law of Contraposition

$$(p \to q) \leftrightarrow (\sim q \to \sim p)$$

p	q	$\sim p$	$\sim q$	$p \to q$	$\sim q \to \sim p$	$(p \to q) \leftrightarrow (\sim q \to \sim p)$
T	T	F	F	T	T	T
T	F	F	T	F	F	T
F	T	T	F	T	T	T
F	F	T	T	T	T	T

EXAMPLE 1

p = It is Christmas.
q = The month is December.
r = The grass is green.

A. If it is Christmas, then the month is December if and only if the month is not December, then it is not Christmas.
$(p \to q) \leftrightarrow (\sim q \to \sim p)$

B. If the grass is green, then the month is December if and only if the month is not December, then the grass is not green.
$(r \to q) \leftrightarrow (\sim q \to \sim r)$

The truth table states that both statements must be true, even though p, or q, or r may be false. Do you agree that they are true statements?

EXAMPLE 2

Using p, q, and r from Example 1, we shall examine the relationships implied by columns 5 and 6 ($p \rightarrow q$, $\sim q \rightarrow \sim p$) of the previous truth table.

A. If it is Christmas, then the month is December.
If it is not December, then it is not Christmas.

B. If the grass is green, then the month is December.
If the month is not December, then the grass is not green.

Do these examples show that when $p \rightarrow q$ is true, then $\sim q \rightarrow \sim p$ is true? Also that when $p \rightarrow q$ is false, then $\sim q \rightarrow \sim p$ is false? (Refer to Section 4.11 and 4.12.)

Summary

There are more than twenty commonly used laws of inference whose truth values have been established. The few that we have listed will be more than sufficient to justify the argument forms used in this course.

PROOF OF CONGRUENCE THEOREMS

If one of the basic congruence assumptions, (7.01, 7.02, 7.03) is accepted as a postulate, the others can be proved as theorems. We will accept as our basic postulate, "Two triangles are congruent if two sides and the included angle of one are congruent to two sides and the included angle of the other (s.a.s.)."

The first demonstration establishes the "s.s.s." theorem.

Given: $\triangle ABC$ and $\triangle MNO$, $\overline{AC} \cong \overline{MO}$, $\overline{AB} \cong \overline{MN}$, $\overline{BC} \cong \overline{NO}$

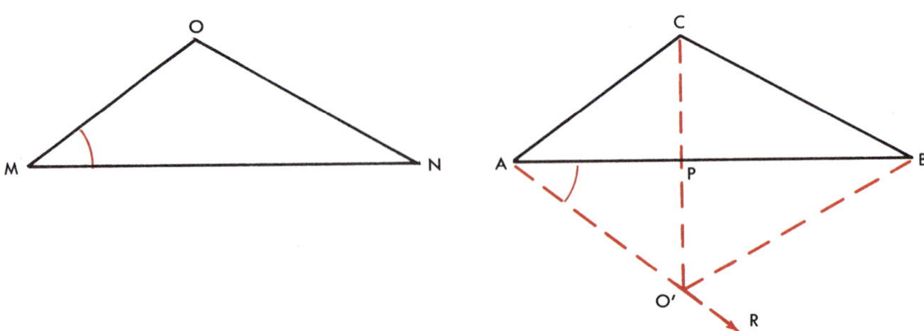

Figure A-1

Conjecture: $\triangle ABC \cong \triangle MNO$

Plan: To use the s.a.s. assumption to prove $\triangle ABC$ and $\triangle MNO$ congruent to a third triangle.

Proof:

Statements	*Reasons*
1. There is a ray $\overrightarrow{AR}$ such that $\angle NMO \cong \angle BAR$	1. 3.01, 3.04
2. There is a segment $\overline{AO'}$ on $\overrightarrow{AR}$ such that $\overline{AO'} \cong \overline{MO}$	2. 2.38
3. $\overline{AB} \cong \overline{MN}$	3. Given
4. $\triangle AO'B \cong \triangle MON$	4. s.a.s.
5. $\overline{AO'} \cong \overline{MO}$. $\overline{BO'} \cong \overline{NO}$	5. c.p.c.t.c.
6. $\overline{AC} \cong \overline{MO}$, $\overline{BC} \cong \overline{NO}$	6. Given
7. $\overline{AO'} \cong \overline{AC}$, $\overline{BO'} \cong \overline{BC}$	7. Transitive Axiom of Congruence
8. $\angle ACO' \cong \angle AO'C$, $\angle BCO' \cong \angle BO'C$	8. Step 7 and 8.07
9. $\angle ACB \cong \angle AO'B$	9. 3.08 and Addition Axiom
10. $\triangle ACB \cong \triangle AO'B$	10. s.a.s.
11. $\triangle ABC \cong \triangle MNO$	11. Steps 4 and 10 and the Transitive Axiom of Congruence

This satisfies the case in which P is between A and B. If P coincides with A, then $\angle ACB \cong \angle AO'B$ by 8.07. Then $\angle C \cong \angle O$ because $\angle O \cong \angle O'$. Therefore $\triangle ABC \cong \triangle MNO$ by s.a.s.

If A is between P and B on $\overleftrightarrow{AB}$, then we show that $\angle ACB \cong \angle AO'B$ by subtracting the measures of the angles in Step 9.

The two remaining cases, where P coincides with B and where B is between A and P are similar to the two cases outlined above.

THEOREM

If three sides of one triangle are congruent, respectively, to the three sides of another triangle, then the triangles are congruent. (7.01)

The following demonstration establishes the "a.s.a." theorem.

Given: $\triangle ABC$ and $\triangle RST$ with $\angle A \cong \angle R$, $\angle B \cong \angle RST$, $\overline{AB} \cong \overline{RS}$

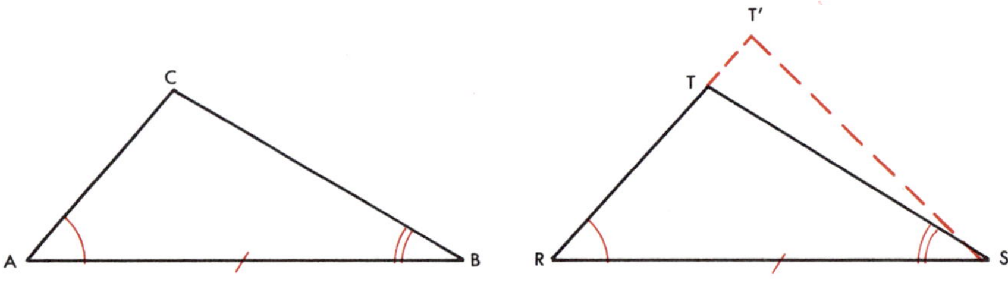

Figure A-2

Conjecture: $\triangle ABC \cong \triangle RST$

Plan: Use s.a.s. postulate by showing that $\overline{AC} \cong \overline{RT}$

Proof: *Statements*	*Reasons*
1. On $\overrightarrow{RT}$ is a point T' such that $\overline{AC} \cong \overline{RT'}$	1. 2.38
2. $\angle A \cong \angle R$, $\overline{AB} \cong \overline{RS}$	2. Given
3. $\therefore \triangle ABC \cong \triangle RST'$	3. s.a.s.
4. Then $\angle ABC \cong \angle RST'$	4. c.p.c.t.c.
5. But $\angle RST \cong \angle ABC$	5. Given
6. So $\angle RST \cong \angle RST'$	6. Steps 4 and 5 and Transitive Axiom of Congruence
7. $\overrightarrow{ST} = \overrightarrow{ST'}$	7. 3.01
8. T coincides with T'	8. 2.15
9. $\overline{AC} \cong \overline{RT}$	9. Steps 1, 8, and 2.38
10. $\triangle ABC \cong \triangle DEF$	10. s.a.s.

THEOREM

If two angles and the included side of one triangle are congruent to the corresponding parts of a second triangle, then the triangles are congruent. (7.03)

The following demonstration establishes the "h.s." theorem.

Given: $\triangle ABC$ and $\triangle DEF$ with right angles A and D, $\overline{CB} \cong \overline{FE}$ and $\overline{AC} \cong \overline{DF}$

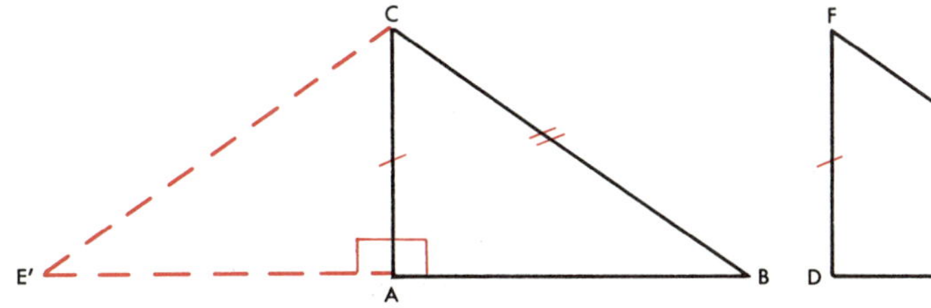

Figure A-3

Conjecture: $\triangle ABC \cong \triangle DEF$

Plan: To show that $\angle B \cong \angle E$, then use s.a.a., (7.04)

Proof:

Statements	*Reasons*
1. On $\overrightarrow{BA}$ is a point E' such that $\overline{AE'} \cong \overline{DE}$.	1. 2.38
2. $\overline{E'C}$ is determined by E' and C.	2. 2.10
3. $\angle E'AC$ is a right angle.	3. 3.16
4. $\overline{AC} \cong \overline{DF}$	4. Given
5. $\triangle AE'C \cong \triangle DEF$	5. s.a.s.
6. $\angle E \cong \angle E'$, $\overline{CE'} \cong \overline{FE}$.	6. c.p.c.t.c.
7. $\overline{CB} \cong \overline{FE}$	7. Given
8. $\overline{CE'} \cong \overline{CB}$	8. Steps 6 and 7, Transitive Axiom of Congruence
9. $\angle E' \cong \angle B$	9. 8.07, step 8
10. $\angle B \cong \angle E$	10. Steps 6 and 9, Transitive Axiom of Congruence
11. $\angle A \cong \angle B$	11. Given, 3.13
12. $\triangle ABC \cong \triangle DEF$	12. s.a.a., Steps 4, 10, 11

THEOREM

If the hypotenuse and a leg of one right triangle are congruent respectively to the hypotenuse and a leg of another right triangle, the triangles are congruent. (7.05)

EUCLID'S PYTHAGOREAN THEOREM

Given: Right $\triangle ABC$, with squares I, II, III on the sides

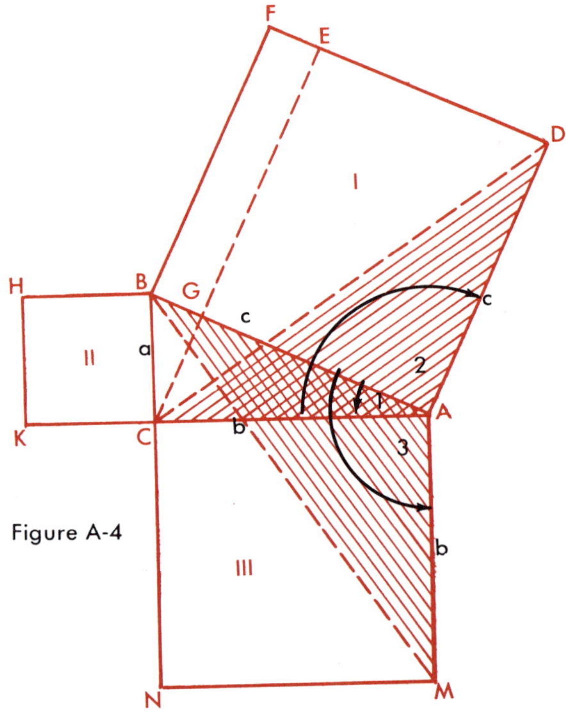

Figure A-4

Prove: The area of square I = area of square II + area of square III (i.e., $a^2 + b^2 = c^2$).

Construction: Draw $\overline{CE} \perp \overline{DF}$, intersecting $\overline{AB}$ at G. Draw $\overline{BM}$ and $\overline{CD}$.

Analysis:

1. $ADEG$ and $GEFB$ are rectangles. Why?
2. $\angle MAB \cong \angle CAD$. Why?
3. $\triangle MAB \cong \triangle CAD$. Why?
4. area of $\triangle MAB = \frac{1}{2}$ area of square III $\left(\text{area} = \frac{1}{2}\, b \cdot b\right)$
5. area of $\triangle CAD = \frac{1}{2}$ area of rectangle $ADEG$. Why?
6. area of Square III = area of rectangle $ADEG$. Why?
   By a similar proof, drawing $\overline{AH}$ and $\overline{CF}$, area of square II = area of rectangle $GEFB$.
7. $\therefore$ area of square I = area of square II + area of square III, or $c^2 = a^2 + b^2$.

LIST OF FORMULAS USED IN GEOMETRY

The sum of the interior angles of any convex polygon $(n - 2)\,180°$

One angle of any regular polygon . . . . $\dfrac{(n - 2)\,180°}{n}$

Circumference of circle . . . . $C = 2\pi r$

Pythagorean Theorem for right triangle . . . . . . $c^2 = a^2 + b^2$

Area of regions of plane figures

Area of rectangular region . . . . $A = lw$

Area of parallelogram region . . . . $A = bh$

Area of triangular region . . . . $A = \dfrac{bh}{2}$

Area of trapezoidal region . . . . $A = \dfrac{1}{2}h(b + b')$

Area of regular polygon. . . . . $A = \dfrac{1}{2}ap$

Area of circular region . . . . $A = \pi r^2$

Area and volume of the regions of geometric solids and surfaces

Lateral Area

Cone of revolution . . . . $L = \pi rs$

Cylinder of revolution. $L = 2\pi rh$

Frustum of cone of revolution. . . . . $L = \dfrac{1}{2}s(C + C')$

Frustum of regular pyramid . . . . $L = \dfrac{1}{2}s(p + p')$

Prism. . . . . $L = ep$

Regular pyramid . . . . $L = \dfrac{1}{2}sp$

Total Area

Cone of revolution . . . . $T = \pi r(r + s)$

Cylinder of revolution. $T = 2\pi r(r + h)$

Sphere . . . . $S = 4\pi r^2$

Volume

Circular cone	$V = \frac{1}{3}\pi r^2 h$
Circular cylinder	$V = Bh$
Cube	$V = e^3$
Cylinder of revolution	$V = \pi r^2 h$
Frustum of circular cone.	$V = \frac{1}{3}\pi h(r^2 + r'^2 + rr')$
Frustum of a pyramid.	$V = \frac{1}{3}h(B + B' + \sqrt{BB'})$
Prism	$V = Bh$
Pyramid	$V = \frac{1}{3}Bh$
Rectangular solid	$V = lwh$
Sphere	$V = \frac{4}{3}\pi r^3$

SYMBOLS AND ABBREVIATIONS

Students will find the following list of symbols and abbreviations useful in their study of geometry.

Symbols

$=$	(is) equal to; equals; equal
$\neq$	is not equal to
$>$	is greater than
$<$	is less than
$\sim$	(is) similar to; similar
$\cong$	(is) congruent to; congruent
$\parallel$	(is) parallel to; parallel
$\perp$	(is) perpendicular to; perpendicular
$x', x'', \ldots$	x-prime, x-double prime, . . .
$x_1, x_2, \ldots$	x-sub-one, x-sub-two, . . .
$\angle$	angle
∟	right angle
$\frown$	arc
$\odot$	circle
$\therefore$	therefore
$\triangle$	triangle
$\overrightarrow{AB}$	ray AB
$\overleftrightarrow{AB}$	line AB
$\overline{AB}$	segment AB

The plural of any symbol representing a noun is formed by adding the letter *s*. Thus ⊙s means *circles*; ∠s means *angles*.

Abbreviations

adj.	= adjacent	**Iden.**	= identity
alt.	= alternate	**int.**	= interior
Ax.	= axiom	**isos.**	= isosceles
comp.	= complementary	**opp.**	= opposite
Const.	= construction	**rt.**	= right
Cor.	= corollary	**sin**	= sine
corr.	= corresponding	**st.**	= straight
cos	= cosine	**supp.**	= supplementary
ext.	= exterior	**tan**	= tangent

s.a.s. = If two triangles have two sides and the included angle of one congruent respectively to two sides and the included angle of the other, the triangles are congruent.

a.s.a. = If two triangles have two angles and the included side of one congruent respectively to two angles and the included side of the other, the triangles are congruent.

s.s.s. = If two triangles have the three sides of one congruent respectively to the three sides of the other, the triangles are congruent.

TABLE 1

Approximate Square Roots and Cube Roots

N	$\sqrt{N}$	$\sqrt{10N}$	$\sqrt[3]{N}$	N	$\sqrt{N}$	$\sqrt{10N}$	$\sqrt[3]{N}$
1	1.000	3.162	1.000	51	7.141	22.58	3.708
2	1.414	4.472	1.260	52	7.211	22.80	3.733
3	1.732	5.477	1.442	53	7.280	23.02	3.756
4	2.000	6.325	1.587	54	7.348	23.24	3.780
5	2.236	7.071	1.710	55	7.416	23.45	3.803
6	2.449	7.746	1.817	56	7.483	23.66	3.826
7	2.646	8.367	1.913	57	7.550	23.87	3.848
8	2.828	8.944	2.000	58	7.616	24.08	3.871
9	3.000	9.487	2.080	59	7.681	24.29	3.893
10	3.162	10.00	2.154	60	7.746	24.49	3.915
11	3.317	10.49	2.224	61	7.810	24.70	3.936
12	3.464	10.95	2.289	62	7.874	24.90	3.958
13	3.606	11.40	2.351	63	7.937	25.10	3.979
14	3.742	11.83	2.410	64	8.000	25.30	4.000
15	3.873	12.25	2.466	65	8.062	25.50	4.021
16	4.000	12.65	2.520	66	8.124	25.69	4.041
17	4.123	13.04	2.571	67	8.185	25.88	4.062
18	4.243	13.42	2.621	68	8.246	26.08	4.082
19	4.359	13.78	2.668	69	8.307	26.27	4.102
20	4.472	14.14	2.714	70	8.367	26.46	4.121
21	4.583	14.49	2.759	71	8.426	26.65	4.141
22	4.690	14.83	2.802	72	8.485	26.83	4.160
23	4.796	15.17	2.844	73	8.544	27.02	4.179
24	4.899	15.49	2.884	74	8.602	27.20	4.198
25	5.000	15.81	2.924	75	8.660	27.39	4.217
26	5.099	16.12	2.962	76	8.718	27.57	4.236
27	5.196	16.43	3.000	77	8.775	27.75	4.254
28	5.292	16.73	3.037	78	8.832	27.93	4.273
29	5.385	17.03	3.072	79	8.888	28.11	4.291
30	5.477	17.32	3.107	80	8.944	28.28	4.309
31	5.568	17.61	3.141	81	9.000	28.46	4.327
32	5.657	17.89	3.175	82	9.055	28.64	4.344
33	5.745	18.17	3.208	83	9.110	28.81	4.362
34	5.831	18.44	3.240	84	9.165	28.98	4.380
35	5.916	18.71	3.271	85	9.220	29.15	4.397
36	6.000	18.97	3.302	86	9.274	29.33	4.414
37	6.083	19.24	3.332	87	9.327	29.50	4.431
38	6.164	19.49	3.362	88	9.381	29.66	4.448
39	6.245	19.75	3.391	89	9.434	29.83	4.465
40	6.325	20.00	3.420	90	9.487	30.00	4.481
41	6.403	20.25	3.448	91	9.539	30.17	4.498
42	6.481	20.49	3.476	92	9.592	30.33	4.514
43	6.557	20.74	3.503	93	9.644	30.50	4.531
44	6.633	20.98	3.530	94	9.695	30.66	4.547
45	6.708	21.21	3.557	95	9.747	30.82	4.563
46	6.782	21.45	3.583	96	9.798	30.98	4.579
47	6.856	21.68	3.609	97	9.849	31.14	4.595
48	6.928	21.91	3.634	98	9.899	31.30	4.610
49	7.000	22.14	3.659	99	9.950	31.46	4.626
50	7.071	22.36	3.684	100	10.000	31.62	4.642

TABLE 2

Trigonometric Ratios

Angle	sin	cos	tan	Angle	sin	cos	tan
0°	.0000	1.0000	.0000	45°	.7071	.7071	1.0000
1°	.0175	.9998	.0175	46°	.7193	.6947	1.0355
2°	.0349	.9994	.0349	47°	.7314	.6820	1.0724
3°	.0523	.9986	.0524	48°	.7431	.6691	1.1106
4°	.0698	.9976	.0699	49°	.7547	.6561	1.1504
5°	.0872	.9962	.0875	50°	.7660	.6428	1.1918
6°	.1045	.9945	.1051	51°	.7771	.6293	1.2349
7°	.1219	.9925	.1228	52°	.7880	.6157	1.2799
8°	.1392	.9903	.1405	53°	.7986	.6018	1.3270
9°	.1564	.9877	.1584	54°	.8090	.5878	1.3764
10°	.1736	.9848	.1763	55°	.8192	.5736	1.4281
11°	.1908	.9816	.1944	56°	.8290	.5592	1.4826
12°	.2079	.9781	.2126	57°	.8387	.5446	1.5399
13°	.2250	.9744	.2309	58°	.8480	.5299	1.6003
14°	.2419	.9703	.2493	59°	.8572	.5150	1.6643
15°	.2588	.9659	.2679	60°	.8660	.5000	1.7321
16°	.2756	.9613	.2867	61°	.8746	.4848	1.8040
17°	.2924	.9563	.3057	62°	.8829	.4695	1.8807
18°	.3090	.9511	.3249	63°	.8910	.4540	1.9626
19°	.3256	.9455	.3443	64°	.8988	.4384	2.0503
20°	.3420	.9397	.3640	65°	.9063	.4226	2.1445
21°	.3584	.9336	.3839	66°	.9135	.4067	2.2460
22°	.3746	.9272	.4040	67°	.9205	.3907	2.3559
23°	.3907	.9205	.4245	68°	.9272	.3746	2.4751
24°	.4067	.9135	.4452	69°	.9336	.3584	2.6051
25°	.4226	.9063	.4663	70°	.9397	.3420	2.7475
26°	.4384	.8988	.4877	71°	.9455	.3256	2.9042
27°	.4540	.8910	.5095	72°	.9511	.3090	3.0777
28°	.4695	.8829	.5317	73°	.9563	.2924	3.2709
29°	.4848	.8746	.5543	74°	.9613	.2756	3.4874
30°	.5000	.8660	.5774	75°	.9659	.2588	3.7321
31°	.5150	.8572	.6009	76°	.9703	.2419	4.0108
32°	.5299	.8480	.6249	77°	.9744	.2250	4.3315
33°	.5446	.8387	.6494	78°	.9781	.2079	4.7046
34°	.5592	.8290	.6745	79°	.9816	.1908	5.1446
35°	.5736	.8192	.7002	80°	.9848	.1736	5.6713
36°	.5878	.8090	.7265	81°	.9877	.1564	6.3138
37°	.6018	.7986	.7536	82°	.9903	.1392	7.1154
38°	.6157	.7880	.7813	83°	.9925	.1219	8.1443
39°	.6293	.7771	.8098	84°	.9945	.1045	9.5144
40°	.6428	.7660	.8391	85°	.9962	.0872	11.4301
41°	.6561	.7547	.8693	86°	.9976	.0698	14.3007
42°	.6691	.7431	.9004	87°	.9986	.0523	19.0811
43°	.6820	.7314	.9325	88°	.9994	.0349	28.6363
44°	.6947	.7193	.9657	89°	.9998	.0175	57.2900
45°	.7071	.7071	1.0000	90°	1.0000	.0000	—

INDEX

D

Q

R